Igor V. Nikolaev
Noncommutative Geometry

De Gruyter Studies in Mathematics

Volume 66

Igor V. Nikolaev

Noncommutative Geometry

A Functorial Approach

2nd edition

DE GRUYTER

Mathematics Subject Classification 2020
11F72, 11G15, 11G45, 11J81, 14A22, 14F42, 14G15, 14H10, 14H52, 14H55, 18Dxx, 16R10, 46L85, 55S35, 57M27, 58F10

Author
Prof. Dr. Igor V. Nikolaev
Department of Mathematics and Computer
Science, St. John's University, 8000 Utopia
Parkway,
New York, NY 11439
USA
igor.v.nikolaev@gmail.com

ISBN 978-3-11-078860-0
e-ISBN (PDF) 978-3-11-078870-9
e-ISBN (EPUB) 978-3-11-078881-5
ISSN 0179-0986

Library of Congress Control Number: 2022936127

Bibliographic information published by the Deutsche Nationalbibliothek
The Deutsche Nationalbibliothek lists this publication in the Deutsche Nationalbibliografie; detailed bibliographic data are available on the Internet at http://dnb.dnb.de.

Typesetting: VTeX UAB, Lithuania
Printing and binding: CPI books GmbH, Leck

www.degruyter.com

To @louanhu

Foreword

The noncommutative geometry (NCG) is a branch of mathematics dealing with an interplay between geometry and noncommutative algebras (NCA). The following naive but vital questions arise:

> What is the purpose of NCG? What is it good for? Can the NCG solve open problems of mathematics inaccessible otherwise? In other words, is it useful? Can the NCG benefit a topologist, an algebraic geometer or a number theorist? What is it anyway?

As a model example, consider the noncommutative torus $\mathcal{A}_\theta$, i. e., a C^*-algebra on two generators u and v satisfying the relation $vu = e^{2\pi i\theta}uv$ for a real constant θ. It turns out that $\mathcal{A}_\theta$ behaves much like a coordinate ring of a nonsingular elliptic curve. In other words, we deal with a functor from the category of elliptic curves to a category of the NCA. Such a functor can be used to construct generators of the abelian extensions of real quadratic fields (Hilbert's 12th problem) and to prove that $e^{2\pi i\theta+\log\log\varepsilon}$ is an algebraic number, where θ and ε are quadratic irrational numbers. This and similar functors are at the heart of our book.

What is the NCG anyway? By such we understand functors defined on the topological, algebraic or arithmetic spaces with values in the NCA. The NCG studies geometry of such spaces using algebraic invariants of the corresponding functors. Our approach is distinct from, but complementary to, a program of classification of the NCA outlined in [57].

Boston – New York

Igor V. Nikolaev

https://doi.org/10.1515/9783110788709-201

Foreword to the second edition

Since the first edition in 2017, I got a courteous offer from the Walter de Gruyter's Senior Publishing Editor Mathematics Steven Elliot to extend and revise the existing volume. The idea was supported by Niels Jacob, the Editor of the Studies in Mathematics Series. I express my deep gratitude to both, and also to the Content Editor Nadja Schedensack, for this excellent opportunity. I thank my colleagues from the Department of Mathematics at St. John's University for their unwaivering support.

The second edition contains two new chapters on the Arithmetic Topology and Quantum Arithmetic. All other parts remain intact.

New York, April 2022 Igor V. Nikolaev

https://doi.org/10.1515/9783110788709-202

Introduction

The book has three parts. Part I is preparatory: Chapter 1 deals with the simplest functors arising in algebraic geometry, number theory, and topology; the functors take value in a category of the C^*-algebras called noncommutative tori. Using these functors, one gets a set of noncommutative invariants of elliptic curves and Anosov's automorphisms of the two-dimensional torus. Chapter 2 is a brief introduction to the categories, functors, and natural transformations. Chapter 3 reviews the C^*-algebras and their K-theory. We describe certain important classes of such algebras: the noncommutative tori, the AF-algebras, the UHF-algebras, and the Cuntz–Krieger algebras.

Part II deals with the noncommutative invariants corresponding to functors on the classical spaces. Chapter 4 studies functors on the topological spaces with values in the category of the stationary AF-algebras; the corresponding invariant is the Handelman triple $(\Lambda, [I], K)$, where Λ is an order in a real algebraic number field K and $[I]$ an equivalence class of the ideals of Λ. We construct a representation of the braid group in a cluster C^*-algebra; this result is used to extend the Jones and HOMFLY polynomials to the multivariable Laurent polynomials. Chapter 5 deals with a functor arising in algebraic geometry. We use the functor to settle the Harvey conjecture on the linearity of the mapping class groups. Finally, Chapter 6 covers functors in number theory. These functors are used to construct generators of the maximal abelian extension of a real quadratic field (Hilbert's 12th problem) and to prove that the complex number $e^{2\pi i\theta+\log\log\varepsilon}$ is algebraic if θ and ε are quadratic irrational numbers.

Part III is a brief survey of the NCG covering finite, von Neumann and Connes geometries, Index Theory, Jones polynomials, quantum groups, noncommutative algebraic geometry, and trends in noncommutative geometry. A bibliography is attached at the end of each chapter; we owe an apology to the authors of many important but omitted works.

There exist a handful of excellent textbooks on the NCG. The first and foremost is the monograph [57] and its extended English edition [58]. The books [153, 100, 280, 133] treat particular aspects of Connes' monograph. A different approach to the NCG is taken in a small but instructive textbook [154]. The specialized volumes [156] and [59] cover links to number theory and physics. Finally, the books [144] and [269] serve the readers with background in physics. None of these books treat the NCG as a functor [178].

I thank the organizers, participants and sponsors of the Spring Institute on Noncommutative Geometry and Operator Algebras (NCGOA) held annually at the Vanderbilt University in Nashville, Tennessee; this book grew from an effort to understand what's going on there. (I still have no answer.) I am grateful to folks who helped me with the project; among them are P. Baum, D. Bisch, B. Blackadar, late O. Bratteli, A. Connes, J. Cuntz, S. Echterhoff, G. Elliott, K. Goodearl, D. Handelman, N. Higson, B. Hughes, V. F. R. Jones, M. Kapranov, M. Khalkhali, W. Krieger, G. Landi, Yu. Manin,

https://doi.org/10.1515/9783110788709-203

V. Manuilov, M. Marcolli, V. Mathai, A. Mishchenko, D. Mundici, S. Novikov, M. Rieffel, R. Schiffler, late W. Thurston, V. Troitsky, G. Yu, and others.

I thank my colleagues in the St. John's University (New York) and the staff of the Walter de Gruyter GmbH for a support.

Contents

Part III: **Brief survey of NCG**

Part I: **Basics**

1 Model examples

We consider the simplest functors arising in algebraic geometry, number theory, and topology; these functors take value in a category of the C^*-algebras called noncommutative tori. We believe that a handful of elegant examples tells more than a general theory; we encourage the reader to keep these examples in mind for the rest of the book. No special knowledge of the C^*-algebras, elliptic curves, or Anosov automorphisms (beyond an intuitive level) is required at this point; the reader can look up the missing definitions in the standard literature given at the end of each section.

1.1 Noncommutative tori

The noncommutative torus $\mathcal{A}_\theta$ is an algebra over the complex numbers on a pair of generators u and v satisfying the commutation relation $vu = e^{2\pi i\theta}uv$ for a real constant θ. A geometric definition of the $\mathcal{A}_\theta$ involves deformation of the commutative algebra $C^\infty(T^2)$ of the smooth complex-valued functions on a two-dimensional torus T^2; we start with this definition, since it clarifies the origin and notation for such algebras.

1.1.1 Geometric definition

Let $C^\infty(T^2)$ be the commutative algebra of infinitely differentiable complex-valued functions on T^2 endowed with the usual pointwise sum and product of two functions. The idea is to replace the commutative product $f(x)g(x)$ of functions $f, g \in C^\infty(T^2)$ by a noncommutative product $f(x) *_\hbar g(x)$ depending on a continuous deformation parameter $\hbar$, so that $\hbar = 0$ corresponds to the commutative product $f(x)g(x)$; the product $f(x) *_\hbar g(x)$ must be associative for each value of $\hbar$. To achieve the goal, it is sufficient to construct the Poisson bracket $\{f, g\}$ on $C^\infty(T^2)$, i. e., a binary operation satisfying the identities $\{f, f\} = 0$ and $\{f, \{g, h\}\} + \{h, \{f, g\}\} + \{g, \{h, f\}\} = 0$; this is a special case of Kontsevich's Theorem for the Poisson manifolds, see Section 14.3. The algebra $C^\infty_\hbar(T^2)$ equipped with the usual sum $f(x) + g(x)$ and a noncommutative associative product $f(x) *_\hbar g(x)$ is called a *deformation quantization* of the algebra $C^\infty(T^2)$.

The Poisson bracket can be constructed as follows. For a real number θ, define a bracket on $C^\infty(T^2)$ by the formula

$$\{f, g\}_\theta := \theta\left(\frac{\partial f}{\partial x}\frac{\partial g}{\partial y} - \frac{\partial f}{\partial y}\frac{\partial g}{\partial x}\right).$$

The reader is encouraged to verify that the bracket satisfies the identities $\{f, f\}_\theta = 0$ and $\{f, \{g, h\}_\theta\}_\theta + \{h, \{f, g\}_\theta\}_\theta + \{g, \{h, f\}_\theta\}_\theta = 0$, i. e., is a Poisson bracket. Kontsevich Theorem says that there exists an associative product $f *_\hbar g$ on $C^\infty(T^2)$ obtained from the bracket $\{f, g\}_\theta$. Namely, let φ and ψ denote the Fourier transform of functions f

https://doi.org/10.1515/9783110788709-001

and g, respectively; one can define an $\hbar$-family of products between the Fourier transforms according to the formula

$$(\varphi *_\hbar \psi)(p) = \sum_{q\in\mathbf{Z}^2} \varphi(q)\psi(p-q)e^{-\pi i\hbar\, k(p,q)},$$

where $k(p,q) = \theta(pq - qp)$ is the kernel of the Fourier transform of the Poisson bracket $\{f,g\}_\theta$, i. e., an expression defined by the formula

$$\{\varphi,\psi\}_\theta = -4\pi^2 \sum_{q\in\mathbf{Z}^2} \varphi(q)\psi(p-q)k(q,p-q).$$

The product $f *_\hbar g$ is defined as a pullback of the product $(\varphi *_\hbar \psi)$; the resulting associative algebra $C^\infty_{\hbar,\theta}(T^2)$ is called the deformation quantization of $C^\infty(T^2)$ in the direction θ defined by the Poisson bracket $\{f,g\}_\theta$, see Fig. 1.1.

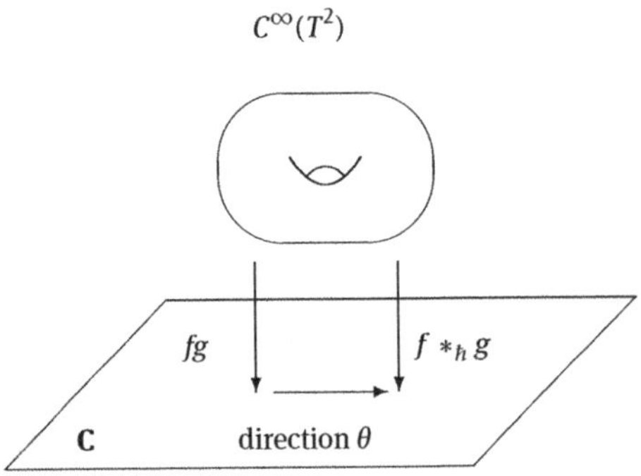

Figure 1.1: Deformation of the algebra $C^\infty(T^2)$.

Remark 1.1.1. The algebra $C^\infty_{\hbar,\theta}(T^2)$ is endowed with the natural involution coming from the complex conjugation on $C^\infty(T^2)$, so that $\varphi^*(p) := \bar{\varphi}(-p)$ for all $p \in \mathbf{Z}^2$. A norm on $C^\infty_{\hbar,\theta}(T^2)$ comes from the operator norm on the Schwartz functions $\varphi,\psi \in \mathcal{S}(\mathbf{Z}^2)$ on the Hilbert space $\ell^2(\mathbf{Z}^2)$.

Definition 1.1.1. By a noncommutative torus $\mathcal{A}^{\text{geometric}}_\theta$ one understands the C^*-algebra obtained from the norm closure of the *-algebra $C^\infty_{1,\theta}(T^2)$.

1.1.2 Analytic definition

The analytic definition involves bounded linear operators acting on a Hilbert space. Namely, let S^1 be the unit circle; denote by $L^2(S^1)$ the Hilbert space of the square-integrable complex valued functions on S^1. Fix a real number $\theta \in [0,1)$; for every

$f(e^{2\pi it}) \in L^2(S^1)$, we consider a pair of bounded linear operators U and V acting by the formula

$$\begin{cases} U[f(e^{2\pi it})] = f(e^{2\pi i(t-\theta)}), \\ V[f(e^{2\pi it})] = e^{2\pi it} f(e^{2\pi it}). \end{cases}$$

It is verified directly that

$$\begin{cases} VU = e^{2\pi i\theta} UV, \\ UU^* = U^* U = E, \\ VV^* = V^* V = E, \end{cases}$$

where U^* and V^* are the adjoint operators of U and V, respectively, and E is the identity operator.

Definition 1.1.2. By a noncommutative torus $\mathcal{A}_\theta^{\text{analytic}}$ one understands the C^*-algebra generated by the operators U and V acting on the Hilbert space $L^2(S^1)$.

1.1.3 Algebraic definition

An algebraic definition is the shortest; it involves the universal algebra, i. e., an associative algebras given by the generators and relations. Namely, let $\mathbf{C}\langle x_1, x_2, x_3, x_4\rangle$ be the polynomial ring in four noncommuting variables x_1, x_2, x_3, and x_4. Consider a two-sided ideal, I_θ, generated by the relations

$$\begin{cases} x_3 x_1 = e^{2\pi i\theta} x_1 x_3, \\ x_1 x_2 = x_2 x_1 = e, \\ x_3 x_4 = x_4 x_3 = e. \end{cases}$$

Definition 1.1.3. By the noncommutative torus $\mathcal{A}_\theta^{\text{algebraic}}$ one understands a C^*-algebra given by the norm closure of the $*$-algebra $\mathbf{C}\langle x_1, x_2, x_3, x_4\rangle / I_\theta$, where the *-involution acts on the generators according to the formula $x_1^* = x_2$ and $x_3^* = x_4$.

1.1.4 Abstract noncommutative torus

Theorem 1.1.1. $\mathcal{A}_\theta^{\text{geometric}} \cong \mathcal{A}_\theta^{\text{analytic}} \cong \mathcal{A}_\theta^{\text{algebraic}}$.

Proof. The isomorphism $\mathcal{A}_\theta^{\text{analytic}} \cong \mathcal{A}_\theta^{\text{algebraic}}$ is obvious, since one can write $x_1 = U$, $x_2 = U^*$, $x_3 = V$, and $x_4 = V^*$. The isomorphism $\mathcal{A}_\theta^{\text{geometric}} \cong \mathcal{A}_\theta^{\text{analytic}}$ is established by the identification of functions $t \mapsto e^{2\pi itp}$ of $\mathcal{A}_\theta^{\text{geometric}}$ with the unitary operators U_p for each $p \in \mathbf{Z}^2$; then the generators of $\mathbf{Z}^2$ will correspond to the operators U and V. □

Definition 1.1.4. We shall write $\mathcal{A}_\theta$ to denote an abstract noncommutative torus independent of its geometric, analytic, or algebraic realization.

1.1.5 First properties of the $\mathcal{A}_\theta$

The algebra $\mathcal{A}_\theta$ has a plethora of features; we shall focus on the following two: the Morita equivalence and real multiplication. The first property is a basic equivalence relation in the category of noncommutative tori; such a relation tells us that the algebras $\mathcal{A}_\theta$ and $\mathcal{A}_{\theta'}$ are the same from the standpoint of noncommutative geometry. The second property is rare – only a countable set of $\mathcal{A}_\theta$ has real multiplication; it means that the algebra $\mathcal{A}_\theta$ has unusually many endomorphisms, i. e., the corresponding ring exceeds **Z**.

1.1.5.1 Morita equivalence

Definition 1.1.5. The noncommutative torus $\mathcal{A}_\theta$ is said to be Morita equivalent (stably isomorphic) to a noncommutative torus $\mathcal{A}_{\theta'}$ if $\mathcal{A}_\theta \otimes \mathcal{K} \cong \mathcal{A}_{\theta'} \otimes \mathcal{K}$, where $\mathcal{K}$ is the C^*-algebra of compact operators.

The Morita equivalence of the algebras A and A' means that they have the same category of projective modules $\mathbf{Mod}(A) \cong \mathbf{Mod}(A')$. In general, it is very hard to describe the Morita equivalence in intrinsic terms. (If $A \cong A'$ are isomorphic, then, of course, A is Morita equivalent to A'.) The following remarkable result provides an elegant solution to the Morita equivalence of the noncommutative tori.

Theorem 1.1.2 ([238]). *The noncommutative tori $\mathcal{A}_\theta$ and $\mathcal{A}_{\theta'}$ are Morita equivalent (stably isomorphic) if and only if*

$$\theta' = \frac{a\theta + b}{c\theta + d} \quad \textit{for a matrix} \begin{pmatrix} a & b \\ c & d \end{pmatrix} \in \mathrm{SL}_2(\mathbf{Z}).$$

Proof. In outline, the proof proceeds by establishing a bijection between the isomorphism classes of the algebras $\mathcal{A}_\theta \otimes \mathcal{K}$ and the range of a canonical trace $\tau : \mathcal{A}_\theta \otimes \mathcal{K} \to \mathbf{R}$ on the projections of the algebra $\mathcal{A}_\theta \otimes \mathcal{K}$. Such a range is proved to be a subgroup $\Lambda = \mathbf{Z} + \mathbf{Z}\theta$ of $\mathbf{R}$. A new set of generators in Λ corresponds to the transformation $\theta' = \frac{a\theta+b}{c\theta+d}$, where $a, b, c, d \in \mathbf{Z}$ and $ad - bc = 1$. Thus $\mathcal{A}_{\theta'} \otimes \mathcal{K} \cong \mathcal{A}_\theta \otimes \mathcal{K}$ if and only if $\theta' = \theta \bmod \mathrm{SL}_2(\mathbf{Z})$. □

1.1.5.2 Real multiplication

Denote by $\mathfrak{Q}$ the set of all quadratic irrational numbers, i. e., the irrational roots of quadratic polynomials with integer coefficients.

Theorem 1.1.3 ([155]). *The endomorphism ring of a noncommutative torus $\mathcal{A}_\theta$ is given by the formula*

$$\operatorname{End}(\mathcal{A}_\theta) \cong \begin{cases} \mathbf{Z}, & \text{if } \theta \in \mathbf{R} - (\mathcal{Q} \cup \mathbf{Q}), \\ \mathbf{Z} + fO_k, & \text{if } \theta \in \mathcal{Q}, \end{cases}$$

where integer $f \geq 1$ is the conductor of an order in the ring of integers O_k of the real quadratic field $k = \mathbf{Q}(\sqrt{D})$.

Proof. The endomorphisms of $\mathcal{A}_\theta$ correspond to multiplications by a real number α preserving the subgroup $\Lambda = \mathbf{Z} + \mathbf{Z}\theta$ of $\mathbf{R}$. In other words, condition $\alpha\Lambda \subseteq \Lambda$ implies that either $\alpha \in \mathbf{Z}$ or $\theta = \frac{a\theta+b}{c\theta+d}$ for $a, b, c, d \in \mathbf{Z}$ and $ad - bc = \pm 1$. In the latter case $\theta \in \mathcal{Q}$ and $\alpha \in \mathbf{Z} + fO_k$, where $k = \mathbf{Q}(\theta)$. □

Definition 1.1.6. The noncommutative torus $\mathcal{A}_\theta$ is said to have real multiplication if $\operatorname{End}(\mathcal{A}_\theta)$ is bigger than $\mathbf{Z}$, i. e., θ is a quadratic irrationality. We denote by $\mathcal{A}_{RM}^{(D,f)}$ a noncommutative torus with real multiplication by an order of conductor f in the quadratic field $\mathbf{Q}(\sqrt{D})$.

Remark 1.1.2. The reader can verify that real multiplication is preserved by the Morita equivalence of the algebra $\mathcal{A}_\theta$.

Guide to the literature

A good introduction to the noncommutative tori is a survey [239]. The $\mathcal{A}_\theta$ is also known as an irrational rotation algebra, see [230] and [238]. The real multiplication has been introduced in [155].

1.2 Elliptic curves

This subject is so fundamental that it hardly needs an introduction; many key facts of the complex analysis, algebraic geometry, and number theory can be recast in terms of the elliptic curves. Perhaps such curves form the oldest well-studied part of mathematics which yet hides the deepest open problems, e. g., the Birch and Swinnerton-Dyer Conjecture.

1.2.1 Weierstrass and Jacobi normal forms

We deal with the ground field of complex numbers $\mathbf{C}$. An *elliptic curve* is the subset of the complex projective plane of the form

$$\mathcal{E}(\mathbf{C}) = \{(x, y, z) \in \mathbf{C}P^2 \mid y^2 z = 4x^3 + axz^2 + bz^3\},$$

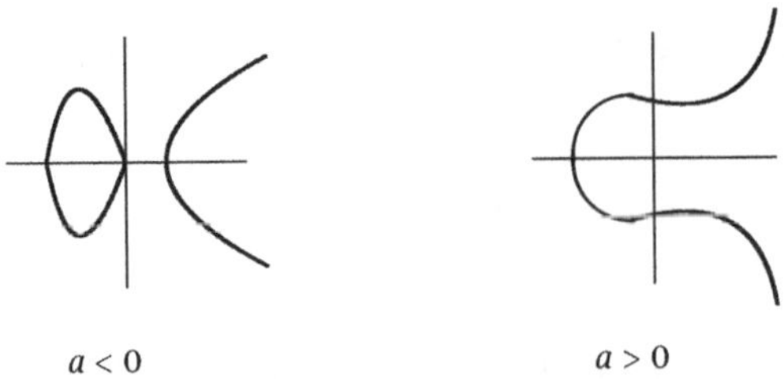

Figure 1.2: The real points of an affine elliptic curve $y^2 = 4x^3 + ax$.

where a and b are some constant complex numbers. The real points of $\mathcal{E}(\mathbf{C})$ are shown in Fig. 1.2. The above equation is called the *Weierstrass normal form* of $\mathcal{E}(\mathbf{C})$.

Remark 1.2.1. The curve $\mathcal{E}(\mathbf{C})$ is isomorphic to the set of points of intersection of a pair of the quadric surfaces in the complex projective space $\mathbf{C}P^3$ given by the system of homogeneous equations

$$\begin{cases} u^2 + v^2 + w^2 + z^2 = 0, \\ Av^2 + Bw^2 + z^2 = 0, \end{cases}$$

where A and B are some constant complex numbers and $(u, v, w, z) \in \mathbf{C}P^3$. The above system of quadratic equations is called the *Jacobi form* of the $\mathcal{E}(\mathbf{C})$.

1.2.2 Complex tori

Definition 1.2.1. By a complex torus one understands the space $\mathbf{C}/(\mathbf{Z}\omega_1 + \mathbf{Z}\omega_2)$, where ω_1 and ω_2 are linearly independent vectors in the complex plane $\mathbf{C}$. The ratio $\tau = \omega_2/\omega_1$ is called a complex modulus, see Fig. 1.3.

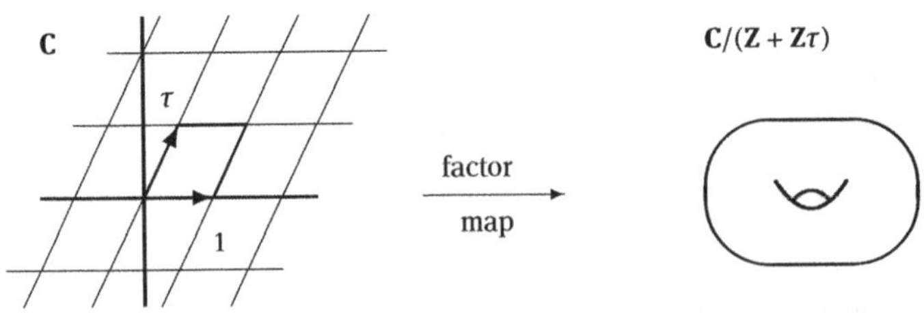

Figure 1.3: Complex torus $\mathbf{C}/(\mathbf{Z} + \mathbf{Z}\tau)$.

Remark 1.2.2. Two complex tori $\mathbf{C}/(\mathbf{Z}+\mathbf{Z}\tau)$ and $\mathbf{C}/(\mathbf{Z}+\mathbf{Z}\tau')$ are isomorphic if and only if

$$\tau' = \frac{a\tau + b}{c\tau + d} \quad \text{for a matrix} \begin{pmatrix} a & b \\ c & d \end{pmatrix} \in SL_2(\mathbf{Z}).$$

(We leave the proof to the reader. Hint: notice that $z \mapsto \alpha z$ is an invertible holomorphic map for each $\alpha \in \mathbf{C} - \{0\}$.)

1.2.3 Weierstrass uniformization

One may ask if the complex manifold $\mathbf{C}/(\mathbf{Z} + \mathbf{Z}\tau)$ can be embedded into an n-dimensional projective space as a variety; it turns out that the answer is positive even for the case $n = 2$. The following classical result relates the complex torus $\mathbf{C}/(\mathbf{Z}+\mathbf{Z}\tau)$ with an elliptic curve $\mathcal{E}(\mathbf{C})$ in the projective plane $\mathbf{C}P^2$.

Theorem 1.2.1 (Weierstrass). *There exists a holomorphic embedding*

$$\mathbf{C}/(\mathbf{Z} + \mathbf{Z}\tau) \hookrightarrow \mathbf{C}P^2$$

given by the formula

$$z \mapsto \begin{cases} (\wp(z), \wp'(z), 1) & \text{for } z \notin L_\tau := \mathbf{Z} + \mathbf{Z}\tau, \\ (0, 1, 0) & \text{for } z \in L_\tau, \end{cases}$$

which is an isomorphism between the complex torus $\mathbf{C}/(\mathbf{Z} + \mathbf{Z}\tau)$ and the elliptic curve

$$\mathcal{E}(\mathbf{C}) = \{(x, y, z) \in \mathbf{C}P^2 \mid y^2 z = 4x^3 + axz^2 + bz^3\},$$

where $\wp(z)$ is the Weierstrass function defined by the convergent series

$$\wp(z) = \frac{1}{z^2} + \sum_{\omega \in L_\tau - \{0\}} \left(\frac{1}{(z-\omega)^2} - \frac{1}{\omega^2} \right)$$

and

$$\begin{cases} a = -60 \sum_{\omega \in L_\tau - \{0\}} \frac{1}{\omega^4}, \\ b = -140 \sum_{\omega \in L_\tau - \{0\}} \frac{1}{\omega^6}. \end{cases}$$

Remark 1.2.3. Weierstrass Theorem identifies elliptic curves $\mathcal{E}(\mathbf{C})$ and complex tori $\mathbf{C}/(\mathbf{Z}\omega_1 + \mathbf{Z}\omega_2)$. Therefore we write $\mathcal{E}_\tau$ to denote elliptic curve corresponding to the complex torus of modulus τ.

1.2.4 Complex multiplication

Remark 1.2.4. The endomorphism ring

$$\mathrm{End}(\mathcal{E}_\tau) := \{\alpha \in \mathbf{C} : \alpha L_\tau \subseteq L_\tau\}$$

of elliptic curve $\mathcal{E}_\tau = \mathbf{C}/L_\tau$ is given by the formula

$$\mathrm{End}(\mathcal{E}_\tau) \cong \begin{cases} \mathbf{Z}, & \text{if } \tau \in \mathbf{C} - \mathcal{Q}, \\ \mathbf{Z} + fO_k, & \text{if } \tau \in \mathcal{Q}, \end{cases}$$

where $\mathcal{Q}$ is the set of all imaginary quadratic numbers and integer $f \geq 1$ is conductor of an order in the ring of integers O_k of the imaginary quadratic number field $k = \mathbf{Q}(\sqrt{-D})$. (The proof is left to the reader as an exercise.)

Definition 1.2.2. Elliptic curve $\mathcal{E}_\tau$ is said to have complex multiplication if the endomorphism ring of $\mathcal{E}_\tau$ is bigger than $\mathbf{Z}$, i. e., τ is a quadratic irrationality. We write $\mathcal{E}_{CM}^{(-D,f)}$ to denote elliptic curves with complex multiplication by an order of conductor f in the quadratic field $\mathbf{Q}(\sqrt{-D})$.

Guide to the literature

Among many excellent textbooks on the elliptic curves are [122, 136, 138, 258–260], and others. More advanced topics are covered in the survey papers [49, 158, 274].

1.3 Functor $\mathcal{E}_\tau \longrightarrow \mathcal{A}_\theta$

Observation 1.3.1. Comparing $\mathcal{E}_\tau$ and $\mathcal{A}_\theta$, one cannot fail to observe the following striking fact:

> An isomorphism of the elliptic curve $\mathcal{E}_\tau$ acts on τ by the formula $\tau \mapsto \frac{a\tau+b}{c\tau+d}$; likewise, the Morita equivalence of a noncommutative torus $\mathcal{A}_\theta$ acts on θ by the formula $\theta \mapsto \frac{a\theta+b}{c\theta+d}$. Can this phenomenon be part of a categorical correspondence? In other words, can one construct a functor F between the two categories shown in Fig. 1.4?

Theorem 1.3.1. *There exists a covariant functor F from the category of elliptic curves $\mathcal{E}_\tau$ to the category of noncommutative tori $\mathcal{A}_\theta$ such that, if $\mathcal{E}_\tau$ is isomorphic to $\mathcal{E}_{\tau'}$ then $\mathcal{A}_\theta = F(\mathcal{E}_\tau)$ is Morita equivalent (stably isomorphic) to $\mathcal{A}_{\theta'} = F(\mathcal{E}_{\tau'})$.*

Proof. We shall give an algebraic proof of this fact based on the notion of a Sklyanin algebra; there exists a geometric proof using the notion of measured foliations and the Teichmüller theory, see Section 5.1.2.

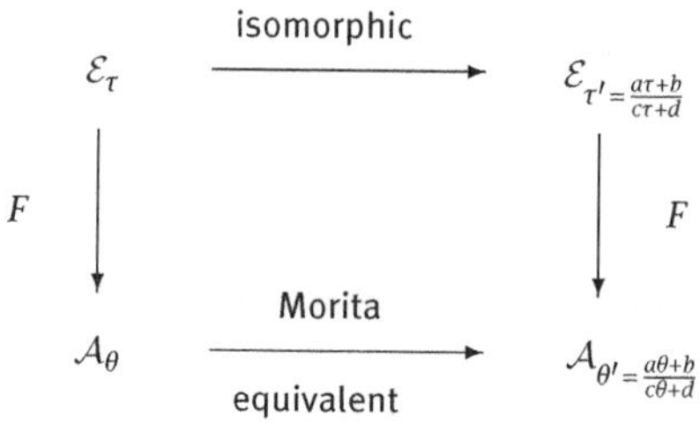

Figure 1.4: An observation.

The *Sklyanin algebra* $S(\alpha,\beta,\gamma)$ is a free $\mathbf{C}$-algebra on four generators $x_1,\dots,x_4$ and six quadratic relations:

$$\begin{cases} x_1x_2 - x_2x_1 = \alpha(x_3x_4 + x_4x_3), \\ x_1x_2 + x_2x_1 = x_3x_4 - x_4x_3, \\ x_1x_3 - x_3x_1 = \beta(x_4x_2 + x_2x_4), \\ x_1x_3 + x_3x_1 = x_4x_2 - x_2x_4, \\ x_1x_4 - x_4x_1 = \gamma(x_2x_3 + x_3x_2), \\ x_1x_4 + x_4x_1 = x_2x_3 - x_3x_2, \end{cases}$$

where $\alpha+\beta+\gamma+\alpha\beta\gamma = 0$ [264, p. 260]. The algebra $S(\alpha,\beta,\gamma)$ is isomorphic to a (twisted homogeneous) coordinate ring of elliptic curve $\mathcal{E}_\tau \subset \mathbf{C}P^3$ having the Jacobi form

$$\begin{cases} u^2 + v^2 + w^2 + z^2 = 0, \\ \frac{1-\alpha}{1+\beta}v^2 + \frac{1+\alpha}{1-\gamma}w^2 + z^2 = 0. \end{cases}$$

The latter means that the algebra $S(\alpha,\beta,\gamma)$ satisfies an isomorphism of Serre

$$\mathbf{Mod}(S(\alpha,\beta,\gamma))/\mathbf{Tors} \cong \mathbf{Coh}(\mathcal{E}_\tau),$$

where **Coh** is the category of quasicoherent sheaves on $\mathcal{E}_\tau$, **Mod** the category of graded left modules over the graded ring $S(\alpha,\beta,\gamma)$, and **Tors** the full subcategory of **Mod** consisting of the torsion modules [249]. The algebra $S(\alpha,\beta,\gamma)$ comes with a natural automorphism $\sigma : \mathcal{E}_\tau \to \mathcal{E}_\tau$ [265, p. 173]. Take this automorphism to have order 4, i. e., $\sigma^4 = 1$; in this case $\beta = 1$, $\gamma = -1$, and it is known that system of quadratic relations for the Sklyanin algebra $S(\alpha,\beta,\gamma)$ can be brought to the skew symmetric form

$$\begin{cases} x_3x_1 = \mu e^{2\pi i\theta} x_1x_3, \\ x_4x_2 = \frac{1}{\mu} e^{2\pi i\theta} x_2x_4, \\ x_4x_1 = \mu e^{-2\pi i\theta} x_1x_4, \\ x_3x_2 = \frac{1}{\mu} e^{-2\pi i\theta} x_2x_3, \\ x_2x_1 = x_1x_2, \\ x_4x_3 = x_3x_4, \end{cases}$$

where $\theta = \mathrm{Arg}(q)$ and $\mu = |q|$ for some complex number $q \in \mathbf{C} - \{0\}$ [84, Remark 1].

On the other hand, the noncommutative torus $\mathcal{A}_\theta$ is defined by the relations

$$\begin{cases} x_3x_1 = e^{2\pi i\theta} x_1x_3, \\ x_4x_2 = e^{2\pi i\theta} x_2x_4, \\ x_4x_1 = e^{-2\pi i\theta} x_1x_4, \\ x_3x_2 = e^{-2\pi i\theta} x_2x_3, \\ x_2x_1 = x_1x_2 = e, \\ x_4x_3 = x_3x_4 = e, \end{cases}$$

which are equivalent to the relations of Section 1.1.3. (We leave the proof to the reader as an exercise in noncommutative algebra.) Comparing these relations with the skew symmetric relations for the Sklyanin algebra $S(\alpha, 1, -1)$, one concludes that they are almost identical; to pin down the difference, we shall add two extra relations,

$$x_1x_3 = x_3x_4 = \frac{1}{\mu} e,$$

to relations of the Sklyanin algebra and bring it (by multiplication and cancellations) to the following equivalent form:

$$\begin{cases} x_3x_1x_4 = e^{2\pi i\theta} x_1, \\ \quad x_4 = e^{2\pi i\theta} x_2x_4x_1, \\ x_4x_1x_3 = e^{-2\pi i\theta} x_1, \\ \quad x_2 = e^{-2\pi i\theta} x_4x_2x_3, \\ x_2x_1 = x_1x_2 = \frac{1}{\mu} e, \\ x_4x_3 = x_3x_4 = \frac{1}{\mu} e. \end{cases}$$

Applying the same equivalent transformations to the system of relations for the noncommutative torus, one brings the system to the form

$$\begin{cases} x_3x_1x_4 = e^{2\pi i\theta}x_1, \\ x_4 = e^{2\pi i\theta}x_2x_4x_1, \\ x_4x_1x_3 = e^{-2\pi i\theta}x_1, \\ x_2 = e^{-2\pi i\theta}x_4x_2x_3, \\ x_2x_1 = x_1x_2 = e, \\ x_4x_3 = x_3x_4 = e. \end{cases}$$

Thus the only difference between relations for the Sklyanin algebra (modulo the ideal I_μ generated by relations $x_1x_3 = x_3x_4 = \frac{1}{\mu}e$) and such for the noncommutative torus $\mathcal{A}_\theta$ is a scaling of the unit $\frac{1}{\mu}e$. Thus one obtains the following isomorphism:

$$\mathcal{A}_\theta \cong S(\alpha, 1, -1) \,/\, I_\mu.$$

The functor F is obtained as a quotient map of the Serre isomorphism

$$I_\mu \backslash \mathbf{Coh}(\mathcal{E}_\tau) \cong \mathbf{Mod}(I_\mu \backslash S(\alpha, 1, -1))/\mathbf{Tors} \cong \mathbf{Mod}(\mathcal{A}_\theta)/\mathbf{Tors}$$

and the fact that the isomorphisms in category $\mathbf{Mod}(\mathcal{A}_\theta)$ correspond to the Morita equivalence (stable isomorphisms) in category $\mathcal{A}_\theta$. □

Remark 1.3.1 (The $\mathcal{A}_\theta$ as a coordinate ring of $\mathcal{E}_\tau$). The formula $\mathcal{A}_\theta \cong S(\alpha, 1, -1) \,/\, I_\mu$ says that the noncommutative torus $\mathcal{A}_\theta$ is a coordinate ring of elliptic curve $\mathcal{E}_\tau$ modulo the ideal I_μ.

Guide to the literature

The noncommutative tori as coordinate rings of elliptic curves were studied in [176] and [179]; the higher genus curves were considered in [180] and [193]. A desingularization of the moduli space of $\mathcal{A}_\theta$ was introduced in [177].

1.4 Ranks of elliptic curves

Functor $F : \mathcal{E}_\tau \to \mathcal{A}_\theta$ is intertwined with the arithmetic of elliptic curves. Below we relate the ranks of elliptic curves to an invariant of the algebra $\mathcal{A}_{RM}$.

1.4.1 Symmetry of complex and real multiplication

We consider a restriction of F to the elliptic curves $\mathcal{E}_{CM}^{(-D,f)}$. Theorem below describes the restriction in terms of the noncommutative tori with real multiplication.

Theorem 1.4.1 ([181, 189]). *$F(\mathcal{E}_{CM}^{(-D,f)}) = \mathcal{A}_{RM}^{(D,f')}$, where f' is the least integer satisfying the equation $|\mathrm{Cl}(\mathbf{Z} + f'O_{\mathbf{Q}(\sqrt{D})})| = |\mathrm{Cl}(\mathbf{Z} + fO_{\mathbf{Q}(\sqrt{-D})})|$ and $\mathrm{Cl}(R)$ is the class group of the ring R.*

1.4.2 Arithmetic complexity of the $\mathcal{A}_{RM}$

Let $k = \mathbf{Q}(\sqrt{-D})$ be an imaginary quadratic number field and let $j(\mathcal{E}_{CM}^{(-D,f)})$ be the j-invariant of elliptic curve $\mathcal{E}_{CM}^{(-D,f)}$; it is well known from complex multiplication that $\mathcal{E}_{CM}^{(-D,f)} \cong \mathcal{E}(K)$, where $K = k(j(\mathcal{E}_{CM}^{(-D,f)}))$ is the Hilbert class field of k, see, e. g., [259, p. 95]. The Mordell–Weil theorem says that the set of the K-rational points of $\mathcal{E}_{CM}^{(-D,f)}$ is a finitely generated abelian group, see [274, p. 192]; the rank of such a group will be denoted by $\mathrm{rk}(\mathcal{E}_{CM}^{(-D,f)})$. For the sake of simplicity, we further restrict to the following class of curves. If $(\mathcal{E}_{CM}^{(-D,f)})^\sigma$, $\sigma \in \mathrm{Gal}(K|\mathbf{Q})$ is the Galois conjugate of the curve $\mathcal{E}_{CM}^{(-D,f)}$, then by a **Q**-*curve* one understands elliptic curve $\mathcal{E}_{CM}^{(-D,f)}$ such that there exists an isogeny between $(\mathcal{E}_{CM}^{(-D,f)})^\sigma$ and $\mathcal{E}_{CM}^{(-D,f)}$ for each $\sigma \in \mathrm{Gal}(K|\mathbf{Q})$, see, e. g., [101]. Let $\mathfrak{P}_{3 \bmod 4}$ be the set of all primes $p = 3 \bmod 4$; it is known that $\mathcal{E}_{CM}^{(-p,1)}$ is a **Q**-curve whenever $p \in \mathfrak{P}_{3 \bmod 4}$, see [101, p. 33]. The rank of $\mathcal{E}_{CM}^{(-p,1)}$ is always divisible by $2h_k$, where h_k is the class number of field $k = \mathbf{Q}(\sqrt{-p})$, see [101, p. 49]; by a **Q**-*rank* of $\mathcal{E}_{CM}^{(-p,1)}$ one understands the integer

$$\mathrm{rk}_{\mathbf{Q}}(\mathcal{E}_{CM}^{(-p,1)}) := \frac{1}{2h_k}\,\mathrm{rk}(\mathcal{E}_{CM}^{(-p,1)}).$$

Definition 1.4.1. By an arithmetic complexity $c(\mathcal{A}_{RM}^{(D,f)})$ of the algebra $\mathcal{A}_{RM}^{(D,f)}$ one understands the number of independent entries a_i in the continued fraction $\sqrt{f^2D} = [a_0, \overline{a_1, a_2, \dots, a_2, a_1, 2a_0}]$, see Section 6.3.2 for the details.

Theorem 1.4.2 ([205]). *$\mathrm{rk}_{\mathbf{Q}}(\mathcal{E}_{CM}^{(-p,1)}) + 1 = c(\mathcal{A}_{RM}^{(p,1)})$, whenever $p = 3 \bmod 4$.*

Remark 1.4.1. There are infinitely many pairwise nonisomorphic **Q**-curves, see, e. g., [101]. All pairwise nonisomorphic **Q**-curves $\mathcal{E}_{CM}^{(-p,1)}$ with $p < 100$ and their invariant $c(\mathcal{A}_{RM}^{(p,1)})$ are calculated in Fig. 1.5.

Guide to the literature

D. Hilbert counted complex multiplication as not only the most beautiful part of mathematics but also of entire science; it surely does as it links complex analysis and num-

$p \equiv 3 \bmod 4$	$\mathrm{rk}_{\mathbb{Q}}(\mathcal{E}_{CM}^{(-p,1)})$	$\sqrt{p}$	$c(\mathcal{A}_{RM}^{(p,1)})$
3	1	$[1,\overline{1,2}]$	2
7	0	$[2,\overline{1,1,1,4}]$	1
11	1	$[3,\overline{3,6}]$	2
19	1	$[4,\overline{2,1,3,1,2,8}]$	2
23	0	$[4,\overline{1,3,1,8}]$	1
31	0	$[5,\overline{1,1,3,5,3,1,1,10}]$	1
43	1	$[6,\overline{1,1,3,1,5,1,3,1,1,12}]$	2
47	0	$[6,\overline{1,5,1,12}]$	1
59	1	$[7,\overline{1,2,7,2,1,14}]$	2
67	1	$[8,\overline{5,2,1,1,7,1,1,2,5,16}]$	2
71	0	$[8,\overline{2,2,1,7,1,2,2,16}]$	1
79	0	$[8,\overline{1,7,1,16}]$	1
83	1	$[9,\overline{9,18}]$	2

Figure 1.5: The $\mathbb{Q}$-curves $\mathcal{E}_{CM}^{(-p,1)}$ with $p < 100$.

ber theory. One cannot beat [250] for an introduction, but more comprehensive [259, Chapter 2] is the must. The problem of ranks dates back to [234, p. 493]. It was proved in [165] that the rank of any rational elliptic curve is always a finite number. The result was extended in [172] to the elliptic curves over any number field K. The ranks of individual elliptic curves are calculated by the method of descent, see, e. g., [49, p. 205]. An analytic approach uses the *Hasse–Weil L-function* $L(\mathcal{E}_\tau, s)$; it was conjectured by B. J. Birch and H. P. F. Swinnerton-Dyer that the order of zero of such a function at $s = 1$ is equal to the rank of $\mathcal{E}_\tau$, see, e. g., [274, p. 198]. Real multiplication has been introduced in [155]. A rank conjecture involving invariants of the noncommutative tori was formulated in [181]. Torsion points were studied in [186].

1.5 Classification of surface automorphisms

One of the main topics of topology are invariants of continuous maps $f : X \to Y$ of the topological spaces X and Y. We call f an automorphism if it is invertible and $X = Y$. The automorphisms $f, f' : X \to X$ are said to be conjugate if there exists an automorphism $h : X \to X$ such that $f' = h \circ f \circ h^{-1}$, where $f \circ f'$ means the composition of f and f'; the conjugation means a change of coordinate system for the topological space X and each property of f invariant under the conjugation is intrinsic, i. e., a topological invariant of f. The automorphism $f : X \to X$ is said to have an infinite order if $f^n \neq \mathrm{Id}$ for each $n \in \mathbf{Z} - \{0\}$. The conjugation problem for infinite order automorphisms is unsolved even when X is a compact two-dimensional manifold, i. e., a surface; such a solution would imply a topological classification of the three-dimensional manifolds, see, e. g., [114].

1.5.1 Anosov automorphisms

We shall focus on the case $X = T^2$, i. e., the two-dimensional torus. Because $T^2 \cong \mathbf{R}^2/\mathbf{Z}^2$, each automorphism of T^2 can be given by an isomorphism of the lattice $\mathbf{Z}^2 \subset \mathbf{R}^2$, see Fig. 1.6; therefore the automorphism $f : T^2 \to T^2$ can be written in the matrix form

$$A_f = \begin{pmatrix} a_{11} & a_{12} \\ a_{21} & a_{22} \end{pmatrix} \in \mathrm{GL}(2, \mathbf{Z}).$$

Definition 1.5.1. An infinite order automorphism $f : T^2 \to T^2$ is called *Anosov* if its matrix form A_f satisfies the inequality $|a_{11} + a_{22}| > 2$.

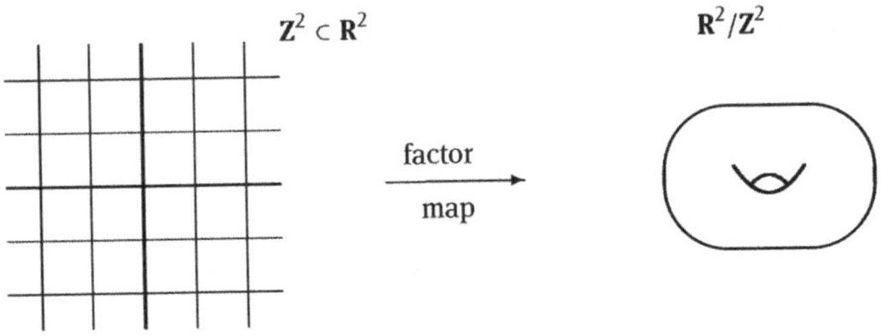

Figure 1.6: Topological torus $T^2 \cong \mathbf{R}^2/\mathbf{Z}^2$.

1.5.2 Functor *F*

To study topological invariants, we shall construct a functor F on the set of all Anosov automorphisms with the values in a category of the noncommutative tori such that the diagram in Fig. 1.7 is commutative; in other words, if f and f' are conjugate Anosov automorphisms, then the corresponding noncommutative tori $\mathcal{A}_\theta$ and $\mathcal{A}_{\theta'}$ are Morita equivalent. The required map $F : A_f \mapsto \mathcal{A}_\theta$ can be constructed as follows. For simplicity, we shall assume that $a_{11} + a_{22} > 2$; the case $a_{11} + a_{22} < -2$ is treated similarly. Moreover, we can assume that A_f is a positive matrix since each class of conjugation of the Anosov automorphism f contains such a representative. Denote by λ_{A_f} the Perron–Frobenius eigenvalue of positive matrix A_f. The noncommutative torus $\mathcal{A}_\theta = F(A_f)$ is defined by the normalized Perron–Frobenius eigenvector $(1, \theta)$ of the matrix A_f, i. e.,

$$A_f \begin{pmatrix} 1 \\ \theta \end{pmatrix} = \lambda_{A_f} \begin{pmatrix} 1 \\ \theta \end{pmatrix}.$$

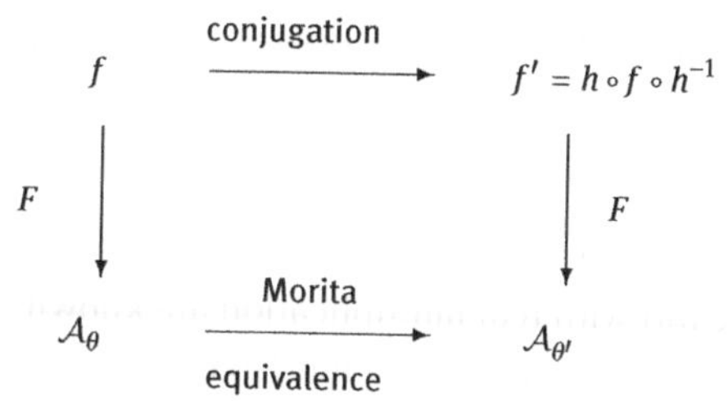

Figure 1.7: Functor F.

Remark 1.5.1. We leave it to the reader to prove that if f is Anosov's, then θ is a quadratic irrationality given by the formula

$$\theta = \frac{a_{22} - a_{11} + \sqrt{(a_{11} + a_{22})^2 - 4}}{2a_{12}}.$$

It follows from the above formula that map F takes values in the noncommutative tori with real multiplication. The following theorem says that our map $F : A_f \mapsto \mathcal{A}_\theta$ is actually a functor.

Theorem 1.5.1 ([185]). *If f and f' are conjugate Anosov automorphisms, then the noncommutative torus $\mathcal{A}_\theta = F(A_f)$ is Morita equivalent (stably isomorphic) to $\mathcal{A}_{\theta'} = F(A_{f'})$.*

1.5.3 Handelman's invariant

In view of Theorem 1.5.1, the problem of conjugation for the Anosov automorphisms can be recast in terms of the noncommutative tori; namely, one needs to calculate invariants of the Morita equivalence of a noncommutative torus with real multiplication. Such a noncommutative invariant was found in [105]. Namely, consider the eigenvalue problem for a matrix $A_f \in GL(2, \mathbf{Z})$, i. e., $A_f v_A = \lambda_{A_f} v_A$, where $\lambda_{A_f} > 1$ is the Perron–Frobenius eigenvalue and $v_A = (v_A^{(1)}, v_A^{(2)})$ the corresponding eigenvector with the positive entries normalized so that $v_A^{(i)} \in K = \mathbf{Q}(\lambda_{A_f})$. Denote by $\mathfrak{m} = \mathbf{Z}v_A^{(1)} + \mathbf{Z}v_A^{(2)}$ a $\mathbf{Z}$-module in the number field K. Recall that the coefficient ring, Λ, of module $\mathfrak{m}$ consists of the elements $\alpha \in K$ such that $\alpha\mathfrak{m} \subseteq \mathfrak{m}$. It is known that Λ is an order in K (i. e., a subring of K containing 1) and, with no restriction, one can assume that $\mathfrak{m} \subseteq \Lambda$. It follows from the definition, that $\mathfrak{m}$ coincides with an ideal, I, whose equivalence class in Λ we shall denote by $[I]$.

Theorem 1.5.2 (Handelman's noncommutative invariant). *The triple*

$$(\Lambda, [I], K)$$

Exercises

1. Show that the bracket on $C^\infty(T^2)$ defined by the formula

$$\{f,g\}_\theta := \theta\left(\frac{\partial f}{\partial x}\frac{\partial g}{\partial y} - \frac{\partial f}{\partial y}\frac{\partial g}{\partial x}\right)$$

is a Poisson bracket, i. e., satisfies the identities $\{f,f\}_\theta = 0$ and $\{f,\{g,h\}_\theta\}_\theta + \{h,\{f,g\}_\theta\}_\theta + \{g,\{h,f\}_\theta\}_\theta = 0$.

2. Prove that real multiplication is an invariant of the stable isomorphism (Morita equivalence) class of noncommutative torus $\mathcal{A}_\theta$.
3. Prove that complex tori $\mathbf{C}/(\mathbf{Z}+\mathbf{Z}\tau)$ and $\mathbf{C}/(\mathbf{Z}+\mathbf{Z}\tau')$ are isomorphic if and only if $\tau' = \frac{a\tau+b}{c\tau+d}$ for some matrix $\left(\begin{smallmatrix} a & b \\ c & d\end{smallmatrix}\right) \in SL_2(\mathbf{Z})$. (Hint: Notice that $z \mapsto \alpha z$ is an invertible holomorphic map for each $\alpha \in \mathbf{C} - \{0\}$.)
4. Prove that the system of relations

$$\begin{cases} x_3x_1 = e^{2\pi i\theta}x_1x_3, \\ x_1x_2 = x_2x_1 = e, \\ x_3x_4 = x_4x_3 = e, \end{cases}$$

involved in the algebraic definition of noncommutative torus $\mathcal{A}_\theta$ is equivalent to the following system of quadratic relations:

$$\begin{cases} x_3x_1 = e^{2\pi i\theta}x_1x_3, \\ x_4x_2 = e^{2\pi i\theta}x_2x_4, \\ x_4x_1 = e^{-2\pi i\theta}x_1x_4, \\ x_3x_2 = e^{-2\pi i\theta}x_2x_3, \\ x_2x_1 = x_1x_2 = e, \\ x_4x_3 = x_3x_4 = e. \end{cases}$$

5. Prove that elliptic curve $\mathcal{E}_\tau$ has complex multiplication if and only if the complex modulus $\tau \in \mathbf{Q}(\sqrt{-D})$ is an imaginary quadratic number. (Hint: Let $\alpha \in \mathbf{C}$ be such that $\alpha(\mathbf{Z}+\mathbf{Z}\tau) \subseteq \mathbf{Z}+\mathbf{Z}\tau$. That is, there exist $m,n,r,s \in \mathbf{Z}$ such that

$$\begin{cases} \alpha = m + n\tau, \\ \alpha\tau = r + s\tau. \end{cases}$$

One can divide the second equation by the first, so that one gets $\tau = \frac{r+s\tau}{m+n\tau}$. Thus $n\tau^2 + (m-s)\tau - r = 0$; in other words, τ is an imaginary quadratic number. Conversely, if τ is an imaginary quadratic number, then it is easy to see that $\mathrm{End}(\mathbf{C}/(\mathbf{Z}+\mathbf{Z}\tau))$ is nontrivial.)

6. Prove that isomorphisms over the field K of an elliptic curve $\mathcal{E}_{CM} \cong \mathcal{E}(K)$ correspond to isomorphisms of the algebra $\mathcal{A}_{RM} = F(\mathcal{E}_{CM})$, while isomorphisms over the field $\bar{K} \cong \mathbf{C}$ of $\mathcal{E}_{CM}$ correspond to the Morita equivalences of the algebra $\mathcal{A}_{RM}$. In other words, all twists of the curve $\mathcal{E}_{CM}$ are Morita equivalent to the algebra $\mathcal{A}_{RM} = F(\mathcal{E}_{CM})$. (Hint: The category $\mathbf{Mod}(\mathcal{A}_{RM})$ is equivalent to a category of the Galois extensions of the field K.)
7. Prove that the rational integer $\Delta(\mathfrak{m})$ is a numerical invariant of Anosov's automorphisms. (Hint: $\Delta(\mathfrak{m})$ does not depend on the basis of module $\mathfrak{m} = \mathbf{Z}v_A^{(1)} + \mathbf{Z}v_A^{(2)}$.)
8. Prove that the matrices

$$A = \begin{pmatrix} 5 & 2 \\ 2 & 1 \end{pmatrix} \quad \text{and} \quad B = \begin{pmatrix} 5 & 1 \\ 4 & 1 \end{pmatrix}$$

are not similar by using the *Gauss method*, i. e., the method of continued fractions. (Hint: Find the fixed points $Ax = x$ and $Bx = x$ which give us $x_A = 1 + \sqrt{2}$ and $x_B = \frac{1+\sqrt{2}}{2}$, respectively. Then one unfolds the fixed points into a periodic continued fraction, which gives us $x_A = [2, 2, 2, \dots]$ and $x_B = [1, 4, 1, 4, \dots]$. Since the period $(\overline{2})$ of x_A differs from the period $(\overline{1, 4})$ of B, one concludes that matrices A and B belong to different similarity classes in $GL(2, \mathbf{Z})$.)
9. Repeat the exercise for matrices

$$A = \begin{pmatrix} 4 & 3 \\ 5 & 4 \end{pmatrix} \quad \text{and} \quad B = \begin{pmatrix} 4 & 15 \\ 1 & 4 \end{pmatrix}.$$

2 Categories and functors

The notions of a category, functor, and natural transformation were introduced in [73]. We refer the reader to the monograph [48] for the basics of homological algebra; for a modern exposition, see [93]. Categories are designed to extend methods of the algebraic topology to the rest of mathematics. They can be described as a formal but amazingly helpful "calculus of arrows" between certain "objects." We briefly review categories in Section 2.1, functors in Section 2.2, and natural transformations in Section 2.3; our exposition follows the book [152]. A set of exercises can be found at the end of the chapter.

2.1 Categories

The category theory deals with certain diagrams representing sets and maps between the sets; the maps can be composed with each other, i. e., they constitute a monoid with the operation of composition. The sets are represented by points of the diagram; the maps between the sets correspond to the arrows (i. e., directed edges) of the diagram.

The diagram is called commutative if for any choice of sets A_i and maps $f_i : A_{i-1} \to A_i$ the resulting composition map from A_0 to A_n is the same. The examples of commutative diagrams are given in Fig. 2.1.

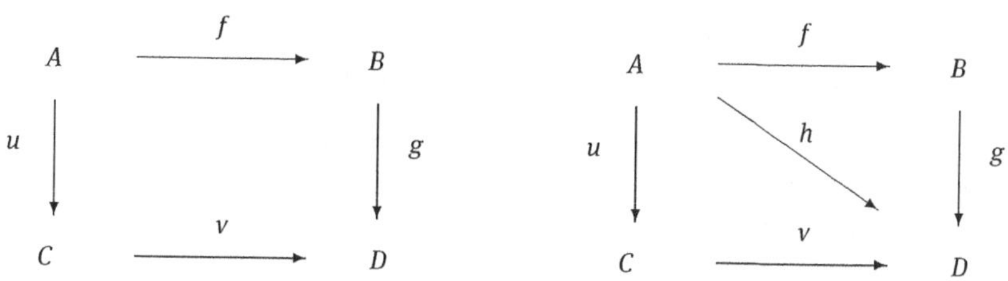

(a) (b)

Figure 2.1: Two examples: (a) $vu = gf$ and (b) $h = vu = gf$.

Example 2.1.1. The commutative diagrams allow proving that functions on the cartesian product of two sets X and Y are uniquely determined by such on the sets X and Y. Indeed, consider the cartesian product $X \times Y$ of two sets, consisting o as usual of all ordered pairs (x, y) of elements $x \in X$ and $y \in Y$. The projections $(x, y) \to x$ and $(x, y) \to y$ of the product define the functions $p : X \times Y \to X$ and $q : X \times Y \to Y$. Any function $h : W \to X \times Y$ from a third set W is uniquely determined by the composites $p \circ h$ and $q \circ h$. Conversely, given W and two functions f and g, there is a unique function h which fits in the commutative diagram of Fig. 2.2.

https://doi.org/10.1515/9783110788709-002

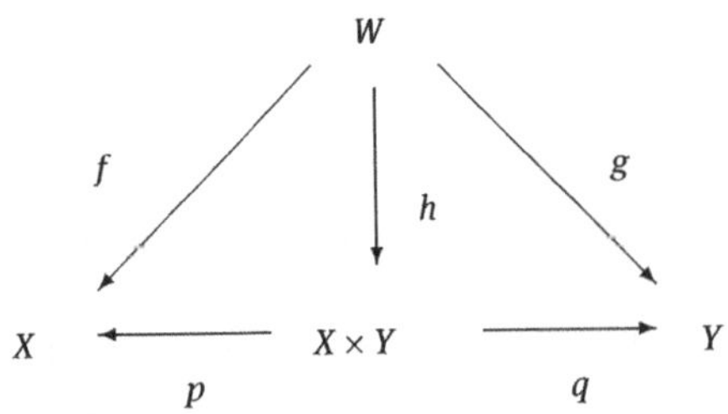

Figure 2.2: The universal property of cartesian product.

Remark 2.1.1. The construction of a cartesian product is called a *functor* because it applies not only to sets but also to the functions between them. Indeed, two functions $f : X \to X'$ and $g : Y \to Y'$ define a function $f \times g$ as their cartesian product according to the formula

$$f \times g : X \times Y \to X' \times Y', \quad (x, y) \mapsto (gx, fy).$$

Below are examples illustrating the universal property of the cartesian product.

Example 2.1.2. Let X, Y, and W be the topological spaces and p, q, f, and g be continuous maps between them. Then the space $X \times Y$ is the product of topological spaces X and Y. Such a product satisfies the universal property of the cartesian product.

Example 2.1.3. Let X, Y, and W be the abelian groups and p, q, f, and g be the homomorphisms between the groups. Consider the direct sum (product) $X \oplus Y$ of the abelian groups X and Y. In view of the canonical embeddings $X \to X \oplus Y$ and $Y \to X \oplus Y$, the direct sum satisfies the universal property of the cartesian product.

Remark 2.1.2. The direct sum of non-abelian groups does not satisfy the universal property of the cartesian product. Indeed, suppose that W has two subgroups isomorphic to X and Y, respectively, but whose elements do not commute with one another; let f and g be isomorphisms of X and Y with these subgroups. Since elements $x \in X$ and $y \in Y$ commute in the direct sum $X \oplus Y$, the diagram for the cartesian product will not be commutative for any homomorphism h.

Example 2.1.4. Let X, Y, and W be any groups and $X * Y$ be the free product of groups X and Y. It is left as an exercise to the reader to prove that the product $X * Y$ satisfies the universal property of the cartesian product.

The above examples involve the sets and certain maps between the sets. Notice that we do not need to consider what kind of elements our sets are made of, or how these elements transform under the maps. We need only that the maps can be composed with one another and that such maps can be arranged into a commutative diagram. Such a standpoint can be axiomatized in the notion of a category.

Definition 2.1.1. A category $\mathcal{C}$ consists of the following data:
(i) a set $\operatorname{Ob}\mathcal{C}$ whose elements are called the objects of $\mathcal{C}$;
(ii) for any $A, B \in \operatorname{Ob}\mathcal{C}$, a set $H(A, B)$ whose elements are called the morphisms of $\mathcal{C}$ from A to B;
(iii) for any $A, B, C \in \operatorname{Ob}\mathcal{C}$, and any $f \in H(A, B)$ and $g \in H(B, C)$, a morphism $h \in H(A, C)$ is defined and which is called the composite $g \circ f$ of g and f;
(iv) for any $A \in \operatorname{Ob}\mathcal{C}$, a morphism $1_A \in H(A, A)$ is defined and called the identity morphism, so that $f \circ 1_A = 1_B \circ f = f$ for each $f \in H(A, B)$;
(v) morphisms are associative, i. e., $h \circ (g \circ f) = (h \circ g) \circ f$ for all $f \in H(A, B)$, $g \in H(B, C)$, and $h \in H(C, D)$.

The morphisms in a category $\mathcal{C}$ are often called arrows, since they are represented by such in the commutative diagrams. Notice that since the objects in a category correspond to its identity arrows, it is possible to omit the objects and deal only with the arrows. The data for an arrows-only category $\mathcal{C}$ consists of arrows, certain ordered pairs (g, f) called the composable pairs of arrows, and an operation assigning to each composable pair (g, f) an arrow $g \circ f$ called the composite. With these data, one defines an identity of $\mathcal{C}$ to be an arrow u such that $f \circ u = f$ whenever the composite $f \circ u$ is defined and $u \circ g = g$ whenever $u \circ g$ is defined. The data is required to satisfy the following three axioms: (i) the composite $(k \circ g) \circ f$ is defined if and only if $k \circ (g \circ f)$ is defined, and we write it as $k \circ g \circ f$, (ii) the triple composite $k \circ g \circ f$ is defined whenever both composites $k \circ g$ and $g \circ f$ are defined, and (iii) for each arrow g of $\mathcal{C}$, there exist identity arrows u and u' of $\mathcal{C}$ such that $u' \circ g$ and $g \circ u$ are defined.

Example 2.1.5. An *empty* category **0** contains no objects and no arrows.

Example 2.1.6. A category is *discrete* when every arrow is an identity arrow. Every set X is the set of objects of a discrete category and every discrete category is determined by its set of objects. Thus, discrete categories are sets.

Example 2.1.7. Consider the category **Ind** whose objects are arbitrary subsets of a given set X and whose arrows are the inclusion maps between them so that $H(A, B)$ is either empty or consists of a single element.

Example 2.1.8. The category **Top** whose objects are topological spaces and whose arrows are continuous maps between them.

Example 2.1.9. The category **Grp** whose objects are all groups and whose arrows are homomorphisms between the groups.

Example 2.1.10. The category **Ab** whose objects are additive abelian groups and arrows are homomorphisms of the abelian groups.

Example 2.1.11. The category **Rng** whose objects are rings and arrows are homomorphisms between the rings preserving the unit.

Example 2.1.12. The category **CRng** whose objects are commutative rings and whose arrows are homomorphisms between the rings.

Example 2.1.13. The category $\mathbf{Mod}_R$ whose objects are modules over a given ring R, and whose arrows are homomorphisms between them. The category $\mathbf{Mod}_\mathbb{Z}$ of abelian groups is denoted by **Ab**.

Guide to the literature

Categories were introduced in [73]. We refer the interested reader to the monograph [48] for the basics of homological algebra; for a modern exposition, see [93]. Our exposition follows the book [152].

2.2 Functors

The invariant (or natural) construction is a powerful tool in mathematics. A formalization of such a construction leads to the notion of a functor. The functor is a morphism of the categories.

Definition 2.2.1. A covariant functor from a category $\mathcal{C}$ to a category $\mathcal{D}$ consists of two maps (denoted by the same letter F):
(i) $F : \operatorname{Ob}\mathcal{C} \to \operatorname{Ob}\mathcal{D}$ and
(ii) $F : H(A, B) \to H(F(A), F(B))$ for all $A, B \in \operatorname{Ob}\mathcal{C}$,

which satisfy the following conditions:
(i) $F(1_A) = 1_{F(A)}$ for all $A \in \operatorname{Ob}\mathcal{C}$;
(ii) $F(f \circ g) = F(f) \circ F(g)$, whenever $f \circ g$ is defined in $\mathcal{C}$.

Remark 2.2.1. A *contravariant functor* is also given by a map $F : \operatorname{Ob}\mathcal{C} \to \operatorname{Ob}\mathcal{D}$ but it defines a reverse order map,

$$F : H(A, B) \to H(F(B), F(A)),$$

for all $A, B \in \operatorname{Ob}\mathcal{C}$; the latter must satisfy the reverse order conditions

$$F(1_A) = 1_{F(A)} \quad \text{and} \quad F(f \circ g) = F(g) \circ F(f).$$

A functor, like a category, can be described in the "arrows-only" mode. In this case F is a function from arrows f of $\mathcal{C}$ to arrows Ff of $\mathcal{D}$ carrying each identity of $\mathcal{C}$ to an identity of $\mathcal{D}$ and each composable pair (g, f) in $\mathcal{C}$ to a composable pair (Fg, Ff) in $\mathcal{D}$ with $Fg \circ Ff = F(g \circ f)$.

Example 2.2.1. The singular n-dimensional homology assigns to each topological space X an abelian group $H_n(X)$ called the nth homology group. Each continuous map

$f : X \to Y$ between the topological spaces X and Y gives rise to a homomorphism $f_* : H_n(X) \to H_n(Y)$ between the corresponding nth homology groups. Therefore, the nth singular homology is a covariant functor $H_n : \mathbf{Top} \to \mathbf{Ab}$ between the category of topological spaces and the category of abelian groups.

Example 2.2.2. The nth homotopy groups $\pi_n(X)$ of the topological space X can be regarded as functors. Since $\pi_n(X)$ depend on the choice of a base point in X, they are functors $\mathbf{Top}_* \to \mathbf{Grp}$ between the category of topological spaces with a distinguished point and the category of groups. Note that the groups are abelian unless $n = 1$.

Example 2.2.3. Let X be a set, e. g., a Hausdorff topological space or the homogeneous space of a Lie group G. Denote by $\mathcal{F}(X, \mathbf{C})$ the space of continuous complex-valued functions on X. Since any map $f : X \to Y$ takes the function $\varphi \in \mathcal{F}(X, \mathbf{C})$ into a function $\varphi' \in \mathcal{F}(X, \mathbf{C})$, it follows that $\mathcal{F}(X, \mathbf{C})$ is a contravariant functor from category of sets to the category of vector spaces (usually infinite-dimensional). The functor maps any transformation of X into an invertible linear operator on the space $\mathcal{F}(X, \mathbf{C})$. In particular, if $X = G$ is a Lie group, one can consider the action of G on itself by the left translations; thus one gets the regular representation of G by the linear operators on $\mathcal{F}(X, \mathbf{C})$.

Example 2.2.4. Let R be a commutative ring. Consider the multiplicative group $\mathrm{GL}_n(R)$ of all nonsingular $n \times n$ matrices with entries in R. Because each homomorphism $f : R \to R'$ produces in the evident way a group homomorphism $\mathrm{GL}_n(R) \to \mathrm{GL}_n(R')$, one gets a functor $\mathrm{GL}_n : \mathbf{CRng} \to \mathbf{Grp}$ from the category of commutative rings to the category of groups.

Remark 2.2.2. Functors can be composed. Namely, given functors

$$\mathcal{C} \xrightarrow{T} \mathcal{B} \xrightarrow{S} \mathcal{A}$$

between categories $\mathcal{A}$, $\mathcal{B}$, and $\mathcal{C}$, the composite functions

$$c \mapsto S(Tc), \quad f \mapsto S(Tf)$$

on the objects $c \in \mathrm{Ob}\, \mathcal{C}$ and arrows of $\mathcal{C}$ define a functor $S \circ T : \mathcal{C} \to \mathcal{A}$ called the *composite* of S with T. The composition is associative. For each category $\mathcal{B}$, there is an identity functor $I_{\mathcal{B}} : \mathcal{B} \to \mathcal{B}$ which acts as an identity for this composition. Thus one gets a category of all categories endowed with the composition arrow.

Definition 2.2.2. A functor F is called forgetful if F forgets some or all of the structure of an algebraic object.

Example 2.2.5. The functor $F : \mathbf{Rng} \to \mathbf{Ab}$ assigns to each ring R the additive abelian group of R; such a functor is a forgetful functor because F forgets the multiplicative structure of the ring R.

Definition 2.2.3. A functor $F : \mathcal{C} \to \mathcal{B}$ is called full when to every pair of objects $C, C' \in \operatorname{Ob}\mathcal{C}$ and every arrow $g : FC \to FC'$ of $\mathcal{B}$, there is an arrow $f : C \to C'$ of $\mathcal{C}$ with $g = Ff$.

Definition 2.2.4. A functor $F : \mathcal{C} \to \mathcal{B}$ is called faithful (or an embedding) when to every pair $C, C' \in \operatorname{Ob}\mathcal{C}$ and to every pair $f_1, f_2 \in H(C, C')$ of parallel arrows of $\mathcal{C}$ the equality $Ff_1 = Ff_2 : F(C) \to F(C')$ implies $f_1 = f_2$.

Definition 2.2.5. An isomorphism $F : \mathcal{C} \to \mathcal{B}$ of categories is a functor F which is a bijection both on the objects and arrows of the respective categories.

Remark 2.2.3. Every full and faithful functor $F : \mathcal{C} \to \mathcal{B}$ is an isomorphism of categories $\mathcal{C}$ and $\mathcal{B}$ provided there are no objects of $\mathcal{B}$ not in the image of F.

Guide to the literature

Functors were introduced and studied in [73]. We refer the interested reader to the classic monograph [48] for the basics of homological algebra; for a modern exposition, see [93]. Our exposition of categories follows the classical book [152].

2.3 Natural transformations

Let $S, T : \mathcal{C} \to \mathcal{B}$ be two functors between the same categories $\mathcal{C}$ and $\mathcal{B}$. The natural transformation τ is a set of arrows of $\mathcal{B}$ which translates the "picture S" of category $\mathcal{C}$ to the "picture T" of $\mathcal{C}$. In other words, a natural transformation is a morphism of functors.

Definition 2.3.1. Given two functors $S, T : \mathcal{C} \to \mathcal{B}$, a natural transformation $\tau : S \dot{\to} T$ is a function which assigns to each object $c \in \mathcal{C}$ an arrow $\tau_c : Sc \to Tc$ of $\mathcal{B}$ in such a way that every arrow $f : c \to c'$ in $\mathcal{C}$ yields the commutative diagram shown in Fig. 2.3.

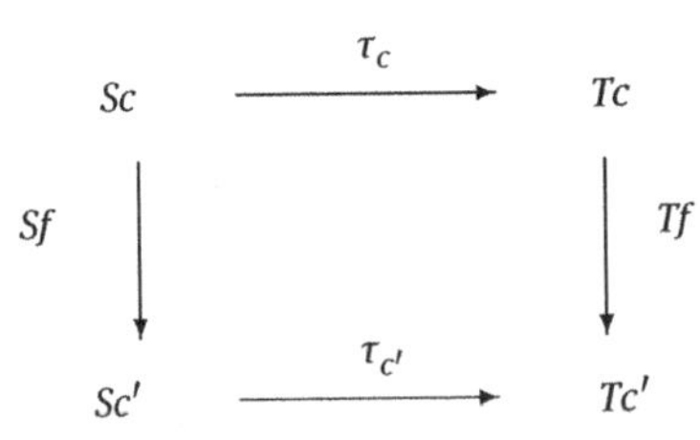

Figure 2.3: The natural transformation.

Remark 2.3.1. A natural transformation τ with every component τ_c invertible in category $\mathcal{B}$ is called a *natural isomorphism* of functors S and T; such an isomorphism is denoted by $\tau : S \cong T$.

Example 2.3.1. Let R be a commutative ring. Denote by R^* the group of units, i. e., the invertible elements of R. Let $S : R \to R^*$ be a functor **CRng** $\to$ **Grp** which is the embedding of R^* into R. Recall that there exists another functor GL_n : **CRng** $\to$ **Grp**, which assigns to R the group $\mathrm{GL}_n(R)$ consisting of the $n \times n$ matrices with the entries in the ring R; we shall put $T = \mathrm{GL}_n$. A natural transformation $\tau : S \xrightarrow{\bullet} T$ is defined by the function $\det_R : M_n(R) \to R^*$ in category **Grp**, which assigns to each matrix $M_n(R)$ its determinant; since $M_n(R)$ is a nonsingular matrix, $\det_R$ takes values in R^*. Thus one gets the commutative diagram shown in Fig. 2.4.

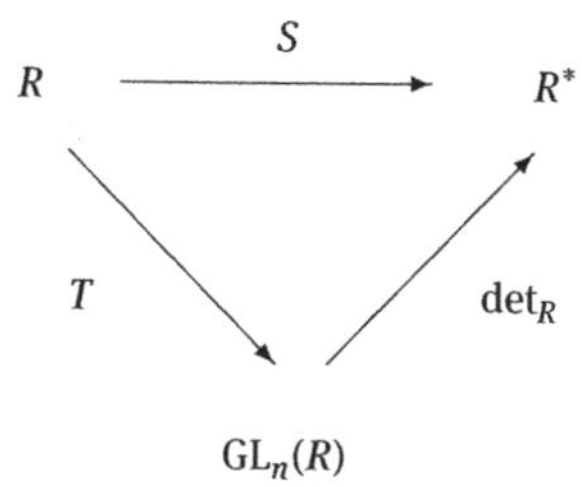

Figure 2.4: Construction of the natural transformation $\det_R$.

Guide to the literature

Natural transformations were introduced in [73]. We refer the reader to the monograph [48] for the basics of homological algebra; for a modern exposition, see [93]. Our exposition of categories follows the book [152].

Exercises

1. Prove that the map from the set of all integral domains to their quotient fields is a functor.
2. Prove that the correspondence between the Lie groups and their Lie algebras is a functor.
3. Prove that the correspondence between affine algebraic varieties and their coordinate rings is a functor. (Hint: For missing definitions, facts and a proof, see, e. g., [108, Chapter 1].)
4. Show that the category of all finite-dimensional vector spaces over a field F is equivalent to the category of matrices over F.

3 C^*-algebras

Such algebras are native to quantum mechanics and representation theory of the locally compact groups; we refer the reader to the seminal paper [171]. The axioms of the C^*-algebras were written in [92]. We recommend the introductory books [13, 32, 64, 69, 86, 170, 243, 286]. Some basic definitions can be found in Section 3.1. The crossed products are reviewed in Section 3.2. The K-theory of C^*-algebras is discussed in Section 3.3. The n-dimensional noncommutative tori are defined in Section 3.4. For a brief review of the AF-algebras, we refer the reader to Section 3.5. The UHF-algebras are covered in Section 3.6. Finally, the Cuntz–Krieger algebras are reviewed in Section 3.7. A bibliography can be found at the end of each section.

3.1 Basic definitions

Definition 3.1.1. A C^*-algebra A is an algebra over $\mathbf{C}$ with a norm $a \mapsto \|a\|$ and an involution $a \mapsto a^*$ on $a \in A$, such that A is complete with respect to the norm and such that for every $a, b \in A$:
(i) $\|ab\| \leq \|a\|\|b\|$;
(ii) $\|a^*a\| = \|a\|^2$.

The C^*-algebra A is called unital if it has a multiplicative identity.

Remark 3.1.1. The C^*-algebras is an axiomatization of the algebras generated by bounded linear operators acting on a Hilbert space $\mathcal{H}$. The space $\mathcal{H}$ comes with a scalar product and, therefore, one gets the norm $\| \bullet \|$ and involution $a \mapsto a^*$ on the operators $a \in A$.

Definition 3.1.2. An element p in a C^*-algebra A is called a projection if $p = p^* = p^2$; two projections p and q are said to be orthogonal if $pq = 0$, and we write $p \perp q$ in this case. The element u in a unital C^*-algebra A is said to be unitary if $uu^* = u^*u = 1$. An element s in A is called a partial isometry when s^*s is a projection.

Example 3.1.1. The field of complex numbers $\mathbf{C}$ is a C^*-algebra with involution given by the complex conjugation $z \mapsto \bar{z}$ of complex numbers $z \in \mathbf{C}$.

Example 3.1.2. The matrix algebra $M_n(\mathbf{C})$ of all $n \times n$ matrices (z_{ij}) with complex entries z_{ij} endowed with the usual matrix norm and involution $(z_{ij})^* = (\bar{z}_{ij})$ is a C^*-algebra. It is a finite-dimensional C^*-algebra since it is isomorphic to the algebra of bounded linear operators on the finite-dimensional Hilbert space $\mathbf{C}^n$. The C^*-algebra $M_n(\mathbf{C})$ is noncommutative unless $n = 1$.

https://doi.org/10.1515/9783110788709-003

Example 3.1.3. If $\mathcal{H}$ is a Hilbert space, then the algebra of all compact linear operators on $\mathcal{H}$ is a C^*-algebra denoted by $\mathcal{K}$; the C^*-algebra $\mathcal{K}$ can be viewed as the limit of finite-dimensional C^*-algebras $M_n(\mathbf{C})$ when n tends to infinity.

Example 3.1.4. If $\mathcal{H}$ is a Hilbert space, then the algebra of all bounded linear operators on $\mathcal{H}$ is a C^*-algebra denoted by $B(\mathcal{H})$.

Example 3.1.5. Let A be a C^*-algebra. Consider an algebra $M_n(A)$ consisting of all $n \times n$ matrices (a_{ij}) with entries $a_{ij} \in A$; the involution on $M_n(A)$ is defined by the formula $(a_{ij})^* = (a_{ji}^*)$. There exists a unique norm on $M_n(A)$ such that $M_n(A)$ is a C^*-algebra.

Remark 3.1.2. The C^*-algebra $M_n(A)$ is isomorphic to a tensor product C^*-algebra $M_n(\mathbf{C}) \otimes A$; since the algebra $\mathcal{K} = \lim_{n\to\infty} M_n(\mathbf{C})$, one can talk about the tensor product C^*-algebra $A \otimes \mathcal{K}$.

Definition 3.1.3. The C^*-algebras A and A' are said to be Morita equivalent (stably isomorphic), if

$$A \otimes \mathcal{K} \cong A' \otimes \mathcal{K}.$$

Let A be a Banach algebra, i. e., a complete normed algebra over $\mathbf{C}$. When A is commutative, a nonzero homomorphism $\tau : A \to \mathbf{C}$ is called a *character* of A; we shall denote by $\Omega(A)$ the space of all characters on A endowed with the weak topology. Define a function $\hat{a} : \Omega(A) \to \mathbf{C}$ by the formula $\tau \mapsto \tau(a)$.

Theorem 3.1.1 (Gelfand). *If A is a commutative Banach algebra, then:*
(i) *$\Omega(A)$ is a locally compact Hausdorff topological space, which is compact whenever A is unital;*
(ii) *the map $A \mapsto C_0(\Omega(A))$ given by the formula $a \mapsto \hat{a}$ is a norm-decreasing homomorphism.*

Corollary 3.1.1. *Each commutative C^*-algebra is isomorphic to the algebra $C_0(X)$ of continuous complex-valued functions vanishing at the infinity of a locally compact Hausdorff topological space X.*

For the noncommutative C^*-algebras, no general classification is known; however, the Gelfand–Naimark–Segal (GNS) construction implies that each C^*-algebra has a concrete realization as a C^*-subalgebra of the algebra $B(\mathcal{H})$ for some Hilbert space $\mathcal{H}$. Below is a brief account of the GNS construction.

A *representation* of a C^*-algebra A is a pair $(\mathcal{H}, \varphi)$, where $\mathcal{H}$ is a Hilbert space and $\varphi : A \to B(\mathcal{H})$ is a $*$-homomorphism. The representation $(\mathcal{H}, \varphi)$ is said to be *faithful* if φ is injective. One can sum up the representations as follows. If $(\mathcal{H}_\lambda, \varphi_\lambda)_{\lambda\in\Lambda}$ is a family of representation of the C^*-algebra A, a *direct sum* is the representation $(\mathcal{H}, \varphi)$ got by setting $\mathcal{H} = \bigoplus_{\lambda\in\Lambda} \mathcal{H}_\lambda$ and $\varphi(a)((x_\lambda)_\lambda) = (\varphi_\lambda(a)(x_\lambda))_\lambda$ for all $a \in A$ and $(x_\lambda)_\lambda \in \mathcal{H}$.

For each positive linear functional $\tau : A \to C$ on the C^*-algebra A, there is an associated representation of A. Indeed, let $N_\tau = \{a \in A \mid \tau(a^*a) = 0\}$ be a subset of A; it is easy to verify that N_τ is a closed left ideal of A. We shall define a map $(A/N_\tau, A/N_\tau) \to \mathbf{C}$ by the formula

$$(a + N_\tau, b + N_\tau) \mapsto \tau(b^*a);$$

such a map is a well-defined inner product on the inner product space A/N_τ. We denote by $\mathcal{H}_\tau$ the Hilbert space completion of A/N_τ. If $a \in A$, then one can define an operator $\varphi(a) \in B(\mathcal{H}_\tau)$ by setting

$$\varphi(a)(b + N_\tau) = ab + N_\tau.$$

Thus one gets a $*$-homomorphism

$$\varphi_\tau : A \to B(\mathcal{H}_\tau)$$

given by the formula $a \mapsto \varphi_\tau(a)$; the representation $(\mathcal{H}_\tau, \varphi_\tau)$ of A is called the *GNS representation* associated to the positive linear functional τ. A *universal* representation of A is defined as the direct sum of representations $(\mathcal{H}_\tau, \varphi_\tau)$ as τ ranges over the space $S(A)$ of all positive linear functionals on A.

Theorem 3.1.2 (Gelfand–Naimark). *If A is a C^*-algebra, then it has a faithful representation in the space $B(\mathcal{H})$ of bounded linear operators on a Hilbert space $\mathcal{H}$; such a representation coincides with the universal representation of A.*

Guide to the literature

The axioms of the C^*-algebra are due to [92]. We refer the reader to the textbooks [13, 64, 69, 86, 170, 243, 286].

3.2 Crossed products

Let A be a C^*-algebra and G a locally compact group. We shall consider a continuous homomorphism α from G to the group $\operatorname{Aut} A$ of $*$-automorphisms of A endowed with the topology of pointwise norm-convergence. Roughly speaking, the idea of the crossed product construction is to embed A into a larger C^*-algebra in which the automorphism becomes the inner automorphism. We shall pass to a detailed description of the crossed product construction.

A *covariant representation* of the triple (A, G, α) is a pair of representations (π, ρ) of A and G on the same Hilbert space $\mathcal{H}$, such that

$$\rho(g)\pi(a)\rho(g)^* = \pi(\alpha_g(a))$$

for all $a \in A$ and $g \in G$. Each covariant representation of (A, G, α) gives rise to a convolution algebra $C(G, A)$ of continuous functions from G to A; the completion of $C(G, A)$ in the norm topology is a C^*-algebra $A \rtimes_\alpha G$ called a *crossed product* of A by G. If α is a single automorphism of A, one gets an action of $\mathbf{Z}$ on A; the crossed product in this case is called simply the crossed product of A by α.

Example 3.2.1. Let $A \cong \mathbf{C}$ the field of complex numbers. The nondegenerate representations ρ of $C(G)$ correspond to unitary representations U of group G via the formula

$$\rho(f) = \int_G f(s) U_s ds,$$

for all $f \in C(G)$. The completion of $C(G)$ in the operator norm is called the *group C^*-algebra* and denoted by $C^*(G)$. Thus $\mathbf{C} \rtimes_{\mathrm{Id}} G \cong C^*(G)$.

Example 3.2.2. Let $A \cong C(S^1)$ be the commutative C^*-algebra of continuous complex-valued functions on the unit circle S^1. Let $\mathbf{Z}/n\mathbf{Z}$ be the cyclic group of order n acting on S^1 by rotation through the rational angle $\frac{2\pi}{n}$. In this case it is known that

$$C(S^1) \rtimes \mathbf{Z}/n\mathbf{Z} \cong M_n(C(S^1)),$$

where $M_n(C(S^1))$ is the C^*-algebra of all $n \times n$ matrices with the entries in the C^*-algebra $C(S^1)$.

Example 3.2.3. Let $A \cong C(S^1)$ be the commutative C^*-algebra of continuous complex-valued functions on the unit circle S^1. Let $\theta \in \mathbf{R} - \mathbf{Q}$ be an irrational number and consider an automorphism of $C(S^1)$ generated by the rotation of S^1 by the irrational angle $2\pi\theta$. In this case $G \cong \mathbf{Z}$ and it is known that

$$C(S^1) \rtimes \mathbf{Z} \cong \mathcal{A}_\theta,$$

where $\mathcal{A}_\theta$ is called an *irrational rotation C^*-algebra*.

One can recover the C^*-algebra A from its crossed product by the action of a locally compact abelian group G; the corresponding construction is known as the *Landstadt–Takai duality*. Namely, let (A, G, α) be a C^*-dynamical system with G locally compact abelian group; let $\hat{G}$ be the dual of G. For each $\gamma \in \hat{G}$, one can define a map $\hat{a}_\gamma : C(G, A) \to C(G, A)$ given by the formula

$$\hat{a}_\gamma(f)(s) = \bar{\gamma}(s) f(s), \quad \forall s \in G.$$

In fact, $\hat{a}_\gamma$ is a $*$-homomorphism, since it respects the convolution product and involution on $C_c(G, A)$ [287]. Because the crossed product $A \rtimes_\alpha G$ is the closure of $C(G, A)$, one gets an extension of $\hat{a}_\gamma$ to an element of $\mathrm{Aut}(A \rtimes_\alpha G)$ and, therefore, a homomorphism

$$\hat{\alpha} : \hat{G} \to \mathrm{Aut}(A \rtimes_\alpha G).$$

Theorem 3.2.1 (Landstadt–Takai duality). *If G is a locally compact abelian group, then*

$$(A \rtimes_\alpha G) \rtimes_{\hat{\alpha}} \hat{G} \cong A \otimes \mathcal{K}(L^2(G)),$$

where $\mathcal{K}(L^2(G))$ is the algebra of compact operators on the Hilbert space $L^2(G)$.

Guide to the literature

The crossed products were introduced in [171]. A detailed account can be found in the monograph [287].

3.3 K-theory of the C^*-algebras

We review the covariant functors

$$\begin{cases} K_0^+ : \mathbf{C^*\text{-}Alg} \to \mathbf{Ab\text{-}Semi}, \\ K_0 : \mathbf{C^*\text{-}Alg} \to \mathbf{Ab}, \\ K_1 : \mathbf{C^*\text{-}Alg} \to \mathbf{Ab}, \end{cases}$$

where **C*-Alg** is the category of C^*-algebras and homomorphisms between them, **Ab-Semi** the category of abelian semigroups and homomorphisms between them, and **Ab** the category of abelian groups and the respective homomorphisms. Let A be a unital C^*-algebra; the definition of K_0-group of A requires simultaneous consideration of all matrix algebras with the entries in A.

Definition 3.3.1. By $M_\infty(A)$ one understands the algebraic direct limit of the C^*-algebras $M_n(A)$ under the embedding $a \mapsto \mathrm{diag}(a, 0)$.

Remark 3.3.1. The direct limit $M_\infty(A)$ can be thought of as the C^*-algebra of infinite-dimensional matrices whose entries are all zero except for a finite number of the nonzero entries taken from the C^*-algebra A.

Definition 3.3.2. The projections $p, q \in M_\infty(A)$ are equivalent if there exists an element $v \in M_\infty(A)$ such that

$$p = v^* v \quad \text{and} \quad q = vv^*.$$

The equivalence relation is denoted by $\sim$ and the corresponding equivalence class of projection p is denoted by $[p]$.

Definition 3.3.3. We shall write $V(A)$ to denote all equivalence classes of projections in the C^*-algebra $M_\infty(A)$; in other words,

$$V(A) := \{[p] : p = p^* = p^2 \in M_\infty(A)\}.$$

Remark 3.3.2. The set $V(A)$ has the natural structure of an abelian semigroup with the addition operation defined by the formula

$$[p] + [q] := [\mathrm{diag}(p, q)] = [p' \oplus q'],$$

where $p' \sim p$, $q' \sim q$ and $p' \perp q'$. The identity of the semigroup $V(A)$ is given by $[0]$, where 0 is the zero projection.

Definition 3.3.4. The semigroup $V(A)$ is said to have the cancellation property if the equality $[p] + [r] = [q] + [r]$ implies $[p] = [q]$ for any elements $[p], [q], [r] \in V(A)$.

Example 3.3.1. Let $A \cong \mathbf{C}$ be the field of complex numbers. It is not hard to see that $V(\mathbf{C})$ has the cancellation property and $V(\mathbf{C}) \cong \mathbf{N} \cup \{0\}$ is the additive semigroup of natural numbers with the zero.

Theorem 3.3.1 (Functor K_0^+). *If $h : A \to A'$ is a homomorphism of the C^*-algebras A and A', then the induced map $h_* : V(A) \to V(A')$ given by the formula $[p] \mapsto [h(p)]$ is a well-defined homomorphism of the abelian semigroups. In other words, the correspondence*

$$K_0^+ : A \to V(A)$$

is a covariant functor from the category **C*-Alg** *to the category* **Ab-Semi**.

Let $(S, +)$ be an abelian semigroup. One can associate to every $(S, +)$ an abelian group as follows. Define an equivalence relation $\sim$ on $S \times S$ by $(x_1, y_1) \sim (x_2, y_2)$ if there exists $z \in S$ such that $x_1 + y_2 + z = x_2 + y_1 + z$. We shall write

$$G(S) := S \times S / \sim$$

and let $[x, y]$ be the equivalence class of (x, y). The operation

$$[x_1, y_1] + [x_2, y_2] = [x_1 + x_2, y_1 + y_2]$$

is well defined and turns the semigroup $(S, +)$ into an abelian group $G(S)$.

Definition 3.3.5. The abelian group $G(S)$ is called the Grothendieck group of the abelian semigroup $(S, +)$. The map $\gamma_S : S \to G(S)$ given by the formula

$$x \mapsto [x + y, y]$$

is independent of the choice of $y \in S$ and called the Grothendieck map.

Remark 3.3.3. The Grothendieck map $\gamma_S : S \to G(S)$ is injective if and only if the semigroup $(S, +)$ has the cancellation property. Without cancellation, it is hard to recover the initial semigroup $(S, +)$ from the Grothendieck group.

Definition 3.3.6. By the K_0-group $K_0(A)$ of the unital C^*-algebra A one understands the Grothendieck group of the abelian semigroup $V(A)$.

Example 3.3.2. Let $A \cong \mathbf{C}$ be the field of complex numbers. It is not hard to see that $K_0(\mathbf{C}) \cong \mathbf{Z}$ is the infinite cyclic group.

Example 3.3.3. If $A \cong M_n(\mathbf{C})$, then $K_0(M_n(\mathbf{C})) \cong \mathbf{Z}$.

Example 3.3.4. If $A \cong C(X)$, then $K_0(C(X)) \cong \mathbf{Z}$, whenever X is a contractible topological space.

Example 3.3.5. If $A \cong \mathcal{K}$, then $K_0(\mathcal{K}) \cong \mathbf{Z}$.

Example 3.3.6. If $A \cong B(\mathcal{H})$, then $K_0(B(\mathcal{H})) \cong 0$.

Example 3.3.7. If A is a C^*-algebra, then $K_0(M_n(A)) \cong K_0(A)$.

Theorem 3.3.2 (Functor K_0). *If $h : A \to A'$ is a homomorphism of the C^*-algebras A and A', then the induced map $h_* : K_0(A) \to K_0(A')$ given by the formula $[p] \mapsto [h(p)]$ is a well-defined homomorphism of the abelian groups. In other words, the correspondence*

$$K_0 : A \to K_0(A)$$

is a covariant functor from the category **C*-Alg** *to the category* **Ab**.

Let A be a unital C^*-algebra. Consider the multiplicative group $\mathrm{GL}_n(A)$ consisting of all invertible $n \times n$ matrices with the entries in A; the $\mathrm{GL}_n(A)$ is a topological group, since A is endowed with a norm. By $\mathrm{GL}_\infty(A)$, one understands the direct limit

$$\mathrm{GL}_\infty(A) := \lim_{n \to \infty} \mathrm{GL}_n(A),$$

defined by the embedding $\mathrm{GL}_n(A) \hookrightarrow \mathrm{GL}_{n+1}$ given by the formula $x \to \mathrm{diag}(x, 1)$. (Note that this embedding is the "exponential" of the embedding of $M_n(A)$ into $M_{n+1}(A)$ used in the construction of the K_0-group of A.) Let $\mathrm{GL}^0_\infty(A)$ be the connected component of $\mathrm{GL}_\infty(A)$, which contains the unit of the group $\mathrm{GL}_\infty(A)$.

Definition 3.3.7. By the K_1-group $K_1(A)$ of the unital C^*-algebra A one understands the multiplicative abelian group consisting of the connected components of the group $\mathrm{GL}_\infty(A)$, i. e.,

$$K_1(A) := \mathrm{GL}_\infty(A)/\mathrm{GL}^0_\infty(A).$$

Example 3.3.8. Let $A \cong \mathbf{C}$ be the field of complex numbers. It is not hard to see that $K_1(\mathbf{C}) \cong 0$ is the infinite cyclic group.

Example 3.3.9. If $A \cong M_n(\mathbf{C})$, then $K_1(M_n(\mathbf{C})) \cong 0$.

Example 3.3.10. If $A \cong C(X)$, then $K_1(C(X)) \cong 0$, whenever X is a contractible topological space.

Example 3.3.11. If $A \cong \mathcal{K}$, then $K_1(\mathcal{K}) \cong 0$.

Example 3.3.12. If $A \cong B(\mathcal{H})$, then $K_1(B(\mathcal{H})) \cong 0$.

Example 3.3.13. If A is a C^*-algebra, then $K_1(M_n(A)) \cong K_1(A)$.

Theorem 3.3.3 (Functor K_1). *If $h : A \to A'$ is a homomorphism of the C^*-algebras A and A', then the induced map $h_* : K_0(A) \to K_0(A')$ is a homomorphism of the multiplicative abelian groups. In other words, the correspondence*

$$K_1 : A \to K_0(A)$$

is a covariant functor from the category **C*-Alg** *to the category* **Ab**.

Remark 3.3.4. The functors K_0 and K_1 are related to each other by the formula

$$K_1(A) \cong K_0(C_0(\mathbf{R}) \otimes A),$$

where $C_0(\mathbf{R})$ is the commutative C^*-algebra of continuous complex-valued functions with the compact support on $\mathbf{R}$; the tensor product $S(A) = C_0(\mathbf{R}) \otimes A$ is called a *suspension* C^*-algebra of A. The suspension can be regarded as a natural transformation implementing the morphism between two functors K_0 and K_1, because the diagram shown in Fig. 3.1 is commutative.

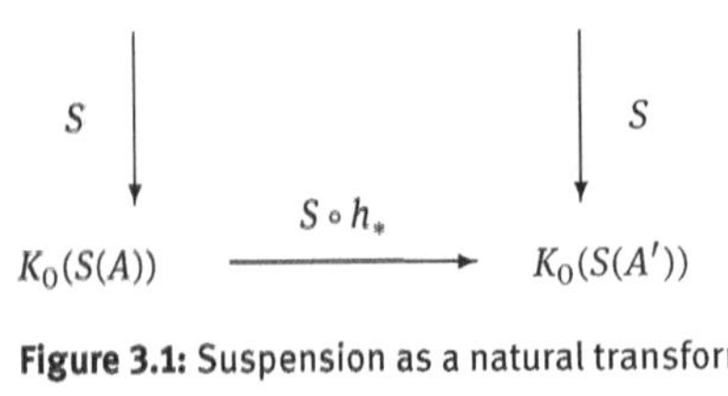

Figure 3.1: Suspension as a natural transformation.

For a calculation of the K_0 and K_1-groups of the crossed product C^*-algebras, the following result is extremely useful.

Theorem 3.3.4 (Pimsner–Voiculescu). *Let A be a C^*-algebra and $\alpha \in \mathrm{Aut}(A)$; consider the crossed product C^*-algebra $A \rtimes_\alpha \mathbf{Z}$ and let $i : A \to A \rtimes_\alpha \mathbf{Z}$ be the canonical embedding. Then there exists a cyclic six-term exact sequence of the abelian groups shown in Fig. 3.2.*

$$\begin{array}{ccccc} K_0(A) & \xrightarrow{1-\alpha_*} & K_0(A) & \xrightarrow{i_*} & K_0(A \rtimes_\alpha \mathbf{Z}) \\ \uparrow & & & & \downarrow \\ K_1(A \rtimes_\alpha \mathbf{Z}) & \xleftarrow{i_*} & K_1(A) & \xleftarrow{1-\alpha_*} & K_1(A) \end{array}$$

Figure 3.2: The Pimsner–Voiculescu exact sequence.

Guide to the literature

The K-theory is a powerful tool in the C^*-algebra theory. The K-theory alone classifies certain types of the C^*-algebras and is linked with the geometry and topology of manifolds, see a review of the Index Theory in Chapter 10. An excellent account of the K-theory for C^*-algebras is given in [32]. For a friendly approach, we refer the reader to [286]. The textbooks [86] and [243] give an encyclopedic and detailed accounts, respectively.

3.4 Noncommutative tori

An n-dimensional noncommutative torus is a C^*-algebra given by n generators and $\frac{n(n-1)}{2}$ quadratic relations. The n-dimensional tori are reviewed in Section 3.4.1. The 2-dimensional tori are treated in Section 3.4.2. We refer the reader to the survey [239].

3.4.1 *n*-dimensional noncommutative tori

Let

$$\Theta = \begin{pmatrix} 0 & \theta_{12} & \dots & \theta_{1n} \\ -\theta_{12} & 0 & \dots & \theta_{2n} \\ \vdots & \vdots & \ddots & \vdots \\ -\theta_{1n} & -\theta_{2n} & \dots & 0 \end{pmatrix}$$

be a skew-symmetric matrix with $\theta_{ij} \in \mathbf{R}$.

Definition 3.4.1. An n-dimensional noncommutative torus $\mathcal{A}_\Theta$ is the universal C^*-algebra generated by n unitary operators $u_1, \ldots, u_n$ satisfying the commutation relations

$$u_j u_i = e^{2\pi i \theta_{ij}} u_i u_j, \quad 1 \le i, j \le n.$$

Definition 3.4.2. Let A, B, C, D be the $n \times n$ matrices with integer entries; by $\mathrm{SO}(n, n|\mathbf{Z})$ we shall understand a subgroup of the matrix group $\mathrm{GL}_{2n}(\mathbf{Z})$ consisting of all matrices of the form

$$\begin{pmatrix} A & B \\ C & D \end{pmatrix},$$

such that the matrices $A, B, C, D \in \mathrm{GL}_k(\mathbf{Z})$ satisfy the conditions

$$A^T D + C^T B = I, \quad A^T C + C^T A = 0 = B^T D + D^T B,$$

where A^T, B^T, C^T, and D^T are the transpose of the matrices A, B, C, and D, respectively, and I is the identity matrix.

Remark 3.4.1. The group $\mathrm{SO}(n, n|\mathbf{Z})$ can be equivalently defined as a subgroup of the group $\mathrm{SO}(n, n|\mathbf{R})$ consisting of linear transformations of the space $\mathbf{R}^{2n}$, which preserve the quadratic form

$$x_1 x_{n+1} + x_2 x_{n+2} + \cdots + x_n x_{2n}.$$

Theorem 3.4.1 (Rieffel–Schwarz). *The n-dimensional noncommutative tori $\mathcal{A}_\Theta$ and $\mathcal{A}_{\Theta'}$ are Morita equivalent (stably isomorphic) if the matrices Θ and Θ' belong to the same orbit of the group $\mathrm{SO}(n, n \mid \mathbf{Z})$ acting on Θ by the formula*

$$\Theta' = \frac{A\Theta + B}{C\Theta + D},$$

where A, B, C, and D are integer matrices.

Theorem 3.4.2. $K_0(\mathcal{A}_\Theta) \cong K_1(\mathcal{A}_\Theta) \cong \mathbf{Z}^{2^{n-1}}$.

Proof. Let T^{n-1} be an $(n-1)$-dimensional topological torus. Because $\mathcal{A}_\Theta \cong C(T^{n-1}) \rtimes_\Theta \mathbf{Z}$ is the crossed product C^*-algebra, one can apply the Pimsner–Voiculescu exact sequence and get the diagram of Fig. 3.3.

It is known that for $i \in \{0, 1\}$ one has $K_i(C(T^{n-1})) \cong K^{i-1}(T^{n-1}) \cong \mathbf{Z}^{2^{n-2}}$, where $K^\bullet(T^{n-1})$ are the topological K-groups of torus T^{n-1}. Thus one gets two split exact sequences

$$0 \to \mathbf{Z}^{2^{n-2}} \to K_i(\mathcal{A}_\Theta) \to \mathbf{Z}^{2^{n-2}} \to 0.$$

Hence $K_i(\mathcal{A}_\Theta) \cong \mathbf{Z}^{2^{n-2}} \oplus \mathbf{Z}^{2^{n-2}} \cong \mathbf{Z}^{2^{n-1}}$. □

$$\begin{array}{ccccc} K_0(C(T^{n-1})) & \xrightarrow{1-\Theta_*} & K_0(C(T^{n-1})) & \xrightarrow{i_*} & K_0(\mathcal{A}_\Theta) \\ \uparrow & & & & \downarrow \\ K_1(\mathcal{A}_\Theta) & \xleftarrow{i_*} & K_1(C(T^{n-1})) & \xleftarrow{1-\Theta_*} & K_1(C(T^{n-1})) \end{array}$$

Figure 3.3: The Pimsner–Voiculescu exact sequence for $\mathcal{A}_\Theta$.

Remark 3.4.2. Unless $n < 5$, the abelian semigroup $V(\mathcal{A}_\Theta)$ does not have cancellation and, therefore, $V(\mathcal{A}_\Theta)$ cannot be embedded into the group $K_0(\mathcal{A}_\Theta) \cong \mathbf{Z}^{2^{n-1}}$. However, for $n = 2$, the semigroup $V(\mathcal{A}_\Theta)$ does embed into $K_0(\mathcal{A}_\Theta) \cong \mathbf{Z}^2$ turning $K_0(\mathcal{A}_\Theta)$ into a totally ordered abelian group; in this case the functor

$$\mathcal{A}_\Theta \longrightarrow V(\mathcal{A}_\Theta)$$

is an equivalence of the categories of 2-dimensional noncommutative tori and totally ordered abelian groups of rank 2.

3.4.2 2-dimensional noncommutative tori

If $n = 2$, then

$$\Theta = \begin{pmatrix} 0 & \theta_{12} \\ -\theta_{12} & 0 \end{pmatrix}.$$

Therefore the noncommutative torus $\mathcal{A}_\Theta$ depends on a single parameter $\theta_{12} := \theta$; we shall simply write $\mathcal{A}_\theta$ in this case. Thus one gets the following

Definition 3.4.3. A noncommutative torus $\mathcal{A}_\theta$ is the universal C^*-algebra on the generators u, u^*, v, v^* and relations

$$\begin{cases} vu = e^{2\pi i\theta} uv, \\ u^*u = uu^* = 1, \\ v^*v = vv^* = 1. \end{cases}$$

Theorem 3.4.3. *The noncommutative tori $\mathcal{A}_\theta$ and $\mathcal{A}_{\theta'}$ are Morita equivalent (stably isomorphic) if and only if*

$$\theta' = \frac{a\theta + b}{c\theta + d},$$

where $a, b.c, d \in \mathbf{Z}$ are such that $ad - bc = \pm 1$.

Proof. We shall use the general formula involving group $\mathrm{SO}(n, n|\mathbf{Z})$. The skew-symmetric matrix Θ has the form

$$\Theta = \begin{pmatrix} 0 & \theta \\ -\theta & 0 \end{pmatrix}.$$

The 2×2 integer matrices A, B, C, D take the form

$$\begin{cases} A = \begin{pmatrix} a & 0 \\ 0 & a \end{pmatrix}, \\ B = \begin{pmatrix} 0 & b \\ -b & 0 \end{pmatrix}, \\ C = \begin{pmatrix} 0 & -c \\ c & 0 \end{pmatrix}, \\ D = \begin{pmatrix} d & 0 \\ 0 & d \end{pmatrix}. \end{cases}$$

The reader can verify that A, B, C, and D satisfy the conditions

$$A^T D + C^T B = I, \quad A^T C + C^T A = 0 = B^T D + D^T B.$$

Therefore, the matrix

$$\begin{pmatrix} A & B \\ C & D \end{pmatrix} \in \mathrm{SO}(2, 2|\mathbf{Z}).$$

A direct computation gives the following equations:

$$\Theta' = \frac{A\Theta + B}{C\Theta + D} = \begin{pmatrix} 0 & \frac{a\theta+b}{c\theta+d} \\ -\frac{a\theta+b}{c\theta+d} & 0 \end{pmatrix}.$$

In other words, one gets $\theta' = \frac{a\theta+b}{c\theta+d}$. □

Theorem 3.4.4. $K_0(\mathcal{A}_\theta) \cong K_1(\mathcal{A}_\theta) \cong \mathbf{Z}^2$.

Proof. Follows from the general formula for $n = 2$. □

Theorem 3.4.5 (Pimsner–Voiculescu–Rieffel). *The abelian semigroup $V(\mathcal{A}_\theta)$ embeds into the group $K_0(\mathcal{A}_\theta) \cong \mathbf{Z}^2$; the image of the injective map*

$$V(\mathcal{A}_\theta) \longrightarrow K_0(\mathcal{A}_\theta)$$

is given by the formula

$$\{(m, n) \in \mathbf{Z}^2 \mid m + n\theta > 0\}.$$

Corollary 3.4.1. *The abelian group $K_0(\mathcal{A}_\theta)$ gets an order structure $\geq$ coming from the semigroup $V(\mathcal{A}_\theta)$ and given by the formula*

$$z_1 \geq z_2 \quad \textit{iff} \quad z_1 - z_2 \in V(\mathcal{A}_\theta) \subset K_0(\mathcal{A}_\theta).$$

If **NTor** *is the category of 2-dimensional noncommutative tori whose arrows are stable homomorphisms of the tori and* **Ab-Ord** *is the category of ordered abelian groups of rank 2 whose arrows are order-homomorphisms of the groups, then the functor*

$$K_0^+ : \mathbf{NTor} \to \mathbf{Ab\text{-}Ord}$$

is full and faithful, i. e., an isomorphism of the categories.

Guide to the literature

We encourage the reader to start with the survey paper [239]. The noncommutative torus is also known as the irrational rotation algebra, see [230] and [238]. The n-dimensional noncommutative tori and their K-theory were considered in [240].

3.5 AF-algebras

Such algebras were introduced in [41] as a generalization of the UHF-algebras studied in [94]. The AF-algebras are classified by their K-theory [76]. Section 3.5.1 covers generic AF-algebras and Section 3.5.2 covers stationary AF-algebras.

3.5.1 Generic AF-algebras

The Approximately Finite-dimensional (AF) algebra is a limit of the finite-dimensional C^*-algebras.

Theorem 3.5.1 (Bratteli–Elliott). *Any finite-dimensional C^*-algebra A is isomorphic to*

$$M_{n_1}(\mathbf{C}) \oplus M_{n_2}(\mathbf{C}) \oplus \cdots \oplus M_{n_k}(\mathbf{C})$$

for some positive integers $n_1, n_2, \ldots, n_k$. Moreover,

$$K_0(A) \cong \mathbf{Z}^k \quad \textit{and} \quad K_1(A) \cong 0.$$

Definition 3.5.1. By an AF-algebra one understands a C^*-algebra $\mathbb{A}$ which is the norm closure of an ascending sequence of the embeddings φ_i of the finite-dimensional C^*-algebras

$$A_1 \overset{\varphi_1}{\hookrightarrow} A_2 \overset{\varphi_2}{\hookrightarrow} \cdots .$$

The set-theoretic limit $\mathbb{A} = \lim A_i$ is endowed with a natural algebraic structure given by the formula $a_m + b_k \to a + b$; here $a_m \to a$, $b_k \to b$ for the sequences $a_m \in A_m$, $b_k \in A_k$.

Remark 3.5.1. To keep track of all embeddings $\varphi_i : A_i \hookrightarrow A_{i+1}$, it is convenient to arrange them into a graph called the Bratteli diagram. Let $A_i = M_{i_1} \oplus \cdots \oplus M_{i_k}$ and $A_{i'} = M_{i'_1} \oplus \cdots \oplus M_{i'_k}$ be the semisimple C^*-algebras and $\varphi_i : A_i \to A_{i'}$ the homomorphism. One has the two sets of vertices $V_{i_1}, \ldots, V_{i_k}$ and $V_{i'_1}, \ldots, V_{i'_k}$ joined by the a_{rs} edges, whenever the summand M_{i_r} contains a_{rs} copies of the summand $M_{i'_s}$ under the embedding φ_i. As i varies, one obtains an infinite graph; the graph is called a *Bratteli diagram* of the AF-algebra $\mathbb{A}$. The Bratteli diagram defines a unique AF-algebra, but a converse is false.

Example 3.5.1. The embeddings

$$M_1(\mathbf{C}) \overset{\text{Id}}{\hookrightarrow} M_2(\mathbf{C}) \overset{\text{Id}}{\hookrightarrow} \cdots$$

define the C^*-algebra $\mathcal{K}$ of all compact operators on a Hilbert space $\mathcal{H}$. The corresponding Bratteli diagram is shown in Fig. 3.4.

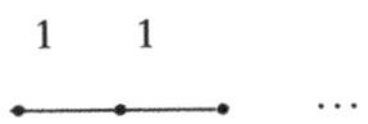

Figure 3.4: Bratteli diagram of the AF-algebra $\mathcal{K}$.

Example 3.5.2. Let $\theta \in \mathbf{R}$ be an irrational number and consider the regular continued fraction

$$\theta = a_0 + \cfrac{1}{a_1 + \cfrac{1}{a_2 + \cdots}} := [a_0, a_1, a_2, \ldots].$$

One can define an AF-algebra $\mathbb{A}_\theta$ by the Bratteli diagram shown in Fig. 3.5; the AF-algebra is called an *Effros–Shen algebra*.

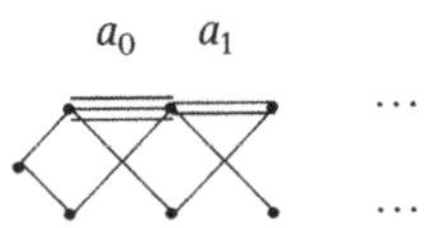

Figure 3.5: The Effros–Shen algebra $\mathbb{A}_\theta$.

Remark 3.5.2. Because $K_1(A_i) \cong 0$ and the K_1-functor is continuous on the inductive limits, one gets

$$K_1(\mathbb{A}) \cong 0,$$

for any AF-algebra $\mathbb{A}$. The group $K_0(\mathbb{A})$ is more interesting because each embedding $\varphi_i : A_i \hookrightarrow A_{i+1}$ induces a homomorphism $(\varphi_i)_* : K_0(A_i) \to K_0(A_{i+1})$ of the abelian groups $K_0(A_i)$.

We need the following

Definition 3.5.2. By a dimension group one understands an ordered abelian group (G, G^+) which is the limit of the sequence of ordered abelian groups $\mathbf{Z}^{n_i}$ and positive homomorphisms $(\varphi_i)_* : \mathbf{Z}^{n_i} \to \mathbf{Z}^{n_{i+1}}$:

$$\mathbf{Z}^{n_1} \stackrel{(\varphi_1)_*}{\to} \mathbf{Z}^{n_2} \stackrel{(\varphi_2)_*}{\to} \cdots,$$

where the positive cone of $\mathbf{Z}^{n_i}$ is defined by the formula

$$(\mathbf{Z}^{n_i})^+ = \{(x_1, \dots, x_{n_i}) \in \mathbf{Z}^{n_i} : x_j \geq 0\}.$$

Theorem 3.5.2 (Elliott). *If $\mathbb{A}$ is an AF-algebra, then*

$$K_0(\mathbb{A}) \cong G \quad \text{and} \quad V(\mathbb{A}) \cong G^+.$$

Moreover, if **AF-Alg** *is the category of the AF-algebras and homomorphisms between them and* **Ab-Ord** *is the category of ordered abelian groups (with the scaled units) and order-preserving homomorphisms between them, then the functor*

$$K_0^+ : \textbf{AF-Alg} \to \textbf{Ab-Ord}$$

is full and faithful, i. e., an isomorphism of the categories.

Example 3.5.3. If $\mathbb{A} \cong \mathcal{K}$, then $K_0(\mathcal{K}) \cong \mathbf{Z}$ and $V(\mathcal{K}) \cong \mathbf{Z}^+$, where $\mathbf{Z}^+$ is an additive semigroup of the nonnegative integers.

Example 3.5.4. If $\mathbb{A}_\theta$ is an Effros–Shen algebra, then

$$K_0(\mathbb{A}_\theta) \cong \mathbf{Z}^2 \quad \text{and} \quad V(\mathbb{A}_\theta) \cong \{(m,n) \in \mathbf{Z}^2 \mid m + n\theta > 0\}.$$

Theorem 3.5.3 (Pimsner–Voiculescu). *Let $\mathcal{A}_\theta$ be a two-dimensional noncommutative torus and $\mathbb{A}_\theta$ the corresponding Effros–Shen algebra. There exists an embedding*

$$\mathcal{A}_\theta \hookrightarrow \mathbb{A}_\theta,$$

which induces an order-isomorphism of the corresponding semigroups $V(\mathcal{A}_\theta) \cong V(\mathbb{A}_\theta)$.

3.5.2 Stationary AF-algebras

Definition 3.5.3. By a stationary AF-algebra $\mathbb{A}_\varphi$ one understands an AF-algebra for which the embedding homomorphisms $\varphi_1 = \varphi_2 = \cdots = \text{const}$; in other words, the stationary AF-algebra has the form:

$$A_1 \overset{\varphi}{\hookrightarrow} A_2 \overset{\varphi}{\hookrightarrow} \cdots,$$

where A_i are some finite-dimensional C^*-algebras.

Remark 3.5.3. The corresponding dimension group $(G, G^+_{\varphi_*})$ can be written as

$$\mathbf{Z}^k \overset{\varphi_*}{\longrightarrow} \mathbf{Z}^k \overset{\varphi_*}{\longrightarrow} \mathbf{Z}^k \overset{\varphi_*}{\longrightarrow} \cdots,$$

where φ_* is a matrix with nonnegative integer entries; one can take a minimal power of φ_* to obtain a strictly positive integer matrix B.

Example 3.5.5. Let $\theta \in \mathbf{R}$ be a quadratic irrationality called the *golden mean*:

$$\frac{1+\sqrt{5}}{2} = 1 + \cfrac{1}{1 + \cfrac{1}{1+\cdots}} = [1, 1, 1, \ldots].$$

The corresponding Effros–Shen algebra has a periodic Bratteli diagram shown in Fig. 3.6; the dimension group (G, G^+_B) is given by the following sequence of positive homomorphisms:

$$\mathbf{Z}^2 \overset{\left(\begin{smallmatrix}1&1\\1&0\end{smallmatrix}\right)}{\longrightarrow} \mathbf{Z}^2 \overset{\left(\begin{smallmatrix}1&1\\1&0\end{smallmatrix}\right)}{\longrightarrow} \mathbf{Z}^2 \overset{\left(\begin{smallmatrix}1&1\\1&0\end{smallmatrix}\right)}{\longrightarrow} \cdots.$$

Figure 3.6: Stationary Effros–Shen algebra $\mathbb{A}_{\frac{1+\sqrt{5}}{2}}$.

Definition 3.5.4. If $\mathbb{A}_\varphi$ is a stationary AF-algebra, consider an order-automorphism of its dimension group $(G, G^+_{\varphi_*})$ generated by the 1-shift of the diagram shown in Fig. 3.7

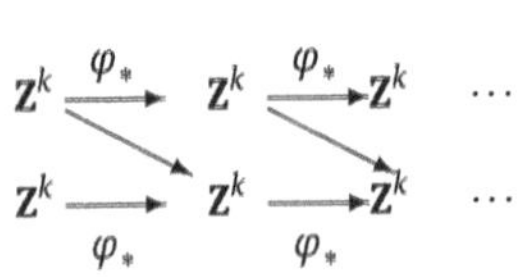

Figure 3.7: Shift automorphism.

and let $\sigma_\varphi : \mathbb{A}_\varphi \to \mathbb{A}_\varphi$ be the corresponding automorphism of $\mathbb{A}_\varphi$; then σ_φ is called the shift automorphism.

To classify all stationary AF-algebras, one needs the following set of invariants. Let $B \in M_k(\mathbf{Z})$ be a matrix with the strictly positive entries corresponding to a stationary dimension group (G, G_B^+), namely

$$\mathbf{Z}^k \xrightarrow{B} \mathbf{Z}^k \xrightarrow{B} \mathbf{Z}^k \xrightarrow{B} \cdots.$$

By the Perron–Frobenius theory, matrix B has a real eigenvalue $\lambda_B > 1$, which exceeds the absolute values of other roots of the characteristic polynomial of B. Note that λ_B is an algebraic integer. Consider the real algebraic number field $K = \mathbf{Q}(\lambda_B)$ obtained as an extension of the field of the rational numbers by the algebraic number λ_B. Let $(v_B^{(1)}, \ldots, v_B^{(k)})$ be the eigenvector corresponding to the eigenvalue λ_B. One can normalize the eigenvector so that $v_B^{(i)} \in K$. Consider the $\mathbf{Z}$-module $\mathfrak{m} = \mathbf{Z}v_B^{(1)} + \cdots + \mathbf{Z}v_B^{(k)}$. The module $\mathfrak{m}$ brings in two new arithmetic objects: (i) the ring Λ of the endomorphisms of $\mathfrak{m}$ and (ii) an ideal I in the ring Λ such that $I = \mathfrak{m}$ after a scaling. The ring Λ is an order in the algebraic number field K and therefore it belongs to an ideal class in Λ. The ideal class of I we denote by $[I]$.

Theorem 3.5.4 (Handelman). *The triple $(\Lambda, [I], K)$ is an invariant of the Morita equivalence (stable isomorphism) class of the stationary AF-algebra $\mathbb{A}_\varphi$.*

Guide to the literature

The AF-algebras were introduced in [41]; it is a generalization of the UHF-algebras studied in [94]. The dimension groups of the AF-algebras were introduced in [76]; he showed that such groups classify the AF-algebras [76]. The Effros–Shen algebras were introduced in [75]. An generalization of such algebras was considered in [194]. The stationary AF-algebras are covered in [74, Chapter 6]. The arithmetic invariant $(\Lambda, [I], K)$ of stationary AF-algebras was introduced in [105].

3.6 UHF-algebras

The uniformly hyperfinite C^*-algebras (the *UHF-algebras*, for brevity) is a special type of the AF-algebras; they were the first AF-algebras studied and classified by an invariant called the supernatural number.

Definition 3.6.1. The UHF-algebra is an AF-algebra which is isomorphic to the inductive limit of the sequence of finite-dimensional C^*-algebras of the form

$$M_{k_1}(\mathbf{C}) \to M_{k_1}(\mathbf{C}) \otimes M_{k_2}(\mathbf{C}) \to M_{k_1}(\mathbf{C}) \otimes M_{k_2}(\mathbf{C}) \otimes M_{k_3}(\mathbf{C}) \to \cdots,$$

where $M_{k_i}(\mathbf{C})$ is a matrix C^*-algebra and $k_i \in \{1, 2, 3, \ldots\}$; we shall denote the UHF-algebra by $M_{\mathbf{k}}$, where $\mathbf{k} = (k_1, k_2, k_3, \ldots)$.

Example 3.6.1. Let p be a prime number and consider the UHF-algebra

$$M_{p^\infty} := M_p(\mathbf{C}) \otimes M_p(\mathbf{C}) \otimes \cdots .$$

For $p = 2$, the algebra M_{2^∞} is known as a Canonical Anticommutation Relations C^*-algebra (the *CAR* or *Fermion algebra*); its Bratteli diagram is shown in Fig. 3.8.

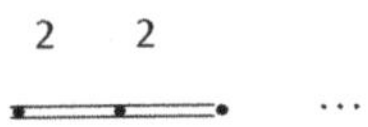

Figure 3.8: Bratteli diagram of the CAR algebra M_{2^∞}.

To classify the UHF-algebras up to a stable isomorphism, one needs the following construction. Let p be a prime number and $n = \sup\{0 \le j \le \infty : p^j \mid \prod_{i=1}^{\infty} k_i\}$; denote by $\mathbf{n} = (n_1, n_2, \ldots)$ an infinite sequence of n_i as p_i runs through the ordered set of all primes.

Definition 3.6.2. By $\mathbf{Q}(\mathbf{n})$ we understand an additive subgroup of $\mathbf{Q}$ consisting of rational numbers whose denominators divide the "supernatural number" $p_1^{n_1} p_2^{n_2} \cdots$, where each n_j belongs to the set $\{0, 1, 2, \ldots, \infty\}$.

Remark 3.6.1. The $\mathbf{Q}(\mathbf{n})$ is a dense subgroup of $\mathbf{Q}$ and every dense subgroup of $\mathbf{Q}$ containing $\mathbf{Z}$ is given by $\mathbf{Q}(\mathbf{n})$ for some $\mathbf{n}$.

Theorem 3.6.1 (Glimm). $K_0(M_{\mathbf{k}}) \cong \mathbf{Q}(\mathbf{n})$.

Example 3.6.2. For the CAR algebra M_{2^∞}, one gets

$$K_0(M_{2^\infty}) \cong \mathbf{Z}\left[\frac{1}{2}\right],$$

where $\mathbf{Z}[\frac{1}{2}]$ are the dyadic rationals.

Theorem 3.6.2 (Glimm). *The UHF-algebras $M_{\mathbf{k}}$ and $M_{\mathbf{k}'}$ are Morita equivalent (stably isomorphic) if and only if $r\mathbf{Q}(\mathbf{n}) = s\mathbf{Q}(\mathbf{n}')$ for some positive integers r and s.*

Guide to the literature

The UHF-algebras were introduced and classified in [94]. The supernatural numbers are covered in [74, p. 28] and [243, Section 7.4].

3.7 Cuntz–Krieger algebras

A Cuntz–Krieger algebra is a C^*-algebra on $2n$ generators and n quadratic relations. The K-groups of such an algebra are finite abelian groups. A Cuntz–Krieger algebra is the crossed product of a stationary AF-algebra by the shift automorphism. (Such a K-theory and the crossed product structure have an application in the noncommutative localization; we refer the reader to Section 6.5.) Denote by

$$B = \begin{pmatrix} b_{11} & b_{12} & \dots & b_{1n} \\ b_{21} & b_{22} & \dots & b_{2n} \\ \vdots & \vdots & \ddots & \vdots \\ b_{n1} & b_{n2} & \dots & b_{nn} \end{pmatrix}$$

a square matrix such that $b_{ij} \in \{0, 1, 2, \dots\}$.

Definition 3.7.1. By a Cuntz–Krieger algebra $\mathcal{O}_B$ one understands the C^*-algebra generated by the partial isometries $s_1, \dots, s_n$ which satisfy the relations

$$\begin{cases} s_1^* s_1 = b_{11} s_1 s_1^* + b_{12} s_2 s_2^* + \dots + b_{1n} s_n s_n^*, \\ s_2^* s_2 = b_{21} s_1 s_1^* + b_{22} s_2 s_2^* + \dots + b_{2n} s_n s_n^*, \\ \vdots \\ s_n^* s_n = b_{n1} s_1 s_1^* + b_{n2} s_2 s_2^* + \dots + b_{nn} s_n s_n^*. \end{cases}$$

Example 3.7.1. Let

$$B = \begin{pmatrix} 1 & 1 & \dots & 1 \\ 1 & 1 & \dots & 1 \\ \vdots & \vdots & \ddots & \vdots \\ 1 & 1 & \dots & 1 \end{pmatrix}.$$

Then $\mathcal{O}_B$ is called a *Cuntz algebra* and denoted by $\mathcal{O}_n$.

Remark 3.7.1. It is known that the C^*-algebra $\mathcal{O}_B$ is simple whenever matrix B is irreducible, i. e., a certain power of B is a strictly positive integer matrix.

Theorem 3.7.1 (Cuntz–Krieger). *If $\mathcal{O}_B$ is a Cuntz–Krieger algebra, then*

$$\begin{cases} K_0(\mathcal{O}_B) \cong \dfrac{\mathbf{Z}^n}{(I - B^T)\mathbf{Z}^n}, \\ K_1(\mathcal{O}_B) \cong \operatorname{Ker}(I - B^T), \end{cases}$$

where B^T is a transpose of the matrix B.

Remark 3.7.2. It is not difficult to see that if $\det(I - B^T) \neq 0$, then $K_0(\mathcal{O}_B)$ is a finite abelian group and $K_1(\mathcal{O}_B) = 0$. Such groups are invariants of the Morita equivalence (stable isomorphism) class of the Cuntz–Krieger algebra.

Theorem 3.7.2 (Cuntz–Krieger). *Let $\mathbb{A}_\varphi$ be a stationary AF-algebra such that $\varphi_* \cong B$ and let $\sigma_\varphi : \mathbb{A}_\varphi \to \mathbb{A}_\varphi$ be the corresponding shift automorphism of $\mathbb{A}_\varphi$. Then*

$$\mathcal{O}_B \otimes \mathcal{K} \cong \mathbb{A}_\varphi \rtimes_{\sigma_\varphi} \mathbf{Z},$$

where the crossed product is taken by the shift automorphism σ_φ.

Guide to the literature

The Cuntz–Krieger algebras were studied in [63]. Such algebras generalize a class of algebras introduced in [62].

Exercises

1. Recall that *spectrum* $\mathbf{Sp}\,(a)$ of an element a of the C^*-algebra A is the set of complex numbers λ such that $a - \lambda e$ is not invertible. Show that if $p \in A$ is a projection, then $\mathbf{Sp}\,(p) \subseteq \{0, 1\}$.
2. Show that if $u \in A$ is a unitary element, then $\mathbf{Sp}\,(u) \subseteq \{z \in \mathbf{C} : |z| = 1\}$.
3. Show that $\|p - q\| \leq 1$ for every pair of projections in the C^*-algebra A.
4. Show that $\|u - v\| \leq 2$ for every pair of unitary elements in the C^*-algebra A.
5. Recall that projections $p, q \in A$ are *orthogonal*, $p \perp q$, if $pq = 0$. Show that the following three conditions are equivalent:
   (i) $p \perp q$;
   (ii) $p + q$ is a projection;
   (iii) $p + q \leq 1$.
6. Recall that v is a *partial isometry* if v^*v is a projection. Show that $v = vv^*v$ and conclude that vv^* is a projection.
7. Prove that if $A \cong \mathbf{C}$, then $K_0(\mathbf{C}) \cong \mathbf{Z}$ is the infinite cyclic group.
8. Prove that if $A \cong M_n(\mathbf{C})$, then $K_0(M_n(\mathbf{C})) \cong \mathbf{Z}$.
9. Prove that if $A \cong \mathbf{C}$, then $K_1(\mathbf{C}) \cong 0$.
10. Prove that if $A \cong M_n(\mathbf{C})$, then $K_1(M_n(\mathbf{C})) \cong 0$.
11. Calculate the K-theory of noncommutative torus $\mathcal{A}_\theta$ using its crossed product structure. (Hint: Use the Pimsner–Voiculscu exact sequence for crossed products.)
12. Calculate the Handelman invariant $(\Lambda, [I], K)$ of the stationary Effros–Shen algebra $\mathbb{A}_{\frac{1+\sqrt{5}}{2}}$ (golden mean algebra).

13. Calculate the K-theory of the Cuntz–Krieger algebra $\mathcal{O}_B$, where

$$B = \begin{pmatrix} 5 & 1 \\ 4 & 1 \end{pmatrix}.$$

(Hint: Use the formulas $K_0(\mathcal{O}_B) \cong \frac{\mathbf{Z}^n}{(I-B^T)\mathbf{Z}^n}$ and $K_1(\mathcal{O}_B) \cong \operatorname{Ker}(I - B^T)$; bring the matrix $I - B^T$ to the *Smith normal form*, see, e. g., [151].)

14. Repeat the exercise for matrix

$$B = \begin{pmatrix} 5 & 2 \\ 2 & 1 \end{pmatrix}.$$

Part II: **Noncommutative invariants**

4 Topology

We construct functors arising in topology of the surface automorphisms, fiber bundles, knots and links, etc. These functors range in a category of the AF-algebras, Cuntz–Krieger algebras, cluster C^*-algebras, etc. The functors define a set of homotopy invariants of the corresponding topological space. The invariants are new but some were known before, e. g., torsion in the fiber bundles, the Jones and HOMFLY polynomials, etc. A reference to the topological facts can be found at the end of each section.

4.1 Classification of the surface automorphisms

We assume that X is a compact oriented surface of genus $g \geq 1$; we shall be interested in the continuous invertible self-maps (automorphisms) of X, i. e.,

$$\phi : X \to X.$$

As it was shown in the model example for $X \cong T^2$, there exists a functor on the set of all Anosov's maps ϕ with values in the category of noncommutative tori with real multiplication; the functor sends the conjugate Anosov's maps to the Morita equivalent (stably isomorphic) noncommutative tori $\mathcal{A}_\theta$. In this section we extend this result to the surfaces of genus $g \geq 2$.

4.1.1 Pseudo-Anosov automorphisms of a surface

Let $\mathrm{Mod}(X)$ be the mapping class group of a compact surface X, i. e., the group of orientation-preserving automorphisms of X modulo the trivial ones. Recall that $\phi, \phi' \in \mathrm{Mod}(X)$ are conjugate automorphisms whenever $\phi' = h \circ \phi \circ h^{-1}$ for an $h \in \mathrm{Mod}(X)$. It is not hard to see that conjugation is an equivalence relation which splits the mapping class group into disjoint classes of conjugate automorphisms. The construction of invariants of the conjugacy classes in $\mathrm{Mod}(X)$ is an important and difficult problem studied in [114, 167], and others; it is important to understand that any knowledge of such invariants leads to a topological classification of three-dimensional manifolds [275]. It is known that any $\phi \in \mathrm{Mod}(X)$ is isotopic to an automorphism ϕ' such that either (i) ϕ' has a finite order or (ii) ϕ' is a *pseudo-Anosov* (aperiodic) automorphism, or else (iii) ϕ' is reducible by a system of curves Γ surrounded by the small tubular neighborhoods $N(\Gamma)$ such that on $X \setminus N(\Gamma)$ automorphism ϕ' satisfies either (i) or (ii). Let ϕ be a representative of the equivalence class of a pseudo-Anosov automorphism. Then there exist a pair consisting of the stable $\mathcal{F}_s$ and unstable $\mathcal{F}_u$ mutually orthogonal measured foliations on the surface X such that $\phi(\mathcal{F}_s) = \frac{1}{\lambda_\phi}\mathcal{F}_s$ and $\phi(\mathcal{F}_u) = \lambda_\phi \mathcal{F}_u$, where $\lambda_\phi > 1$ is called a dilatation of ϕ. The foliations $\mathcal{F}_s, \mathcal{F}_u$ are minimal, uniquely ergodic,

https://doi.org/10.1515/9783110788709-004

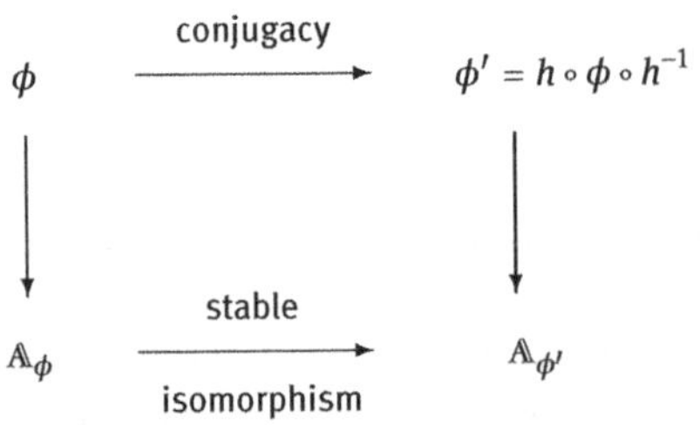

Figure 4.1: Conjugation of the pseudo-Anosov maps.

and describe the automorphism ϕ up to a power. In the sequel, we shall focus on the conjugacy problem for the pseudo-Anosov automorphisms of a surface X; we shall try to solve the problem using functors with values in the NCG. Namely, we shall assign to each pseudo-Anosov map ϕ an AF-algebra, $\mathbb{A}_\phi$, so that for every $h \in \mathrm{Mod}(X)$ the diagram in Fig. 4.1 is commutative. In words, if ϕ and ϕ' are conjugate pseudo-Anosov automorphisms, then the AF-algebras $\mathbb{A}_\phi$ and $\mathbb{A}_{\phi'}$ are Morita equivalent (stably isomorphic). For the sake of clarity, we shall consider an example illustrating the idea in the case $X \cong T^2$.

Example 4.1.1 (Case $X \cong T^2$). Let $\phi \in \mathrm{Mod}(T^2)$ be the Anosov automorphism given by a nonnegative matrix $A_\phi \in SL(2, \mathbf{Z})$. Consider a stationary AF-algebra, $\mathbb{A}_\phi$, given by the periodic Bratteli diagram shown in Fig. 4.2, where a_{ij} indicate the multiplicity of the respective edges of the graph. (We encourage the reader to verify that $F : \phi \mapsto \mathbb{A}_\phi$ is a well-defined function on the set of Anosov automorphisms given by the hyperbolic matrices with nonnegative entries.) Let us show that if $\phi, \phi' \in \mathrm{Mod}(T^2)$ are conjugate Anosov automorphisms, then $\mathbb{A}_\phi, \mathbb{A}_{\phi'}$ are Morita equivalent (stably isomorphic) AF-algebras. Indeed, let $\phi' = h \circ \phi \circ h^{-1}$ for an $h \in \mathrm{Mod}(X)$. Then $A_{\phi'} = TA_\phi T^{-1}$ for a matrix $T \in SL_2(\mathbf{Z})$. Note that $(A'_\phi)^n = (TA_\phi T^{-1})^n = TA_\phi^n T^{-1}$, where $n \in \mathbf{N}$. We shall use the following criterion: the AF-algebras $\mathbb{A}, \mathbb{A}'$ are Morita equivalent (stably isomorphic) if and only if their Bratteli diagrams contain a common block of an arbitrary length, see [74, Theorem 2.3] and recall that an order-isomorphism mentioned in the theorem is equivalent to the condition that the corresponding Bratteli diagrams have the same infinite tails, i. e., a common block of infinite length. Consider the following sequences of matrices:

$$\begin{cases} \underbrace{A_\phi A_\phi \dots A_\phi}_{n}, \\ T\underbrace{A_\phi A_\phi \dots A_\phi}_{n} T^{-1}, \end{cases}$$

which mimic the Bratteli diagrams of $\mathbb{A}_\phi$ and $\mathbb{A}_{\phi'}$. Letting $n \to \infty$, we conclude that $\mathbb{A}_\phi \otimes \mathcal{K} \cong \mathbb{A}_{\phi'} \otimes \mathcal{K}$.

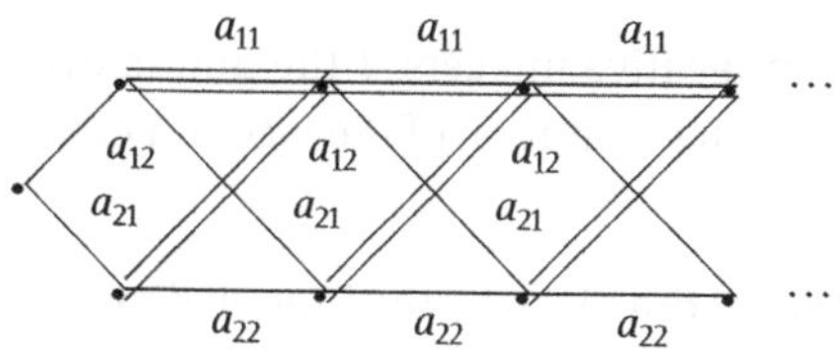

$$A_\phi = \begin{pmatrix} a_{11} & a_{12} \\ a_{21} & a_{22} \end{pmatrix},$$

Figure 4.2: The AF-algebra $\mathbb{A}_\phi$.

Remark 4.1.1 (Handelman's invariant of the AF-algebra $\mathbb{A}_\phi$). One can reformulate the conjugacy problem for the automorphisms $\phi : T^2 \to T^2$ in terms of the AF-algebras $\mathbb{A}_\phi$; namely, one needs to find invariants of the stable isomorphism (Morita equivalence) classes of the stationary AF-algebras $\mathbb{A}_\phi$. One such invariant was introduced in Section 1.4; let us recall its definition and properties. Consider an eigenvalue problem for the matrix $A_\phi \in SL(2, \mathbf{Z})$, i. e., $A_\phi v_A = \lambda_A v_A$, where $\lambda_A > 1$ is the Perron–Frobenius eigenvalue and $v_A = (v_A^{(1)}, v_A^{(2)})$ the corresponding eigenvector with the positive entries normalized so that $v_A^{(i)} \in K = \mathbf{Q}(\lambda_A)$. Denote by $\mathfrak{m} = \mathbf{Z}v_A^{(1)} + \mathbf{Z}v_A^{(2)}$ the $\mathbf{Z}$-module in the number field K. The coefficient ring, Λ, of module $\mathfrak{m}$ consists of the elements $\alpha \in K$ such that $\alpha\mathfrak{m} \subseteq \mathfrak{m}$. It is known that Λ is an order in K (i. e., a subring of K containing 1) and, with no restriction, one can assume that $\mathfrak{m} \subseteq \Lambda$. It follows from the definition that $\mathfrak{m}$ coincides with an ideal, I, whose equivalence class in Λ we shall denote by $[I]$. The triple $(\Lambda, [I], K)$ is an arithmetic invariant of the stable isomorphism class of $\mathbb{A}_\phi$: the $\mathbb{A}_\phi$, $\mathbb{A}_{\phi'}$ are Morita equivalent (stably isomorphic) AF-algebras if and only if $\Lambda = \Lambda'$, $[I] = [I']$ and $K = K'$, see [105].

4.1.2 Functors and invariants

Denote by $\mathcal{F}_\phi$ the stable foliation of a pseudo Anosov automorphism $\phi \in \mathrm{Mod}(X)$. For brevity, we assume that $\mathcal{F}_\phi$ is an oriented foliation given by the trajectories of a closed 1-form $\omega \in H^1(X; \mathbf{R})$. Let $v^{(i)} = \int_{\gamma_i} \omega$, where $\{\gamma_1, \dots, \gamma_n\}$ is a basis in the relative homology $H_1(X, \mathbf{Sing}\,\mathcal{F}_\phi; \mathbf{Z})$, such that $\theta = (\theta_1, \dots, \theta_{n-1})$ is a vector with positive coordinates $\theta_i = v^{(i+1)}/v^{(1)}$.

Remark 4.1.2. The constants θ_i depend on a basis in the homology group, but the $\mathbf{Z}$-module generated by the θ_i does not.

Consider the infinite Jacobi–Perron continued fraction of θ,

$$\begin{pmatrix} 1 \\ \theta \end{pmatrix} = \lim_{k\to\infty} \begin{pmatrix} 0 & 1 \\ I & b_1 \end{pmatrix} \cdots \begin{pmatrix} 0 & 1 \\ I & b_k \end{pmatrix} \begin{pmatrix} 0 \\ \mathbb{I} \end{pmatrix},$$

where $b_i = (b_1^{(i)}, \dots, b_{n-1}^{(i)})^T$ is a vector of the nonnegative integers, I the unit matrix and $\mathbb{I} = (0, \dots, 0, 1)^T$; we refer the reader to [23] for the definition of the Jacobi–Perron algorithm and related continued fractions.

Definition 4.1.1. By $\mathbb{A}_\phi$ one understands the AF-algebra given by the Bratteli diagram defined by the incidence matrices $B_k = \left(\begin{smallmatrix} 0 & 1 \\ I & b_k \end{smallmatrix}\right)$ for $k = 1, \dots, \infty$.

Remark 4.1.3. We encourage the reader to verify that $\mathbb{A}_\phi$ coincides with the one for the Anosov maps. (Hint: The Jacobi–Perron fractions of dimension $n = 2$ coincide with the regular continued fractions.)

Definition 4.1.2. For a matrix $A \in GL_n(\mathbf{Z})$ with positive entries, we shall denote by λ_A the Perron–Frobenius eigenvalue and let $(v_A^{(1)}, \dots, v_A^{(n)})$ be the corresponding normalized eigenvector such that $v_A^{(i)} \in K = \mathbf{Q}(\lambda_A)$. The coefficient (endomorphism) ring of the module $\mathfrak{m} = \mathbf{Z}v_A^{(1)} + \cdots + \mathbf{Z}v_A^{(n)}$ will shall write as Λ; the equivalence class of ideal I in Λ will be written as $[I]$. We shall denote by $\Delta = \mathbf{det}(a_{ij})$ and Σ the determinant and signature of the symmetric bilinear form $q(x,y) = \sum_{i,j}^n a_{ij}x_ix_j$, where $a_{ij} = \mathbf{Tr}(v_A^{(i)}v_A^{(j)})$ and $\mathbf{Tr}(\bullet)$ is the trace function.

Theorem 4.1.1. *The $\mathbb{A}_\phi$ is a stationary AF-algebra.*

Let Φ be a category of all pseudo-Anosov (resp. Anosov) automorphisms of a surface of the genus $g \geq 2$ (resp. $g = 1$); the arrows (morphisms) are conjugations between the automorphisms. Likewise, let $\mathcal{A}$ be the category of all stationary AF-algebras $\mathbb{A}_\phi$, where ϕ runs over the set Φ; the arrows of $\mathcal{A}$ are stable isomorphisms among the algebras $\mathbb{A}_\phi$.

Theorem 4.1.2 (Functor on pseudo-Anosov maps). *Let $F : \Phi \to \mathcal{A}$ be a map given by the formula $\phi \mapsto \mathbb{A}_\phi$. Then:*

(i) *F is a functor which maps conjugate pseudo-Anosov automorphisms to Morita equivalent (stably isomorphic) AF-algebras;*
(ii) *$\operatorname{Ker} F = [\phi]$, where $[\phi] = \{\phi' \in \Phi \mid (\phi')^m = \phi^n,\ m, n \in \mathbf{N}\}$ is the commensurability class of the pseudo-Anosov automorphism ϕ.*

Corollary 4.1.1 (Noncommutative invariants). *The following are invariants of the conjugacy classes of the pseudo-Anosov automorphisms:*

(i) *triples $(\Lambda, [I], K)$;*
(ii) *integers Δ and Σ.*

Theorems 4.1.1, 4.1.2 and Corollary 4.1.1 will be proved in Section 4.1.5; the necessary background is developed in the sections below.

4.1.3 Jacobian of measured foliations

Let $\mathcal{F}$ be a measured foliation on a compact surface X [276]. For the sake of brevity, we shall always assume that $\mathcal{F}$ is an oriented foliation, i. e., given by the trajectories of a closed 1-form ω on X. (The assumption is not a restriction – each measured foliation is oriented on a surface $\widetilde{X}$, which is a double cover of X ramified at the singular points of the half-integer index of the nonoriented foliation [120].) Let $\{\gamma_1, \dots, \gamma_n\}$ be a basis in the relative homology group $H_1(X, \mathbf{Sing}\,\mathcal{F}; \mathbf{Z})$, where $\mathbf{Sing}\,\mathcal{F}$ is the set of singular points of the foliation $\mathcal{F}$. It is well known that $n = 2g + m - 1$, where g is the genus of X and $m = |\mathbf{Sing}(\mathcal{F})|$. The periods of ω in the above basis will be written as

$$\lambda_i = \int_{\gamma_i} \omega.$$

The real numbers λ_i are coordinates of $\mathcal{F}$ in the space of all measured foliations on X with the fixed set of singular points, see, e. g., [71].

Definition 4.1.3. By a jacobian $\mathrm{Jac}(\mathcal{F})$ of the measured foliation $\mathcal{F}$ we understand a $\mathbf{Z}$-module $\mathfrak{m} = \mathbf{Z}\lambda_1 + \cdots + \mathbf{Z}\lambda_n$ regarded as a subset of the real line $\mathbf{R}$.

An importance of the jacobians stems from an observation that although the periods λ_i depend on the choice of basis in $H_1(X, \mathbf{Sing}\,\mathcal{F}; \mathbf{Z})$, the jacobian does not. Moreover, up to a scalar multiple, the jacobian is an invariant of the equivalence class of the foliation $\mathcal{F}$. We formalize these observations in the following two lemmas.

Lemma 4.1.1. *The $\mathbf{Z}$-module $\mathfrak{m}$ is independent of the choice of a basis in $H_1(X, \mathbf{Sing}\,\mathcal{F}; \mathbf{Z})$ and depends solely on the foliation $\mathcal{F}$.*

Proof. Indeed, let $A = (a_{ij}) \in \mathrm{GL}(n, \mathbf{Z})$ and let

$$\gamma_i' = \sum_{j=1}^{n} a_{ij}\gamma_j$$

be a new basis in $H_1(X, \mathbf{Sing}\,\mathcal{F}; \mathbf{Z})$. Then using the integration rules, we get

$$\lambda_i' = \int_{\gamma_i'} \omega = \int_{\sum_{j=1}^{n} a_{ij}\gamma_j} \omega = \sum_{j=1}^{n} \int_{\gamma_j} \omega = \sum_{j=1}^{n} a_{ij}\lambda_j.$$

To prove that $\mathfrak{m} = \mathfrak{m}'$, consider the following equations:

$$\mathfrak{m}' = \sum_{i=1}^{n} \mathbf{Z}\lambda_i' = \sum_{i=1}^{n} \mathbf{Z} \sum_{j=1}^{n} a_{ij}\lambda_j = \sum_{j=1}^{n} \left(\sum_{i=1}^{n} a_{ij}\mathbf{Z} \right) \lambda_j \subseteq \mathfrak{m}.$$

Let $A^{-1} = (b_{ij}) \in \mathrm{GL}(n, \mathbf{Z})$ be an inverse to the matrix A. Then $\lambda_i = \sum_{j=1}^{n} b_{ij}\lambda'_j$ and

$$\mathfrak{m} = \sum_{i=1}^{n} \mathbf{Z}\lambda_i = \sum_{i=1}^{n} \mathbf{Z} \sum_{j=1}^{n} b_{ij}\lambda'_j = \sum_{j=1}^{n} \left(\sum_{i=1}^{n} b_{ij}\mathbf{Z} \right) \lambda'_j \subseteq \mathfrak{m}'.$$

Since both $\mathfrak{m}' \subseteq \mathfrak{m}$ and $\mathfrak{m} \subseteq \mathfrak{m}'$, we conclude that $\mathfrak{m}' = \mathfrak{m}$. Lemma 4.1.1 follows. □

Definition 4.1.4. Two measured foliations $\mathcal{F}$ and $\mathcal{F}'$ are said to *equivalent* if there exists an automorphism $h \in \mathrm{Mod}(X)$ which sends the leaves of the foliation $\mathcal{F}$ to the leaves of the foliation $\mathcal{F}'$.

Remark 4.1.4. The equivalence relation involves the topological foliations, i. e., projective classes of the measured foliations, see [276] for the details.

Lemma 4.1.2. *Let $\mathcal{F}, \mathcal{F}'$ be the equivalent measured foliations on a surface X. Then*

$$\mathrm{Jac}(\mathcal{F}') = \mu \,\mathrm{Jac}(\mathcal{F}),$$

where $\mu > 0$ is a real number.

Proof. Let $h : X \to X$ be an automorphism of the surface X. Denote by h_* its action on $H_1(X, \mathbf{Sing}(\mathcal{F}); \mathbf{Z})$ and by h^* on $H^1(X; \mathbf{R})$ connected by the formula

$$\int_{h_*(\gamma)} \omega = \int_{\gamma} h^*(\omega), \quad \forall \gamma \in H_1(X, \mathbf{Sing}(\mathcal{F}); \mathbf{Z}), \ \forall \omega \in H^1(X; \mathbf{R}).$$

Let $\omega, \omega' \in H^1(X; \mathbf{R})$ be the closed 1-forms whose trajectories define the foliations $\mathcal{F}$ and $\mathcal{F}'$, respectively. Since $\mathcal{F}, \mathcal{F}'$ are equivalent measured foliations,

$$\omega' = \mu h^*(\omega)$$

for a $\mu > 0$.

Let $\mathrm{Jac}(\mathcal{F}) = \mathbf{Z}\lambda_1 + \cdots + \mathbf{Z}\lambda_n$ and $\mathrm{Jac}(\mathcal{F}') = \mathbf{Z}\lambda'_1 + \cdots + \mathbf{Z}\lambda'_n$. Then

$$\lambda'_i = \int_{\gamma_i} \omega' = \mu \int_{\gamma_i} h^*(\omega) = \mu \int_{h_*(\gamma_i)} \omega, \quad 1 \le i \le n.$$

By Lemma 4.1.1, it holds that

$$\mathrm{Jac}(\mathcal{F}) = \sum_{i=1}^{n} \mathbf{Z} \int_{\gamma_i} \omega = \sum_{i=1}^{n} \mathbf{Z} \int_{h_*(\gamma_i)} \omega.$$

Therefore,

$$\mathrm{Jac}(\mathcal{F}') = \sum_{i=1}^{n} \mathbf{Z} \int_{\gamma_i} \omega' = \mu \sum_{i=1}^{n} \mathbf{Z} \int_{h_*(\gamma_i)} \omega = \mu \, \mathrm{Jac}(\mathcal{F}).$$

Lemma 4.1.2 follows. □

4.1.4 Equivalent foliations

Recall that, for a measured foliation $\mathcal{F}$, we constructed an AF-algebra, $\mathbb{A}_{\mathcal{F}}$. Our goal is to prove commutativity of the diagram in Fig. 4.3.; in other words, two equivalent measured foliations map to the Morita equivalent (stably isomorphic) AF-algebras $\mathbb{A}_{\mathcal{F}}$.

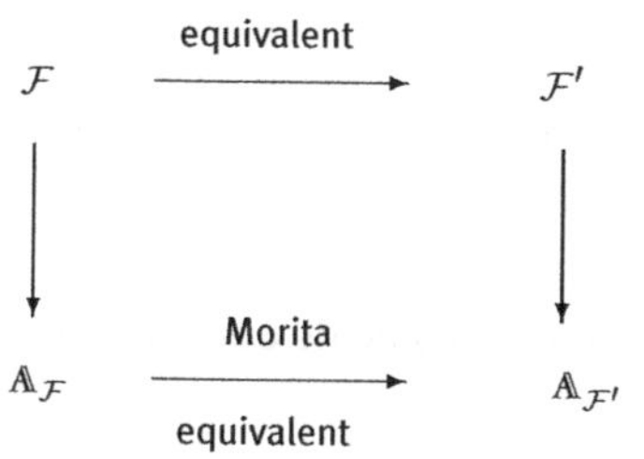

Figure 4.3: Functor on measured foliations.

Lemma 4.1.3 (Perron). *Let* $\mathfrak{m} = \mathbf{Z}\lambda_1 + \cdots + \mathbf{Z}\lambda_n$ *and* $\mathfrak{m}' = \mathbf{Z}\lambda_1' + \cdots + \mathbf{Z}\lambda_n'$ *be two* **Z***-modules such that* $\mathfrak{m}' = \mu\mathfrak{m}$ *for a* $\mu > 0$. *Then the Jacobi–Perron continued fractions of the vectors* λ *and* λ' *coincide except, possibly, at a finite number of terms.*

Proof. Let $\mathfrak{m} = \mathbf{Z}\lambda_1 + \cdots + \mathbf{Z}\lambda_n$ and $\mathfrak{m}' = \mathbf{Z}\lambda_1' + \cdots + \mathbf{Z}\lambda_n'$. Since $\mathfrak{m}' = \mu\mathfrak{m}$, where μ is a positive real, one gets the following identity of the **Z**-modules:

$$\mathbf{Z}\lambda_1' + \cdots + \mathbf{Z}\lambda_n' = \mathbf{Z}(\mu\lambda_1) + \cdots + \mathbf{Z}(\mu\lambda_n).$$

One can always assume that λ_i and λ_i' are positive reals. For obvious reasons, there exists a basis $\{\lambda_1'', \dots, \lambda_n''\}$ of the module $\mathfrak{m}'$ such that

$$\begin{cases} \lambda'' = A(\mu\lambda), \\ \lambda'' = A'\lambda', \end{cases}$$

where $A, A' \in GL^+(n, \mathbf{Z})$ are the matrices whose entries are nonnegative integers. In view of [21, Proposition 3], we have

$$\begin{cases} A = \begin{pmatrix} 0 & 1 \\ I & b_1 \end{pmatrix} \cdots \begin{pmatrix} 0 & 1 \\ I & b_k \end{pmatrix}, \\ A' = \begin{pmatrix} 0 & 1 \\ I & b'_1 \end{pmatrix} \cdots \begin{pmatrix} 0 & 1 \\ I & b'_l \end{pmatrix}, \end{cases}$$

where b_i, b'_i are nonnegative integer vectors. Since the Jacobi–Perron continued fractions for the vectors λ and $\mu\lambda$ coincide for any $\mu > 0$ (see, e. g., [23]), we conclude that

$$\begin{cases} \begin{pmatrix} 1 \\ \theta \end{pmatrix} = \begin{pmatrix} 0 & 1 \\ I & b_1 \end{pmatrix} \cdots \begin{pmatrix} 0 & 1 \\ I & b_k \end{pmatrix} \begin{pmatrix} 0 & 1 \\ I & a_1 \end{pmatrix} \begin{pmatrix} 0 & 1 \\ I & a_2 \end{pmatrix} \cdots \begin{pmatrix} 0 \\ \mathbb{I} \end{pmatrix}, \\ \begin{pmatrix} 1 \\ \theta' \end{pmatrix} = \begin{pmatrix} 0 & 1 \\ I & b'_1 \end{pmatrix} \cdots \begin{pmatrix} 0 & 1 \\ I & b'_l \end{pmatrix} \begin{pmatrix} 0 & 1 \\ I & a_1 \end{pmatrix} \begin{pmatrix} 0 & 1 \\ I & a_2 \end{pmatrix} \cdots \begin{pmatrix} 0 \\ \mathbb{I} \end{pmatrix}, \end{cases}$$

where

$$\begin{pmatrix} 1 \\ \theta'' \end{pmatrix} = \lim_{i\to\infty} \begin{pmatrix} 0 & 1 \\ I & a_1 \end{pmatrix} \cdots \begin{pmatrix} 0 & 1 \\ I & a_i \end{pmatrix} \begin{pmatrix} 0 \\ \mathbb{I} \end{pmatrix}.$$

In other words, the continued fractions of the vectors λ and λ' coincide but at a finite number of terms. Lemma 4.1.3 follows. □

Lemma 4.1.4 (Basic lemma). *Let $\mathcal{F}$ and $\mathcal{F}'$ be equivalent measured foliations on a surface X. Then the AF-algebras $\mathbb{A}_{\mathcal{F}}$ and $\mathbb{A}_{\mathcal{F}'}$ are Morita equivalent (stably isomorphic).*

Proof. Notice that Lemma 4.1.2 implies that equivalent measured foliations $\mathcal{F}, \mathcal{F}'$ have proportional jacobians, i. e., $\mathfrak{m}' = \mu\mathfrak{m}$ for a $\mu > 0$. On the other hand, by Lemma 4.1.3, the continued fraction expansions of the basis vectors of the proportional jacobians must coincide, except for a finite number of terms. Thus, the AF-algebras $\mathbb{A}_{\mathcal{F}}$ and $\mathbb{A}_{\mathcal{F}'}$ are given by the Bratteli diagrams, which are identical, except for a finite part of the diagram. It is well known (see, e. g., [74, Theorem 2.3]) that the AF-algebras, which have such a property, are Morita equivalent (stably isomorphic). Lemma 4.1.4 follows. □

4.1.5 Proofs

4.1.5.1 Proof of Theorem 4.1.1

Let $\phi \in Mod(X)$ be a pseudo-Anosov automorphism of the surface X. Denote by $\mathcal{F}_\phi$ the invariant foliation of ϕ. By definition of such a foliation, $\phi(\mathcal{F}_\phi) = \lambda_\phi \mathcal{F}_\phi$, where $\lambda_\phi > 1$ is the dilatation of ϕ. Consider the jacobian $Jac(\mathcal{F}_\phi) = \mathfrak{m}_\phi$ of foliation $\mathcal{F}_\phi$. Since $\mathcal{F}_\phi$ is an invariant foliation of the pseudo-Anosov automorphism ϕ, one gets the following equality of the **Z**-modules:

$$\mathfrak{m}_\phi = \lambda_\phi \mathfrak{m}_\phi, \quad \lambda_\phi \neq \pm 1.$$

Let $\{v^{(1)}, \dots, v^{(n)}\}$ be a basis in module $\mathfrak{m}_\phi$ such that $v^{(i)} > 0$; from the above equation, one obtains the following system of linear equations:

$$\begin{cases} \lambda_\phi v^{(1)} = a_{11}v^{(1)} + a_{12}v^{(2)} + \dots + a_{1n}v^{(n)}, \\ \lambda_\phi v^{(2)} = a_{21}v^{(1)} + a_{22}v^{(2)} + \dots + a_{2n}v^{(n)}, \\ \quad \vdots \\ \lambda_\phi v^{(n)} = a_{n1}v^{(1)} + a_{n2}v^{(2)} + \dots + a_{nn}v^{(n)}, \end{cases}$$

where $a_{ij} \in \mathbf{Z}$. The matrix $A = (a_{ij})$ is invertible. Indeed, since foliation $\mathcal{F}_\phi$ is minimal, real numbers $v^{(1)}, \dots, v^{(n)}$ are linearly independent over **Q**. So are numbers $\lambda_\phi v^{(1)}, \dots, \lambda_\phi v^{(n)}$, which therefore can be taken for a basis of the module $\mathfrak{m}_\phi$. Thus, there exists an integer matrix $B = (b_{ij})$ such that $v^{(j)} = \sum_{i,j} w^{(i)}$, where $w^{(i)} = \lambda_\phi v^{(i)}$. Clearly, B is an inverse to matrix A. Therefore, $A \in GL(n, \mathbf{Z})$.

Moreover, without loss of the generality, one can assume that $a_{ij} \geq 0$. Indeed, if it is not yet the case, consider the conjugacy class $[A]$ of the matrix A. Since $v^{(i)} > 0$, there exists a matrix $A^+ \in [A]$ whose entries are nonnegative integers. One has to replace basis $v = (v^{(1)}, \dots, v^{(n)})$ in the module $\mathfrak{m}_\phi$ by a basis Tv, where $A^+ = TAT^{-1}$. It will be further assumed that $A = A^+$.

Lemma 4.1.5. *Vector $(v^{(1)}, \dots, v^{(n)})$ is the limit of a periodic Jacobi–Perron continued fraction.*

Proof. It follows from the discussion above that there exists a nonnegative integer matrix A such that $Av = \lambda_\phi v$. In view of [21, Proposition 3], matrix A admits the unique factorization

$$A = \begin{pmatrix} 0 & 1 \\ I & b_1 \end{pmatrix} \cdots \begin{pmatrix} 0 & 1 \\ I & b_k \end{pmatrix},$$

where $b_i = (b_1^{(i)}, \dots, b_n^{(i)})^T$ are vectors of the nonnegative integers. Let us consider the periodic Jacobi–Perron continued fraction

$$\mathrm{Per}\,\overline{\begin{pmatrix} 0 & 1 \\ I & b_1 \end{pmatrix} \cdots \begin{pmatrix} 0 & 1 \\ I & b_k \end{pmatrix}} \begin{pmatrix} 0 \\ \mathbb{I} \end{pmatrix}.$$

According to Perron [224, Satz XII], the above fraction converges to vector $w = (w^{(1)}, \dots, w^{(n)})$, such that w satisfies equation $(B_1 B_2 \cdots B_k)w = Aw = \lambda_\phi w$. In view of equation $Av = \lambda_\phi v$, we conclude that vectors v and w are collinear. Therefore, the Jacobi–Perron continued fractions of v and w must coincide. Lemma 4.1.5 follows. □

It is easy to see that the AF-algebra attached to foliation $\mathcal{F}_\phi$ is stationary. Indeed, by Lemma 4.1.5, the vector of periods $v^{(i)} = \int_{\gamma_i} \omega$ unfolds into a periodic Jacobi–Perron continued fraction. By definition, the Bratteli diagram of the AF-algebra $\mathbb{A}_\phi$ is periodic as well. In other words, the AF-algebra $\mathbb{A}_\phi$ is stationary. Theorem 4.1.1 is proved.

4.1.5.2 Proof of Theorem 4.1.2

(i) Let us prove the first statement. For the sake of completeness, let us give a proof of the following well-known lemma.

Lemma 4.1.6. *If ϕ and ϕ' are conjugate pseudo-Anosov automorphisms of a surface X, then their invariant measured foliations $\mathcal{F}_\phi$ and $\mathcal{F}_{\phi'}$ are equivalent.*

Proof. Let $\phi, \phi' \in \mathrm{Mod}(X)$ be conjugate, i. e., $\phi' = h \circ \phi \circ h^{-1}$ for an automorphism $h \in \mathrm{Mod}(X)$. Since ϕ is the pseudo-Anosov automorphism, there exists a measured foliation $\mathcal{F}_\phi$ such that $\phi(\mathcal{F}_\phi) = \lambda_\phi \mathcal{F}_\phi$. Let us evaluate the automorphism ϕ' on the foliation $h(\mathcal{F}_\phi)$:

$$\phi'(h(\mathcal{F}_\phi)) = h\phi h^{-1}(h(\mathcal{F}_\phi)) = h\phi(\mathcal{F}_\phi) = h\lambda_\phi \mathcal{F}_\phi = \lambda_\phi(h(\mathcal{F}_\phi)).$$

Thus, $\mathcal{F}_{\phi'} = h(\mathcal{F}_\phi)$ is the invariant foliation for the pseudo-Anosov automorphism ϕ' and $\mathcal{F}_\phi, \mathcal{F}_{\phi'}$ are equivalent foliations. Note also that the pseudo-Anosov automorphism ϕ' has the same dilatation as the automorphism ϕ. Lemma 4.1.6 follows. □

To finish the proof of claim (i), suppose that ϕ and ϕ' are conjugate pseudo-Anosov automorphisms. Functor F acts by the formulas $\phi \mapsto \mathbb{A}_\phi$ and $\phi' \mapsto \mathbb{A}_{\phi'}$, where $\mathbb{A}_\phi, \mathbb{A}_{\phi'}$ are the AF-algebras corresponding to invariant foliations $\mathcal{F}_\phi, \mathcal{F}_{\phi'}$. In view of Lemma 4.1.6, $\mathcal{F}_\phi$ and $\mathcal{F}_{\phi'}$ are equivalent measured foliations. Then, by Lemma 4.1.4, the AF-algebras $\mathbb{A}_\phi$ and $\mathbb{A}_{\phi'}$ are Morita equivalent (stably isomorphic) AF-algebras. Claim (i) follows.

(ii) Let us prove the second statement. We start with an elementary observation. Let $\phi \in \mathrm{Mod}(X)$ be a pseudo-Anosov automorphism. Then there exists a unique measured foliation $\mathcal{F}_\phi$ such that $\phi(\mathcal{F}_\phi) = \lambda_\phi \mathcal{F}_\phi$, where $\lambda_\phi > 1$ is an algebraic integer. Let us evaluate automorphism $\phi^2 \in \mathrm{Mod}(X)$ on the foliation $\mathcal{F}_\phi$:

$$\phi^2(\mathcal{F}_\phi) = \phi(\phi(\mathcal{F}_\phi)) = \phi(\lambda_\phi \mathcal{F}_\phi) = \lambda_\phi \phi(\mathcal{F}_\phi) = \lambda_\phi^2 \mathcal{F}_\phi = \lambda_{\phi^2} \mathcal{F}_\phi,$$

where $\lambda_{\phi^2} := \lambda_\phi^2$. Thus, foliation $\mathcal{F}_\phi$ is an invariant foliation for the automorphism ϕ^2 as well. By induction, one concludes that $\mathcal{F}_\phi$ is an invariant foliation of the automorphism ϕ^n for any $n \geq 1$.

Even more is true. Suppose that $\psi \in \mathrm{Mod}(X)$ is a pseudo-Anosov automorphism, such that $\psi^m = \phi^n$ for some $m \geq 1$ and $\psi \neq \phi$. Then $\mathcal{F}_\phi$ is an invariant foliation for the automorphism ψ. Indeed, $\mathcal{F}_\phi$ is invariant foliation of the automorphism ψ^m.

If there exists $\mathcal{F}' \neq \mathcal{F}_\phi$ such that the foliation $\mathcal{F}'$ is an invariant foliation of ψ, then the foliation $\mathcal{F}'$ is also an invariant foliation of the pseudo-Anosov automorphism ψ^m. Thus, by the uniqueness, $\mathcal{F}' = \mathcal{F}_\phi$. We have just proved the following lemma.

Lemma 4.1.7. *If $[\phi]$ is the set of all pseudo-Anosov automorphisms ψ of X, such that $\psi^m = \phi^n$ for some positive integers m and n, then the pseudo-Anosov foliation $\mathcal{F}_\phi$ is an invariant foliation for every pseudo-Anosov automorphism $\psi \in [\phi]$.*

In view of Lemma 4.1.7, one gets the following identities for the AF-algebras

$$\mathbb{A}_\phi = \mathbb{A}_{\phi^2} = \cdots = \mathbb{A}_{\phi^n} = \mathbb{A}_{\psi^m} = \cdots = \mathbb{A}_{\psi^2} = \mathbb{A}_\psi.$$

Thus, functor F is not an injective functor: the preimage, $\operatorname{Ker} F$, of algebra $\mathbb{A}_\phi$ consists of a countable set of the pseudo-Anosov automorphisms $\psi \in [\phi]$, commensurable with the automorphism ϕ. Theorem 4.1.2 is proved.

4.1.5.3 Proof of Corollary 4.1.1

(i) Theorem 4.1.1 says that $\mathbb{A}_\phi$ is a stationary AF-algebra. An arithmetic invariant of the stable isomorphism classes of the stationary AF-algebras has been found by D. Handelman in [106]. Summing up his results, the invariant is as follows. Let $A \in GL(n, \mathbf{Z})$ be a matrix with the strictly positive entries, such that A is equal to the minimal period of the Bratteli diagram of the stationary AF-algebra. (In case the matrix A has zero entries, it is necessary to take a proper minimal power of the matrix A.) By the Perron–Frobenius theory, matrix A has a real eigenvalue $\lambda_A > 1$, which exceeds the absolute values of other roots of the characteristic polynomial of A. Note that λ_A is an invertible algebraic integer (the unit). Consider the real algebraic number field $K = \mathbf{Q}(\lambda_A)$ obtained as an extension of the field of the rational numbers by the algebraic number λ_A. Let $(v_A^{(1)}, \ldots, v_A^{(n)})$ be the eigenvector corresponding to the eigenvalue λ_A. One can normalize the eigenvector so that $v_A^{(i)} \in K$. The departure point of Handelman's invariant is the $\mathbf{Z}$-module $\mathfrak{m} = \mathbf{Z}v_A^{(1)} + \cdots + \mathbf{Z}v_A^{(n)}$. The module $\mathfrak{m}$ brings in two new arithmetic objects: (i) the ring Λ of the endomorphisms of $\mathfrak{m}$ and (ii) an ideal I in the ring Λ, such that $I = \mathfrak{m}$ after a scaling, see, e. g., [38, Lemma 1, p. 88]. The ring Λ is an order in the algebraic number field K, and therefore one can talk about the ideal classes in Λ. The ideal class of I is denoted by $[I]$. Omitting the embedding question for the field K, the triple $(\Lambda, [I], K)$ is an invariant of the stable isomorphism class of the stationary AF-algebra $\mathbb{A}_\phi$, see [106, § 5]. Claim (i) follows.

(ii) Numerical invariants of the stable isomorphism classes of the stationary AF-algebras can be derived from the triple $(\Lambda, [I], K)$. These invariants are the rational integers – called the determinant and signature – which can be obtained as follows. Let $\mathfrak{m}, \mathfrak{m}'$ be the full $\mathbf{Z}$-modules in an algebraic number field K. It follows from (i) that if $\mathfrak{m} \neq \mathfrak{m}'$ are distinct as the $\mathbf{Z}$-modules, then the corresponding AF-algebras cannot be Morita equivalent (stably isomorphic). We wish to find the numerical invariants, which discern the case $\mathfrak{m} \neq \mathfrak{m}'$. It is assumed that a **Z**-module is given by the set of generators $\{\lambda_1, \ldots, \lambda_n\}$. Therefore, the problem can be formulated as follows: find a number attached to the set of generators $\{\lambda_1, \ldots, \lambda_n\}$, which does not change on the set of generators $\{\lambda_1', \ldots, \lambda_n'\}$ of the same **Z**-module. One such invariant is associated with the trace function on the algebraic number field K. Recall that

$$\mathbf{Tr} : K \to \mathbf{Q}$$

is a linear function on K such that $\mathbf{Tr}(\alpha + \beta) = \mathbf{Tr}(\alpha) + \mathbf{Tr}(\beta)$ and $\mathbf{Tr}(a\alpha) = a\mathbf{Tr}(\alpha)$ for $\forall \alpha, \beta \in K$ and $\forall a \in \mathbf{Q}$. Let $\mathfrak{m}$ be a full **Z**-module in the field K. The trace function defines a symmetric bilinear form $q(x,y) : \mathfrak{m} \times \mathfrak{m} \to \mathbf{Q}$ by the formula

$$(x,y) \longmapsto \mathbf{Tr}(xy), \quad \forall x, y \in \mathfrak{m}.$$

The form $q(x,y)$ depends on the basis $\{\lambda_1, \ldots, \lambda_n\}$ in the module $\mathfrak{m}$

$$q(x,y) = \sum_{j=1}^{n} \sum_{i=1}^{n} a_{ij} x_i y_j, \quad \text{where } a_{ij} = \mathbf{Tr}(\lambda_i \lambda_j).$$

However, the general theory of the bilinear forms (over the fields **Q**, **R**, **C**, or the ring of rational integers **Z**) tells us that certain numerical quantities will not depend on the choice of such a basis.

Definition 4.1.5. By a determinant of the bilinear form $q(x,y)$ one understands the rational integer number

$$\Delta = \mathbf{det}(\mathbf{Tr}(\lambda_i \lambda_j)).$$

Lemma 4.1.8. *The determinant $\Delta(\mathfrak{m})$ is independent of the choice of the basis $\{\lambda_1, \ldots, \lambda_n\}$ in the module $\mathfrak{m}$.*

Proof. Consider a symmetric matrix A corresponding to the bilinear form $q(x,y)$, i. e.,

$$A = \begin{pmatrix} a_{11} & a_{12} & \ldots & a_{1n} \\ a_{12} & a_{22} & \ldots & a_{2n} \\ \vdots & \vdots & \ddots & \vdots \\ a_{1n} & a_{2n} & \ldots & a_{nn} \end{pmatrix}.$$

It is known that matrix A, written in a new basis, will take the form $A' = U^T A U$, where $U \in GL(n, \mathbf{Z})$. Then $\mathbf{det}(A') = \mathbf{det}(U^T A U) = \mathbf{det}(U^T)\mathbf{det}(A)\mathbf{det}(U) = \mathbf{det}(A)$. Therefore, the rational integer number

$$\Delta = \mathbf{det}(\mathbf{Tr}(\lambda_i \lambda_j))$$

does not depend on the choice of the basis $\{\lambda_1, \dots, \lambda_n\}$ in the module $\mathfrak{m}$. Lemma 4.1.8 follows. □

Remark 4.1.5 (p-adic invariants). Lemma 4.1.8 says that determinant $\Delta(\mathfrak{m})$ discerns two distinct modules, i. e., $\mathfrak{m} \neq \mathfrak{m}'$. Note that if $\Delta(\mathfrak{m}) = \Delta(\mathfrak{m}')$ for the modules $\mathfrak{m}$ and $\mathfrak{m}'$, one cannot conclude that $\mathfrak{m} = \mathfrak{m}'$. The problem of equivalence of the symmetric bilinear forms over $\mathbf{Q}$ (i. e., the existence of a linear substitution over $\mathbf{Q}$, which transforms one form to the other) is a fundamental question of number theory. The Minkowski–Hasse theorem says that two such forms are equivalent if and only if they are equivalent over the p-adic field $\mathbf{Q}_p$ for every prime number p and over the field $\mathbf{R}$. Clearly, the resulting p-adic quantities will give new invariants of the stable isomorphism classes of the AF-algebras. The question is much similar to the Minkowski units attached to knots, see, e. g., [236].

Definition 4.1.6. By a signature of the bilinear form $q(x,y)$ one understands the rational integer $\Sigma = (\#a_i^+) - (\#a_i^-)$, where a_i^+ are the positive and a_i^- the negative entries in the diagonal form

$$a_1x_1^2 + a_2x_2^2 + \cdots + a_nx_n^2$$

of $q(x,y)$; recall that each $q(x,y)$ can be brought by an integer linear transformation to the diagonal form.

Lemma 4.1.9. *The signature $\Sigma(\mathfrak{m})$ is independent of the choice of basis in the module $\mathfrak{m}$ and, therefore, $\Sigma(\mathfrak{m}) \neq \Sigma(\mathfrak{m}')$ implies $\mathfrak{m} \neq \mathfrak{m}'$.*

Proof. The claim follows from the Law of Inertia for the signature of the bilinear form $q(x,y)$. □

Corollary 4.1.1 follows from Lemmas 4.1.8 and 4.1.9.

4.1.6 Anosov maps of the torus

We shall calculate the noncommutative invariants $\Delta(\mathfrak{m})$ and $\Sigma(\mathfrak{m})$ for the Anosov automorphisms of the two-dimensional torus; we construct concrete examples of Anosov automorphisms with the same Alexander polynomial $\Delta(t)$ but different invariant $\Delta(\mathfrak{m})$, i. e., showing that $\Sigma(\mathfrak{m})$ is *finer* than $\Delta(t)$. Recall that isotopy classes of the orientation-preserving diffeomorphisms of the torus T^2 are bijective with the 2×2 matrices with integer entries and determinant $+1$, i. e., $\mathrm{Mod}(T^2) \cong SL(2,\mathbf{Z})$. Under the identification, the aperiodic automorphisms correspond to the matrices $A \in SL(2,\mathbf{Z})$ with $|\mathbf{Tr}\,A| > 2$. Let $K = \mathbf{Q}(\sqrt{d})$ be a quadratic extension of the field of rational numbers $\mathbf{Q}$. Further we suppose that d is a positive square free integer. Let

$$\omega = \begin{cases} \frac{1+\sqrt{d}}{2} & \text{if } d \equiv 1 \mod 4, \\ \sqrt{d} & \text{if } d \equiv 2,3 \mod 4. \end{cases}$$

Remark 4.1.6. Recall that if f is a positive integer then every order in K has the form $\Lambda_f = \mathbf{Z} + (f\omega)\mathbf{Z}$, where f is the conductor of Λ_f, see, e. g., [38, pp. 130–132]. This formula allows classifying the similarity classes of the full modules in the field K. Indeed, there exist a finite number of $\mathfrak{m}_f^{(1)}, \dots, \mathfrak{m}_f^{(s)}$ of the nonsimilar full modules in the field K whose coefficient ring is the order Λ_f, see [38, Chapter 2.7, Theorem 3]. Thus one gets a finite-to-one classification of the similarity classes of full modules in the field K.

4.1.6.1 Numerical invariants of the Anosov maps

Let Λ_f be an order in K with the conductor f. Under the addition operation, the order Λ_f is a full module, which we denote by $\mathfrak{m}_f$. Let us evaluate the invariants $q(x,y)$, Δ, and Σ on the module $\mathfrak{m}_f$. To calculate $(a_{ij}) = \mathbf{Tr}(\lambda_i\lambda_j)$, we let $\lambda_1 = 1, \lambda_2 = f\omega$. Then

$$a_{11} = 2, \quad a_{12} = a_{21} = f, \quad a_{22} = \frac{1}{2}f^2(d+1) \quad \text{if } d \equiv 1 \mod 4,$$
$$a_{11} = 2, \quad a_{12} = a_{21} = 0, \quad a_{22} = 2f^2 d \qquad \text{if } d \equiv 2,3 \mod 4,$$

and

$$q(x,y) = 2x^2 + 2fxy + \frac{1}{2}f^2(d+1)y^2 \quad \text{if } d \equiv 1 \mod 4,$$
$$q(x,y) = 2x^2 + 2f^2 dy^2 \qquad \text{if } d \equiv 2,3 \mod 4.$$

Therefore

$$\Delta(\mathfrak{m}_f) = \begin{cases} f^2 d & \text{if } d \equiv 1 \mod 4, \\ 4f^2 d & \text{if } d \equiv 2,3 \mod 4, \end{cases}$$

and

$$\Sigma(\mathfrak{m}_f) = +2.$$

Example 4.1.2. Consider the Anosov maps $\phi_A, \phi_B : T^2 \to T^2$ given by matrices

$$A = \begin{pmatrix} 5 & 2 \\ 2 & 1 \end{pmatrix} \quad \text{and} \quad B = \begin{pmatrix} 5 & 1 \\ 4 & 1 \end{pmatrix},$$

respectively. The reader can verify that the Alexander polynomials of ϕ_A and ϕ_B are identical and equal to $\Delta_A(t) = \Delta_B(t) = t^2 - 6t + 1$; yet ϕ_A and ϕ_B are *not* conjugate.

Indeed, the Perron–Frobenius eigenvector of matrix A is $v_A = (1, \sqrt{2} - 1)$ while that of the matrix B is $v_B = (1, 2\sqrt{2} - 2)$. The bilinear forms for the modules $\mathfrak{m}_A = \mathbf{Z} + (\sqrt{2} - 1)\mathbf{Z}$ and $\mathfrak{m}_B = \mathbf{Z} + (2\sqrt{2} - 2)\mathbf{Z}$ can be written as

$$q_A(x,y) = 2x^2 - 4xy + 6y^2, \quad q_B(x,y) = 2x^2 - 8xy + 24y^2,$$

respectively. The modules $\mathfrak{m}_A, \mathfrak{m}_B$ are not similar in the number field $K = \mathbf{Q}(\sqrt{2})$, since their determinants $\Delta(\mathfrak{m}_A) = 8$ and $\Delta(\mathfrak{m}_B) = 32$ are not equal. Therefore, matrices A and B are not similar in the group $\mathrm{SL}(2, \mathbf{Z})$. Note that the class number $h_K = 1$ for the field K.

Remark 4.1.7 (Gauss method). The reader can verify that A and B are nonsimilar by using the *method of periods*, which dates back to C.-F. Gauss. According to the algorithm, we have to find the fixed points $Ax = x$ and $Bx = x$, which gives us $x_A = 1 + \sqrt{2}$ and $x_B = \frac{1+\sqrt{2}}{2}$, respectively. Then one unfolds the fixed points into a periodic continued fraction, which gives us $x_A = [2, 2, 2, \ldots]$ and $x_B = [1, 4, 1, 4, \ldots]$. Since the period $(\overline{2})$ of x_A differs from the period $(\overline{1, 4})$ of B, the matrices A and B belong to different similarity classes in $\mathrm{SL}(2, \mathbf{Z})$.

Example 4.1.3. Consider the Anosov maps $\phi_A, \phi_B : T^2 \to T^2$ given by matrices

$$A = \begin{pmatrix} 4 & 3 \\ 5 & 4 \end{pmatrix} \quad \text{and} \quad B = \begin{pmatrix} 4 & 15 \\ 1 & 4 \end{pmatrix},$$

respectively. The Alexander polynomials of ϕ_A and ϕ_B are identical $\Delta_A(t) = \Delta_B(t) = t^2 - 8t + 1$; yet the automorphisms ϕ_A and ϕ_B are not conjugate. Indeed, the Perron–Frobenius eigenvector of matrix A is $v_A = (1, \frac{1}{3}\sqrt{15})$ while that of the matrix B is $v_B = (1, \frac{1}{15}\sqrt{15})$. The corresponding modules are $\mathfrak{m}_A = \mathbf{Z} + (\frac{1}{3}\sqrt{15})\mathbf{Z}$ and $\mathfrak{m}_B = \mathbf{Z} + (\frac{1}{15}\sqrt{15})\mathbf{Z}$; therefore

$$q_A(x,y) = 2x^2 + 18y^2, \quad q_B(x,y) = 2x^2 + 450y^2,$$

respectively. The modules $\mathfrak{m}_A, \mathfrak{m}_B$ are not similar in the number field $K = \mathbf{Q}(\sqrt{15})$, since the module determinants $\Delta(\mathfrak{m}_A) = 36$ and $\Delta(\mathfrak{m}_B) = 900$ are not equal. Therefore, matrices A and B are not similar in the group $\mathrm{SL}(2, \mathbf{Z})$.

Example 4.1.4 ([107, p. 12]). Let a, b be a pair of positive integers satisfying the Pell equation $a^2 - 8b^2 = 1$; the latter has infinitely many solutions, e. g., $a = 3, b = 1$, etc. Denote by $\phi_A, \phi_B : T^2 \to T^2$ the Anosov maps given by matrices

$$A = \begin{pmatrix} a & 4b \\ 2b & a \end{pmatrix} \quad \text{and} \quad B = \begin{pmatrix} a & 8b \\ b & a \end{pmatrix},$$

respectively. The Alexander polynomials of ϕ_A and ϕ_B are identical $\Delta_A(t) = \Delta_B(t) = t^2 - 2at + 1$; yet maps ϕ_A and ϕ_B are *not* conjugate. Indeed, the Perron–Frobenius eigenvector of matrix A is $v_A = (1, \frac{1}{4b}\sqrt{a^2 - 1})$ while that of matrix B is $v_B = (1, \frac{1}{8b}\sqrt{a^2 - 1})$.

The corresponding modules are $\mathfrak{m}_A = \mathbf{Z} + (\frac{1}{4b}\sqrt{a^2-1})\mathbf{Z}$ and $\mathfrak{m}_B = \mathbf{Z} + (\frac{1}{8b}\sqrt{a^2-1})\mathbf{Z}$. It is easy to see that the discriminant $d = a^2 - 1 \equiv 3 \bmod 4$ for all $a \geq 2$. Indeed, $d = (a-1)(a+1)$ and, therefore, integer $a \not\equiv 1; 3 \bmod 4$; hence $a \equiv 2 \bmod 4$ so that $a - 1 \equiv 1 \bmod 4$ and $a + 1 \equiv 3 \bmod 4$ and, thus, $d = a^2 - 1 \equiv 3 \bmod 4$. Therefore the corresponding conductors are $f_A = 4b$ and $f_B = 8b$, and

$$q_A(x,y) = 2x^2 + 32b^2(a^2-1)y^2, \quad q_B(x,y) = 2x^2 + 128b^2(a^2-1)y^2,$$

respectively. The modules $\mathfrak{m}_A, \mathfrak{m}_B$ are not similar in the number field $K = \mathbf{Q}(\sqrt{a^2-1})$, since their determinants $\Delta(\mathfrak{m}_A) = 64\,b^2(a^2-1)$ and $\Delta(\mathfrak{m}_B) = 256\,b^2(a^2-1)$ are not equal. Therefore, matrices A and B are not similar in the group $SL(2,\mathbf{Z})$.

Guide to the literature

Topology of the surface automorphisms is the oldest part of geometric topology dating back to the works of J. Nielsen [175] and M. Dehn [66]. W. Thurston proved that there are only three types of such automorphisms: they are either of finite order or pseudo-Anosov, or else a mixture of the two, see [276]; the topological classification of pseudo-Anosov automorphisms is the next biggest problem after the *Geometrization Conjecture* proved by G. Perelman, see [275]. An excellent introduction to the subject are the books [81] and [51]. The measured foliations on compact surfaces were introduced in 1970s by W. Thurston [276] and covered in [120]. The Jacobi–Perron algorithm can be found in [224] and [23]. Noncommutative invariants of the pseudo-Anosov automorphisms were constructed in [185].

4.2 Torsion in the torus bundles

We assume that M_α is a torus bundle, i. e., an $(n+1)$-dimensional manifold fibering over the circle with monodromy $\alpha : T^n \to T^n$, where T^n is the n-dimensional torus. We construct a covariant functor on such bundles with the values in a category of the Cuntz–Krieger algebras. Such a functor maps homeomorphic bundles M_α to the Morita equivalent (stably isomorphic) Cuntz–Krieger algebras.

4.2.1 Cuntz–Krieger functor

Definition 4.2.1. If T^n is a torus of dimension $n \geq 1$, then by a torus bundle one understands an $(n+1)$-dimensional manifold

$$M_\alpha = \{T^n \times [0,1] \mid (T^n, 0) = (\alpha(T^n), 1)\},$$

where $\alpha : T^n \to T^n$ is an automorphism of T^n.

Remark 4.2.1. The torus bundles M_α and $M_{\alpha'}$ are homeomorphic if and only if the automorphisms α and α' are conjugate, i. e., $\alpha' = \beta \circ \alpha \circ \beta^{-1}$ for an automorphism $\beta : T^n \to T^n$.

Let $H_1(T^n; \mathbf{Z}) \cong \mathbf{Z}^n$ be the first homology of the torus; consider the group $\mathrm{Aut}(T^n)$ of (homotopy classes of) automorphisms of T^n. Any $\alpha \in \mathrm{Aut}(T^n)$ induces a linear transformation of $H_1(T^n; \mathbf{Z})$, given by an invertible $n \times n$ matrix A with the integer entries; conversely, each $A \in GL(n, \mathbf{Z})$ defines an automorphism $\alpha : T^n \to T^n$. In this matrix representation, the conjugate automorphisms α and α' define similar matrices $A, A' \in GL(n, \mathbf{Z})$, i. e., such that $A' = BAB^{-1}$ for a matrix $B \in GL(n, \mathbf{Z})$. Each class of matrices, similar to a matrix $A \in GL(n, \mathbf{Z})$ and such that $\mathrm{tr}(A) \geq 0$ ($\mathrm{tr}(A) \leq 0$), contains a matrix with only the nonnegative (nonpositive) entries. We always assume, that our bundle M_α is given by a nonnegative matrix A; the matrices with $\mathrm{tr}(A) \leq 0$ can be reduced to this case by switching the sign (from negative to positive) in the respective nonpositive representative.

Definition 4.2.2. Denote by $\mathcal{M}$ a category of torus bundles (of fixed dimension) endowed with homeomorphisms between the bundles; denote by $\mathcal{A}$ a category of the Cuntz–Krieger algebras $\mathcal{O}_A$ with $\det(A) = \pm 1$, endowed with stable isomorphisms between the algebras. By a Cuntz–Krieger map $F : \mathcal{M} \to \mathcal{A}$ one understands the map given by the formula

$$M_\alpha \mapsto \mathcal{O}_A.$$

Theorem 4.2.1 (Functor on torus bundles). *The map F is a covariant functor, which induces an isomorphism between the abelian groups*

$$H_1(M_\alpha; \mathbf{Z}) \cong \mathbf{Z} \oplus K_0(F(M_\alpha)).$$

Remark 4.2.2. Functor $F : M_\alpha \mapsto \mathcal{O}_A$ can be obtained from functor $\mathbf{F} : \alpha \mapsto \mathbb{A}_\alpha$ on automorphisms $\alpha : T^n \to T^n$ with values in the stationary AF-algebras $\mathbb{A}_\alpha$, see Section 4.1; the correspondence between F and $\mathbf{F}$ comes from the canonical isomorphism

$$\mathcal{O}_A \otimes \mathcal{K} \cong \mathbb{A}_\alpha \rtimes_\sigma \mathbf{Z},$$

where σ is the shift automorphism of $\mathbb{A}_\alpha$, see Section 3.7. Thus, one can interpret the invariant $\mathbf{Z} \oplus K_0(\mathcal{O}_A)$ as "abelianized" Handelman's invariant $(\Lambda, [I], K)$ of algebra $\mathbb{A}_\alpha$; here we assume that $(\Lambda, [I], K)$ is an analog of the fundamental group $\pi_1(M_\alpha)$.

4.2.2 Proof of Theorem 4.2.1

The idea of proof consists in a reduction of the conjugacy problem for the automorphisms of T^n to the Cuntz–Krieger theorem on the flow equivalence of the subshifts of finite type, see, e. g., [151]. For a different proof of Theorem 4.2.1, see [241].

(i) The main reference to the subshifts of finite type (SFT) is [151]. Recall that a *full* Bernoulli n-shift is the set X_n of biinfinite sequences $x = \{x_k\}$, where x_k is a symbol taken from a set S of cardinality n. The set X_n is endowed with the product topology, making X_n a Cantor set. The shift homeomorphism $\sigma_n : X_n \to X_n$ is given by the formula $\sigma_n(\dots x_{k-1}x_k x_{k+1}\dots) = (\dots x_k x_{k+1} x_{k+2}\dots)$ The homeomorphism defines a (discrete) dynamical system $\{X_n, \sigma_n\}$ given by the iterations of σ_n. Let A be an $n\times n$ matrix, whose entries $a_{ij} := a(i,j)$ are 0 or 1. Consider a subset X_A of X_n consisting of the biinfinite sequences, which satisfy the restriction $a(x_k, x_{k+1}) = 1$ for all $-\infty < k < \infty$. (It takes a moment to verify that X_A is indeed a subset of X_n and $X_A = X_n$ if and only if all the entries of A are 1's.) By definition, $\sigma_A = \sigma_n|X_A$ and the pair $\{X_A, \sigma_A\}$ is called an *SFT*. A standard edge shift construction described in [151] allows extending the notion of SFT to any matrix A with nonnegative entries. It is well known that the SFTs $\{X_A, \sigma_A\}$ and $\{X_B, \sigma_B\}$ are topologically conjugate (as dynamical systems) if and only if the matrices A and B are *strongly shift equivalent* (SSE), see [151] for the corresponding definition. The SSE of two matrices is a difficult algorithmic problem, which motivates the consideration of a weaker equivalence between the matrices called a *shift equivalence* (SE). Recall, that the matrices A and B are said to be shift equivalent (over $\mathbf{Z}^+$) when there exist nonnegative matrices R and S and a positive integer k (a lag), satisfying the equations $AR = RB$, $BS = SA$, $A^k = RS$, and $SR = B^k$. Finally, the SFTs $\{X_A, \sigma_A\}$ and $\{X_B, \sigma_B\}$ (and the matrices A and B) are said to be *flow equivalent* (FE), if the suspension flows of the SFTs act on the topological spaces, which are homeomorphic under a homeomorphism that respects the orientation of the orbits of the suspension flow. We shall use the following implications

$$\text{SSE} \Rightarrow \text{SE} \Rightarrow \text{FE}.$$

Remark 4.2.3. The first implication is rather classical, while for the second we refer the reader to [151, p. 456].

We further restrict to the SFTs given by the matrices with determinant ± 1. In view of [281, Corollary 2.13], the matrices A and B with $\det(A) = \pm 1$ and $\det(B) = \pm 1$ are SE (over $\mathbf{Z}^+$) if and only if matrices A and B are similar in $GL_n(\mathbf{Z})$. Let now α and α' be a pair of conjugate automorphisms of T^n. Since the corresponding matrices A and A' are similar in $GL_n(\mathbf{Z})$, one concludes that the SFTs $\{X_A, \sigma_A\}$ and $\{X_{A'}, \sigma_{A'}\}$ are SE. In particular, the SFTs $\{X_A, \sigma_A\}$ and $\{X_{A'}, \sigma_{A'}\}$ are FE. One can now apply the known result due to Cuntz and Krieger; it says that the C^*-algebra $\mathcal{O}_A \otimes \mathcal{K}$ is an invariant of the flow equivalence of the irreducible SFTs, see [63, p. 252] and its proof in Section 4 of the same work. Thus, the map F sends the conjugate automorphisms of T^n into the stably isomorphic Cuntz–Krieger algebras, i. e., F is a functor.

Let us show that F is a covariant functor. Consider the commutative diagram in Fig. 4.4, where $A, B \in GL_n(\mathbf{Z})$ and $\mathcal{O}_A, \mathcal{O}_{BAB^{-1}} \in \mathcal{A}$. Let g_1, g_2 be the arrows (similarity of matrices) in the upper category and $F(g_1), F(g_2)$ the corresponding arrows (stable

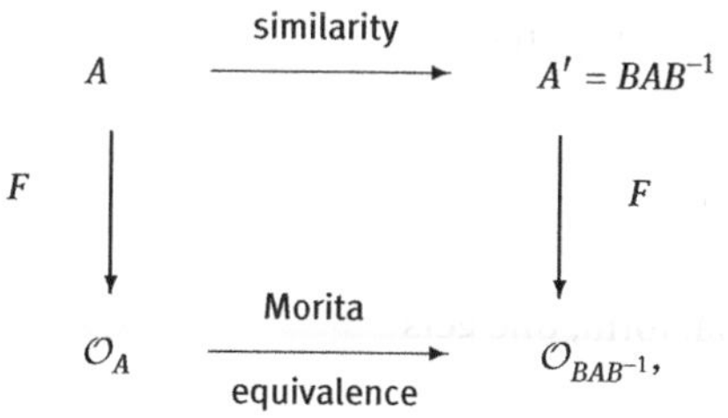

Figure 4.4: Cuntz–Krieger functor.

isomorphisms) in the lower category. In view of the diagram, we have the following identities:

$$F(g_1g_2) = \mathcal{O}_{B_2B_1AB_1^{-1}B_2^{-1}} = \mathcal{O}_{B_2(B_1AB_1^{-1})B_2^{-1}} = \mathcal{O}_{B_2A'B_2^{-1}} = F(g_1)F(g_2),$$

where $F(g_1)(\mathcal{O}_A) = \mathcal{O}_{A'}$ and $F(g_2)(\mathcal{O}_{A'}) = \mathcal{O}_{A''}$. Thus, F does not reverse the arrows and is, therefore, a covariant functor. The first statement of Theorem 4.2.1 is proved.

(ii) Let M_α be a torus bundle with a monodromy, given by the matrix $A \in GL(n, \mathbf{Z})$. It can be calculated, e. g., using the Leray spectral sequence for the fiber bundles, that $H_1(M_\alpha; \mathbf{Z}) \cong \mathbf{Z} \oplus \mathbf{Z}^n/(A-I)\mathbf{Z}^n$. Comparing this calculation with the K-theory of the Cuntz–Krieger algebra, one concludes that $H_1(M_\alpha; \mathbf{Z}) \cong \mathbf{Z} \oplus K_0(\mathcal{O}_A)$, where $\mathcal{O}_A = F(M_\alpha)$. The second statement of Theorem 4.2.1 follows.

4.2.3 Noncommutative invariants of torus bundles

To illustrate Theorem 4.2.1, we shall consider concrete examples of the torus bundles and calculate the noncommutative invariant $K_0(\mathcal{O}_A)$ for them. The reader can see that in some cases $K_0(\mathcal{O}_A)$ is a complete invariant of a family of the torus bundles. We compare $K_0(\mathcal{O}_A)$ with the corresponding Alexander polynomial $\Delta(t)$.

Example 4.2.1. Consider a three-dimensional torus bundle

$$A_1^n = \begin{pmatrix} 1 & n \\ 0 & 1 \end{pmatrix}, \quad n \in \mathbf{Z}.$$

Using the reduction of matrix to its Smith normal form (see, e. g., [151]), one can easily calculate

$$K_0(\mathcal{O}_{A_1^n}) \cong \mathbf{Z} \oplus \mathbf{Z}_n.$$

Remark 4.2.4. The Cuntz–Krieger invariant $K_0(\mathcal{O}_{A_1^n}) \cong \mathbf{Z} \oplus \mathbf{Z}_n$ is a complete topological invariant of the family of bundles $M_{\alpha_1^n}$; thus, such an invariant solves the classification problem for such bundles.

Example 4.2.2. Consider a three-dimensional torus bundle

$$A_2 = \begin{pmatrix} 5 & 2 \\ 2 & 1 \end{pmatrix}.$$

Using the reduction of matrix to its Smith normal form, one gets

$$K_0(\mathcal{O}_{A_2}) \cong \mathbf{Z}_2 \oplus \mathbf{Z}_2.$$

Example 4.2.3. Consider a three-dimensional torus bundle

$$A_3 = \begin{pmatrix} 5 & 1 \\ 4 & 1 \end{pmatrix}.$$

Using the reduction of matrix to its Smith normal form, one obtains

$$K_0(\mathcal{O}_{A_3}) \cong \mathbf{Z}_4.$$

Remark 4.2.5 ($K_0(\mathcal{O}_A)$ versus the Alexander polynomial). Note that for the bundles M_{α_2} and M_{α_3} the Alexander polynomial is

$$\Delta_{A_2}(t) = \Delta_{A_3}(t) = t^2 - 6t + 1.$$

Therefore, the Alexander polynomial cannot distinguish between the bundles M_{α_2} and M_{α_3}; however, since $K_0(\mathcal{O}_{A_2}) \not\cong K_0(\mathcal{O}_{A_3})$, Theorem 4.2.1 says that the torus bundles M_{α_2} and M_{α_3} are topologically distinct. Thus the noncommutative invariant $K_0(\mathcal{O}_A)$ is *finer* than the Alexander polynomial.

Remark 4.2.6. According to the Thurston Geometrization Theorem, the torus bundle $M_{\alpha_1^n}$ is a nilmanifold for any n, while torus bundles M_{α_2} and M_{α_3} are solvmanifolds [275].

Guide to the literature

For an excellent introduction to the subshifts of finite type, we refer the reader to the book [151] and the survey [281]. The Cuntz–Krieger algebras $\mathcal{O}_A$, the abelian group $K_0(\mathcal{O}_A)$ and their connection to the subshifts of finite type were introduced in [63]. Note that Theorem 4.2.1 follows from the results [241]; however, our argument is different and the proof is more direct and shorter than in the above cited work. The Cuntz–Krieger functor was constructed in [182].

4.3 Obstruction theory for Anosov's bundles

We construct a covariant functor F form the category of mapping tori of the Anosov diffeomorphisms $\varphi : M \to M$ of a smooth manifold M (Anosov's bundles) to a category of stationary AF-algebras. The functor sends each continuous map between the bundles to a stable homomorphism between the corresponding AF-algebras. We develop an obstruction theory for continuous maps between Anosov's bundles. Such a theory exploits noncommutative invariants derived from the Handelman triple $(\Lambda, [I], K)$ attached to a stationary AF-algebra. We illustrate the obstruction theory by concrete examples of dimension 2, 3 and 4.

4.3.1 Fundamental AF-algebra

By a q-dimensional, class C^r foliation of an m-dimensional manifold M one understands a decomposition of M into a union of disjoint connected subsets $\{\mathcal{L}_\alpha\}_{\alpha\in A}$, called the *leaves*, see, e. g., [149]. They must satisfy the following property: each point in M has a neighborhood U and a system of local class C^r coordinates $x = (x^1, \dots, x^m) : U \to \mathbf{R}^m$ such that for each leaf $\mathcal{L}_\alpha$, the components of $U \cap \mathcal{L}_\alpha$ are described by the equations $x^{q+1} = \mathrm{const}, \dots, x^m = \mathrm{const}$. Such a foliation is denoted by $\mathcal{F} = \{\mathcal{L}_\alpha\}_{\alpha\in A}$. The number $k = m - q$ is called the codimension of the foliation. An example of a codimension k foliation $\mathcal{F}$ is given by a closed k-form ω on M: the leaves of $\mathcal{F}$ are tangent to a plane defined by the normal vector $\omega(p) = 0$ at each point p of M. The C^r-foliations $\mathcal{F}_0$ and $\mathcal{F}_1$ of codimension k are said to be C^s-conjugate $(0 \le s \le r)$ if there exists an (orientation-preserving) diffeomorphism of M, of class C^s, which maps the leaves of $\mathcal{F}_0$ onto the leaves of $\mathcal{F}_1$; when $s = 0$, $\mathcal{F}_0$ and $\mathcal{F}_1$ are *topologically conjugate*. Denote by $f : N \to M$ a map of class C^s $(1 \le s \le r)$ of a manifold N into M; the map f is said to be *transverse* to $\mathcal{F}$, if for all $x \in N$ it holds $T_y(M) = \tau_y(\mathcal{F}) + f_* T_x(N)$, where $\tau_y(\mathcal{F})$ are the vectors of $T_y(M)$ tangent to $\mathcal{F}$ and $f_* : T_x(N) \to T_y(M)$ is the linear map on tangent vectors induced by f, where $y = f(x)$. If map $f : N \to M$ is transverse to a foliation $\mathcal{F}' = \{\mathcal{L}\}_{\alpha\in A}$ on M, then f induces a class C^s foliation $\mathcal{F}$ on N, where the leaves are defined as $f^{-1}(\mathcal{L}_\alpha)$ for all $\alpha \in A$; it is immediate, that $\mathrm{codim}(\mathcal{F}) = \mathrm{codim}(\mathcal{F}')$. We shall call $\mathcal{F}$ an *induced foliation*. When f is a submersion, it is transverse to any foliation of M; in this case, the induced foliation $\mathcal{F}$ is correctly defined for all $\mathcal{F}'$ on M, see [149, p. 373]. Notice, that for $M = N$ the above definition corresponds to topologically conjugate foliations $\mathcal{F}$ and $\mathcal{F}'$. To introduce measured foliations, denote by P and Q two k-dimensional submanifolds of M, which are everywhere transverse to a foliation $\mathcal{F}$ of codimension k. Consider a collection of C^r homeomorphisms between subsets of P and Q induced by a return map along the leaves of $\mathcal{F}$. The collection of all such homeomorphisms between subsets of all possible pairs of transverse manifolds generates a *holonomy pseudogroup* of $\mathcal{F}$ under composition of the homeomorphisms, see [231, p. 329]. A foliation $\mathcal{F}$ is said to have measure preserving holonomy, if its holonomy pseudogroup has a nontrivial invariant measure, which is finite on compact sets; for brevity, we call $\mathcal{F}$ a *measured foliation*. An example of measured foliation is a foliation determined by closed k-form ω; the restriction of ω to a transverse k-dimensional manifold determines a volume element, which gives a positive invariant measure on open sets. Each measured foliation $\mathcal{F}$ defines an element of the cohomology group $H^k(M;\mathbf{R})$, see [231]; in the case of $\mathcal{F}$ given by a closed k-form ω, such an element coincides with the de Rham cohomology class of ω [231]. In view of the isomorphism $H^k(M;\mathbf{R}) \cong \mathrm{Hom}(H_k(M),\mathbf{R})$, foliation $\mathcal{F}$ defines a linear map h from the kth homology group $H_k(M)$ to $\mathbf{R}$.

Definition 4.3.1. By a Plante group $P(\mathcal{F})$ of measured foliation $\mathcal{F}$ one understand the finitely generated abelian subgroup $h(H_k(M)/\mathrm{Tors}) \subset \mathbf{R}$.

Remark 4.3.1. If $\{\gamma_i\}$ is a basis of the homology group $H_k(M)$, then the periods $\lambda_i = \int_{\gamma_i} \omega$ are generators of the group $P(\mathcal{F})$, see [231].

Let $\lambda = (\lambda_1, \dots, \lambda_n)$ be a basis of the Plante group $P(\mathcal{F})$ of a measured foliation $\mathcal{F}$ such that $\lambda_i > 0$. Take a vector $\theta = (\theta_1, \dots, \theta_{n-1})$ with $\theta_i = \lambda_{i+1}/\lambda_1$; the Jacobi–Perron continued fraction of vector $(1,\theta)$ (or projective class of vector λ) is given by the formula

$$\begin{pmatrix} 1 \\ \theta \end{pmatrix} = \lim_{i\to\infty} \begin{pmatrix} 0 & 1 \\ I & b_1 \end{pmatrix} \cdots \begin{pmatrix} 0 & 1 \\ I & b_i \end{pmatrix} \begin{pmatrix} 0 \\ \mathbb{I} \end{pmatrix} = \lim_{i\to\infty} B_i \begin{pmatrix} 0 \\ \mathbb{I} \end{pmatrix},$$

where $b_i = (b_1^{(i)}, \dots, b_{n-1}^{(i)})^T$ is a vector of nonnegative integers, I the unit matrix and $\mathbb{I} = (0, \dots, 0, 1)^T$, see [23, p. 13]; the b_i are obtained from θ by the Euclidean algorithm [23, pp. 2–3].

Definition 4.3.2. An AF-algebra given by the Bratteli diagram with the incidence matrices B_i will be called associated to the measured foliation $\mathcal{F}$; we shall denote such an algebra by $\mathbb{A}_{\mathcal{F}}$.

Remark 4.3.2. Taking another basis of the Plante group $P(\mathcal{F})$ gives an AF-algebra which is Morita equivalent (stably isomorphic) to $\mathbb{A}_{\mathcal{F}}$; this is an algebraic recast of the main property of the Jacobi–Perron fractions.

If $\mathcal{F}'$ is a measured foliation on a manifold M and $f : N \to M$ is a submersion, then induced foliation $\mathcal{F}$ on N is a measured foliation. We shall denote by **MFol** the category of all manifolds with measured foliations (of fixed codimension), whose arrows are submersions of the manifolds; by $\mathbf{MFol}_0$ we understand a subcategory of **MFol**, consisting of manifolds, whose foliations have a unique transverse measure. Let **AF-alg** be a category of the (isomorphism classes of) AF-algebras given by *convergent* Jacobi–Perron fractions, so that the arrows of **AF-alg** are stable homomorphisms of the AF-algebras. By $F : \mathbf{MFol}_0 \to$ **AF-alg** we denote a map given by the formula $\mathcal{F} \mapsto \mathbb{A}_{\mathcal{F}}$. Notice, that F is correctly defined since foliations with the unique measure have the convergent Jacobi–Perron fractions; this assertion follows from [21]. The following result will be proved in Section 4.3.2.

Theorem 4.3.1. *The map F : $\mathbf{MFol}_0 \to$ **AF-alg** is a functor, which sends any pair of induced foliations to a pair of stably homomorphic AF-algebras.*

Let M be an m-dimensional manifold and $\varphi : M \to M$ a diffeomorphism of M; recall that an orbit of point $x \in M$ is the subset $\{\varphi^n(x) \mid n \in \mathbf{Z}\}$ of M. The finite orbits $\varphi^m(x) = x$ are called periodic; when $m = 1$, x is a *fixed point* of diffeomorphism φ. The fixed point p is *hyperbolic* if the eigenvalues λ_i of the linear map $D\varphi(p) : T_p(M) \to T_p(M)$ do not lie on the unit circle. If $p \in M$ is a hyperbolic fixed point of a diffeomorphism $\varphi : M \to M$, denote by $T_p(M) = V^s + V^u$ the corresponding decomposition of the tangent space under the linear map $D\varphi(p)$, where V^s (V^u) is the eigenspace of $D\varphi(p)$ corresponding to $|\lambda_i| > 1$ ($|\lambda_i| < 1$). For a submanifold $W^s(p)$, there exists a contraction $g : W^s(p) \to W^s(p)$ with fixed point p_0 and an injective equivariant immersion $J : W^s(p) \to M$ such that $J(p_0) = p$ and $DJ(p_0) : T_{p_0}(W^s(p)) \to T_p(M)$ is an isomorphism; the image of J defines an immersed submanifold $W^s(p) \subset M$ called a *stable manifold* of φ at p. Clearly, $\dim(W^s(p)) = \dim(V^s)$.

Definition 4.3.3 ([3]). A diffeomorphism $\varphi : M \to M$ is called Anosov if there exists a splitting of the tangent bundle $T(M)$ into a continuous Whitney sum $T(M) = E^s + E^u$ invariant under $D\varphi : T(M) \to T(M)$, so that the map $D\varphi : E^s \to E^s$ is contracting and $D\varphi : E^u \to E^u$ is expanding.

Remark 4.3.3. The Anosov diffeomorphism imposes a restriction on the topology of manifold M in the sense that not every manifold can support such a diffeomorphism; however, if one Anosov diffeomorphism exists on M, there are infinitely many (conjugacy classes of) such diffeomorphisms on M. It is an open problem of S. Smale to describe which M can carry an Anosov diffeomorphism; so far, it is proved that the hyperbolic diffeomorphisms of m-dimensional tori and certain automorphisms of the nilmanifolds are Anosov, see, e. g., [263].

Let p be a fixed point of the Anosov diffeomorphism $\varphi : M \to M$ and $W^s(p)$ its stable manifold. Since $W^s(p)$ cannot have self-intersections or limit compacta, $W^s(p) \to M$ is a dense immersion, i. e., the closure of $W^s(p)$ is the entire M. Moreover, if q is a periodic point of φ of period n, then $W^s(q)$ is a translate of $W^s(p)$, i. e., locally they look like two parallel lines. Consider a foliation $\mathcal{F}$ of M, whose leaves are the translates of $W^s(p)$; the $\mathcal{F}$ is a continuous foliation, which is invariant under the action of diffeomorphism φ on its leaves, i. e., φ moves leaves of $\mathcal{F}$ to the leaves of $\mathcal{F}$, see [263, p. 760]. The holonomy of $\mathcal{F}$ preserves the Lebesgue measure and, therefore, $\mathcal{F}$ is a measured foliation; we shall call it an *invariant measured foliation* and denote by $\mathcal{F}_\varphi$.

Definition 4.3.4. By a fundamental AF-algebra we understand the AF-algebra of foliation $\mathcal{F}_\varphi$, where $\varphi : M \to M$ is an Anosov diffeomorphism of a manifold M; the fundamental AF-algebra will be denoted by $\mathbb{A}_\varphi$.

Theorem 4.3.2. *The $\mathbb{A}_\varphi$ is a stationary AF-algebra.*

Consider the mapping torus of the Anosov diffeomorphism φ, i. e., a manifold

$$M_\varphi := M \times [0,1]/\sim, \quad \text{where } (x,0) \sim (\varphi(x),1),\ \forall x \in M.$$

Let **AnoBnd** be a category of the mapping tori of all Anosov diffeomorphisms; the arrows of **AnoBnd** are continuous maps between the mapping tori. Likewise, let **Fund-AF** be a category of all fundamental AF-algebras; the arrows of **Fund-AF** are stable homomorphisms between the fundamental AF-algebras. By F : **AnoBnd** $\to$ **Fund-AF** we understand a map given by the formula $M_\varphi \mapsto \mathbb{A}_\varphi$, where $M_\varphi \in$ **AnoBnd** and $\mathbb{A}_\varphi \in$ **Fund-AF**. The following theorem says that F is a functor.

Theorem 4.3.3 (Functor on Anosov bundles). *The map F is a covariant functor, which sends each continuous map $N_\psi \to M_\varphi$ to a stable homomorphism $\mathbb{A}_\psi \to \mathbb{A}_\varphi$ of the corresponding fundamental AF-algebras.*

Remark 4.3.4 (Obstruction theory). Theorem 4.3.3 can be used, e. g., in the obstruction theory because stable homomorphisms of the fundamental AF-algebras are easier to detect than continuous maps between manifolds N_ψ and M_φ; such homomorphisms are bijective with the inclusions of certain **Z**-modules belonging to a real algebraic number field. Often it is possible to prove that no inclusion is possible and, thus, draw a topological conclusion about the maps, see Section 4.3.3.

4.3.2 Proofs

4.3.2.1 Proof of Theorem 4.3.1

Let $\mathcal{F}'$ be measured foliation on M given by a closed form $\omega' \in H^k(M;\mathbf{R})$; let $\mathcal{F}$ be measured foliation on N induced by a submersion $f : N \to M$. Roughly speaking, we have to prove that diagram in Fig. 4.5 is commutative; the proof amounts to the fact that the periods of form ω' are among the periods of form $\omega \in H^k(N;\mathbf{R})$ corresponding to the foliation $\mathcal{F}$. The map f defines a homomorphism $f_* : H_k(N) \to H_k(M)$ of the kth

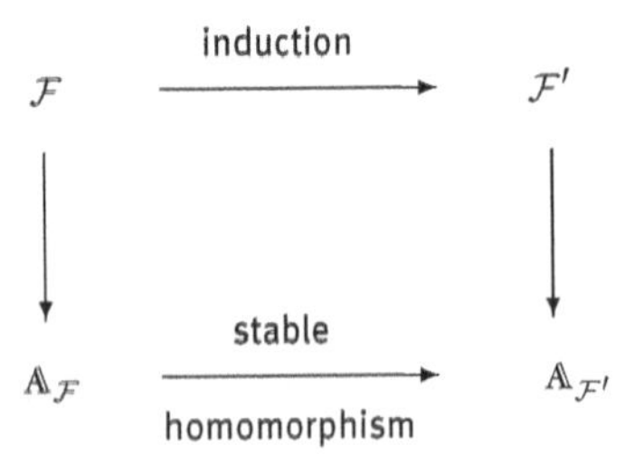

Figure 4.5: Map F : **MFol$_0$** $\to$ **AF-alg**.

homology groups; let $\{e_i\}$ and $\{e_i'\}$ be bases in $H_k(N)$ and $H_k(M)$, respectively. Since $H_k(M) = H_k(N)/\ker(f_*)$, we shall denote by $[e_i] := e_i + \ker(f_*)$ a coset representative of e_i; these can be identified with the elements $e_i \notin \ker(f_*)$. The integral $\int_{e_i} \omega$ defines a scalar product $H_k(N) \times H^k(N;\mathbf{R}) \to \mathbf{R}$, so that f_* is a linear self-adjoint operator; thus, we can write

$$\lambda_i' = \int_{e_i'} \omega' = \int_{e_i'} f^*(\omega) = \int_{f_*^{-1}(e_i')} \omega = \int_{[e_i]} \omega \in P(\mathcal{F}),$$

where $P(\mathcal{F})$ is the Plante group (the group of periods) of foliation $\mathcal{F}$. Since λ_i' are generators of $P(\mathcal{F}')$, we conclude that $P(\mathcal{F}') \subseteq P(\mathcal{F})$. Note that $P(\mathcal{F}') = P(\mathcal{F})$ if and only if f_* is an isomorphism.

One can apply a criterion of the stable homomorphism of AF-algebras; namely, $\mathbb{A}_{\mathcal{F}}$ and $\mathbb{A}_{\mathcal{F}'}$ are stably homomorphic, if and only if, there exists a positive homomorphism $h : G \to H$ between their dimension groups G and H, see [74, p. 15]. But $G \cong P(\mathcal{F})$ and $H \cong P(\mathcal{F}')$, while $h = f_*$. Thus, $\mathbb{A}_{\mathcal{F}}$ and $\mathbb{A}_{\mathcal{F}'}$ are stably homomorphic.

The functor F is compatible with the composition; indeed, let $f : N \to M$ and $f' : L \to N$ be submersions. If $\mathcal{F}$ is a measured foliation of M, one gets the induced foliations $\mathcal{F}'$ and $\mathcal{F}''$ on N and L, respectively; these foliations fit the diagram $(L, \mathcal{F}'') \xrightarrow{f'} (N, \mathcal{F}') \xrightarrow{f} (M, \mathcal{F})$ and the corresponding Plante groups are included, $P(\mathcal{F}'') \supseteq P(\mathcal{F}') \supseteq P(\mathcal{F})$. Thus, $F(f' \circ f) = F(f') \circ F(f)$, since the inclusion of the Plante groups corresponds to the composition of homomorphisms; Theorem 4.3.1 is proved.

4.3.2.2 Proof of Theorem 4.3.2

Let $\varphi : M \to M$ be an Anosov diffeomorphism; we proceed by showing that invariant foliation $\mathcal{F}_\varphi$ is given by form $\omega \in H^k(M;\mathbf{R})$, which is an eigenvector of the linear map $[\varphi] : H^k(M;\mathbf{R}) \to H^k(M;\mathbf{R})$ induced by φ. Indeed, let $0 < c < 1$ be contracting constant of the stable subbundle E^s of diffeomorphism φ and Ω the corresponding volume element; by definition, $\varphi(\Omega) = c\Omega$. Note that Ω is given by restriction of form ω to a k-dimensional manifold, transverse to the leaves of $\mathcal{F}_\varphi$. The leaves of $\mathcal{F}_\varphi$ are fixed by φ and, therefore, $\varphi(\Omega)$ is given by a multiple $c\omega$ of form ω. Since $\omega \in H^k(M;\mathbf{R})$ is a vector whose coordinates define $\mathcal{F}_\varphi$ up to a scalar, we conclude, that $[\varphi](\omega) = c\omega$, i. e., ω is an eigenvector of the linear map $[\varphi]$. Let $(\lambda_1, \dots, \lambda_n)$ be a basis of the Plante group $P(\mathcal{F}_\varphi)$ such that $\lambda_i > 0$. Notice, that φ acts on λ_i as multiplication by constant c; indeed, since $\lambda_i = \int_{\gamma_i} \omega$, we have

$$\lambda_i' = \int_{\gamma_i} [\varphi](\omega) = \int_{\gamma_i} c\omega = c \int_{\gamma_i} \omega = c\lambda_i,$$

where $\{\gamma_i\}$ is a basis in $H_k(M)$. Since φ preserves the leaves of $\mathcal{F}_\varphi$, one concludes that $\lambda_i' \in P(\mathcal{F}_\varphi)$; therefore, $\lambda_j' = \sum b_{ij}\lambda_i$ for a nonnegative integer matrix $B = (b_{ij})$. According to Bauer [21], matrix B can be written as a finite product,

$$B = \begin{pmatrix} 0 & 1 \\ \mathbf{I} & b_1 \end{pmatrix} \cdots \begin{pmatrix} 0 & 1 \\ \mathbf{I} & b_p \end{pmatrix} := B_1 \cdots B_p,$$

where $b_i = (b_1^{(i)}, \dots, b_{n-1}^{(i)})^T$ is a vector of nonnegative integers and $\mathbf{I}$ the unit matrix. Let $\lambda = (\lambda_1, \dots, \lambda_n)$. Consider a purely periodic Jacobi–Perron continued fraction

$$\lim_{i\to\infty} \overline{B_1 \cdots B_p} \begin{pmatrix} 0 \\ \mathbb{I} \end{pmatrix},$$

where $\mathbb{I} = (0, \dots, 0, 1)^T$; by a basic property of such fractions, it converges to an eigenvector $\lambda' = (\lambda'_1, \dots, \lambda'_n)$ of matrix $B_1 \cdots B_p$, see [23, Chapter 3]. But $B_1 \cdots B_p = B$ and λ is an eigenvector of matrix B; therefore, vectors λ and λ' are collinear. The collinear vectors are known to have the same continued fractions; thus, we have

$$\begin{pmatrix} 1 \\ \theta \end{pmatrix} = \lim_{i\to\infty} \overline{B_1 \cdots B_p} \begin{pmatrix} 0 \\ \mathbb{I} \end{pmatrix},$$

where $\theta = (\theta_1, \dots, \theta_{n-1})$ and $\theta_i = \lambda_{i+1}/\lambda_1$. Since vector $(1, \theta)$ unfolds into a periodic Jacobi–Perron continued fraction, we conclude that the AF-algebra $\mathbb{A}_\varphi$ is stationary. Theorem 4.3.2 is proved.

4.3.2.3 Proof of Theorem 4.3.3

Let $\psi : N \to N$ and $\varphi : M \to M$ be a pair of Anosov diffeomorphisms; denote by $(N, \mathcal{F}_\psi)$ and $(M, \mathcal{F}_\varphi)$ the corresponding invariant foliations of manifolds N and M, respectively. In view of Theorem 4.3.1, it is sufficient to prove that the diagram in Fig. 4.6 is commutative. We shall split the proof in a series of lemmas.

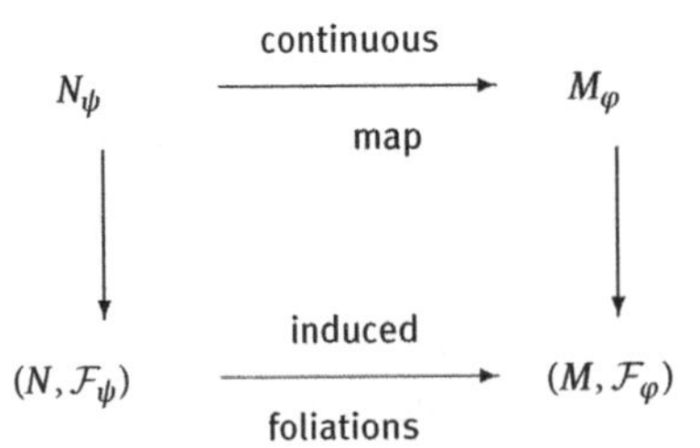

Figure 4.6: Mapping tori and invariant foliations.

Lemma 4.3.1. *There exists a continuous map $N_\psi \to M_\varphi$, whenever $f \circ \varphi = \psi \circ f$ for a submersion $f : N \to M$.*

Proof. (i) Suppose that $h : N_\psi \to M_\varphi$ is a continuous map; let us show that there exists a submersion $f : N \to M$ such that $f \circ \varphi = \psi \circ f$. Both N_ψ and M_φ fiber over the circle S^1 with the projection map p_ψ and p_φ, respectively; therefore, the diagram in Fig. 4.7 is commutative. Let $x \in S^1$; since $p_\psi^{-1} = N$ and $p_\varphi^{-1} = M$, the restriction of h to x defines a submersion $f : N \to M$, i. e., $f = h_x$. Moreover, since ψ and φ are monodromy maps of the bundle, it holds that

$$\begin{cases} p_\psi^{-1}(x + 2\pi) = \psi(N), \\ p_\varphi^{-1}(x + 2\pi) = \varphi(M). \end{cases}$$

From the diagram in Fig. 4.7, we get $\psi(N) = p_\psi^{-1}(x+2\pi) = f^{-1}(p_\varphi^{-1}(x+2\pi)) = f^{-1}(\varphi(M)) = f^{-1}(\varphi(f(N)))$; thus, $f \circ \psi = \varphi \circ f$. The necessary condition of Lemma 4.3.1 follows.

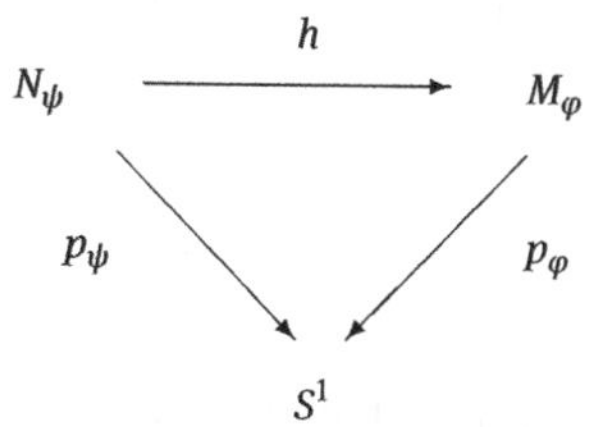

Figure 4.7: The fiber bundles N_ψ and M_φ over S^1.

(ii) Suppose that $f : N \to M$ is a submersion such that $f \circ \varphi = \psi \circ f$; we have to construct a continuous map $h : N_\psi \to M_\varphi$. Recall that

$$\begin{cases} N_\psi = \{N \times [0,1] \mid (x,0) \sim (\psi(x),1)\}, \\ M_\varphi = \{M \times [0,1] \mid (y,0) \sim (\varphi(y),1)\}. \end{cases}$$

We shall identify the points of N_ψ and M_φ using the substitution $y = f(x)$; it remains to verify that such an identification will satisfy the gluing condition $y \sim \varphi(y)$. In view of condition $f \circ \varphi = \psi \circ f$, we have

$$y = f(x) \sim f(\psi(x)) = \varphi(f(x)) = \varphi(y).$$

Thus, $y \sim \varphi(y)$ and, therefore, the map $h : N_\psi \to M_\varphi$ is continuous. The sufficient condition of Lemma 4.3.1 is proved. □

Lemma 4.3.2. *If a submersion $f : N \to M$ satisfies the condition $f \circ \varphi = \psi \circ f$ for the Anosov diffeomorphisms $\psi : N \to N$ and $\varphi : M \to M$, then the invariant foliations $(N, \mathcal{F}_\psi)$ and $(M, \mathcal{F}_\varphi)$ are induced by f.*

Proof. The invariant foliations $\mathcal{F}_\psi$ and $\mathcal{F}_\varphi$ are measured; we shall denote by $\omega_\psi \in H^k(N; \mathbf{R})$ and $\omega_\varphi \in H^k(M; \mathbf{R})$ the corresponding cohomology classes, respectively. The

$$\begin{array}{ccc} H^k(N;\mathbf{R}) & \xrightarrow{[\psi]} & H^k(N,\mathbf{R}) \\ {\scriptstyle [f]}\downarrow & & \downarrow{\scriptstyle [f]} \\ H^k(M,\mathbf{R}) & \xrightarrow{[\varphi]} & H^k(M,\mathbf{R}) \end{array}$$

Figure 4.8: The linear maps $[\psi]$, $[\varphi]$, and $[f]$.

linear maps on $H^k(N;\mathbf{R})$ and $H^k(M;\mathbf{R})$, induced by ψ and φ, we shall denote by $[\psi]$ and $[\varphi]$; the linear map between $H^k(N;\mathbf{R})$ and $H^k(M;\mathbf{R})$, induced by f, we write as $[f]$. Notice that $[\psi]$ and $[\varphi]$ are isomorphisms, while $[f]$ is generally a homomorphism. It was shown earlier that ω_ψ and ω_φ are eigenvectors of linear maps $[\psi]$ and $[\varphi]$, respectively; in other words, we have

$$\begin{cases} [\psi]\omega_\psi = c_1\omega_\psi, \\ [\varphi]\omega_\varphi = c_2\omega_\varphi, \end{cases}$$

where $0 < c_1 < 1$ and $0 < c_2 < 1$. Consider a diagram in Fig. 4.8 which involves the linear maps $[\psi]$, $[\varphi]$, and $[f]$; the diagram is commutative since the condition $f \circ \varphi = \psi \circ f$ implies that $[\varphi] \circ [f] = [f] \circ [\psi]$. Take the eigenvector ω_ψ and consider its image under the linear map $[\varphi] \circ [f]$,

$$[\varphi] \circ [f](\omega_\psi) = [f] \circ [\psi](\omega_\psi) = [f](c_1\omega_\psi) = c_1([f](\omega_\psi)).$$

Therefore, vector $[f](\omega_\psi)$ is an eigenvector of the linear map $[\varphi]$; let us compare it with the eigenvector ω_φ:

$$\begin{cases} [\varphi]([f](\omega_\psi)) = c_1([f](\omega_\psi)), \\ \qquad\qquad [\varphi]\omega_\varphi = c_2\omega_\varphi. \end{cases}$$

We conclude, therefore, that ω_φ and $[f](\omega_\psi)$ are collinear vectors such that $c_1^m = c_2^n$ for some integers $m, n > 0$; a scaling gives us $[f](\omega_\psi) = \omega_\varphi$. The latter is an analytic formula, which says that the submersion $f : N \to M$ induces foliation $(N, \mathcal{F}_\psi)$ from the foliation $(M, \mathcal{F}_\varphi)$. Lemma 4.3.2 is proved. □

To finish the proof of Theorem 4.3.3, let $N_\psi \to M_\varphi$ be a continuous map; by Lemma 4.3.1, there exists a submersion $f : N \to M$ such that $f \circ \varphi = \psi \circ f$. Lemma 4.3.2 says that in this case the invariant measured foliations $(N, \mathcal{F}_\psi)$ and $(M, \mathcal{F}_\varphi)$ are induced. On the other hand, from Theorem 4.3.2 we know that the Jacobi–Perron continued fraction connected to foliations $\mathcal{F}_\psi$ and $\mathcal{F}_\varphi$ are periodic and hence convergent, see, e. g., [23]; therefore, one can apply Theorem 4.3.1 which says that the AF-algebra

$\mathbb{A}_\psi$ is stably homomorphic to the AF-algebra $\mathbb{A}_\varphi$. The latter are, by definition, the fundamental AF-algebras of the Anosov diffeomorphisms ψ and φ, respectively. Theorem 4.3.3 is proved.

4.3.3 Obstruction theory

Let $\mathbb{A}_\psi$ be a fundamental AF-algebra and B its primitive incidence matrix, i. e., B is not a power of some positive integer matrix. Suppose that the characteristic polynomial of B is irreducible and let K_ψ be its splitting field; then K_ψ is a Galois extension of $\mathbf{Q}$.

Definition 4.3.5. We call $\mathrm{Gal}(\mathbb{A}_\psi) := \mathrm{Gal}(K_\psi|\mathbf{Q})$ the Galois group of the fundamental AF-algebra $\mathbb{A}_\psi$; such a group is determined by the AF-algebra $\mathbb{A}_\psi$.

The second algebraic field is connected to the Perron–Frobenius eigenvalue λ_B of the matrix B; we shall denote this field by $\mathbf{Q}(\lambda_B)$. Note that $\mathbf{Q}(\lambda_B) \subseteq K_\psi$ and $\mathbf{Q}(\lambda_B)$ is not, in general, a Galois extension of $\mathbf{Q}$, the reason being that the polynomial char(B) may have complex roots, and if there are no such roots then $\mathbf{Q}(\lambda_B) = K_\psi$. There is still a group $\mathrm{Aut}(\mathbf{Q}(\lambda_B))$ of automorphisms of $\mathbf{Q}(\lambda_B)$ fixing the field $\mathbf{Q}$ and $\mathrm{Aut}(\mathbf{Q}(\lambda_B)) \subseteq \mathrm{Gal}(K_\psi)$ is a subgroup inclusion.

Lemma 4.3.3. *If $h : \mathbb{A}_\psi \to \mathbb{A}_\varphi$ is a stable homomorphism, then $\mathbf{Q}(\lambda_{B'}) \subseteq K_\psi$ is a field inclusion.*

Proof. Notice that the nonnegative matrix B becomes strictly positive when a proper power of it is taken; we always assume B positive. Let $\lambda = (\lambda_1, \dots, \lambda_n)$ be a basis of the Plante group $P(\mathcal{F}_\psi)$. Following the proof of Theorem 4.3.2, one concludes that $\lambda_i \in K_\psi$; indeed, $\lambda_B \in K_\psi$ is the Perron–Frobenius eigenvalue of B, while λ is the corresponding eigenvector. The latter can be scaled so that $\lambda_i \in K_\psi$. Any stable homomorphism $h : \mathbb{A}_\psi \to \mathbb{A}_\varphi$ induces a positive homomorphism of their dimension groups $[h] : G \to H$; but $G \cong P(\mathcal{F}_\psi)$ and $H \cong P(\mathcal{F}_\varphi)$. From inclusion $P(\mathcal{F}_\varphi) \subseteq P(\mathcal{F}_\psi)$, one gets $\mathbf{Q}(\lambda_{B'}) \cong P(\mathcal{F}_\varphi) \otimes \mathbf{Q} \subseteq P(\mathcal{F}_\psi) \otimes \mathbf{Q} \cong \mathbf{Q}(\lambda_B) \subseteq K_f$ and, therefore, $\mathbf{Q}(\lambda_{B'}) \subseteq K_\psi$. Lemma 4.3.3 follows. □

Corollary 4.3.1. *If $h : \mathbb{A}_\psi \to \mathbb{A}_\varphi$ is a stable homomorphism, then $\mathrm{Aut}(\mathbf{Q}(\lambda_{B'}))$ (or $\mathrm{Gal}(\mathbb{A}_\varphi)$) is a subgroup (or a normal subgroup) of $\mathrm{Gal}(\mathbb{A}_\psi)$.*

Proof. The (Galois) subfields of the Galois field K_ψ are bijective with the (normal) subgroups of the group $\mathrm{Gal}(K_\psi)$, see, e. g., [163]. □

Let $T^m \cong \mathbf{R}^m/\mathbf{Z}^m$ be an m-dimensional torus; let ψ_0 be an $m \times m$ integer matrix with $\det(\psi_0) = 1$ such that it is similar to a positive matrix. The matrix ψ_0 defines a linear transformation of $\mathbf{R}^m$, which preserves the lattice $L \cong \mathbf{Z}^m$ of points with integer coordinates. There is an induced diffeomorphism ψ of the quotient $T^m \cong \mathbf{R}^m/\mathbf{Z}^m$ onto itself; this diffeomorphism $\psi : T^m \to T^m$ has a fixed point p corresponding to the origin of $\mathbf{R}^m$. Suppose that ψ_0 is hyperbolic, i. e., there are no eigenvalues of ψ_0 on the unit circle; then p is a hyperbolic fixed point of ψ and the stable manifold $W^s(p)$ is the image of the corresponding eigenspace of ψ_0 under the projection $\mathbf{R}^m \to T^m$. If codim $W^s(p) = 1$, the hyperbolic linear transformation ψ_0 (and the diffeomorphism ψ) will be called *tight*.

Lemma 4.3.4. *The tight hyperbolic matrix ψ_0 is similar to the matrix B of the fundamental AF-algebra $\mathbb{A}_\psi$.*

Proof. Since $H_k(T^m;\mathbf{R}) \cong \mathbf{R}^{\frac{m!}{k!(m-k)!}}$, one gets $H_{m-1}(T^m;\mathbf{R}) \cong \mathbf{R}^m$; in view of the Poincaré duality, $H^1(T^m;\mathbf{R}) = H_{m-1}(T^m;\mathbf{R}) \cong \mathbf{R}^m$. Since codim $W^s(p) = 1$, measured foliation $\mathcal{F}_\psi$ is given by a closed form $\omega_\psi \in H^1(T^m;\mathbf{R})$ such that $[\psi]\omega_\psi = \lambda_\psi\omega_\psi$, where λ_ψ is the eigenvalue of the linear transformation $[\psi] : H^1(T^m;\mathbf{R}) \to H^1(T^m;\mathbf{R})$. It is easy to see that $[\psi] = \psi_0$ because $H^1(T^m;\mathbf{R}) \cong \mathbf{R}^m$ is the universal cover for T^m, where the eigenspace $W^u(p)$ of ψ_0 is the span of the eigenform ω_ψ. On the other hand, from the proof of Theorem 4.3.2, we know that the Plante group $P(\mathcal{F}_\psi)$ is generated by the coordinates of vector ω_ψ; the matrix B is nothing but the matrix ψ_0 written in a new basis of $P(\mathcal{F}_\psi)$. Each change of basis in the $\mathbf{Z}$-module $P(\mathcal{F}_\psi)$ is given by an integer invertible matrix S; therefore, $B = S^{-1}\psi_0 S$. Lemma 4.3.4 follows. □

Let $\psi : T^m \to T^m$ be a hyperbolic diffeomorphism; the mapping torus T^m_ψ will be called a (hyperbolic) *torus bundle* of dimension m. Let $k = |\mathrm{Gal}(\mathbb{A}_\psi)|$; it follows from the Galois theory, that $1 < k \le m!$. Denote t_i the cardinality of a subgroup $G_i \subseteq \mathrm{Gal}(\mathbb{A}_\psi)$.

Corollary 4.3.2. *There are no (nontrivial) continuous maps $T^m_\psi \to T^{m'}_\varphi$, whenever $t'_i \nmid k$ for all $G'_i \subseteq \mathrm{Gal}(\mathbb{A}_\varphi)$.*

Proof. If $h : T^m_\psi \to T^{m'}_\varphi$ were a continuous map to a torus bundle of dimension $m' < m$, then, by Theorem 4.3.3 and Corollary 4.3.1, $\mathrm{Aut}(\mathbf{Q}(\lambda_{B'}))$ (or $\mathrm{Gal}(\mathbb{A}_\varphi)$) would be a nontrivial subgroup (or normal subgroup) of the group $\mathrm{Gal}(\mathbb{A}_\psi)$; since $k = |\mathrm{Gal}(\mathbb{A}_\psi)|$, one concludes that one of t'_i divides k. This contradicts our assumption. □

Definition 4.3.6. The torus bundle T^m_ψ is called robust if there exists $m' < m$ such that no continuous map $T^m_\psi \to T^{m'}_\varphi$ is possible.

4.3.3.1 Case $m = 2$

This case is trivial; ψ_0 is a hyperbolic matrix and always tight. Also $\mathrm{char}(\psi_0) = \mathrm{char}(B)$ is an irreducible quadratic polynomial with two real roots; $\mathrm{Gal}(\mathbb{A}_\psi) \cong \mathbf{Z}_2$ and, therefore, $|\mathrm{Gal}(\mathbb{A}_\psi)| = 2$. Formally, T^2_ψ is robust since no torus bundle of a smaller dimension is defined.

4.3.3.2 Case $m = 3$

If ψ_0 is hyperbolic; it is always tight, since one root of $\mathrm{char}(\psi_0)$ is real and isolated inside or outside the unit circle.

Corollary 4.3.3. *Let*

$$\psi_0(b,c) = \begin{pmatrix} -b & 1 & 0 \\ -c & 0 & 1 \\ -1 & 0 & 0 \end{pmatrix}$$

be such that $\mathrm{char}(\psi_0(b,c)) = x^3+bx^2+cx+1$ *is irreducible and* $-4b^3+b^2c^2+18bc-4c^3-27$ *is the square of an integer; then* T^3_ψ *admits no continuous map to any* T^2_φ.

Proof. We have $\mathrm{char}(\psi_0(b,c)) = x^3+bx^2+cx+1$ and the discriminant $D = -4b^3+b^2c^2+18bc-4c^3-27$. By Morandi [163, Theorem 13.1], we have $\mathrm{Gal}(\mathbb{A}_\psi) \cong \mathbf{Z}_3$ and, therefore, $k = |\mathrm{Gal}(\mathbb{A}_\psi)| = 3$. For $m' = 2$, it was shown that $\mathrm{Gal}(\mathbb{A}_\varphi) \cong \mathbf{Z}_2$ and, therefore, $t'_1 = 2$. Since $2 \nmid 3$, Corollary 4.3.2 says that no continuous map $T^3_\psi \to T^2_\varphi$ can be constructed. Corollary 4.3.3 follows. □

Example 4.3.1. There are infinitely many matrices $\psi_0(b,c)$ satisfying the assumptions of Corollary 4.3.3; below are a few numerical examples of robust bundles:

$$\begin{pmatrix} 0 & 1 & 0 \\ 3 & 0 & 1 \\ -1 & 0 & 0 \end{pmatrix}, \quad \begin{pmatrix} 1 & 1 & 0 \\ 2 & 0 & 1 \\ -1 & 0 & 0 \end{pmatrix}, \quad \begin{pmatrix} 2 & 1 & 0 \\ 1 & 0 & 1 \\ -1 & 0 & 0 \end{pmatrix}, \quad \begin{pmatrix} 3 & 1 & 0 \\ 0 & 0 & 1 \\ -1 & 0 & 0 \end{pmatrix}.$$

Remark 4.3.5. Notice that the above matrices are not pairwise similar; this can be gleaned from their traces; thus they represent topologically distinct torus bundles.

4.3.3.3 Case $m = 4$

Let $p(x) = x^4+ax^3+bx^2+cx+d$ be a quartic. Consider the associated cubic polynomial $r(x) = x^3 - bx^2 + (ac-4d)x + 4bd - a^2d - c^2$; denote by L the splitting field of $r(x)$.

Corollary 4.3.4. *Let*

$$\psi_0(a,b,c) = \begin{pmatrix} -a & 1 & 0 & 0 \\ -b & 0 & 1 & 0 \\ -c & 0 & 0 & 1 \\ -1 & 0 & 0 & 0 \end{pmatrix}$$

be tight and such that $\mathrm{char}(\psi_0(a,b,c)) = x^4 + ax^3 + bx^2 + cx + 1$ *is irreducible and one of the following holds:* (i) $L = \mathbf{Q}$; (ii) $r(x)$ *has a unique root* $t \in \mathbf{Q}$ *and* $h(x) = (x^2 - tx + 1)[x^2 + ax + (b-t)]$ *splits over* L; (iii) $r(x)$ *has a unique root* $t \in \mathbf{Q}$ *and* $h(x)$ *does not split over* L. *Then* T^4_ψ *admits no continuous map to any* T^3_φ *with* $D > 0$.

Proof. According to Morandi [163, Theorem 13.4], $\mathrm{Gal}(\mathbb{A}_\psi) \cong \mathbf{Z}_2 \oplus \mathbf{Z}_2$ in case (i); $\mathrm{Gal}(\mathbb{A}_\psi) \cong \mathbf{Z}_4$ in case (ii); and $\mathrm{Gal}(\mathbb{A}_\psi) \cong D_4$ (the dihedral group) in case (iii). Therefore, $k = |\mathbf{Z}_2 \oplus \mathbf{Z}_2| = |\mathbf{Z}_4| = 4$ or $k = |D_4| = 8$. On the other hand, for $m' = 3$ with $D > 0$ (all roots are real), we have $t_1' = |\mathbf{Z}_3| = 3$ and $t_2' = |S_3| = 6$. Since $3; 6 \nmid 4; 8$, Corollary 4.3.2 says that having a continuous map $T_\psi^4 \to T_\varphi^3$ is impossible. Corollary 4.3.4 follows. □

Example 4.3.2. There are infinitely many matrices ψ_0 which satisfy the assumptions of Corollary 4.3.4; indeed, consider a family

$$\psi_0(a,c) = \begin{pmatrix} -2a & 1 & 0 & 0 \\ -a^2 - c^2 & 0 & 1 & 0 \\ -2c & 0 & 0 & 1 \\ -1 & 0 & 0 & 0 \end{pmatrix},$$

where $a, c \in \mathbf{Z}$. The associated cubic becomes $r(x) = x[x^2 - (a^2 + c^2)x + 4(ac - 1)]$, so that $t = 0$ is a rational root; then $h(x) = (x^2 + 1)[x^2 + 2ax + a^2 + c^2]$. The matrix $\psi_0(a,c)$ satisfies one of the conditions (i)–(iii) of Corollary 4.3.4 for each $a, c \in \mathbf{Z}$; it remains to eliminate the (nongeneric) matrices, which are not tight or irreducible. Thus, $\psi_0(a,c)$ defines a family of topologically distinct robust bundles.

Guide to the literature

The Anosov diffeomorphisms were introduced and studied in [3]; for a classical account of the differentiable dynamical systems see [263]. An excellent survey of foliations has been compiled in [149]. The Galois theory is covered in the textbook [163]. The obstruction theory for Anosov bundles can be found in [188].

4.4 Cluster C^*-algebras and knot polynomials

We construct a representation of the braid groups in the cluster C^*-algebra associated to a triangulation of the Riemann surface S with one or two cusps. It is proved that the Laurent polynomials coming from the K-theory of such an algebra are topological invariants of the closure of braids. The Jones and HOMFLY polynomials are special case of our construction corresponding to the S being a sphere with two cusps and a torus with one cusp, respectively.

4.4.1 Invariant Laurent polynomials

The trace invariant $V_L(t)$ of a link L was introduced in [125]. The $V_L(t)$ is a Laurent polynomial $\mathbf{Z}[t^{\pm\frac{1}{2}}]$ obtained from a representation of the braid group B_k in a W^*-algebra A_k, see Chapter 11. The canonical trace on A_k times a multiple $\frac{t}{(1+t)^2}$ is invariant of the

Markov move of type II of a braid $b \in B_k$. The Jones polynomial $V_L(t)$ is a powerful topological invariant of a link L obtained by the closure of b. The algebra A_k itself comes from an analog of the Galois theory for the W^*-algebras [126, Section 2.6].

Cluster algebras are a class of commutative rings introduced in [87] having deep roots in hyperbolic geometry and Teichmüller theory [288]. Such an algebra $\mathcal{A}(\mathbf{x}, B)$ is a subring of the field of rational functions in n variables depending on a cluster $\mathbf{x} = (x_1, \dots, x_n; y_1, \dots, y_m)$ of *mutable* variables x_i and *frozen* variables y_i and a skew-symmetric matrix $B = (b_{ij}) \in M_n(\mathbf{Z})$; the pair $(\mathbf{x}, B)$ is called a *seed*. In terms of the coefficients c_i from a semifield $(\mathbb{P}, \oplus, \bullet)$, a new cluster $\mathbf{x}' = (x_1, \dots, x_k', \dots, x_n; c_1, \dots, c_j', \dots, c_n)$ and a new skew-symmetric matrix $B' = (b_{ij}')$ are obtained from $(\mathbf{x}, B)$ by the exchange relations:

$$b_{ij}' = \begin{cases} -b_{ij} & \text{if } i = k \text{ or } j = k, \\ b_{ij} + \frac{|b_{ik}|b_{kj} + b_{ik}|b_{kj}|}{2} & \text{otherwise}, \end{cases}$$

$$c_j' = \begin{cases} \frac{1}{c_k} & \text{if } j = k, \\ \frac{c_j c_k^{\max(b_{kj},0)}}{(c_k \oplus 1)^{b_{kj}}} & \text{otherwise}, \end{cases}$$

$$x_k' = \frac{c_k \prod_{i=1}^n x_i^{\max(b_{ik},0)} + \prod_{i=1}^n x_i^{\max(-b_{ik},0)}}{(c_k \oplus 1) x_k},$$

see [288, Definition 2.22] for the details. The seed $(\mathbf{x}', B')$ is said to be a *mutation* of $(\mathbf{x}, B)$ in direction k, where $1 \le k \le n$; the algebra $\mathcal{A}(\mathbf{x}, B)$ is generated by cluster variables $\{x_i\}_{i=1}^\infty$ obtained from the initial seed $(\mathbf{x}, B)$ by the iteration of mutations in all possible directions k. The *Laurent phenomenon* proved in [87] says that $\mathcal{A}(\mathbf{x}, B) \subset \mathbf{Z}[\mathbf{x}^{\pm 1}]$, where $\mathbf{Z}[\mathbf{x}^{\pm 1}]$ is the ring of the Laurent polynomials in variables $\mathbf{x} = (x_1, \dots, x_n)$; in other words, each generator x_i of algebra $\mathcal{A}(\mathbf{x}, B)$ can be written as a Laurent polynomial in n variables with the integer coefficients. (Note that the Laurent phenomenon turns $\mathcal{A}(\mathbf{x}, B)$ into an additive abelian group with an order coming from the semigroup of the Laurent polynomials with positive coefficients.) In what follows, we deal with a cluster algebra $\mathcal{A}(\mathbf{x}, S_{g,n})$ coming from a triangulation of the Riemann surface $S_{g,n}$ of genus g with n cusps, see [88] for the details.

Cluster C^*-algebras are a class of noncommutative rings $\mathbb{A}(\mathbf{x}, B)$ such that $K_0(\mathbb{A}(\mathbf{x}, B)) \cong \mathcal{A}(\mathbf{x}, B)$ [199]; here $K_0(\mathbb{A}(\mathbf{x}, B))$ is the K_0-group of a C^*-algebra $\mathbb{A}(\mathbf{x}, B)$ and $\cong$ is an isomorphism of the additive abelian groups with order [32]. Ring $\mathbb{A}(\mathbf{x}, B)$ is an *AF-algebra* given by the Bratteli diagram [41]; such a diagram can be obtained from a mutation tree of the initial seed $(\mathbf{x}, B)$ modulo an equivalence relation between the seeds lying at the same level. It is known that the *AF*-algebras are characterized by their K-theory [76]. Equivalently, $\mathbb{A}(\mathbf{x}, B)$ is an algebra over the complex numbers generated by a series of projections $\{e_i\}_{i=1}^\infty$.

In this section we construct a representation of the braid group $B_k = \{\sigma_1, \dots, \sigma_{k-1} \mid \sigma_i\sigma_{i+1}\sigma_i = \sigma_{i+1}\sigma_i\sigma_{i+1},\ \sigma_i\sigma_j = \sigma_j\sigma_i \text{ if } |i-j| \geq 2\}$ into an algebra $\mathbb{A}(\mathbf{x}, S_{g,n})$, so that the Laurent phenomenon in $K_0(\mathbb{A}(\mathbf{x}, S_{g,n}))$ corresponds to the polynomial invariants of the closure of braids $b \in B_k$. In particular, if $g = 0$ and $n = 2$ or $g = n = 1$, one recovers the Jones invariant $V_L(t)$ or the HOMFLY polynomials of knots [90], respectively. If $g \geq 2$, one gets new topological invariants generalizing the Jones and HOMFLY polynomials to an arbitrary (but finite) number of variables. The *AF*-algebra $\mathbb{A}(\mathbf{x}, S_{g,n})$ itself can be viewed as an analog of the tower $\bigcup_{k=1}^{\infty} A_k$ of von Neumann algebras A_k arising from the basic construction [126, Section 3.4]. Unlike [125], we exploit the phenomenon of cluster algebras and the *Birman–Hilden Theorem* relating the braid groups B_{2g+n} with the mapping class group of surface $\{S_{g,n} \mid n = 1; 2\}$ [26]. Our main result can be formulated as follows.

Theorem 4.4.1. *The formula $\sigma_i \mapsto e_i + 1$ defines a representation*

$$\rho : \begin{cases} B_{2g+1} \to \mathbb{A}(\mathbf{x}, S_{g,1}), \\ B_{2g+2} \to \mathbb{A}(\mathbf{x}, S_{g,2}). \end{cases}$$

If $b \in B_{2g+1}$ (resp. $b \in B_{2g+2}$) is a braid, there exists a Laurent polynomial $[\rho(b)] \in K_0(\mathbb{A}(\mathbf{x}, S_{g,1}))$ (resp. $[\rho(b)] \in K_0(\mathbb{A}(\mathbf{x}, S_{g,2}))$) with the integer coefficients depending on $2g$ (resp. $2g+1$) variables such that $[\rho(b)]$ is a topological invariant of the closure of b.

4.4.2 Birman–Hilden Theorem

Let $S_{g,n}$ be a Riemann surface of genus $g \geq 0$ with $n \geq 1$ cusps and such that $2g-2+n > 0$; denote by $T_{g,n} \cong \mathbf{R}^{6g-6+2n}$ the (decorated) Teichmüller space of $S_{g,n}$, i. e., a collection of all Riemann surfaces of genus g with n cusps endowed with the natural topology. By $\operatorname{Mod} S_{g,n}$ we understand the mapping class group of surface $S_{g,n}$, i. e., a group of the homotopy classes of orientation preserving automorphisms of $S_{g,n}$ fixing all cusps. It is well known that two Riemann surfaces $S, S' \in T_{g,n}$ are isomorphic if and only if there exists a $\varphi \in \operatorname{Mod} S_{g,n}$ such that $S' = \varphi(S)$; thus each $\varphi \in \operatorname{Mod} S_{g,n}$ corresponds to a homeomorphism of the Teichmüller space $T_{g,n}$. If γ is a simple closed curve on $S_{g,n}$, let $D_\gamma \in \operatorname{Mod} S_{g,n}$ be the Dehn twist around γ; a pair of the Dehn twists D_{γ_i} and D_{γ_j} satisfy the braid relations:

$$\begin{cases} D_{\gamma_i}D_{\gamma_j}D_{\gamma_i} = D_{\gamma_j}D_{\gamma_i}D_{\gamma_j}, & \text{if } \gamma_i \cap \gamma_j = \{\text{single point}\}, \\ D_{\gamma_i}D_{\gamma_j} = D_{\gamma_j}D_{\gamma_i}, & \text{if } \gamma_i \cap \gamma_j = \emptyset. \end{cases}$$

A system of simple closed curves $\{\gamma_i\}$ on $S_{g,n}$ is called a *chain*, if

$$\begin{cases} \gamma_i \cap \gamma_{i+1} = \{\text{single point}\}, \\ \gamma_i \cap \gamma_j = \emptyset \quad \text{otherwise}. \end{cases}$$

Consider a chain $\{\gamma_1, \dots, \gamma_{2g+1}\}$ shown in [82, Fig. 2.7]. The fundamental domain of $S_{g,n}$ obtained by a cut along the chain is a $(4g+2)$-gon with the opposite sides identified [82, Fig. 2.2]. A *hyperelliptic involution* $\iota \in \operatorname{Mod} S_{g,n}$ is a rotation by the angle π of the $(4g+2)$-gon; clearly, the chain $\{\gamma_1, \dots, \gamma_{2g+1}\}$ is an invariant of the involution ι.

In what follows, we focus on the case $n = 1$ or 2. Notice that $S_{g,1}$ (resp. $S_{g,2}$) can be replaced by a surface of the same genus with one (resp. two) boundary components and no cusps; since the mapping class group preserves the boundary components, we work with this new surface while keeping the old notation. It is known that $S_{g,1}$ (resp. $S_{g,2}$) is a double cover of a disk $\mathcal{D}_{2g+1}$ (resp. $\mathcal{D}_{2g+2}$) ramified at the $2g+1$ (resp. $2g+2$) inner points of the disk; we refer the reader to [82, Fig. 9.15] for a picture. It transpires that each automorphism of $\mathcal{D}_{2g+1}$ (resp. $\mathcal{D}_{2g+2}$) pulls back to an automorphism of $S_{g,1}$ (resp. $S_{g,2}$) commuting with the hyperelliptic involution ι. Recall that $\operatorname{Mod} \mathcal{D}_{2g+1} \cong B_{2g+1}$ (resp. $\operatorname{Mod} \mathcal{D}_{2g+2} \cong B_{2g+2}$); a subgroup of $\operatorname{Mod} S_{g,1}$ (resp. $\operatorname{Mod} S_{g,2}$) commuting with ι is called *symmetric* and denoted by $\operatorname{SMod} S_{g,1}$ (resp. $\operatorname{SMod} S_{g,2}$) The following result is critical.

Theorem 4.4.2 ([26]). *There exists an isomorphism*

$$\begin{cases} B_{2g+1} \cong \operatorname{SMod} S_{g,1}, \\ B_{2g+2} \cong \operatorname{SMod} S_{g,2}, \end{cases}$$

given by the formula $\sigma_i \mapsto D_{\gamma_i}$, where σ_i is a generator of the braid group B_{2g+1} (resp. B_{2g+2}) and D_{γ_i} is the Dehn twist around the simple closed curve γ_i of a chain in the $S_{g,1}$ (resp. $S_{g,2}$).

4.4.3 Cluster C^*-algebras

It is well known that the fundamental domain of the Riemann surface $S_{g,n}$ has a triangulation by the $6g-6+3n$ geodesic arcs γ_i, see, e. g., [223]; the endpoint of each γ_i is a cusp at the absolute of Lobachevsky plane $\mathbb{H} = \{x + iy \in \mathbf{C} \mid y > 0\}$. Denote by $l(\gamma_i)$ the $\pm$ hyperbolic length of γ_i between two horocycles around the endpoints of γ_i; consider the $\lambda(\gamma_i) = \exp(\frac{1}{2} l(\gamma_i))$. The following result says that $\lambda(\gamma_i)$ are coordinates in the Teichmüller space $T_{g,n}$.

Theorem 4.4.3 ([223]). *The map $\lambda : \{\gamma_i\}_{i=1}^{6g-6+3n} \to T_{g,n}$ is a homeomorphism.*

Remark 4.4.1. The six-tuples of numbers $\lambda(\gamma_i)$ must satisfy the *Ptolemy relation* $\lambda(\gamma_1)\lambda(\gamma_2) + \lambda(\gamma_3)\lambda(\gamma_4) = \lambda(\gamma_5)\lambda(\gamma_6)$, where $\gamma_1, \dots, \gamma_4$ are pairwise opposite sides and γ_5, γ_6 are the diagonals of a geodesic quadrilateral in $\mathbb{H}$. The n Ptolemy relations reduce the number of independent variables $\lambda(\gamma_i)$ to $6g-6+2n = \dim T_{g,n}$.

Let $T = \{\gamma_i\}_{i=1}^{6g-6+3n}$ be a triangulation of $S_{g,n}$; consider a skew-symmetric matrix $B_T = (b_{ij})$ of rank $6g-6+3n$, where b_{ij} is equal to the number of triangles in T with sides γ_i and γ_j in clockwise order minus the number of triangles in T with sides γ_i and γ_j in the counterclockwise order. (For a quick example of matrix B_T, we refer the reader to Section 4.4.6.)

Theorem 4.4.4 ([88]). *The cluster algebra $\mathcal{A}(\mathbf{x}, B_T)$ does not depend on triangulation T, but only on the surface $S_{g,n}$; namely, a replacement of the geodesic arc γ_k by a new geodesic arc γ_k' (a flip of γ_k) corresponds to a mutation μ_k of the seed $(\mathbf{x}, B_T)$.*

Definition 4.4.1. The algebra $\mathcal{A}(\mathbf{x}, S_{g,n}) := \mathcal{A}(\mathbf{x}, B_T)$ is called a cluster algebra of the Riemann surface $S_{g,n}$.

Recall that an AF-algebra $\mathbb{A}$ is defined as the norm closure of an ascending sequence of finite-dimensional C^*-algebras M_n, where M_n is the C^*-algebra of the $n \times n$ matrices with entries in $\mathbf{C}$; such an algebra is given by an infinite graph called *Bratteli diagram*, see Section 3.5. The dimension group $(K_0(\mathbb{A}), K_0^+(\mathbb{A}), u)$ is a complete isomorphism invariant of the algebra $\mathbb{A}$ [76]. The order-isomorphism class $(K_0(\mathbb{A}), K_0^+(\mathbb{A}))$ is an invariant of the *Morita equivalence* of algebra $\mathbb{A}$, i. e., an isomorphism class in the category of finitely generated projective modules over $\mathbb{A}$. The *scale* Γ is a subset of $K_0^+(\mathbb{A})$ which is generating, hereditary and directed, i. e., (i) for each $a \in K_0^+(\mathbb{A})$ there exist $a_1, \dots, a_r \in \Gamma(\mathbb{A})$, such that $a = a_1 + \cdots + a_r$; (ii) if $0 \le a \le b \in \Gamma$, then $a \in \Gamma$; (iii) given $a, b \in \Gamma$ there exists $c \in \Gamma$ such that $a, b \le c$. If u is an order unit, then the set $\Gamma := \{a \in K_0^+(\mathbb{A}) \mid 0 \le a \le u\}$ is a scale; thus the dimension group of algebra $\mathbb{A}$ can be written in the form $(K_0(\mathbb{A}), K_0^+(\mathbb{A}), \Gamma)$.

Definition 4.4.2 ([199]). By a cluster C^*-algebra $\mathbb{A}(\mathbf{x}, S_{g,n})$ one understands an AF-algebra satisfying an isomorphism of the scaled dimension groups,

$$(K_0(\mathbb{A}(\mathbf{x}, S_{g,n})), K_0^+(\mathbb{A}(\mathbf{x}, S_{g,n})), u) \cong (\mathcal{A}(\mathbf{x}, S_{g,n}), \mathcal{A}^+(\mathbf{x}, S_{g,n}), u'),$$

where $\mathcal{A}^+(\mathbf{x}, S_{g,n})$ is a semigroup of the Laurent polynomials with positive coefficients and u' is an order unit in $\mathcal{A}^+(\mathbf{x}, S_{g,n})$.

Remark 4.4.2. Theorems 4.4.3 and 4.4.4 imply that the algebra $\mathbb{A}(\mathbf{x}, S_{g,n})$ is a noncommutative coordinate ring of the Teichmüller space $T_{g,n}$; in other words, the diagram in Fig. 4.9 must be commutative.

Remark 4.4.3. The $\mathbb{A}(\mathbf{x}, S_{g,n})$ is the norm closure of an algebra of the noncommutative polynomials $\mathbf{C}\langle e_1, e_2, \dots\rangle$, where $\{e_i\}_{i=1}^{\infty}$ are projections in the algebra $\mathbb{A}(\mathbf{x}, S_{g,n})$; this fact follows from the K-theory of $\mathbb{A}(\mathbf{x}, S_{g,n})$. On the other hand, the algebra $\mathcal{A}(\mathbf{x}, S_{g,n})$ is generated by the cluster variables $\{x_i\}_{i=1}^{\infty}$. We shall denote by $\rho(x_i) = e_i$ a natural bijection between the two sets of generators.

$$\begin{array}{ccc} T_{g,n} & \xrightarrow{\text{homeomorphism}} & T_{g,n} \\ \downarrow & & \downarrow \\ \mathbb{A}(\mathbf{x}, S_{g,n}) & \xrightarrow[\text{automorphism}]{\text{inner}} & \mathbb{A}(\mathbf{x}, S_{g,n}) \end{array}$$

Figure 4.9: Coordinate ring of the space $T_{g,n}$.

4.4.4 Jones and HOMFLY polynomials

Recall that a *knot* is a tame embedding of the circle S^1 into the Euclidean space $\mathbf{R}^3$; a *link* with the n components is such an embedding of the union $S^1 \cup \cdots \cup S^1$. (We refer the reader to Section 11.2 for more details.) The classification of knots and links is a difficult open problem of topology. Alexander Theorem says that every knot or link comes from the closure of a braid $b \in B_k = \{\sigma_1, \dots, \sigma_{k-1} | \sigma_i\sigma_{i+1}\sigma_i = \sigma_{i+1}\sigma_i\sigma_{i+1}, \sigma_i\sigma_j = \sigma_j\sigma_i \text{ if } |i-j| \geq 2\}$, i. e., tying the top end of each string of b to the end of a string in the same position at the bottom. Thus the braids can classify knots and links but sadly rather "unrelated" braids $b \in B_k$ and $b' \in B_{k'}$ can produce the same knot. Namely, *Markov Theorem* says that the closure of braids $b, b' \in B_k$ corresponds to the same knot or link if and only if: (i) $b' = gbg^{-1}$ for a braid $g \in B_k$ or (ii) $b' = b\sigma_k^{\pm 1}$ for the generator $\sigma_k \in B_{k+1}$. Thus the *Markov move of type II* always pushes "to infinity" the desired classification, hinting that an asymptotic invariant is required. Let us recall two examples of such invariants; we refer the reader to Chapter 11 for more details.

The *Jones polynomial* of the closure L of a braid $b \in B_k$ is defined by the formula

$$V_L(t) = \left(-\frac{t+1}{\sqrt{t}}\right)^{k-1} \operatorname{tr}(r_t(b)),$$

where r_t is a representation of B_k in a von Neumann algebra A_k and tr is a trace function; $V_L(t) \in \mathbf{Z}[t^{\pm\frac{1}{2}}]$, i. e., a Laurent polynomial in the variable $\sqrt{t}$ [125]. If K is the unknot, then $V_K(t) = 1$ and each polynomial $V_L(t)$ can be calculated from K using the *skein relation*,

$$\frac{1}{t}V_{L^-} - tV_{L^+} = \left(\sqrt{t} - \frac{1}{\sqrt{t}}\right)V_L,$$

where L^+ (resp. L^-) is a link obtained by adding an overpass (resp. underpass) to the link L [125, Theorem 12].

The *HOMFLY polynomial* of a link L is a Laurent polynomial $\rho_L(l, m) \in \mathbf{Z}[l^{\pm 1}, m^{\pm 1}]$; $\rho_L(l, m)$ is defined recursively from the HOMFLY polynomial $\rho_K(l, m) = 1$ of the unknot

K using the skein relation,

$$l\rho_{L^+} + \frac{1}{l}\rho_{L^-} + m\rho_L = 0,$$

where L^+ (resp. L^-) is a link obtained by adding an overpass (resp. underpass) to L [90, Remark 3].

4.4.5 Proof of Theorem 4.4.1

For the sake of clarity, let us outline the main ideas. Roughly speaking, the Birman–Hilden's Theorem 4.4.2 says that a generator $\sigma_i \in B_{2g+1}$ (resp. B_{2g+2}) is given by the Dehn twist $D_{\gamma_i} \in \operatorname{Mod} S_{g,1}$ (resp. $\operatorname{Mod} S_{g,2}$) around a closed curve γ_i. The D_{γ_i} itself is a homeomorphism of the Teichmüller space $T_{g,1}$ (resp. $T_{g,2}$); therefore the D_{γ_i} induced an inner automorphism $x \mapsto u_i x u_i^{-1}$, where u_i is a unit of the cluster C^*-algebra $\mathbb{A}(\mathbf{x}, S_{g,1})$ (resp. $\mathbb{A}(\mathbf{x}, S_{g,2})$). (We refer the reader to Fig. 4.9.) On the other hand, the Fomin–Shapiro–Thurston Theorem 4.4.4 and Remark 4.4.3 imply that units u_i and projections e_i in algebra $\mathbb{A}(\mathbf{x}, S_{g,1})$ (resp. $\mathbb{A}(\mathbf{x}, S_{g,2})$) are bijective. But the minimal degree polynomial in variable e_i corresponding to a unit u_i is a linear polynomial of the form $u_i = ae_i + b$, where a and b are complex constants. (The inverse is given by the formula $u_i^{-1} = -\frac{a}{(a+b)b}e_i + \frac{1}{b}$.) Birman–Hilden Theorem 4.4.2 implies that the u_i satisfy the braid relations $\{u_i u_{i+1} u_i = u_{i+1} u_i u_{i+1},\ u_i u_j = u_j u_i, \text{ if } |i-j| \geq 2\}$; moreover, if one substitutes $u_i = ae_i + b$ in the braid relations, then

$$\begin{cases} e_i^2 = e_i, \\ e_i e_{i\pm 1} e_i = -\dfrac{(a+b)b}{a^2} e_i, \\ e_i e_j = e_j e_i, \qquad \text{if } |i-j| \geq 2. \end{cases}$$

Remark 4.4.4. The above relations are invariant of the involution $e_i^* = e_i$ if and only if $\frac{(a+b)b}{a^2} \in \mathbf{R}$; in this case the algebra $\mathbb{A}(\mathbf{x}, S_{g,1})$ (resp. $\mathbb{A}(\mathbf{x}, S_{g,2})$) contains a finite-dimensional C^*-algebra $\mathbb{A}_{2g}$ (resp. $\mathbb{A}_{2g+1}$) obtained from the norm closure of a self-adjoint representation of a *Temperley–Lieb algebra*.

Thus one gets a representation $\rho : B_{2g+1} \to \mathbb{A}(\mathbf{x}, S_{g,1})$ (resp. $B_{2g+2} \to \mathbb{A}(\mathbf{x}, S_{g,2})$) given by the formula $\sigma_i \mapsto ae_i + b$, where $1 \leq i \leq 2g$ (resp. $1 \leq i \leq 2g+1$). It follows from the above that the set

$$\begin{aligned} \mathcal{E} := \{(e_{i_1} e_{i_1 - 1} \dots e_{j_1}) \dots (e_{i_p} e_{i_p - 1} \dots e_{j_p}) \mid 1 \leq i_1 < \dots < i_p < 2g \text{ (resp. } 2g+1); \\ 1 \leq j_1 < \dots < j_p < 2g \text{ (resp. } 2g+1); \\ j_1 \leq i_1, \dots, j_p \leq i_p\} \end{aligned}$$

is multiplicatively closed; moreover, $|\mathcal{E}| \le \frac{1}{n+1}\binom{2n}{n}$ = nth Catalan number, where $n = 2g$ (resp. $n = 2g+1$) In particular, $\mathbb{A}_{2g}$ (resp. $\mathbb{A}_{2g+1}$) of Remark 4.4.4 is a finite-dimensional C^*-algebra and each element $\varepsilon \in \mathcal{E}$ is equivalent to a projection; the (Murray–von Neumann) equivalence class of the projection will be denoted by $[\varepsilon]$. If $\{b = \sigma_1^{k_1}\dots\sigma_{n-1}^{k_n} \in B_{n+1} \mid k_i \in \mathbf{Z}\}$ is a braid for $n = 2g$ (resp. $n = 2g+1$), then the polynomial $\rho(b) = (ae_1+b)^{k_1}\cdots(ae_{n-1}+b)^{k_n}$ unfolds into a finite sum $\{\sum_{i=1}^{|\mathcal{E}|} a_i\varepsilon_i \mid \varepsilon_i \in \mathcal{E},\ a_i \in \mathbf{Z}\}$. Therefore $[\rho(b)] := \{\sum_{i=1}^{|\mathcal{E}|} a_i[\varepsilon_i] \mid [\varepsilon_i] \in K_0(\mathbb{A}(\mathbf{x}, S_{g,1})),\ a_i \in \mathbf{Z}\}$ is an element of the group $K_0(\mathbb{A}(\mathbf{x}, S_{g,1}))$ (resp. $K_0(\mathbb{A}(\mathbf{x}, S_{g,2}))$). But $K_0(\mathbb{A}(\mathbf{x}, S_{g,n})) \cong \mathcal{A}(\mathbf{x}, S_{g,n}) \subset \mathbf{Z}[\mathbf{x}^{\pm 1}]$, where $n = 1$ (resp. $n = 2$); thus $[\rho(b)] \in K_0(\mathbb{A}(\mathbf{x}, S_{g,n}))$ is a Laurent polynomial with the integer coefficients depending on $2g$ (resp. $2g+1$) variables. The element $[\rho(b)]$ is, in fact, a topological invariant of the closure of b. Indeed, $[\rho(gbg^{-1})] = [\rho(b)]$ for all $b \in B_{2g+1}$ (resp. B_{2g+2}) because the K_0-group and a canonical trace τ on the algebra $\mathbb{A}(\mathbf{x}, S_{g,1})$ (resp. $\mathbb{A}(\mathbf{x}, S_{g,2})$) are related [32, Section 7.3]; since τ is a character of the representation ρ, one gets the formula $[\rho(gbg^{-1})] = [\rho(b)]$. The invariance of the $[\rho(b)]$ with respect to the Markov move of type II is a bit subtler, but follows from a stability of the K-theory [32, Section 5.1]. Namely, the map $\sigma_{2g+1}^{\pm 1} \mapsto 2e_{2g+1} - 1$ (resp. $\sigma_{2g+2}^{\pm 1} \mapsto 2e_{2g+2} - 1$) gives rise to a crossed product C^*-algebra $\mathbb{A}_{2g} \rtimes_\alpha G$ (resp. $\mathbb{A}_{2g+1} \rtimes_\alpha G$), where $G \cong \mathbf{Z}/2\mathbf{Z}$; the crossed product is isomorphic to the algebra $M_2(\mathbb{A}_{2g}^\alpha)$ (resp. $M_2(\mathbb{A}_{2g+1}^\alpha)$), where $\mathbb{A}_{2g}^\alpha$ (resp. $\mathbb{A}_{2g+1}^\alpha$) is the fixed-point algebra of the automorphism α [86, Section 3.8.5]. But $K_0(M_2(\mathbb{A}_{2g}^\alpha)) \cong K_0(\mathbb{A}_{2g}^\alpha)$ by the stability of the K-theory; the crossed product itself consists of the formal sums $\sum_{\gamma \in G} a_\gamma u_\gamma$, and one easily derives that $[\rho(b\sigma_{2g+1}^{\pm 1})] = [\rho(b)]$ for all $b \in B_{2g+1}$ (resp. B_{2g+2}).

We shall pass to a detailed proof of Theorem 4.4.1 by splitting the argument into a series of lemmas.

Lemma 4.4.1. *The map $\{\sigma_i \mapsto e_i + 1 \mid 1 \le i \le 2g\}$ (resp. $\{\sigma_i \mapsto e_i + 1 \mid 1 \le i \le 2g+1\}$) defines a representation $\rho : B_{2g+1} \to \mathbb{A}(\mathbf{x}, S_{g,1})$ (resp. $\rho : B_{2g+2} \to \mathbb{A}(\mathbf{x}, S_{g,2})$) of the braid group with an odd (resp. even) number of strings into a cluster C^*-algebra of a Riemann surface with one cusp (resp. two cusps).*

Proof. We shall prove the case $\rho : B_{2g+1} \to \mathbb{A}(\mathbf{x}, S_{g,1})$ of the braid groups with an odd number of strings; the case of an even number of strings is treated likewise.

Let $\{\gamma_1, \dots, \gamma_{2g}\}$ be a chain of simple closed curves on the surface $S_{g,1}$. The Dehn twists $\{D_{\gamma_1}, \dots, D_{\gamma_{2g}}\}$ around γ_i satisfy the braid relations. The subgroup SMod $S_{g,1}$ of Mod $S_{g,1}$ consisting of the automorphisms of $S_{g,1}$ commuting with the hyperelliptic involution ι is isomorphic to the braid group B_{2g+1} (Birman–Hilden Theorem 4.4.2).

On the other hand, each D_{γ_i} is a homeomorphism of the Teichmüller space $T_{g,1}$; the D_{γ_i} induces an inner automorphism of the cluster C^*-algebra $\mathbb{A}(\mathbf{x}, S_{g,1})$, see Remark 4.4.2. We shall denote such an automorphism by

$$\{u_i x u_i^{-1} \mid u_i \in \mathbb{A}(\mathbf{x}, S_{g,1}),\ \forall x \in \mathbb{A}(\mathbf{x}, S_{g,1})\}.$$

The units $\{u_i \in \mathbb{A}(\mathbf{x}, S_{g,1}) \mid 1 \le i \le 2g\}$ satisfy the braid relations

$$\begin{cases} u_i u_{i+1} u_i = u_{i+1} u_i u_{i+1}, & \\ u_i u_j = u_j u_i, & \text{if } |i-j| \ge 2. \end{cases}$$

Indeed, from the Birman–Hilden Theorem 4.4.2 one gets

$$\begin{aligned} \text{SMod}\, S_{g,1} = \{D_{\gamma_1}, \dots, D_{\gamma_{2g}} \mid D_{\gamma_i} D_{\gamma_{i+1}} D_{\gamma_i} = D_{\gamma_{i+1}} D_{\gamma_i} D_{\gamma_{i+1}}, \\ D_{\gamma_i} D_{\gamma_j} = D_{\gamma_j} D_{\gamma_i} \text{ if } |i-j| \ge 2\}. \end{aligned}$$

If Inn $\mathbb{A}(\mathbf{x}, S_{g,1})$ is a group of the inner automorphisms of the algebra $\mathbb{A}(\mathbf{x}, S_{g,1})$, then $\{u_i \mid 1 \le i \le 2g\}$ are generators of Inn $\mathbb{A}(\mathbf{x}, S_{g,1})$. Using the commutative diagram in Fig. 4.9, one gets from $D_{\gamma_i} D_{\gamma_{i+1}} D_{\gamma_i} = D_{\gamma_{i+1}} D_{\gamma_i} D_{\gamma_{i+1}}$ the equality $u_i u_{i+1} u_i = u_{i+1} u_i u_{i+1}$; similarly, the $D_{\gamma_i} D_{\gamma_j} = D_{\gamma_j} D_{\gamma_i}$ implies the equality $u_i u_j = u_j u_i$ for $|i-j| \ge 2$. In particular, SMod $S_{g,1} \cong$ Inn $\mathbb{A}(\mathbf{x}, S_{g,1})$, where the isomorphism is given by the formula $D_{\gamma_i} \mapsto u_i$.

It remains to express the units u_i in terms of generators e_i of the algebra $\mathbb{A}(\mathbf{x}, S_{g,1})$. The e_i itself is not invertible, but a polynomial $u_i(e_i) = e_i + 1$ has an inverse $u_i^{-1} = -\frac{1}{2} e_i + 1$. From an isomorphism $B_{2g+1} \cong$ SMod $S_{g,1}$ given by the formula $\sigma_i \mapsto D_{\gamma_i}$ one gets a representation $\rho : B_{2g+1} \to \mathbb{A}(\mathbf{x}, S_{g,1})$ given by the formula $\sigma_i \mapsto e_i + 1$. Lemma 4.4.1 is proved. □

Corollary 4.4.1. *The norm closure of a self-adjoint representation of a Temperley–Lieb algebra* $\text{TL}_{2g}(\frac{i\sqrt{2}}{2})$ *(resp.* $\text{TL}_{2g+1}(\frac{i\sqrt{2}}{2})$*) is a finite-dimensional sub-*C^**-algebra* $\mathbb{A}_{2g}$ *(resp.* $\mathbb{A}_{2g+1}$*) of the* $\mathbb{A}(\mathbf{x}, S_{g,1})$ *(resp.* $\mathbb{A}(\mathbf{x}, S_{g,2})$*).*

Proof. Let us substitute $u_i = e_i + 1$ into the braid relations. The reader is encouraged to verify that they are equivalent to the following system of relations:

$$\begin{cases} e_i^2 = e_i, \ e_i^* = e_i, & \\ e_i e_{i\pm 1} e_i = -2 e_i, & \\ e_i e_j = e_j e_i, & \text{if } |i-j| \ge 2. \end{cases}$$

The normalization $e_i' = \frac{i\sqrt{2}}{2} e_i$ brings the above system to the form

$$\begin{cases} e_i^2 = \dfrac{i\sqrt{2}}{2} e_i, & \\ e_i e_{i\pm 1} e_i = e_i, & \\ e_i e_j = e_j e_i, & \text{if } |i-j| \ge 2. \end{cases}$$

These relations are defining relations for a Temperley–Lieb algebra $\text{TL}_{2g}(\frac{i\sqrt{2}}{2})$ (resp. $\text{TL}_{2g+1}(\frac{i\sqrt{2}}{2})$), see, e. g., [126, p. 85]; such an algebra is always finite-dimensional, see the next lemma. Corollary 4.4.1 follows. □

Lemma 4.4.2 ([126, Section 3.5]). *The set*

$$\mathcal{E} := \{(e_{i_1} e_{i_1-1} \dots e_{j_1}) \dots (e_{i_p} e_{i_p-1} \dots e_{j_p}) \mid 1 \le i_1 < \dots < i_p < 2g \text{ (resp. } 2g+1); \\ 1 \le j_1 < \dots < j_p < 2g \text{ (resp. } 2g+1); \\ j_1 \le i_1, \dots, j_p \le i_p\}$$

is multiplicatively closed and

$$|\mathcal{E}| \le \frac{1}{n+1}\binom{2n}{n},$$

where $n = 2g$ (resp. $n = 2g+1$).

Proof. An elegant proof of this fact is based on a representation of the above relations by the diagrams of the noncrossing strings reminiscent of the braid diagrams. □

Corollary 4.4.2. *Each element $e \in \mathcal{E}$ is equivalent to a projection in the cluster C^*-algebra $\mathbb{A}(\mathbf{x}, S_{g,1})$ (resp. $\mathbb{A}(\mathbf{x}, S_{g,2})$).*

Proof. Indeed, if $e \in \mathcal{E}$ then e^2 must coincide with one of the elements of $\mathcal{E}$. But e^2 cannot be any such, except for the e itself. Thus $e^2 = e$, i. e., e is an idempotent. On the other hand, it is well known that each idempotent in a C^*-algebra is (Murray–von Neumann) equivalent to a projection in the same algebra, see, e. g., [32, Proposition 4.6.2]. Corollary 4.4.2 follows. □

Lemma 4.4.3. *If $b \in B_{2g+1}$ (resp. $b \in B_{2g+2}$) is a braid, there exists a Laurent polynomial $[\rho(b)]$ with the integer coefficients depending on $2g$ (resp. $2g+1$) variables such that $[\rho(b)] \in K_0(\mathbb{A}(\mathbf{x}, S_{g,1}))$ (resp. $[\rho(b)] \in K_0(\mathbb{A}(\mathbf{x}, S_{g,2}))$).*

Proof. We shall prove this fact for the braid groups with an odd number of strings; the case of an even number of strings is treated likewise.

Let $\{b = \sigma_1^{k_1} \dots \sigma_{2g}^{k_{2g}} \in B_{2g+1} \mid k_i \in \mathbf{Z}\}$ be a braid. By Lemma 4.4.1, such a braid has a representation $\rho(b)$ in the cluster C^*-algebra $\mathbb{A}(\mathbf{x}, S_{g,1})$ given by the formula

$$\rho(b) = (e_1 + 1)^{k_1} \cdots (e_{2g} + 1)^{k_{2g}} \in \mathbb{A}(\mathbf{x}, S_{g,1}).$$

One can unfold the above product into a sum of the monomials in variables e_i; by Lemma 4.4.2, any such monomial is an element of the set $\mathcal{E}$. In other words, one gets a finite sum

$$\rho(b) = \left\{ \sum_{i=1}^{|\mathcal{E}|} a_i \varepsilon_i \;\middle|\; \varepsilon_i \in \mathcal{E},\ a_i \in \mathbf{Z} \right\}.$$

(Note that, whenever $k_i < 0$, the coefficient a_i is a rational number, but, clearing the denominators, we can assume $a_i \in \mathbf{Z}$.) On the other hand, Corollary 4.4.2 says that each ε_i is a projection; therefore ε_i defines an equivalence class $[\varepsilon_i] \in K_0(\mathbb{A}(\mathbf{x}, S_{g,1}))$ of projections in the cluster C^*-algebra $\mathbb{A}(\mathbf{x}, S_{g,1})$. Thus one gets $[\rho(b)] \in K_0(\mathbb{A}(\mathbf{x}, S_{g,1}))$ given by a finite sum

$$[\rho(b)] = \left\{ \sum_{i=1}^{|\mathcal{E}|} a_i[\varepsilon_i] \;\middle|\; [\varepsilon_i] \in K_0(\mathbb{A}(\mathbf{x}, S_{g,1})),\ a_i \in \mathbf{Z} \right\}.$$

But $K_0(\mathbb{A}(\mathbf{x}, S_{g,1})) \cong \mathcal{A}(\mathbf{x}, S_{g,1}) \subset \mathbf{Z}[\mathbf{x}^{\pm 1}]$; in particular, $[\rho(b)] \in \mathbf{Z}[\mathbf{x}^{\pm 1}]$ is a Laurent polynomial with the integer coefficients.

To calculate the number of variables in $[\rho(b)]$, recall that rank $\mathcal{A}(\mathbf{x}, S_{g,1}) = 6g - 3$; on the other hand, a fundamental domain of the Riemann surface $S_{g,1}$ is a paired $(4g+2)$-gon, where one pair of sides corresponds to a boundary component obtained from the cusp. Since the boundary component contracts to a cusp, one gets a paired $4g$-gon whose triangulation requires $4g-3$ interior geodesic arcs. Thus the cluster $|\mathbf{x}| = 6g-3$ can be written in the form

$$\mathbf{x} = (x_1, \ldots, x_{2g}; y_1, \ldots, y_{4g-3}),$$

where x_i are mutable and y_i are frozen variables [288, Definition 2.6]. One can always assume y_i = const and therefore the Laurent polynomial $[\rho(b)]$ depends on the $2g$ variables x_i. Lemma 4.4.3 follows. □

Lemma 4.4.4. *The Laurent polynomial $[\rho(b)]$ is a topological invariant of the closure of the braid $b \in B_{2g+1}$ (resp. $b \in B_{2g+2}$).*

Proof. Again we shall prove the case $b \in B_{2g+1}$; the case $b \in B_{2g+1}$ can be treated similarly and is left to the reader. To prove that $[\rho(b)]$ is a topological invariant, it is enough to demonstrate that (i) $[\rho(gbg^{-1})] = [\rho(b)]$ for all $g \in B_{2g+1}$; and (ii) $[\rho(b\sigma_{2g+1}^{\pm 1})] = [\rho(b)]$ for the generator $\sigma_{2g+1} \in B_{2g+2}$.

(i) Recall that the K-theory of an AF-algebra $\mathbb{A}(\mathbf{x}, S_{g,1})$ can be recovered from the canonical trace $\tau : \mathbb{A}(\mathbf{x}, S_{g,1}) \to \mathbf{C}$ [32, Section 7.3]. On the other hand, trace τ is a character of the representation $\rho : B_{2g+1} \to \mathbb{A}(\mathbf{x}, S_{g,1})$; in particular, $\tau(\rho(g)\rho(b)\rho(g^{-1})) = \tau(\rho(b))$. In other words, $[\rho(gbg^{-1})] = [\rho(b)]$ for all $g \in B_{2g+1}$. Item (i) follows.

(ii) Let $\mathbb{A}_{2g}$ be a finite-dimensional C^*-algebra of Corollary 4.4.1. Let u_{2g+1} and u_{2g+1}^{-1} be a unit and its inverse in the algebra $\mathbb{A}(\mathbf{x}, S_{g,1})$ given by the formulas

$$u_{2g+1} = u_{2g+1}^{-1} := 2e_{2g+1} - 1.$$

Clearly, the units $u_{2g+1}^{\pm 1} \notin \mathbb{A}_{2g}$, and they make the group $G \cong \mathbf{Z}/2\mathbf{Z}$ under a multiplication. We shall consider an extension $\mathbb{A}_{2g} \rtimes G$ of the algebra $\mathbb{A}_{2g}$ given by the formal sums

$$\mathbb{A}_{2g} \rtimes G := \left\{ \sum_{\gamma \in G} a_\gamma u_\gamma \;\middle|\; u_\gamma \in \mathbb{A}_{2g} \right\}.$$

(The algebra of the above formal sums is isomorphic to a crossed product C^*-algebra of the algebra $\mathbb{A}_{2g}$ by the outer automorphisms α given the elements $u_\gamma \in \{u_{2g+1}^{\pm 1}\}$; hence our notation.) It is well known that

$$\mathbb{A}_{2g} \rtimes G \cong M_2(\mathbb{A}_{2g}^{\alpha}),$$

where $\mathbb{A}_{2g}^{\alpha} \subset \mathbb{A}_{2g}$ is a fixed-point algebra of the automorphism $\alpha : \mathbb{A}_{2g} \to \mathbb{A}_{2g}$, see, e. g., [86, Section 3.8.5]. By the second formula above, one gets $u_i u_{2g+1} = u_{2g+1} u_i$ for all $1 \le i \le 2g-1$; in other words,

$$u_i = u_{2g+1} u_i u_{2g+1}^{-1}$$

for all $1 \le i \le 2g-1$. Since all generators u_i of algebra $\mathbb{A}_{2g-1}$ are fixed by the automorphism α, one gets an isomorphism

$$\mathbb{A}_{2g}^{\alpha} \cong \mathbb{A}_{2g-1}.$$

On the other hand, $K_0(M_2(A)) \cong K_0(A)$ by a stability of the K-theory [32, Section 5.1]; thus the above formulas imply an isomorphism

$$K_0(\mathbb{A}_{2g} \rtimes G) \cong K_0(\mathbb{A}_{2g-1}).$$

It remains to notice that if one maps the generators $\sigma_{2g+1}^{\pm 1} \in B_{2g+2}$ into the units $u_{2g+1}^{\pm 1} \in \mathbb{A}_{2g} \rtimes G$, then above formulas imply the equality

$$[\rho(b\sigma_{2g+1}^{\pm 1})] = [\rho(b)]$$

for all $b \in B_{2g}$. Item (ii) follows from the last equation, and Lemma 4.4.4 is proved. □

Theorem 4.4.1 follows from Lemmas 4.4.1, 4.4.3, and 4.4.4.

4.4.6 Examples

To illustrate Theorem 4.4.1, we shall consider a representation $\rho : B_2 \to \mathbb{A}(\mathbf{x}, S_{0,2})$ (resp. $\rho : B_3 \to \mathbb{A}(\mathbf{x}, S_{1,1})$); it will be shown that for such a representation the Laurent polynomials $[\rho(b)]$ correspond to the Jones (resp. HOMFLY) invariants of knots and links.

4.4.6.1 Jones polynomials

If $g = 0$ and $n = 2$, then $S_{0,2}$ is a sphere with two cusps; $S_{0,2}$ is homotopy equivalent to an annulus $\mathfrak{A} := \{z = u + iv \in \mathbf{C} \mid r \leq |z| \leq R\}$. The Riemann surface $\mathfrak{A}$ has an ideal triangulation T with one marked point on each boundary component given by the matrix

$$B_T = \begin{pmatrix} 0 & 2 \\ -2 & 0 \end{pmatrix},$$

see [88, Example 4.4]. Using the exchange relations, the reader can verify that the cluster C^*-algebra $\mathbb{A}(\mathbf{x}, S_{0,2})$ is given by the Bratteli diagram shown in Fig. 4.10; $\mathbb{A}(\mathbf{x}, S_{0,2})$ coincides with the so-called *GICAR algebra* [41, Section 5.5].

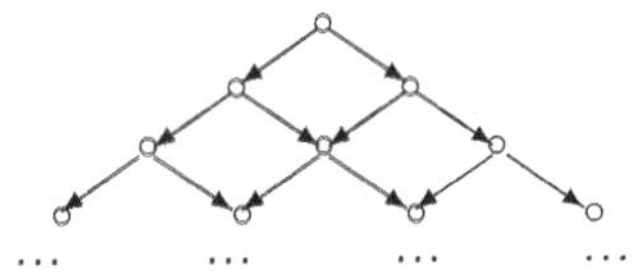

Figure 4.10: Bratteli diagram of the algebra $\mathbb{A}(\mathbf{x}, S_{0,2})$.

The cluster $\mathbf{x} = (x; c)$ consists of a mutable variable x and a coefficient $c \in (\mathbb{P}, \oplus, \bullet)$. Theorem 4.4.1 says that there exists a representation

$$\rho : B_2 \to \mathbb{A}(\mathbf{x}, S_{0,2})$$

such that $[\rho(b)] \in \mathbf{Z}[x^{\pm 1}]$ is a topological invariant of the closure L of $b \in B_2$; since the Laurent polynomial $[\rho(b)]$ depends on x and c, we shall write it as $[\rho(b)](x, c)$. Let $N \geq 1$ be the minimal number of the overpass (resp. underpass) crossings added to the unknot K to get the link L [125, Fig. 2]. The following result compares $[\rho(b)](x, c)$ with the Jones polynomial $V_L(t)$.

Corollary 4.4.3.

$$V_L(t) = \left(-\frac{\sqrt{t}}{t+1}\right)^N [\rho(b)](t, -t^2).$$

Proof. Recall that each polynomial $[\rho(b)](x, c)$ is obtained from an initial seed $(\mathbf{x}, B_T)$ by a finite number of mutations given by the exchange relations; likewise, each polynomial $V_L(t)$ can be obtained from $V_K(t) = 1$ using the skein relation. Roughly speaking, the idea is to show that the mutation and skein relations are equivalent up to a multiple of $-\frac{\sqrt{t}}{t+1}$. Indeed, consider the Laurent polynomials

$$\begin{cases} W_{L^+} = \left(-\frac{t+1}{\sqrt{t}}\right) V_{L^+}, \\ W_{L^-} = \left(-\frac{t+1}{\sqrt{t}}\right) V_{L^-}. \end{cases}$$

The skein relation for the $W_{L^\pm}$ takes the form

$$V_L = \frac{t^2}{t^2-1} W_{L^+} - \frac{1}{t^2-1} W_{L^-}.$$

The substitution

$$\begin{cases} V_L = x_k', \\ W_{L^+} = \frac{1}{x_k} \prod_{i=1}^n x_i^{\max(b_{ik},0)}, \\ W_{L^-} = \frac{1}{x_k} \prod_{i=1}^n x_i^{\max(-b_{ik},0)}, \\ c_k = -t^2 \end{cases}$$

transforms the skein relation to the exchange relations. It remains to observe that an extra crossing added to the unknot K corresponds to a multiplication of $[\rho(b)](x,c)$ by $-\frac{\sqrt{t}}{t+1}$; the minimal number N of such crossings required to get the link L from K gives us the Nth power of the multiple. Corollary 4.4.3 follows. □

4.4.6.2 HOMFLY polynomials

If $g = n = 1$, then $S_{1,1}$ is a torus with a cusp. The matrix B_T associated to an ideal triangulation of the Riemann surface $S_{1,1}$ has the form

$$B_T = \begin{pmatrix} 0 & 2 & -2 \\ -2 & 0 & 2 \\ 2 & -2 & 0 \end{pmatrix},$$

see [88, Example 4.6]. Using the exchange relations the reader can verify that the cluster C^*-algebra $\mathbb{A}(\mathbf{x}, S_{1,1})$ is given by the Bratteli diagram shown in Fig. 4.11; $\mathbb{A}(\mathbf{x}, S_{1,1})$ coincides with the *Mundici algebra* $\mathfrak{m}_1$ [169].

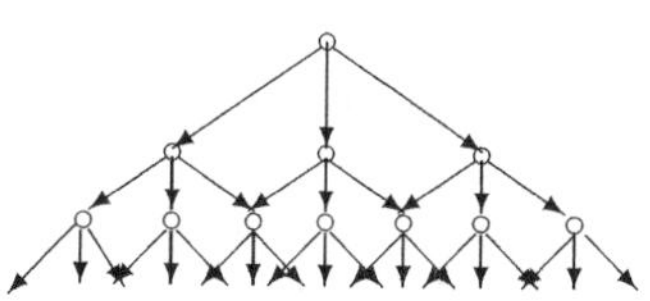

Figure 4.11: Bratteli diagram of the algebra $\mathbb{A}(\mathbf{x}, S_{1,1})$.

The above formulas imply that cluster $\mathbf{x} = (x_1, x_2; y_1)$ consists of two mutable variables x_1, x_2 and a frozen variable y_1. Theorem 4.4.1 says that there exists a representation

$$\rho : B_3 \to \mathbb{A}(\mathbf{x}, S_{1,1})$$

such that $[\rho(b)] \in \mathbf{Z}[x_1^{\pm 1}, x_2^{\pm 1}]$ is a topological invariant of the closure L of $b \in B_3$; since the Laurent polynomial $[\rho(b)]$ depends on two variables x_1, x_2 and two coefficients $c_1, c_2 \in (\mathbb{P}, \oplus, \bullet)$, we shall write it as $[\rho(b)](x_1, x_2; c_1, c_2)$. The following result says that $[\rho(b)](x_1, x_2; c_1, c_2)$ for special values of c_i is related to the HOMFLY polynomial $p_L(l, m)$.

Corollary 4.4.4. *The exchange relations with matrix B imply the skein relation for the HOMFLY polynomial $p_L(l, m)$.*

Proof. Since the variable y_1 is frozen, we consider a reduced matrix

$$\tilde{B}_T = \begin{pmatrix} 0 & 2 \\ -2 & 0 \end{pmatrix},$$

and our seed has the form $(\mathbf{x}, \tilde{B}_T)$, where $\mathbf{x} = (x_1, x_2; c_1, c_2)$. The exchange relations for the variables x_3, x_4, x_5 and the coefficient c_3 imply the following system of equations:

$$\begin{cases} x_3 = \dfrac{c_1 + x_2^2}{(c_1 + 1)x_1}, \\ x_4 = \dfrac{c_2 x_3^2 + 1}{(c_2 + 1)x_2}, \\ x_5 = \dfrac{c_3 + x_4^2}{(c_3 + 1)x_3}, \\ c_3 = \dfrac{1}{c_1}. \end{cases}$$

Clearing the denominators, one gets an expression

$$c_1 x_3 + c_3 x_5 = \frac{c_1 + x_2^2 - x_1 x_3}{x_1} + \frac{c_3 + x_4^2 - x_3 x_5}{x_3}.$$

We exclude $x_4 = \frac{c_2 x_3^2 + 1}{(c_2+1)x_2}$ and $c_3 = \frac{1}{c_1}$ and get the following equality of the Laurent polynomials:

$$c_1 x_3 + \frac{1}{c_1} x_5 + c_2 W = 0,$$

where $W := -c_2[\frac{c_1+x_2^2-x_1x_3}{c_2^2x_1} + \frac{c_1^{-1}-x_3x_5}{c_2^2x_3} + \frac{1}{x_3}(\frac{c_2x_3^2+1}{c_2(c_2+1)x_2})^2]$. The substitution:

$$\begin{cases} c_1 = x_1 = l, \\ c_2 = x_2 = m, \\ x_3 = \rho_{L^+}, \\ x_5 = \rho_{L^-}, \\ W = \rho_L \end{cases}$$

brings the above to the skein relation. Corollary 4.4.4 follows. □

Guide to the literature

Cluster algebras have been introduced in the seminal paper [87]; an excellent survey can be found in [288]. Cluster algebras coming from a triangulation of the Riemann surface were considered in [88]. The cluster C^*-algebras were introduced in [190]. The Birman–Hilden Theorem was proved in [26]; a good review of this and related results can be found in [82]. Cluster C^*-algebras were introduced in [190].

Exercises, problems, and conjectures

1. Verify that $F : \phi \mapsto \mathbb{A}_\phi$ is a well-defined function on the set of all Anosov automorphisms given by the hyperbolic matrices with nonnegative entries.
2. Verify that the definition of the AF-algebra $\mathbb{A}_\phi$ for the pseudo-Anosov maps coincides with the one for the Anosov maps. (Hint: The Jacobi–Perron fractions of dimension $n = 2$ coincide with the regular continued fractions.)
3. ***p*-adic invariants of pseudo-Anosov maps.** Let $\phi \in \mathrm{Mod}(X)$ be pseudo-Anosov automorphism of a surface X. If λ_ϕ is the dilatation of ϕ, then one can consider a $\mathbf{Z}$-module $\mathfrak{m} = \mathbf{Z}v^{(1)} + \cdots + \mathbf{Z}v^{(n)}$ in the number field $K = \mathbf{Q}(\lambda_\phi)$ generated by the normalized eigenvector $(v^{(1)}, \dots, v^{(n)})$ corresponding to the eigenvalue λ_ϕ. The trace function on the number field K gives rise to a symmetric bilinear form $q(x, y)$ on the module $\mathfrak{m}$. The form is defined over the field $\mathbf{Q}$. It has been shown that a pseudo-Anosov automorphism ϕ', conjugate to ϕ, yields a form $q'(x, y)$, equivalent to $q(x, y)$, i. e., $q(x, y)$ can be transformed to $q'(x, y)$ by an invertible linear substitution with the coefficients in $\mathbf{Z}$. It is well known that two rational bilinear forms $q(x, y)$ and $q'(x, y)$ are equivalent whenever the following conditions are satisfied:
   (i) $\Delta = \Delta'$, where Δ is the determinant of the form;
   (ii) for each prime number p (including $p = \infty$) certain p-adic equation between the coefficients of forms q, q' must be satisfied, see, e. g., [38, Chapter 1, § 7.5]. (In fact, only a *finite* number of such equations have to be verified.)

Condition (i) has been already used to discern between the conjugacy classes of the pseudo-Anosov automorphisms. One can use condition (ii) to discern between the pseudo-Anosov automorphisms with $\Delta = \Delta'$; in other words, one gets a problem of *how to define p-adic invariants of the pseudo-Anosov maps.*

4. **The signature of a pseudo-Anosov map.** The signature is an important and well-known invariant connected to the chirality and knotting number of knots and links, see, e. g., [236]. It will be interesting to find a geometric interpretation of the signature Σ for the pseudo-Anosov automorphisms; one can ask the following question: *How to find a geometric interpretation of the invariant Σ?*
5. **The number of conjugacy classes of pseudo-Anosov maps with the same dilatation.** The dilatation λ_ϕ is an invariant of the conjugacy class of the pseudo-Anosov automorphism $\phi \in Mod(X)$. On the other hand, it is known that there exist nonconjugate pseudo-Anosov maps with the same dilatation and the number of such classes is finite, see [276, p. 428]. It is natural to expect that the invariants of operator algebras can be used to evaluate the number; we have the following

Conjecture 4.4.1. *Let $(\Lambda, [I], K)$ be the triple corresponding to a pseudo-Anosov map $\phi \in Mod(X)$. Then the number of the conjugacy classes of the pseudo-Anosov automorphisms with the dilatation λ_ϕ is equal to the class number $h_\Lambda = |\Lambda/[I]|$ of the integral order Λ.*

5 Algebraic geometry

Let **CRng** be the category of the coordinate rings of projective varieties. Consider a functor F,

$$\mathbf{CRng} \xrightarrow{\mathrm{GL}_n} \mathbf{Grp} \hookrightarrow \mathbf{Grp\text{-}Rng},$$

where GL_n is a functor to the category of groups introduced in Example 2.2.4 and $\mathbf{Grp} \hookrightarrow \mathbf{Grp\text{-}Rng}$ is a natural embedding of the **Grp** into the category of group rings. When the **CRng** are the coordinate rings of elliptic curves, the category **Grp-Rng** consists of the noncommutative tori. We prove this important fact in two different ways in Section 5.1. If the **CRng** are coordinate rings of the algebraic curves of genus $g \geq 1$, then the category **Grp-Rng** contains the so-called toric AF-algebras; we refer the reader to Section 5.2. If the **CRng** comprises coordinate rings of projective varieties of dimension $n \geq 1$, then the **Grp-Rng** consists of the *Serre C^*-algebras*; we refer the reader to Section 5.3 for the definition. Elliptic curves over the field of p-adic numbers are considered in Section 5.4; it is proved that in this case the category **Grp-Rng** consists of a family of the UHF-algebras. As an application of the functor F, we show in Section 5.5 that the mapping class groups of genus $g \geq 2$ are linear, i. e., they admit a faithful representation into the matrix group $\mathrm{GL}_{6g-6}(\mathbf{Z})$.

5.1 Elliptic curves

Let **Ell** be the category of all elliptic curves $\mathcal{E}_\tau$; the arrows of **Ell** are identified with the isomorphisms between elliptic curves. Denote by **NC-Tor** the category of all noncommutative tori $\mathcal{A}_\theta$; the arrows of **NC-Tor** are identified with the Morita equivalences (stable isomorphisms) between noncommutative tori.

Theorem 5.1.1 (Functor on elliptic curves). *There exists a covariant functor*

$$F : \mathbf{Ell} \longrightarrow \mathbf{NC\text{-}Tor},$$

which maps isomorphic elliptic curves $\mathcal{E}_\tau$ to the Morita equivalent (stably isomorphic) noncommutative tori $\mathcal{A}_\theta$ as shown in Fig. 1.4. Functor F is noninjective and $\mathrm{Ker}\, F \cong (0, \infty)$.

Theorem 5.1.1 will be proved in Section 5.1.1 using the Sklyanin algebras and in Section 5.1.2 using measured foliations and the Teichmüller theory.

https://doi.org/10.1515/9783110788709-005

5.1.1 Noncommutative tori via Sklyanin algebras

Definition 5.1.1 ([261]). By the Sklyanin algebra $S(\alpha,\beta,\gamma)$ one understands a free **C**-algebra on four generators $x_1,\dots,x_4$ and six quadratic relations

$$\begin{cases} x_1x_2 - x_2x_1 = \alpha(x_3x_4 + x_4x_3), \\ x_1x_2 + x_2x_1 = x_3x_4 - x_4x_3, \\ x_1x_3 - x_3x_1 = \beta(x_4x_2 + x_2x_4), \\ x_1x_3 + x_3x_1 = x_4x_2 - x_2x_4, \\ x_1x_4 - x_4x_1 = \gamma(x_2x_3 + x_3x_2), \\ x_1x_4 + x_4x_1 = x_2x_3 - x_3x_2, \end{cases}$$

where $\alpha+\beta+\gamma+\alpha\beta\gamma = 0$.

Remark 5.1.1 ([264, p. 260]). The algebra $S(\alpha,\beta,\gamma)$ is isomorphic to a twisted homogeneous coordinate ring of elliptic curve $\mathcal{E}_\tau \subset \mathbf{C}P^3$ given in the Jacobi form

$$\begin{cases} u^2 + v^2 + w^2 + z^2 = 0, \\ \frac{1-\alpha}{1+\beta}v^2 + \frac{1+\alpha}{1-\gamma}w^2 + z^2 = 0, \end{cases}$$

i. e., $S(\alpha,\beta,\gamma)$ satisfies an isomorphism

$$\mathbf{Mod}(S(\alpha,\beta,\gamma))/\ \mathbf{Tors} \cong \mathbf{Coh}(\mathcal{E}_\tau),$$

where **Coh** is the category of quasi-coherent sheaves on $\mathcal{E}_\tau$, **Mod** the category of graded left modules over the graded ring $S(\alpha,\beta,\gamma)$, and **Tors** the full subcategory of **Mod** consisting of the torsion modules [249]. The algebra $S(\alpha,\beta,\gamma)$ defines a natural *automorphism* $\sigma : \mathcal{E}_\tau \to \mathcal{E}_\tau$ of the elliptic curve $\mathcal{E}_\tau$, see, e. g., [265, p. 173].

Lemma 5.1.1. *If $\sigma^4 = \mathrm{Id}$, then algebra $S(\alpha,\beta,\gamma)$ is isomorphic to a free algebra $\mathbf{C}\langle x_1, x_2, x_3, x_4\rangle$ modulo an ideal generated by six skew-symmetric quadratic relations*

$$\begin{cases} x_3x_1 = \mu e^{2\pi i\theta}x_1x_3, \\ x_4x_2 = \frac{1}{\mu}e^{2\pi i\theta}x_2x_4, \\ x_4x_1 = \mu e^{-2\pi i\theta}x_1x_4, \\ x_3x_2 = \frac{1}{\mu}e^{-2\pi i\theta}x_2x_3, \\ x_2x_1 = x_1x_2, \\ x_4x_3 = x_3x_4, \end{cases}$$

where $\theta \in S^1$ and $\mu \in (0,\infty)$.

Proof. (i) Since $\sigma^4 = \mathrm{Id}$, the Sklyanin algebra $S(\alpha,\beta,\gamma)$ is isomorphic to a free algebra $\mathbf{C}\langle x_1,x_2,x_3,x_4\rangle$ modulo an ideal generated by the skew-symmetric relations

$$\begin{cases} x_3x_1 = q_{13}x_1x_3, \\ x_4x_2 = q_{24}x_2x_4, \\ x_4x_1 = q_{14}x_1x_4, \\ x_3x_2 = q_{23}x_2x_3, \\ x_2x_1 = q_{12}x_1x_2, \\ x_4x_3 = q_{34}x_3x_4, \end{cases}$$

where $q_{ij} \in \mathbf{C} \setminus \{0\}$, see [84, Remark 1] and [85, § 2] for the proof.

(ii) It is verified directly that above relations are invariant of the involution $x_1^* = x_2, x_3^* = x_4$ if and only if the following restrictions on the constants q_{ij} hold:

$$\begin{cases} q_{13} = (\bar{q}_{24})^{-1}, \\ q_{24} = (\bar{q}_{13})^{-1}, \\ q_{14} = (\bar{q}_{23})^{-1}, \\ q_{23} = (\bar{q}_{14})^{-1}, \\ q_{12} = \bar{q}_{12}, \\ q_{34} = \bar{q}_{34}, \end{cases}$$

where $\bar{q}_{ij}$ means the complex conjugate of $q_{ij} \in \mathbf{C} \setminus \{0\}$.

Remark 5.1.2. The skew-symmetric relations invariant of the involution $x_1^* = x_2, x_3^* = x_4$ define an involution on the Sklyanin algebra; we shall call such an algebra a *Sklyanin $*$-algebra.*

(iii) Consider a one-parameter family $S(q_{13})$ of the Sklyanin $*$-algebras defined by the following additional constraints:

$$\begin{cases} q_{13} = \bar{q}_{14}, \\ q_{12} = q_{34} = 1. \end{cases}$$

It is not hard to see that the $*$-algebras $S(q_{13})$ are pairwise nonisomorphic for different values of complex parameter q_{13}; therefore, family $S(q_{13})$ is a normal form of the Sklyanin $*$-algebra $S(\alpha,\beta,\gamma)$ with $\sigma^4 = \mathrm{Id}$. It remains to notice that one can write the complex parameter q_{13} in the polar form $q_{13} = \mu e^{2\pi i\theta}$, where $\theta = \mathrm{Arg}(q_{13})$ and $\mu = |q_{13}|$. Lemma 5.1.1 follows. □

Lemma 5.1.2. *The system of relations*

$$\begin{cases} x_3x_1 = e^{2\pi i\theta}x_1x_3, \\ x_1x_2 = x_2x_1 = e, \\ x_3x_4 = x_4x_3 = e \end{cases}$$

defining the noncommutative torus $\mathcal{A}_\theta$ is equivalent to the following system of quadratic relations:

$$\begin{cases} x_3x_1 = e^{2\pi i\theta}x_1x_3, \\ x_4x_2 = e^{2\pi i\theta}x_2x_4, \\ x_4x_1 = e^{-2\pi i\theta}x_1x_4, \\ x_3x_2 = e^{-2\pi i\theta}x_2x_3, \\ x_2x_1 = x_1x_2 = e, \\ x_4x_3 = x_3x_4 = e. \end{cases}$$

Proof. Indeed, the first and the last two equations of both systems coincide; we shall proceed stepwise to show the equivalence of the rest of the equations.

(i) Let us prove that equations for $\mathcal{A}_\theta$ imply $x_1x_4 = e^{2\pi i\theta}x_4x_1$. It follows from $x_1x_2 = e$ and $x_3x_4 = e$ that $x_1x_2x_3x_4 = e$. Since $x_1x_2 = x_2x_1$, we can bring the last equation to the form $x_2x_1x_3x_4 = e$ and multiply both sides by the constant $e^{2\pi i\theta}$; thus one gets the equation $x_2(e^{2\pi i\theta}x_1x_3)x_4 = e^{2\pi i\theta}$. But $e^{2\pi i\theta}x_1x_3 = x_3x_1$ and our main equation takes the form $x_2x_3x_1x_4 = e^{2\pi i\theta}$.

We can multiply on the left of both sides of the equation by the element x_1 and thus get the equation $x_1x_2x_3x_1x_4 = e^{2\pi i\theta}x_1$; since $x_1x_2 = e$, one arrives at the equation $x_3x_1x_4 = e^{2\pi i\theta}x_1$.

Again one can multiply on the left of both sides by the element x_4 and thus get the equation $x_4x_3x_1x_4 = e^{2\pi i\theta}x_4x_1$; since $x_4x_3 = e$, one gets the required identity $x_1x_4 = e^{2\pi i\theta}x_4x_1$.

(ii) Let us prove that equations for $\mathcal{A}_\theta$ imply $x_2x_3 = e^{2\pi i\theta}x_3x_2$. As in case (i), it follows from the equations $x_1x_2 = e$ and $x_3x_4 = e$ that $x_3x_4x_1x_2 = e$. Since $x_3x_4 = x_4x_3$, we can bring the last equation to the form $x_4x_3x_1x_2 = e$ and multiply both sides by the constant $e^{-2\pi i\theta}$; thus one gets the equation $x_4(e^{-2\pi i\theta}x_3x_1)x_2 = e^{-2\pi i\theta}$. But $e^{-2\pi i\theta}x_3x_1 = x_1x_3$ and our main equation takes the form $x_4x_1x_3x_2 = e^{-2\pi i\theta}$.

We can multiply on the left of both sides of the equation by the element x_3 and thus get the equation $x_3x_4x_1x_3x_2 = e^{-2\pi i\theta}x_3$; since $x_3x_4 = e$, one arrives at the equation $x_1x_3x_2 = e^{-2\pi i\theta}x_3$.

Again one can multiply on the left of both sides by the element x_2 and thus get the equation $x_2x_1x_3x_2 = e^{-2\pi i\theta}x_2x_3$; since $x_2x_1 = e$, one gets the equation $x_3x_2 =$

$e^{-2\pi i\theta}x_2x_3$. Multiplying both sides by constant $e^{2\pi i\theta}$, we obtain the required identity $x_2x_3 = e^{2\pi i\theta}x_3x_2$.

(iii) Let us prove that equations for $\mathcal{A}_\theta$ imply $x_4x_2 = e^{2\pi i\theta}x_2x_4$. Indeed, it was proved in (i) that $x_1x_4 = e^{2\pi i\theta}x_4x_1$; we shall multiply this equation on the right by the equation $x_2x_1 = e$. Thus one arrives at the equation $x_1x_4x_2x_1 = e^{2\pi i\theta}x_4x_1$.

Notice that in the latter equation one can cancel x_1 on the right thus bringing it to the simpler form $x_1x_4x_2 = e^{2\pi i\theta}x_4$.

We shall multiply on the left both sides of the above equation by the element x_2; one gets therefore $x_2x_1x_4x_2 = e^{2\pi i\theta}x_2x_4$. But $x_2x_1 = e$ and the left-hand side simplifies, giving the required identity $x_4x_2 = e^{2\pi i\theta}x_2x_4$.

Lemma 5.1.2 follows. □

Lemma 5.1.3 (Basic isomorphism). *The system of relations for noncommutative torus $\mathcal{A}_\theta$, namely*

$$\begin{cases} x_3x_1 = e^{2\pi i\theta}x_1x_3, \\ x_4x_2 = e^{2\pi i\theta}x_2x_4, \\ x_4x_1 = e^{-2\pi i\theta}x_1x_4, \\ x_3x_2 = e^{-2\pi i\theta}x_2x_3, \\ x_2x_1 = x_1x_2 = e, \\ x_4x_3 = x_3x_4 = e, \end{cases}$$

is equivalent to the system of relations for the Sklyanin $*$-algebra

$$\begin{cases} x_3x_1 = \mu e^{2\pi i\theta}x_1x_3, \\ x_4x_2 = \frac{1}{\mu}e^{2\pi i\theta}x_2x_4, \\ x_4x_1 = \mu e^{-2\pi i\theta}x_1x_4, \\ x_3x_2 = \frac{1}{\mu}e^{-2\pi i\theta}x_2x_3, \\ x_2x_1 = x_1x_2, \\ x_4x_3 = x_3x_4, \end{cases}$$

modulo the following "scaled unit relation":

$$x_1x_2 = x_3x_4 = \frac{1}{\mu}e.$$

Proof. (i) Using the last two relations, one can bring the noncommutative torus relations to the form

$$\begin{cases} x_3x_1x_4 = e^{2\pi i\theta}x_1, \\ x_4 = e^{2\pi i\theta}x_2x_4x_1, \\ x_4x_1x_3 = e^{-2\pi i\theta}x_1, \\ x_2 = e^{-2\pi i\theta}x_4x_2x_3, \\ x_1x_2 = x_2x_1 = e, \\ x_3x_4 = x_4x_3 = e. \end{cases}$$

(ii) The system of relations for the Sklyanin $*$-algebra complemented by the scaled unit relation, i. e.,

$$\begin{cases} x_3x_1 = \mu e^{2\pi i\theta}x_1x_3, \\ x_4x_2 = \frac{1}{\mu}e^{2\pi i\theta}x_2x_4, \\ x_4x_1 = \mu e^{-2\pi i\theta}x_1x_4, \\ x_3x_2 = \frac{1}{\mu}e^{-2\pi i\theta}x_2x_3, \\ x_2x_1 = x_1x_2 = \frac{1}{\mu}e, \\ x_4x_3 = x_3x_4 = \frac{1}{\mu}e, \end{cases}$$

is equivalent to the system

$$\begin{cases} x_3x_1x_4 = e^{2\pi i\theta}x_1, \\ x_4 = e^{2\pi i\theta}x_2x_4x_1, \\ x_4x_1x_3 = e^{-2\pi i\theta}x_1, \\ x_2 = e^{-2\pi i\theta}x_4x_2x_3, \\ x_2x_1 = x_1x_2 = \frac{1}{\mu}e, \\ x_4x_3 = x_3x_4 = \frac{1}{\mu}e \end{cases}$$

by using multiplication and cancellation involving the last two equations.

(iii) For each $\mu \in (0,\infty)$, consider a *scaled unit* $e' := \frac{1}{\mu}e$ of the Sklyanin $*$-algebra $S(q_{13})$ and the two-sided ideal $I_\mu \subset S(q_{13})$ generated by the relations $x_1x_2 = x_3x_4 = e'$.

Comparing the defining relations for $S(q_{13})$ with those for the noncommutative torus $\mathcal{A}_\theta$, one gets an isomorphism

$$S(q_{13})/I_\mu \cong \mathcal{A}_\theta,$$

see items (i) and (ii). The isomorphism maps generators $x_1, \dots, x_4$ of $*$-algebra $S(q_{13})$ to those of the C^*-algebra $\mathcal{A}_\theta$ and the scaled unit $e' \in S(q_{13})$ to the *ordinary* unit of algebra $\mathcal{A}_\theta$. Lemma 5.1.3 follows. □

To finish the proof of Theorem 5.1.1, recall that the Sklyanin $*$-algebra $S(q_{13})$ satisfies the fundamental isomorphism $\mathbf{Mod}(S(q_{13}))/\mathbf{Tors} \cong \mathbf{Coh}(\mathcal{E}_\tau)$. Using the isomorphism $S(q_{13})/I_\mu \cong \mathcal{A}_\theta$ established in Lemma 5.1.3, we conclude that

$$I_\mu \backslash \mathbf{Coh}(\mathcal{E}_\tau) \cong \mathbf{Mod}(I_\mu \backslash S(q_{13}))/\mathbf{Tors} \cong \mathbf{Mod}(\mathcal{A}_\theta)/\mathbf{Tors}.$$

Thus one gets an isomorphism $\mathbf{Coh}(\mathcal{E}_\tau)/I_\mu \cong \mathbf{Mod}(\mathcal{A}_\theta)/\mathbf{Tors}$, which defines a map $F : \mathbf{Ell} \to \mathbf{NC\text{-}Tor}$. Moreover, map F is a functor because isomorphisms in the category $\mathbf{Mod}$ $(\mathcal{A}_\theta)$ give rise to the stable isomorphisms (Morita equivalences) in the category **NC-Tor**. The second part of Theorem 5.1.1 is due to the fact that F forgets scaling of the unit, i. e., for each $\mu \in (0, \infty)$ we have a constant map

$$S(q_{13}) \ni e' := \frac{1}{\mu} e \longmapsto e \in \mathcal{A}_\theta.$$

Thus $\operatorname{Ker} F \cong (0, \infty)$. Theorem 5.1.1 is proved.

5.1.2 Noncommutative tori via measured foliations

Definition 5.1.2 ([276]). By a measured foliation $\mathcal{F}$ on a surface X one understands a partition of X into the singular points $x_1, \dots, x_n$ of order $k_1, \dots, k_n$ and the regular leaves, i. e., 1-dimensional submanifolds of X; on each open cover U_i of $X \backslash \{x_1, \dots, x_n\}$ there exists a nonvanishing real-valued closed 1-form ϕ_i such that:

(i) $\phi_i = \pm\phi_j$ on $U_i \cap U_j$;

(ii) at each x_i there exists a local chart $(u, v) : V \to \mathbf{R}^2$ such that for $z = u + iv$, it holds that $\phi_i = \operatorname{Im}(z^{\frac{k_i}{2}} dz)$ on $V \cap U_i$ for some branch of $z^{\frac{k_i}{2}}$.

The pair (U_i, ϕ_i) is called an atlas for the measured foliation $\mathcal{F}$. A measure μ is assigned to each segment $(t_0, t) \in U_i$; the measure is transverse to the leaves of $\mathcal{F}$ and is defined by the integral $\mu(t_0, t) = \int_{t_0}^{t} \phi_i$. Such a measure is invariant along the leaves of $\mathcal{F}$, hence the name.

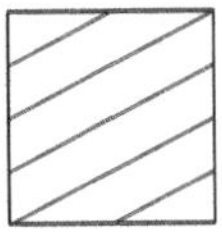

Figure 5.1: A measured foliation on the torus $\mathbf{R}^2/\mathbf{Z}^2$.

Remark 5.1.3. In case $X \cong T^2$ (a torus), each measured foliation is given by a family of parallel lines of a slope $\theta > 0$ as shown in Fig. 5.1.

Let $T(g)$ be the Teichmüller space of surface X of genus $g \geq 1$, i. e., the space of the complex structures on X. Consider the vector bundle $p : Q \to T(g)$ over $T(g)$, whose fiber above a point $S \in T_g$ is the vector space $H^0(S, \Omega^{\otimes 2})$. Given a nonzero $q \in Q$ above S, we can consider the horizontal measured foliation $\mathcal{F}_q \in \Phi_X$ of q, where Φ_X denotes the space of the equivalence classes of the measured foliations on X. If $\{0\}$ is the zero section of Q, the above construction defines a map $Q - \{0\} \longrightarrow \Phi_X$. For any $\mathcal{F} \in \Phi_X$, let $E_{\mathcal{F}} \subset Q - \{0\}$ be the fiber above $\mathcal{F}$. In other words, $E_{\mathcal{F}}$ is a subspace of the holomorphic quadratic forms, whose horizontal trajectory structure coincides with the measured foliation $\mathcal{F}$.

Remark 5.1.4. If $\mathcal{F}$ is a measured foliation with the simple zeroes (a generic case), then $E_{\mathcal{F}} \cong \mathbf{R}^n - 0$ and $T(g) \cong \mathbf{R}^n$, where $n = 6g - 6$ if $g \geq 2$ and $n = 2$ if $g = 1$.

Theorem 5.1.2 ([120]). *The restriction of p to $E_{\mathcal{F}}$ defines a homeomorphism (an embedding) $h_{\mathcal{F}} : E_{\mathcal{F}} \to T(g)$.*

Corollary 5.1.1. *There exists a canonical homeomorphism $h : \Phi_X \to T(g) - \{pt\}$, where $pt = h_{\mathcal{F}}(0)$ and $\Phi_X \cong \mathbf{R}^n - 0$ is the space of equivalence classes of measured foliations $\mathcal{F}'$ on X.*

Proof. Denote by $\mathcal{F}'$ a vertical trajectory structure of q. Since $\mathcal{F}$ and $\mathcal{F}'$ define q, and $\mathcal{F} = \text{const}$ for all $q \in E_{\mathcal{F}}$, one gets a homeomorphism between $T(g) - \{pt\}$ and Φ_X. Corollary 5.1.1 follows. □

Remark 5.1.5. The homeomorphism $h : \Phi_X \to T(g) - \{pt\}$ depends on a foliation $\mathcal{F}$; yet there exists a canonical homeomorphism $h = h_{\mathcal{F}}$ as follows. Let $\mathrm{Sp}(S)$ be the length spectrum of the Riemann surface S and $\mathrm{Sp}(\mathcal{F}')$ be the set of positive reals $\inf \mu(\gamma_i)$, where γ_i runs over all simple closed curves which are transverse to the foliation $\mathcal{F}'$. A canonical homeomorphism $h = h_{\mathcal{F}} : \Phi_X \to T(g) - \{pt\}$ is defined by the formula $\mathrm{Sp}(\mathcal{F}') = \mathrm{Sp}(h_{\mathcal{F}}(\mathcal{F}'))$ for $\forall \mathcal{F}' \in \Phi_X$.

Let $X \cong T^2$; then $T(1) \cong \mathbb{H} := \{z = x + iy \in \mathbf{C} \mid y > 0\}$. Since $q \neq 0$, there are no singular points and each $q \in H^0(S, \Omega^{\otimes 2})$ has the form $q = \omega^2$, where ω is a nowhere zero holomorphic differential on the complex torus S. Note that ω is just a constant times dz, and hence its vertical trajectory structure is just a family of the parallel lines of a slope θ, see, e. g., [268, pp. 54–55]. Therefore, Φ_{T^2} consists of the equivalence classes of the nonsingular measured foliations on the two-dimensional torus. It is well known (the Denjoy theory) that every such foliation is measure equivalent to the foliation of a slope θ and a transverse measure $\mu > 0$ which is invariant along the leaves of the foliation. Thus one obtains a canonical bijection

$$h : \Phi_{T^2} \longrightarrow \mathbb{H} - \{pt\}.$$

Definition 5.1.3 (Category of lattices). By a lattice in the complex plane $\mathbf{C}$ one understands a triple $(\Lambda, \mathbf{C}, j)$, where $\Lambda \cong \mathbf{Z}^2$ and $j : \Lambda \to \mathbf{C}$ is an injective homomorphism with the discrete image. A morphism of lattices $(\Lambda, \mathbf{C}, j) \to (\Lambda', \mathbf{C}, j')$ is the identity $j \circ \psi = \varphi \circ j'$ where φ is a group homomorphism and ψ is a $\mathbf{C}$-linear map. It is not hard to see that any isomorphism class of a lattice contains a representative given by $j : \mathbf{Z}^2 \to \mathbf{C}$ such that $j(1,0) = 1, j(0,1) = \tau \in \mathbb{H}$. The category of lattices $\mathcal{L}$ consists of $\mathrm{Ob}(\mathcal{L})$, which are lattices $(\Lambda, \mathbf{C}, j)$ and morphisms $H(L, L')$ between $L, L' \in \mathrm{Ob}(\mathcal{L})$ which coincide with the morphisms of lattices specified above. For any $L, L', L'' \in \mathrm{Ob}(\mathcal{L})$ and any morphisms $\varphi' : L \to L'$, $\varphi'' : L' \to L''$ a morphism $\phi : L \to L''$ is the composite of φ' and φ'', which we write as $\phi = \varphi''\varphi'$. The identity morphism, 1_L, is a morphism $H(L, L)$.

Remark 5.1.6. The lattices are bijective with the complex tori (and elliptic curves) via the formula $(\Lambda, \mathbf{C}, j) \mapsto \mathbf{C}/j(\Lambda)$; thus $\mathcal{L} \cong \mathbf{Ell}$.

Definition 5.1.4 (Category of pseudo-lattices). By a pseudo-lattice (of rank 2) in the real line $\mathbf{R}$ one understands a triple $(\Lambda, \mathbf{R}, j)$, where $\Lambda \cong \mathbf{Z}^2$ and $j : \Lambda \to \mathbf{R}$ is a homomorphism. A morphism of the pseudo-lattices $(\Lambda, \mathbf{R}, j) \to (\Lambda', \mathbf{R}, j')$ is the identity $j \circ \psi = \varphi \circ j'$, where φ is a group homomorphism and ψ is an inclusion map (i. e., $j'(\Lambda') \subseteq j(\Lambda)$). Any isomorphism class of a pseudo-lattice contains a representative given by $j : \mathbf{Z}^2 \to \mathbf{R}$, such that $j(1,0) = \lambda_1, j(0,1) = \lambda_2$, where λ_1, λ_2 are the positive reals. The pseudo-lattices make up a category, which we denote by $\mathcal{PL}$.

Lemma 5.1.4. *The pseudo-lattices are bijective with the measured foliations on torus via the formula* $(\Lambda, \mathbf{R}, j) \mapsto \mathcal{F}^{\lambda_1}_{\lambda_2/\lambda_1}$, *where* $\mathcal{F}^{\lambda_1}_{\lambda_2/\lambda_1}$ *is a foliation of the slope* $\theta = \lambda_2/\lambda_1$ *and measure* $\mu = \lambda_1$.

Proof. Define a pairing by the formula $(\gamma, \mathrm{Re}\,\omega) \mapsto \int_\gamma \mathrm{Re}\,\omega$, where $\gamma \in H_1(T^2, \mathbf{Z})$ and $\omega \in H^0(S; \Omega)$. The trajectories of the closed differential $\phi := \mathrm{Re}\,\omega$ define a measured foliation on T^2. Thus, in view of the pairing, the linear spaces Φ_{T^2} and $\mathrm{Hom}(H_1(T^2, \mathbf{Z}); \mathbf{R})$ are isomorphic. Notice that the latter space coincides with the space of the pseudo-lattices. To obtain an explicit bijection formula, let us evaluate the integral

$$\int_{\mathbf{Z}\gamma_1 + \mathbf{Z}\gamma_2} \phi = \mathbf{Z}\int_{\gamma_1} \phi + \mathbf{Z}\int_{\gamma_2} \phi = \mathbf{Z}\int_0^1 \mu dx + \mathbf{Z}\int_0^1 \mu dy,$$

where $\{\gamma_1, \gamma_2\}$ is a basis in $H_1(T^2, \mathbf{Z})$. Since $\frac{dy}{dx} = \theta$, one gets

$$\begin{cases} \int_0^1 \mu dx = \mu = \lambda_1, \\ \int_0^1 \mu dy = \int_0^1 \mu\theta dx = \mu\theta = \lambda_2. \end{cases}$$

Thus, $\mu = \lambda_1$ and $\theta = \frac{\lambda_2}{\lambda_1}$. Lemma 5.1.4 follows. □

Remark 5.1.7. It follows from Lemma 5.1.4 and the canonical bijection $h : \Phi_{T^2} \to \mathbb{H} - \{pt\}$ that $\mathcal{L} \cong \mathcal{PL}$ are the equivalent categories.

Definition 5.1.5 (Category of projective pseudo-lattices). By a projective pseudo-lattice (of rank 2) one understands a triple $(\Lambda, \mathbf{R}, j)$, where $\Lambda \cong \mathbf{Z}^2$ and $j : \Lambda \to \mathbf{R}$ is a homomorphism. A morphism of the projective pseudo-lattices $(\Lambda, \mathbf{C}, j) \to (\Lambda', \mathbf{R}, j')$ is the identity $j \circ \psi = \varphi \circ j'$, where φ is a group homomorphism and ψ is an $\mathbf{R}$-linear map. (Notice that unlike the case of the pseudo-lattices, ψ is a scaling map as opposed to an inclusion map. Thus, the two pseudo-lattices can be projectively equivalent, while being distinct in the category $\mathcal{PL}$.) It is not hard to see that any isomorphism class of a projective pseudo-lattice contains a representative given by $j : \mathbf{Z}^2 \to \mathbf{R}$ such that $j(1,0) = 1, j(0,1) = \theta$, where θ is a positive real. The projective pseudo-lattices make up a category, which we shall denote by $\mathcal{PPL}$.

Lemma 5.1.5. *$\mathcal{PPL} \cong$ **NC-Tor**, i. e., projective pseudo-lattices and noncommutative tori are equivalent categories.*

Proof. Notice that projective pseudo-lattices are bijective with the noncommutative tori via the formula $(\Lambda, \mathbf{R}, j) \mapsto \mathcal{A}_\theta$. An isomorphism $\varphi : \Lambda \to \Lambda'$ acts by the formula $1 \mapsto a + b\theta$, $\theta \mapsto c + d\theta$, where $ad - bc = 1$ and $a, b, c, d \in \mathbf{Z}$. Therefore, $\theta' = \frac{c+d\theta}{a+b\theta}$. Thus, isomorphic projective pseudo-lattices map to the Morita equivalent (stably isomorphic) noncommutative tori. Lemma 5.1.5 follows. □

To define a map $F : \mathbf{Ell} \to \mathbf{NC\text{-}Tor}$, we shall consider a composition of the following morphisms

$$\mathbf{Ell} \longrightarrow \mathcal{L} \longrightarrow \mathcal{PL} \xrightarrow{F} \mathcal{PPL} \longrightarrow \mathbf{NC\text{-}Tor},$$

where all the arrows, except for F, have been defined. To define F, let $\mathrm{PL} \in \mathcal{PL}$ be a pseudo-lattice such that $\mathrm{PL} = \mathrm{PL}(\lambda_1, \lambda_2)$, where $\lambda_1 = j(1,0), \lambda_2 = j(0,1)$ are positive reals. Let $\mathrm{PPL} \in \mathcal{PPL}$ be a projective pseudo-lattice such that $\mathrm{PPL} = \mathrm{PPL}(\theta)$, where $j(1,0) = 1$ and $j(0,1) = \theta$ is a positive real. Then $F : \mathcal{PL} \to \mathcal{PPL}$ is given by the formula $\mathrm{PL}(\lambda_1, \lambda_2) \longmapsto \mathrm{PPL}(\frac{\lambda_2}{\lambda_1})$. It is easy to see that $\mathrm{Ker}\, F \cong (0, \infty)$ and F is not an injective map. Since all the arrows, but F, are isomorphisms between the categories, one gets a map

$$F : \mathbf{Ell} \longrightarrow \mathbf{NC\text{-}Tor}.$$

Lemma 5.1.6 (Basic lemma). *The map $F : \mathbf{Ell} \to \mathbf{NC\text{-}Tor}$ is a covariant functor which maps isomorphic complex tori to the Morita equivalent (stably isomorphic) noncommutative tori; the functor is noninjective functor and* $\operatorname{Ker} F \cong (0, \infty)$.

Proof. (i) Let us show that F maps isomorphic complex tori to the stably isomorphic noncommutative tori. Let $\mathbf{C}/(\mathbf{Z}\omega_1 + \mathbf{Z}\omega_2)$ be a complex torus. Recall that the periods $\omega_1 = \int_{\gamma_1} \omega_E$ and $\omega_2 = \int_{\gamma_2} \omega_E$, where $\omega_E = dz$ is an invariant (Néron) differential on the complex torus and $\{\gamma_1, \gamma_2\}$ is a basis in $H_1(T^2, \mathbf{Z})$. The map F can be written as

$$\mathbf{C}/L_{(\int_{\gamma_2} \omega_E)/(\int_{\gamma_1} \omega_E)} \overset{F}{\longmapsto} \mathcal{A}_{(\int_{\gamma_2} \phi)/(\int_{\gamma_1} \phi)},$$

where L_{ω_2/ω_1} is a lattice and $\phi = \operatorname{Re} \omega$ is a closed differential defined earlier. Note that every isomorphism in the category $\mathbf{Ell}$ is induced by an orientation preserving automorphism φ of the torus T^2. The action of φ on the homology basis $\{\gamma_1, \gamma_2\}$ of T^2 is given by the formula

$$\begin{cases} \gamma_1' = a\gamma_1 + b\gamma_2, \\ \gamma_2' = c\gamma_1 + d\gamma_2, \end{cases} \quad \text{where } \begin{pmatrix} a & b \\ c & d \end{pmatrix} \in SL_2(\mathbf{Z}).$$

The functor F acts by the formula

$$\tau = \frac{\int_{\gamma_2} \omega_E}{\int_{\gamma_1} \omega_E} \mapsto \theta = \frac{\int_{\gamma_2} \phi}{\int_{\gamma_1} \phi}.$$

(a) From the left-hand side of the above equation, one obtains

$$\begin{cases} \omega_1' = \int\limits_{\gamma_1'} \omega_E = \int\limits_{a\gamma_1 + b\gamma_2} \omega_E = a \int\limits_{\gamma_1} \omega_E + b \int\limits_{\gamma_2} \omega_E = a\omega_1 + b\omega_2, \\ \omega_2' = \int\limits_{\gamma_2'} \omega_E = \int\limits_{c\gamma_1 + d\gamma_2} \omega_E = c \int\limits_{\gamma_1} \omega_E + d \int\limits_{\gamma_2} \omega_E = c\omega_1 + d\omega_2, \end{cases}$$

and therefore, $\tau' = \frac{\int_{\gamma_2'} \omega_E}{\int_{\gamma_1'} \omega_E} = \frac{c + d\tau}{a + b\tau}$.

(b) From the right-hand side, one obtains

$$\begin{cases} \lambda_1' = \int\limits_{\gamma_1'} \phi = \int\limits_{a\gamma_1+b\gamma_2} \phi = a\int\limits_{\gamma_1}\phi + b\int\limits_{\gamma_2}\phi = a\lambda_1 + b\lambda_2, \\ \lambda_2' = \int\limits_{\gamma_2'} \phi = \int\limits_{c\gamma_1+d\gamma_2} \phi = c\int\limits_{\gamma_1}\phi + d\int\limits_{\gamma_2}\phi = c\lambda_1 + d\lambda_2, \end{cases}$$

and therefore $\theta' = \frac{\int_{\gamma_2'}\phi}{\int_{\gamma_1'}\phi} = \frac{c+d\theta}{a+b\theta}$. Comparing (a) and (b), one concludes that F maps isomorphic complex tori to the stably isomorphic (Morita equivalent) noncommutative tori.

(ii) Let us show that F is a covariant functor, i. e., F does not reverse the arrows. Indeed, it can be verified directly using the above formulas, that $F(\varphi_1\varphi_2) = \varphi_1\varphi_2 = F(\varphi_1)F(\varphi_2)$ for any pair of the isomorphisms $\varphi_1, \varphi_2 \in \mathrm{Aut}(T^2)$.

(iii) Since $F : \mathcal{PL} \to \mathcal{PPL}$ is given by the formula $\mathrm{PL}(\lambda_1, \lambda_2) \longmapsto \mathrm{PPL}(\frac{\lambda_2}{\lambda_1})$, one gets $\mathrm{Ker}\, F \cong (0, \infty)$ and F is not an injective map. Lemma 5.1.6 is proved. □

Theorem 5.1.1 follows from Lemma 5.1.6.

Guide to the literature

The basics of elliptic curves are covered in [122, 136, 138, 258–260], and others. More advanced topics are discussed in the survey papers [49, 158, 274]. The Sklyanin algebras were introduced and studied in [261] and [262]; for a detailed account, see [84] and [85]. The general theory is covered in [265]. The basics of measured foliations and the Teichmüller theory can be found in [276] and [120]. The idea of infinite-dimensional representations of Sklyanin's algebras by the linear operators on a Hilbert space $\mathcal{H}$ belongs to [261, Section 3]. A functor from elliptic curves to noncommutative tori was constructed in [179] using measured foliations and in [200] using Sklyanin's algebras. A desingularization of the moduli space of noncommutative tori was considered in [177].

5.2 Algebraic curves of genus $g \geq 1$

The complex algebraic curve is a subset of the complex projective plane of the form

$$C = \{(x, y, z) \in \mathbf{C}P^2 \mid P(x, y, z) = 0\},$$

where $P(x, y, z)$ is a homogeneous polynomial with complex coefficients. The C is isomorphic to the two-dimensional manifold of genus g endowed with a complex structure, i. e., the Riemann surfaces S. We construct a functor F on a generic set of the Riemann surfaces with values in the category of the toric AF-algebras. Functor F maps isomorphic algebraic curves to the Morita equivalent (stably isomorphic) toric AF-algebras. For $g = 1$, the toric AF-algebras coincide with the Effros–Shen algebras $\mathbb{A}_\theta$; by Theorem 3.5.3, such an algebra contains the noncommutative torus $\mathcal{A}_\theta$. Functor F will be used in Section 5.5 to prove a long standing conjecture on the linearity of the mapping class groups.

5.2.1 Toric AF-algebras

We recall some facts of the Teichmüller theory [120]. Denote by $T_S(g)$ the Teichmüller space of genus $g \geq 1$ (i. e., the space of all complex 2-dimensional manifolds of genus g) endowed with a distinguished point S. Let $q \in H^0(S, \Omega^{\otimes 2})$ be a holomorphic quadratic differential on the Riemann surface S, such that all zeroes of q (if any) are simple. By $\widetilde{S}$ we understand a double cover of S ramified over the zeroes of q and by $H_1^{\mathrm{odd}}(\widetilde{S})$ the odd part of the integral homology of $\widetilde{S}$ relatively the zeroes. Note that $H_1^{\mathrm{odd}}(\widetilde{S}) \cong \mathbf{Z}^n$, where $n = 6g - 6$ if $g \geq 2$ and $n = 2$ if $g = 1$. It is known that

$$T_S(g) \cong \mathrm{Hom}(H_1^{\mathrm{odd}}(\widetilde{S}); \mathbf{R}) - \{0\},$$

where 0 is the zero homomorphism [120]. Denote by $\lambda = (\lambda_1, \ldots, \lambda_n)$ the image of a basis of $H_1^{\mathrm{odd}}(\widetilde{S})$ in the real line $\mathbf{R}$ such that $\lambda_1 \neq 0$.

Remark 5.2.1. The claim $\lambda_1 \neq 0$ is not restrictive because the zero homomorphism is excluded.

We let $\theta = (\theta_1, \ldots, \theta_{n-1})$, where $\theta_i = \lambda_{i-1}/\lambda_1$. Recall that, up to a scalar multiple, vector $(1, \theta) \in \mathbf{R}^n$ is the limit of a generically convergent Jacobi–Perron continued fraction [23]

$$\begin{pmatrix} 1 \\ \theta \end{pmatrix} = \lim_{k \to \infty} \begin{pmatrix} 0 & 1 \\ I & b_1 \end{pmatrix} \cdots \begin{pmatrix} 0 & 1 \\ I & b_k \end{pmatrix} \begin{pmatrix} 0 \\ \mathbb{I} \end{pmatrix},$$

where $b_i = (b_1^{(i)}, \ldots, b_{n-1}^{(i)})^T$ is a vector of nonnegative integers, I the unit matrix, and $\mathbb{I} = (0, \ldots, 0, 1)^T$.

Definition 5.2.1. By a toric AF-algebra $\mathbb{A}_\theta$ one understands the AF-algebra given by the Bratteli diagram in Fig. 5.2, where numbers $b_j^{(i)}$ indicate the multiplicity of edges of the graph.

Remark 5.2.2. Note that in the case $g = 1$, the Jacobi–Perron fraction coincides with the regular continued fraction and $\mathbb{A}_\theta$ becomes the Effros–Shen algebra, see Example 3.5.2.

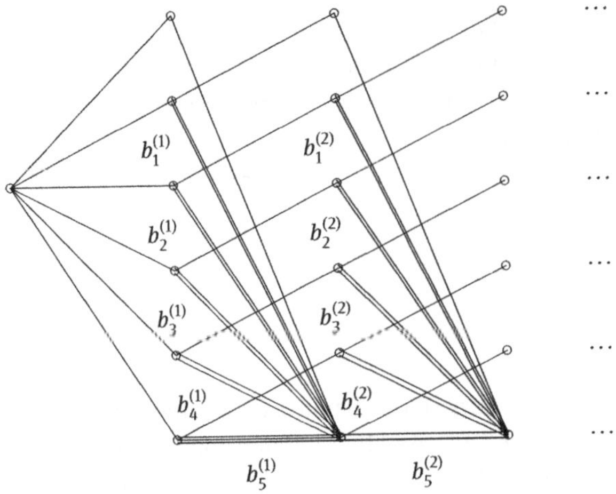

Figure 5.2: Toric AF-algebra $\mathbb{A}_\theta$ (case $g = 2$).

By **Alg-Gen** we understand the maximal subset of $T_S(g)$ such that, for each complex algebraic curve $C \in$ **Alg-Gen**, the corresponding Jacobi–Perron continued fraction is convergent; the arrows of **Alg-Gen** are isomorphisms between complex algebraic curves C. We shall write **AF-Toric** to denote the category of all toric AF-algebras $\mathbb{A}_\theta$; the arrows of **AF-Toric** are stable isomorphisms (Morita equivalences) between toric AF-algebras $\mathbb{A}_\theta$. By F we understand a map given by the formula $C \mapsto \mathbb{A}_\theta$; in other words, we have a map

$$F : \textbf{Alg-Gen} \longrightarrow \textbf{AF-Toric}.$$

Theorem 5.2.1 (Functor on algebraic curves). *The set* **Alg-Gen** *is a generic subset of* $T_S(g)$ *and the map F has the following properties:*

(i) **Alg-Gen** $\cong$ **AF-Toric** $\times (0, \infty)$ *is a trivial fiber bundle, whose projection map* p : **Alg-Gen** $\to$ **AF-Toric** *coincides with F;*
(ii) F : **Alg-Gen** $\to$ **AF-Toric** *is a covariant functor, which maps isomorphic complex algebraic curves* $C, C' \in$ **Alg-Gen** *to the Morita equivalent (stably isomorphic) toric AF-algebras* $\mathbb{A}_\theta, \mathbb{A}_{\theta'} \in$ **AF-Toric.**

5.2.2 Proof of Theorem 5.2.1

Let S be a Riemann surface, and $q \in H^0(S, \Omega^{\otimes 2})$ a holomorphic quadratic differential on S. The lines $\operatorname{Re} q = 0$ and $\operatorname{Im} q = 0$ define a pair of measured foliations on R, which are transversal to each other outside the set of singular points. The set of singular points is common to both foliations and coincides with the zeroes of q. The above measured foliations are said to represent the vertical and horizontal trajectory structure of q, respectively. Denote by $T(g)$ the Teichmüller space of the topological surface X of genus $g \geq 1$, i. e., the space of the complex structures on X. Consider the vector bundle $p : Q \to T(g)$ over $T(g)$ whose fiber above a point $S \in T_g$ is the vector space $H^0(S, \Omega^{\otimes 2})$. Given a nonzero $q \in Q$ above S, we can consider horizontal measured foliation $\mathcal{F}_q \in \Phi_X$ of q, where Φ_X denotes the space of equivalence classes of measured foliations on X. If $\{0\}$ is the zero section of Q, the above construction defines a map $Q - \{0\} \longrightarrow \Phi_X$. For any $\mathcal{F} \in \Phi_X$, let $E_{\mathcal{F}} \subset Q - \{0\}$ be the fiber above $\mathcal{F}$. In other words, $E_{\mathcal{F}}$ is a subspace of the holomorphic quadratic forms whose horizontal trajectory structure coincides with the measured foliation $\mathcal{F}$. If $\mathcal{F}$ is a measured foliation with the simple zeroes (a generic case), then $E_{\mathcal{F}} \cong \mathbf{R}^n - 0$, while $T(g) \cong \mathbf{R}^n$, where $n = 6g-6$ if $g \geq 2$ and $n = 2$ if $g = 1$. The restriction of p to $E_{\mathcal{F}}$ defines a homeomorphism (an embedding)

$$h_{\mathcal{F}} : E_{\mathcal{F}} \to T(g).$$

The above result implies that the measured foliations parametrize the space $T(g) - \{pt\}$, where $pt = h_{\mathcal{F}}(0)$. Indeed, denote by $\mathcal{F}'$ a vertical trajectory structure of q. Since $\mathcal{F}$ and $\mathcal{F}'$ define q, and $\mathcal{F} = \text{const}$ for all $q \in E_{\mathcal{F}}$, one gets a homeomorphism between $T(g) - \{pt\}$ and Φ_X, where $\Phi_X \cong \mathbf{R}^n - 0$ is the space of equivalence classes of the measured foliations $\mathcal{F}'$ on X. Note that the above parametrization depends on a foliation $\mathcal{F}$. However, there exists a unique canonical homeomorphism $h = h_{\mathcal{F}}$ as follows. Let $\mathrm{Sp}(S)$ be the length spectrum of the Riemann surface S and $\mathrm{Sp}(\mathcal{F}')$ be the set of positive reals $\inf \mu(\gamma_i)$, where γ_i runs over all simple closed curves which are transverse to the foliation $\mathcal{F}'$. A canonical homeomorphism $h = h_{\mathcal{F}} : \Phi_X \to T(g) - \{pt\}$ is defined by the formula $\mathrm{Sp}(\mathcal{F}') = \mathrm{Sp}(h_{\mathcal{F}}(\mathcal{F}'))$ for all $\mathcal{F}' \in \Phi_X$. Thus, there exists a canonical homeomorphism

$$h : \Phi_X \to T(g) - \{pt\}.$$

Recall that Φ_X is the space of equivalence classes of measured foliations on the topological surface X. Following [71], we consider the following coordinate system on Φ_X. For clarity, let us make a generic assumption that $q \in H^0(S, \Omega^{\otimes 2})$ is a nontrivial holomorphic quadratic differential with only simple zeroes. We wish to construct a Riemann surface of $\sqrt{q}$, which is a double cover of S with ramification over the zeroes of q. Such a surface, denoted by $\widetilde{S}$, is unique and has an advantage of carrying a holomorphic differential ω such that $\omega^2 = q$. We further denote by $\pi : \widetilde{S} \to S$ the covering projection. The vector space $H^0(\widetilde{S}, \Omega)$ splits into the direct sum $H^0_{\text{even}}(\widetilde{S}, \Omega) \oplus H^0_{\text{odd}}(\widetilde{S}, \Omega)$ in view of the involution π^{-1} of $\widetilde{S}$, and the vector space $H^0(S, \Omega^{\otimes 2}) \cong H^0_{\text{odd}}(\widetilde{S}, \Omega)$. Let $H_1^{\text{odd}}(\widetilde{S})$ be the odd part of the homology of $\widetilde{S}$ relatively the zeroes of q. Consider the pairing $H_1^{\text{odd}}(\widetilde{S}) \times H^0(S, \Omega^{\otimes 2}) \to \mathbf{C}$, defined by the integration $(\gamma, q) \mapsto \int_\gamma \omega$. We shall take the associated map $\psi_q : H^0(S, \Omega^{\otimes 2}) \to \mathrm{Hom}(H_1^{\text{odd}}(\widetilde{S}); \mathbf{C})$ and let $h_q = \mathrm{Re}\, \psi_q$.

Lemma 5.2.1 ([71]). *The map*

$$h_q : H^0(S, \Omega^{\otimes 2}) \longrightarrow \mathrm{Hom}(H_1^{\mathrm{odd}}(\tilde{S}); \mathbf{R})$$

is an **R**-*isomorphism.*

Remark 5.2.3. Since each $\mathcal{F} \in \Phi_X$ is the vertical foliation $\mathrm{Re}\, q = 0$ for a $q \in H^0(S, \Omega^{\otimes 2})$, Lemma 5.2.1 implies that $\Phi_X \cong \mathrm{Hom}(H_1^{\mathrm{odd}}(\tilde{S}); \mathbf{R})$. By formulas for the relative homology, one finds that $H_1^{\mathrm{odd}}(\tilde{S}) \cong \mathbf{Z}^n$, where $n = 6g - 6$ if $g \geq 2$ and $n = 2$ if $g = 1$. Each $h \in \mathrm{Hom}(\mathbf{Z}^n; \mathbf{R})$ is given by the reals $\lambda_1 = h(e_1), \dots, \lambda_n = h(e_n)$, where $(e_1, \dots, e_n)$ is a basis in $\mathbf{Z}^n$. The numbers $(\lambda_1, \dots, \lambda_n)$ are the coordinates in the space Φ_X and, therefore, in the Teichmüller space $T(g)$.

To prove Theorem 5.2.1, we shall consider the following categories: (i) generic complex algebraic curves **Alg-Gen**; (ii) pseudo-lattices $\mathcal{PL}$; (iii) projective pseudo-lattices $\mathcal{PPL}$, and (iv) category **AF-Toric** of the toric AF-algebras. First, we show that **Alg-Gen** $\cong \mathcal{PL}$ are equivalent categories such that isomorphic complex algebraic curves $C, C' \in$ **Alg-Gen** map to isomorphic pseudo-lattices $\mathrm{PL}, \mathrm{PL}' \in \mathcal{PL}$. Next, a noninjective functor $F : \mathcal{PL} \to \mathcal{PPL}$ is constructed. Functor F maps isomorphic pseudo-lattices to isomorphic projective pseudo-lattices and $\mathrm{Ker}\, F \cong (0, \infty)$. Finally, it is shown that a subcategory $U \subseteq \mathcal{PPL}$ and **AF-Toric** are equivalent categories. In other words, we have the following diagram:

$$\textbf{Alg-Gen} \xrightarrow{\alpha} \mathcal{PL} \xrightarrow{F} U \xrightarrow{\beta} \textbf{AF-Toric},$$

where α is an injective map, β is a bijection, and $\mathrm{Ker}\, F \cong (0, \infty)$.

Definition 5.2.2. Let $\mathrm{Mod}\, X$ be the mapping class group of the surface X. A *complex algebraic curve* is a triple (X, C, j), where X is a topological surface of genus $g \geq 1$, $j : X \to C$ is a complex (conformal) parametrization of X and C is a Riemann surface. A *morphism* of complex algebraic curves $(X, C, j) \to (X, C', j')$ is the identity $j \circ \psi = \varphi \circ j'$, where $\varphi \in \mathrm{Mod}\, X$ is a diffeomorphism of X and ψ is an isomorphism of Riemann surfaces. A category of generic complex algebraic curves, **Alg-Gen**, consists of Ob(**Alg-Gen**) which are complex algebraic curves $C \in T_S(g)$ and morphisms $H(C, C')$ between $C, C' \in$ Ob(**Alg-Gen**) which coincide with the morphisms specified above. For any $C, C', C'' \in$ Ob(**Alg-Gen**) and any morphisms $\varphi' : C \to C'$, $\varphi'' : C' \to C''$ a morphism $\phi : C \to C''$ is the composite of φ' and φ'', which we write as $\phi = \varphi'' \varphi'$. The identity morphism, 1_C, is a morphism $H(C, C)$.

Definition 5.2.3. By a pseudo-lattice (of rank n) one understands the triple $(\Lambda, \mathbf{R}, j)$, where $\Lambda \cong \mathbf{Z}^n$ and $j : \Lambda \to \mathbf{R}$ is a homomorphism. A morphism of pseudo-lattices $(\Lambda, \mathbf{R}, j) \to (\Lambda, \mathbf{R}, j')$ is the identity $j \circ \psi = \varphi \circ j'$, where φ is a group homomorphism and ψ is an inclusion map, i. e., $j'(\Lambda') \subseteq j(\Lambda)$. Any isomorphism class of a pseudo-lattice contains a representative given by $j : \mathbf{Z}^n \to \mathbf{R}$ such that $j(1, 0, \dots, 0) = \lambda_1$, $j(0, 1, \dots, 0) =$

$\lambda_2, \dots, j(0,0,\dots,1) = \lambda_n$, where $\lambda_1, \lambda_2, \dots, \lambda_n$ are positive reals. The pseudo-lattices of rank n make up a category, which we denote by $\mathcal{PL}_n$.

Lemma 5.2.2 (Basic lemma). *Let $g \geq 2$ ($g = 1$) and $n = 6g - 6$ ($n = 2$). There exists an injective covariant functor α : **Alg-Gen** $\to \mathcal{PL}_n$, which maps isomorphic complex algebraic curves $C, C' \in$ **Alg-Gen** to the isomorphic pseudo-lattices* $\mathrm{PL}, \mathrm{PL}' \in \mathcal{PL}_n$.

Proof. Recall that we have a map $\alpha : T(g) - \{pt\} \to \mathrm{Hom}(H_1^{\mathrm{odd}}(\widetilde{S}); \mathbf{R}) - \{0\}$; α is a homeomorphism and, therefore, α is injective. Let us find the image $\alpha(\varphi) \in \mathrm{Mor}(\mathcal{PL})$ of $\varphi \in \mathrm{Mor}(\textbf{Alg-Gen})$. Let $\varphi \in \mathrm{Mod}\, X$ be a diffeomorphism of X, and let $\widetilde{X} \to X$ be the ramified double cover of X. We denote by $\widetilde{\varphi}$ the induced mapping on $\widetilde{X}$. Note that $\widetilde{\varphi}$ is a diffeomorphism of $\widetilde{X}$ modulo the covering involution $\mathbf{Z}_2$. Denote by $\widetilde{\varphi}^*$ the action of $\widetilde{\varphi}$ on $H_1^{\mathrm{odd}}(\widetilde{X}) \cong \mathbf{Z}^n$. Since $\widetilde{\varphi} \bmod \mathbf{Z}_2$ is a diffeomorphism of $\widetilde{X}$, $\widetilde{\varphi}^* \in \mathrm{GL}_n(\mathbf{Z})$. Thus, $\alpha(\varphi) = \widetilde{\varphi}^* \in \mathrm{Mor}(\mathcal{PL})$. Let us show that α is a functor. Indeed, let $C, C' \in$ **Alg-Gen** be isomorphic complex algebraic curves such that $C' = \varphi(C)$ for a $\varphi \in \mathrm{Mod}\, X$. Let a_{ij} be the elements of matrix $\widetilde{\varphi}^* \in \mathrm{GL}_n(\mathbf{Z})$. Recall that $\lambda_i = \int_{\gamma_i} \phi$ for a closed 1-form $\phi = \mathrm{Re}\,\omega$ and $\gamma_i \in H_1^{\mathrm{odd}}(\widetilde{X})$. Then $\gamma_j = \sum_{i=1}^n a_{ij}\gamma_i$, $j = 1, \dots, n$ are the elements of a new basis in $H_1^{\mathrm{odd}}(\widetilde{X})$. By integration rules, we have

$$\lambda_j' = \int_{\gamma_j} \phi = \int_{\sum a_{ij}\gamma_i} \phi = \sum_{i=1}^n a_{ij}\lambda_i.$$

Let $j(\Lambda) = \mathbf{Z}\lambda_1 + \cdots + \mathbf{Z}\lambda_n$ and $j'(\Lambda) = \mathbf{Z}\lambda_1' + \cdots + \mathbf{Z}\lambda_n'$. Since $\lambda_j' = \sum_{i=1}^n a_{ij}\lambda_i$ and $(a_{ij}) \in \mathrm{GL}_n(\mathbf{Z})$, we conclude that $j(\Lambda) = j'(\Lambda) \subset \mathbf{R}$. In other words, the pseudo-lattices $(\Lambda, \mathbf{R}, j)$ and $(\Lambda, \mathbf{R}, j')$ are isomorphic. Hence, α : **Alg-Gen** $\to \mathcal{PL}$ maps isomorphic complex algebraic curves to the isomorphic pseudo-lattices, i. e., α is a functor. Let us show that α is a covariant functor. Indeed, let $\varphi_1, \varphi_2 \in \mathrm{Mor}(\textbf{Alg-Gen})$. Then $\alpha(\varphi_1\varphi_2) = (\widetilde{\varphi_1\varphi_2})^* = \widetilde{\varphi}_1^*\widetilde{\varphi}_2^* = \alpha(\varphi_1)\alpha(\varphi_2)$. Lemma 5.2.2 follows. □

Definition 5.2.4. By a projective pseudo-lattice (of rank n) one understands a triple $(\Lambda, \mathbf{R}, j)$, where $\Lambda \cong \mathbf{Z}^n$ and $j : \Lambda \to \mathbf{R}$ is a homomorphism. A morphism of projective pseudo-lattices $(\Lambda, \mathbf{C}, j) \to (\Lambda, \mathbf{R}, j')$ is the identity $j \circ \psi = \varphi \circ j'$, where φ is a group homomorphism and ψ is an $\mathbf{R}$-linear map. It is not hard to see that any isomorphism class of a projective pseudo-lattice contains a representative given by $j : \mathbf{Z}^n \to \mathbf{R}$ such that $j(1,0,\dots,0) = 1$, $j(0,1,\dots,0) = \theta_1, \dots, j(0,0,\dots,1) = \theta_{n-1}$, where θ_i are positive reals. The projective pseudo-lattices of rank n make up a category, which we denote by $\mathcal{PPL}_n$.

Remark 5.2.4. Notice that, unlike in the case of pseudo-lattices, ψ is a scaling map as opposed to an inclusion map. This allows two pseudo-lattices to be projectively equivalent, while being distinct in the category $\mathcal{PL}_n$.

Definition 5.2.5. Finally, the toric AF-algebras $\mathbb{A}_\theta$, modulo stable isomorphism (Morita equivalences), make up a category which we shall denote by **AF-Toric**.

Lemma 5.2.3. *Let $U_n \subseteq \mathcal{PPL}_n$ be a subcategory consisting of the projective pseudo-lattices* $\mathrm{PPL} = \mathrm{PPL}(1,\theta_1,\dots,\theta_{n-1})$ *for which the Jacobi–Perron fraction of the vector* $(1,\theta_1,\dots,\theta_{n-1})$ *converges to the vector. Define a map $\beta : U_n \to$ **AF-Toric** by the formula* $\mathrm{PPL}(1,\theta_1,\dots,\theta_{n-1}) \mapsto \mathbb{A}_\theta$, *where $\theta = (\theta_1,\dots,\theta_{n-1})$. Then β is a bijective functor, which maps isomorphic projective pseudo-lattices to the stably isomorphic toric AF-algebras.*

Proof. It is evident that β is injective and surjective. Let us show that β is a functor. Indeed, according to [74, Corollary 4.7], every totally ordered abelian group of rank n has form $\mathbf{Z}+\theta_1\mathbf{Z}+\dots+\mathbf{Z}\theta_{n-1}$. The latter is a projective pseudo-lattice PPL from the category U_n. On the other hand, by Theorem 3.5.2, the PPL defines a stable isomorphism class of the toric AF-algebra $\mathbb{A}_\theta \in$ **AF-Toric**. Therefore, β maps isomorphic projective pseudo-lattices (from the set U_n) to the stably isomorphic toric AF-algebras, and vice versa. Lemma 5.2.3 follows. □

Lemma 5.2.4. *Let $F : \mathcal{PL}_n \to \mathcal{PPL}_n$ be a map given by formula*

$$\mathrm{PL}(\lambda_1,\lambda_2,\dots,\lambda_n) \mapsto \mathrm{PPL}\left(1,\frac{\lambda_2}{\lambda_1},\dots,\frac{\lambda_n}{\lambda_1}\right),$$

where $\mathrm{PL}(\lambda_1,\lambda_2,\dots,\lambda_n) \in \mathcal{PL}_n$ and $\mathrm{PPL}(1,\theta_1,\dots,\theta_{n-1}) \in \mathcal{PPL}_n$. Then $\operatorname{Ker} F = (0,\infty)$ and F is a functor which maps isomorphic pseudo-lattices to isomorphic projective pseudo-lattices.

Proof. Indeed, F can be thought of as a map from $\mathbf{R}^n$ to $\mathbf{R}P^n$. Hence $\operatorname{Ker} F = \{\lambda_1 : \lambda_1 > 0\} \cong (0,\infty)$. The second part of lemma is evident. Lemma 5.2.4 is proved. □

Theorem 5.2.1 follows from Lemmas 5.2.2–5.2.4 with $n = 6g-6$ ($n = 2$) for $g \geq 2$ ($g = 1$).

Guide to the literature

An excellent introduction to complex algebraic curves is the book [135]. For measured foliations and their relation to the Teichmüller theory the reader is referred to [120]. Functor F : **Alg-Gen** → **AF-Toric** was constructed in [180] and [193]; the term toric AF-algebras was suggested by Yu. I. Manin (private communication).

5.3 Projective varieties of dimension $n \geq 1$

We extend functor F to a category **Proj-Alg** of the projective varieties of complex dimension $n \geq 1$. Namely, we construct a covariant functor

$$F : \textbf{Proj-Alg} \longrightarrow \textbf{C*-Serre},$$

where **C*-Serre** is the category of the *Serre C^*-algebras* $\mathcal{A}_X$ of variety X. In particular, if $n = 1$, one gets that $\mathcal{A}_X$ is a noncommutative torus $\mathcal{A}_\theta$ for $X \cong \mathcal{E}_\tau$ and $\mathcal{A}_X$ is a toric

AF-algebra $\mathbb{A}_\theta$ for $X \cong C$. On the other hand, we prove that $X \cong \mathbf{Irred}(\mathcal{A}_X \rtimes_{\hat{\alpha}} \hat{\mathbf{Z}})$, where $\hat{\alpha}$ is an automorphism of $\mathcal{A}_X$ and **Irred** is the space of all irreducible representations of the crossed product C^*-algebra. The formula $X \cong \mathbf{Irred}(\mathcal{A}_X \rtimes_{\hat{\alpha}} \hat{\mathbf{Z}})$ is illustrated for $\mathcal{A}_X \cong \mathcal{A}_{RM}$.

5.3.1 Serre C^*-algebras

Let X be a projective scheme over a field k, and let $\mathcal{L}$ be the invertible sheaf $\mathcal{O}_X(1)$ of linear forms on X. Recall that the homogeneous coordinate ring of X is a graded k-algebra, which is isomorphic to the algebra

$$B(X, \mathcal{L}) = \bigoplus_{n \geq 0} H^0(X, \mathcal{L}^{\otimes n}).$$

Denote by **Coh** the category of quasi-coherent sheaves on a scheme X and by **Mod** the category of graded left modules over a graded ring B. If $M = \oplus M_n$ and $M_n = 0$ for $n \gg 0$, then the graded module M is called *right bounded*. The direct limit $M = \lim M_\alpha$ is called a *torsion* if each M_α is a right-bounded graded module. Denote by **Tors** the full subcategory of **Mod** of the torsion modules. The following result is the fundamental fact about the graded ring $B = B(X, \mathcal{L})$.

Theorem 5.3.1 ([249]).

$$\mathbf{Mod}(B(X, \mathcal{L}))/\mathbf{Tors} \cong \mathbf{Coh}(X).$$

Definition 5.3.1. Let α be an automorphism of a projective scheme X; the pullback of sheaf $\mathcal{L}$ along α will be denoted by $\mathcal{L}^\alpha$, i. e., $\mathcal{L}^\alpha(U) := \mathcal{L}(\alpha U)$ for every $U \subset X$. We shall set

$$B(X, \mathcal{L}, \alpha) = \bigoplus_{n \geq 0} H^0(X, \mathcal{L} \otimes \mathcal{L}^\alpha \otimes \cdots \otimes \mathcal{L}^{\alpha^n}).$$

The multiplication of sections is defined by the rule

$$ab = a \otimes b^{\alpha^m},$$

whenever $a \in B_m$ and $b \in B_n$. Given a pair (X, α) consisting of a Noetherian scheme X and an automorphism α of X, an invertible sheaf $\mathcal{L}$ on X is called α-ample, if for every coherent sheaf $\mathcal{F}$ on X, the cohomology group $H^q(X, \mathcal{L} \otimes \mathcal{L}^\alpha \otimes \cdots \otimes \mathcal{L}^{\alpha^{n-1}} \otimes \mathcal{F})$ vanishes for $q > 0$ and $n \gg 0$. (Notice that if α is trivial, this definition is equivalent to the usual definition of ample invertible sheaf, see [249].) If $\alpha : X \to X$ is an automorphism of a projective scheme X over k and $\mathcal{L}$ is an α-ample invertible sheaf on X, then $B(X, \mathcal{L}, \alpha)$ is called a twisted homogeneous coordinate ring of X.

Theorem 5.3.2 ([9]).

$$\mathbf{Mod}(B(X,\mathcal{L},\alpha))/\mathbf{Tors} \cong \mathbf{Coh}(X).$$

Remark 5.3.1. Theorem 5.3.2 extends Theorem 5.3.1 to the noncommutative rings; hence the name for ring $B(X,\mathcal{L},\alpha)$. The question of which invertible sheaves are α-ample is fairly subtle, and there is no characterization of the automorphisms α for which such an invertible sheaf exists. However, in many important special cases this problem is solvable, see [9, Corollary 1.6].

Remark 5.3.2. In practice, any twisted homogeneous coordinate ring $B(X,\mathcal{L},\alpha)$ of a projective scheme X can be constructed as follows. Let R be a commutative graded ring such that $X = \mathrm{Spec}(R)$. Consider the ring $B(X,\mathcal{L},\alpha) := R[t,t^{-1};\alpha]$, where $R[t,t^{-1};\alpha]$ is the ring of skew Laurent polynomials defined by the commutation relation $b^{\alpha}t = tb$, for all $b \in R$; here $b^{\alpha} \in R$ is the image of b under automorphism α. The ring $B(X,\mathcal{L},\alpha)$ satisfies the isomorphism $\mathbf{Mod}(B(X,\mathcal{L},\alpha))/\mathbf{Tors} \cong \mathbf{Coh}(X)$, i. e., is the twisted homogeneous coordinate ring of projective scheme X, see Lemma 5.3.1.

Example 5.3.1. Let k be a field and $U_{\infty}(k)$ the algebra of polynomials over k in two noncommuting variables x_1 and x_2, and a quadratic relation $x_1x_2-x_2x_1-x_1^2 = 0$; let $\mathbb{P}^1(k)$ be the projective line over k. Then $B(X,\mathcal{L},\alpha) = U_{\infty}(k)$ and $X = \mathbb{P}^1(k)$ satisfy equation $\mathbf{Mod}(B(X,\mathcal{L},\alpha))/\mathbf{Tors} \cong \mathbf{Coh}(X)$. The ring $U_{\infty}(k)$ corresponds to the automorphism $\alpha(u) = u+1$ of the projective line $\mathbb{P}^1(k)$. Indeed, $u = x_2x_1^{-1} = x_1^{-1}x_2$ and, therefore, α maps x_2 to $x_1 + x_2$; if one substitutes $t = x_1, b = x_2$ and $b^{\alpha} = x_1 + x_2$ into equation $b^{\alpha}t = tb$ (see Remark 5.3.2), then one gets the defining relation $x_1x_2 - x_2x_1 - x_1^2 = 0$ for the algebra $U_{\infty}(k)$.

To get a C^*-algebra from the ring $B(X,\mathcal{L},\alpha)$, we shall consider infinite-dimensional representations of $B(X,\mathcal{L},\alpha)$ by bounded linear operators on a Hilbert space $\mathcal{H}$; as usual, let $\mathcal{B}(\mathcal{H})$ stay for the algebra of all bounded linear operators on $\mathcal{H}$. For a ring of skew Laurent polynomials $R[t,t^{-1};\alpha]$ described in Remark 5.3.2, we shall consider a homomorphism

$$\rho : R[t,t^{-1};\alpha] \longrightarrow \mathcal{B}(\mathcal{H}).$$

Recall that algebra $\mathcal{B}(\mathcal{H})$ is endowed with a $*$-involution; such an involution is the adjoint with respect to the scalar product on the Hilbert space $\mathcal{H}$.

Definition 5.3.2. The representation ρ will be called $*$-coherent if:
(i) $\rho(t)$ and $\rho(t^{-1})$ are unitary operators such that $\rho^*(t) = \rho(t^{-1})$;
(ii) for all $b \in R$, it holds that $(\rho^*(b))^{\alpha(\rho)} = \rho^*(b^{\alpha})$, where $\alpha(\rho)$ is an automorphism of $\rho(R)$ induced by α.

Example 5.3.2. The ring $U_\infty(k)$ has no $*$-coherent representations. Indeed, involution acts on the generators of $U_\infty(k)$ by the formula $x_1^* = x_2$; the latter does not preserve the defining relation $x_1x_2 - x_2x_1 - x_1^2 = 0$.

Definition 5.3.3. By a Serre C^*-algebra $\mathcal{A}_X$ of the projective scheme X one understands the norm closure of an $*$-coherent representation $\rho(B(X, \mathcal{L}, \alpha))$ of the twisted homogeneous coordinate ring $B(X, \mathcal{L}, \alpha) \cong R[t, t^{-1}; \alpha]$ of scheme X.

Example 5.3.3. If $X \cong \mathcal{E}_\tau$ is an elliptic curve, then the ring $R[t, t^{-1}; \alpha]$ is isomorphic to the Sklyanin algebra; for such algebras, there exists a $*$-coherent representation; see Section 5.1.1. The resulting Serre C^*-algebra $\mathcal{A}_X \cong \mathcal{A}_\theta$, where $\mathcal{A}_\theta$ is the noncommutative torus.

Theorem 5.3.3 (Functor on complex projective varieties). *If* **Proj-Alg** *is the category of all complex projective varieties X (of dimension n) and* **C*-Serre** *the category of all Serre C^*-algebras $\mathcal{A}_X$, then the formula $X \mapsto \mathcal{A}_X$ gives rise to a map*

$$F : \textbf{Proj-Alg} \longrightarrow \textbf{C*-Serre}.$$

The map F is a functor which takes isomorphisms between projective varieties to the Morita equivalences (stable isomorphisms) between the corresponding Serre C^*-algebras.

Proof. The proof repeats the argument for elliptic and algebraic curves and is left to the reader. □

Let $\textbf{Spec}(B(X, \mathcal{L}))$ be the space of all prime ideals of the commutative homogeneous coordinate ring $B(X, \mathcal{L})$ of a complex projective variety X [9]. To get an analog of the classical formula

$$X \cong \textbf{Spec}(B(X, \mathcal{L}))$$

for the Serre C^*-algebras $\mathcal{A}_X$, we shall recall that for each continuous homomorphism $\alpha : G \to \mathrm{Aut}(\mathcal{A})$ of a locally compact group G into the group of automorphisms of a C^*-algebra $\mathcal{A}$, there exists a crossed product C^*-algebra $\mathcal{A} \rtimes_\alpha G$; we refer the reader to Section 3.2. Let $G = \mathbf{Z}$ and let $\hat{\mathbf{Z}} \cong S^1$ be its Pontryagin dual. We shall write **Irred** for the set of all irreducible representations of given C^*-algebra.

Theorem 5.3.4. *For each Serre C^*-algebra $\mathcal{A}_X$ there exists an $\hat{\alpha} \in \mathrm{Aut}(\mathcal{A}_X)$ such that*

$$X \cong \textbf{Irred}(\mathcal{A}_X \rtimes_{\hat{\alpha}} \hat{\mathbf{Z}}).$$

5.3.2 Proof of Theorem 5.3.4

Lemma 5.3.1. *$B(X, \mathcal{L}, \alpha) \cong R[t, t^{-1}; \alpha]$, where $X = \mathbf{Spec}(R)$.*

Proof. Let us write the twisted homogeneous coordinate ring $B(X, \mathcal{L}, \alpha)$ of projective variety X in the following form:

$$B(X, \mathcal{L}, \alpha) = \bigoplus_{n \geq 0} H^0(X, \mathfrak{B}_n),$$

where $\mathfrak{B}_n = \mathcal{L} \otimes \mathcal{L}^{\alpha} \otimes \cdots \otimes \mathcal{L}^{\alpha^n}$ and $H^0(X, \mathfrak{B}_n)$ is the zero sheaf cohomology of X, i. e., the space of sections $\Gamma(X, \mathfrak{B}_n)$; compare with [9, formula (3.5)]. If one denotes by $\mathcal{O}$ the structure sheaf of X, then

$$\mathfrak{B}_n = \mathcal{O}t^n$$

can be interpreted as a free left $\mathcal{O}$-module of rank one with basis $\{t^n\}$, see [9, p. 252]. Recall that spaces $B_i = H^0(X, \mathfrak{B}_i)$ have been endowed with the multiplication rule between the sections $a \in B_m$ and $b \in B_n$, see Definition 5.3.1; such a rule translates into the formula

$$at^m bt^n = ab^{\alpha^m} t^{m+n}.$$

One can eliminate a and t^n on both sides of the above equation; this operation gives us the following equation:

$$t^m b = b^{\alpha^m} t^m.$$

First notice that our ring $B(X, \mathcal{L}, \alpha)$ contains a commutative subring R such that $\mathbf{Spec}(R) = X$. Indeed, let $m = 0$ in formula $t^m b = b^{\alpha^m} t^m$; then $b = b^{\mathrm{Id}}$ and, thus, $\alpha = \mathrm{Id}$. We conclude therefore, that $R = B_0$ is a commutative subring of $B(X, \mathcal{L}, \alpha)$, and $\mathbf{Spec}(R) = X$.

Let us show that equations $b^{\alpha} t = tb$ of Remark 5.3.2 and $t^m b = b^{\alpha^m} t^m$ are equivalent. First, let us show that $b^{\alpha} t = tb$ implies $t^m b = b^{\alpha^m} t^m$. Indeed, equation $b^{\alpha} t = tb$ can be written as $b^{\alpha} = tbt^{-1}$. Then

$$\begin{cases} b^{\alpha^2} = tb^{\alpha} t^{-1} = t^2 bt^{-2}, \\ b^{\alpha^3} = tb^{\alpha^2} t^{-1} = t^3 bt^{-3}, \\ \vdots \\ b^{\alpha^m} = tb^{\alpha^{m-1}} t^{-1} = t^m bt^{-m}. \end{cases}$$

The last equation of the above system is equivalent to equation $t^m b = b^{\alpha^m} t^m$. The converse is evident; one sets $m = 1$ in $t^m b = b^{\alpha^m} t^m$ and obtains equation $b^{\alpha} t = tb$. Thus,

$b^{\alpha}t = tb$ and $t^m b = b^{\alpha^m} t^m$ are equivalent equations. It is easy now to establish an isomorphism $B(X, \mathcal{L}, \alpha) \cong R[t, t^{-1}; \alpha]$. For that, take $b \in R \subset B(X, \mathcal{L}, \alpha)$; then $B(X, \mathcal{L}, \alpha)$ coincides with the ring of the skew Laurent polynomials $R[t, t^{-1}; \alpha]$, since the commutation relation $b^{\alpha}t = tb$ is equivalent to equation $t^m b = b^{\alpha^m} t^m$. Lemma 5.3.1 follows. □

Lemma 5.3.2. *$\mathcal{A}_X \cong C(X) \rtimes_{\alpha} \mathbf{Z}$, where $C(X)$ is the C^*-algebra of all continuous complex-valued functions on X and α is a $*$-coherent automorphism of X.*

Proof. By definition of the Serre algebra $\mathcal{A}_X$, the ring of skew Laurent polynomials $R[t, t^{-1}; \alpha]$ is dense in $\mathcal{A}_X$; roughly speaking, one has to show that this property defines a crossed product structure on $\mathcal{A}_X$. We shall proceed in the following steps.

(i) Recall that $R[t, t^{-1}; \alpha]$ consists of the finite sums

$$\sum b_k t^k, \quad b_k \in R,$$

subject to the commutation relation

$$b_k^{\alpha} t = t b_k.$$

Because of the $*$-coherent representation, there is also an involution on $R[t, t^{-1}; \alpha]$, subject to the following rules:

$$\begin{cases} \text{(a)} & t^* = t^{-1}, \\ \text{(b)} & (b_k^*)^{\alpha} = (b_k^{\alpha})^*. \end{cases}$$

(ii) Following [287, p. 47], we shall consider the set $C_c(\mathbf{Z}, R)$ of continuous functions from $\mathbf{Z}$ to R having a compact support; then the finite sums can be viewed as elements of $C_c(\mathbf{Z}, R)$ via the identification

$$k \mapsto b_k.$$

It can be verified that multiplication operation of the finite sums translates into a convolution product of functions $f, g \in C_c(\mathbf{Z}, R)$ given by the formula

$$(fg)(k) = \sum_{l \in \mathbf{Z}} f(l) t^l g(k - l) t^{-l},$$

while involution translates into an involution on $C_c(\mathbf{Z}, R)$ given by the formula

$$f^*(k) = t^k f^*(-k) t^{-k}.$$

It is easy to see that the multiplication given by the convolution product and involution turn $C_c(\mathbf{Z}, R)$ into an $*$-algebra, which is isomorphic to the algebra $R[t, t^{-1}; \alpha]$.

(iii) There exists a standard construction of a norm on $C_c(\mathbf{Z}, R)$; we omit it here referring the reader to [287, Section 2.3]. The completion of $C_c(\mathbf{Z}, R)$ in that norm defines a crossed product C^*-algebra $R \rtimes_\alpha \mathbf{Z}$ [287, Lemma 2.27].

(iv) Since R is a commutative C^*-algebra and $X = \mathbf{Spec}(R)$, one concludes that $R \cong C(X)$. Thus, one obtains $\mathcal{A}_X = C(X) \rtimes_\alpha \mathbf{Z}$. Lemma 5.3.2 follows. □

Remark 5.3.3. It is easy to prove that equations $b_k^\alpha t = tb_k$ and $t^* = t^{-1}$ imply equation $(b_k^*)^\alpha = (b_k^\alpha)^*$; in other words, if involution does not commute with automorphism α, representation ρ cannot be unitary, i. e., $\rho^*(t) \neq \rho(t^{-1})$.

Lemma 5.3.3. *There exists* $\hat{\alpha} \in \mathrm{Aut}(\mathcal{A}_X)$ *such that*

$$X \cong \mathbf{Irred}(\mathcal{A}_X \rtimes_{\hat{\alpha}} \hat{\mathbf{Z}}).$$

Proof. The above formula is an implication of the Takai duality for the crossed products, see, e. g., [287, Section 7.1]; for the sake of clarity, we shall repeat this construction. Let (A, G, α) be a C^*-dynamical system with G locally compact abelian group; let $\hat{G}$ be the dual of G. For each $\gamma \in \hat{G}$, one can define a map $\hat{\alpha}_\gamma : C_c(G, A) \to C_c(G, A)$ given by the formula

$$\hat{\alpha}_\gamma(f)(s) = \bar{\gamma}(s) f(s), \quad \forall s \in G.$$

In fact, $\hat{\alpha}_\gamma$ is a $*$-homomorphism, since it respects the convolution product and involution on $C_c(G, A)$ [287]. Because the crossed product $A \rtimes_\alpha G$ is the closure of $C_c(G, A)$, one gets an extension of $\hat{\alpha}_\gamma$ to an element of $\mathrm{Aut}(A \rtimes_\alpha G)$ and, therefore, a homomorphism

$$\hat{\alpha} : \hat{G} \to \mathrm{Aut}(A \rtimes_\alpha G).$$

The *Takai duality* asserts that

$$(A \rtimes_\alpha G) \rtimes_{\hat{\alpha}} \hat{G} \cong A \otimes \mathcal{K}(L^2(G)),$$

where $\mathcal{K}(L^2(G))$ is the algebra of compact operators on the Hilbert space $L^2(G)$. Let us substitute $A = C_0(X)$ and $G = \mathbf{Z}$ in the above equation; one gets the following isomorphism:

$$(C_0(X) \rtimes_\alpha \mathbf{Z}) \rtimes_{\hat{\alpha}} \hat{\mathbf{Z}} \cong C_0(X) \otimes \mathcal{K}(L^2(\mathbf{Z})).$$

Lemma 5.3.2 asserts that $C_0(X) \rtimes_\alpha \mathbf{Z} \cong \mathcal{A}_X$; therefore, one arrives at the following isomorphism:

$$\mathcal{A}_X \rtimes_{\hat{\alpha}} \hat{\mathbf{Z}} \cong C_0(X) \otimes \mathcal{K}(L^2(\mathbf{Z})).$$

Consider the set of all irreducible representations of the C^*-algebras in the above equation; then one gets the following equality of representations:

$$\mathbf{Irred}(\mathcal{A}_X \rtimes_{\hat{\alpha}} \hat{\mathbf{Z}}) = \mathbf{Irred}(C_0(X) \otimes \mathcal{K}(L^2(\mathbf{Z}))).$$

Let π be a representation of the tensor product $C_0(X) \otimes \mathcal{K}(L^2(\mathbf{Z}))$ on the Hilbert space $\mathcal{H} \otimes L^2(\mathbf{Z})$; then $\pi = \varphi \otimes \psi$, where $\varphi : C_0(X) \to \mathcal{B}(\mathcal{H})$ and $\psi : \mathcal{K} \to \mathcal{B}(L^2(\mathbf{Z}))$. It is known that the only irreducible representation of the algebra of compact operators is the identity representation. Thus, one gets

$$\mathbf{Irred}(C_0(X) \otimes \mathcal{K}(L^2(\mathbf{Z}))) = \mathbf{Irred}(C_0(X)) \otimes \{pt\} = \mathbf{Irred}(C_0(X)).$$

Further, the C^*-algebra $C_0(X)$ is commutative, hence the following equations are true:

$$\mathbf{Irred}(C_0(X)) = \mathbf{Spec}(C_0(X)) = X.$$

Putting together the last three equations, one obtains

$$\mathbf{Irred}(\mathcal{A}_X \rtimes_{\hat{\alpha}} \hat{\mathbf{Z}}) \cong X.$$

The conclusion of Lemma 5.3.3 follows from the above equation. □

Theorem 5.3.4 follows from Lemma 5.3.3.

5.3.3 Example

We illustrate Theorem 5.3.4 for $\mathcal{A}_X \cong \mathcal{A}_{RM}$, i. e., a noncommutative torus with real multiplication; notice that $\mathcal{A}_{RM}$ is the Serre C^*-algebra, see Example 5.3.3.

Theorem 5.3.5.

$$\mathbf{Irred}(\mathcal{A}_{RM} \rtimes_{\hat{\alpha}} \hat{\mathbf{Z}}) \cong \mathcal{E}(K),$$

where $\mathcal{E}(K)$ is a nonsingular elliptic curve defined over a field of algebraic numbers K.

Proof. We shall view the crossed product $\mathcal{A}_{RM} \rtimes_{\hat{\alpha}} \hat{\mathbf{Z}}$ as a C^*-dynamical system $(\mathcal{A}_{RM}, \hat{\mathbf{Z}}, \hat{\alpha})$, see [287] for the details. Recall that the irreducible representations of C^*-dynamical system $(\mathcal{A}_{RM}, \hat{\mathbf{Z}}, \hat{\alpha})$ are in the one-to-one correspondence with the minimal sets of the dynamical system (i. e., closed $\hat{\alpha}$-invariant sub-C^*-algebras of $\mathcal{A}_{RM}$ not containing a smaller object with the same property). To calculate the minimal sets of $(\mathcal{A}_{RM}, \hat{\mathbf{Z}}, \hat{\alpha})$, let θ be quadratic irrationality such that $\mathcal{A}_{RM} \cong \mathcal{A}_\theta$. It is known that every nontrivial sub-C^*-algebra of $\mathcal{A}_\theta$ has the form $\mathcal{A}_{n\theta}$ for some positive integer n, see [238, p. 419]. It is easy to deduce that the *maximal* proper sub-C^*-algebra of $\mathcal{A}_\theta$ has the form $\mathcal{A}_{p\theta}$, where p is a prime number. (Indeed, each composite $n = n_1 n_2$ cannot be maximal since $\mathcal{A}_{n_1n_2\theta} \subset \mathcal{A}_{n_1\theta} \subset \mathcal{A}_\theta$ or $\mathcal{A}_{n_1n_2\theta} \subset \mathcal{A}_{n_2\theta} \subset \mathcal{A}_\theta$, where all inclusions are strict.) We claim that $(\mathcal{A}_{p\theta}, \hat{\mathbf{Z}}, \hat{\alpha}^{\pi(p)})$ is the minimal C^*-dynamical system, where $\pi(p)$ is certain power of the automorphism $\hat{\alpha}$. Indeed, the automorphism $\hat{\alpha}$ of $\mathcal{A}_\theta$ corresponds to multiplication by the fundamental unit, ε, of pseudo-lattice $\Lambda = \mathbf{Z} + \theta\mathbf{Z}$. It is known that certain power, $\pi(p)$, of ε coincides with the fundamental unit of pseudo-lattice $\mathbf{Z} + (p\theta)\mathbf{Z}$, see, e. g., [111, p. 298]. Thus one gets the minimal C^*-dynamical system $(\mathcal{A}_{p\theta}, \hat{\mathbf{Z}}, \hat{\alpha}^{\pi(p)})$, which is defined on the sub-C^*-algebra $\mathcal{A}_{p\theta}$ of $\mathcal{A}_\theta$. Therefore we have an isomorphism

$$\mathbf{Irred}(\mathcal{A}_{RM} \rtimes_{\hat{\alpha}} \hat{\mathbf{Z}}) \cong \bigcup_{p \in \mathcal{P}} \mathbf{Irred}(\mathcal{A}_{p\theta} \rtimes_{\hat{\alpha}^{\pi(p)}} \hat{\mathbf{Z}}),$$

where $\mathcal{P}$ is the set of all (but a finite number) of primes. To simplify the RHS of the above equation, let us introduce some notation. Recall that matrix form of the fundamental unit ε of pseudo-lattice Λ coincides with the matrix A, see above. For each prime $p \in \mathcal{P}$, consider the matrix

$$L_p = \begin{pmatrix} \mathrm{tr}(A^{\pi(p)}) - p & p \\ \mathrm{tr}(A^{\pi(p)}) - p - 1 & p \end{pmatrix},$$

where tr is the trace of matrix. Let us show that

$$\mathcal{A}_{p\theta} \rtimes_{\hat{\alpha}^{\pi(p)}} \hat{\mathbf{Z}} \cong \mathcal{A}_\theta \rtimes_{L_p} \hat{\mathbf{Z}},$$

where L_p is an endomorphism of $\mathcal{A}_\theta$ (of degree p) induced by matrix L_p. Indeed, because $\deg(L_p) = p$ the endomorphism L_p maps pseudo-lattice $\Lambda = \mathbf{Z} + \theta\mathbf{Z}$ to a sublattice of index p; any such can be written in the form $\Lambda_p = \mathbf{Z} + (p\theta)\mathbf{Z}$, see, e. g., [38, p. 131]. Notice that pseudo-lattice Λ_p corresponds to the sub-C^*-algebra $\mathcal{A}_{p\theta}$ of algebra $\mathcal{A}_\theta$ and L_p induces a shift automorphism of $\mathcal{A}_{p\theta}$, see, e. g., [62] the beginning of Section 2.1 for terminology and details of this construction. It is not hard to see that the shift automorphism coincides with $\hat{\alpha}^{\pi(p)}$. Indeed, it is verified directly that $\mathrm{tr}(\hat{\alpha}^{\pi(p)}) = \mathrm{tr}(A^{\pi(p)}) = \mathrm{tr}(L_p)$; thus one gets a bijection between powers of $\hat{\alpha}^{\pi(p)}$ and such of L_p. But $\hat{\alpha}^{\pi(p)}$ corresponds to the fundamental unit of pseudo-lattice Λ_p; therefore the shift automorphism induced by L_p must coincide with $\hat{\alpha}^{\pi(p)}$. The required isomorphism is proved and, therefore, our last formula can be written in the form

$$\mathbf{Irred}(\mathcal{A}_{RM} \rtimes_{\hat{\alpha}} \hat{\mathbf{Z}}) \cong \bigcup_{p \in \mathcal{P}} \mathbf{Irred}(\mathcal{A}_{RM} \rtimes_{L_p} \hat{\mathbf{Z}}).$$

To calculate irreducible representations of the crossed product C^*-algebra $\mathcal{A}_{RM} \rtimes_{L_p} \hat{\mathbf{Z}}$ at the RHS of the above equation, recall that such are in a one-to-one correspondence with the set of invariant measures on a subshift of finite type given by the positive integer matrix L_p, see [40] and [62]; the measures make an abelian group under the addition operation. Such a group is isomorphic to $\mathbf{Z}^2/(I-L_p)\mathbf{Z}^2$, where I is the identity matrix, see [40, Theorem 2.2]. Therefore our last equation can be written in the form

$$\mathbf{Irred}(\mathcal{A}_{RM} \rtimes_{\hat{\alpha}} \hat{\mathbf{Z}}) \cong \bigcup_{p\in\mathcal{P}} \frac{\mathbf{Z}^2}{(I-L_p)\mathbf{Z}^2}.$$

Let $\mathcal{E}(K)$ be a nonsingular elliptic curve defined over the algebraic number field K; let $\mathcal{E}(\mathbb{F}_p)$ be the reduction of $\mathcal{E}(K)$ modulo prime ideal over a "good" prime number p. Recall that $|\mathcal{E}(\mathbb{F}_p)| = \det(I - \mathrm{Fr}_p)$, where Fr_p is an integer two-by-two matrix corresponding to the action of Frobenius endomorphism on the ℓ-adic cohomology of $\mathcal{E}(K)$, see, e. g., [274, p. 187]. Since $|\mathbf{Z}^2/(I-L_p)\mathbf{Z}^2| = \det(I-L_p)$, one can identify Fr_p and L_p and, therefore, one obtains an isomorphism $\mathcal{E}(\mathbb{F}_p) \cong \mathbf{Z}^2/(I-L_p)\mathbf{Z}^2$. Thus our equation can be written in the form

$$\mathbf{Irred}(\mathcal{A}_{RM} \rtimes_{\hat{\alpha}} \hat{\mathbf{Z}}) \cong \bigcup_{p\in\mathcal{P}} \mathcal{E}(\mathbb{F}_p).$$

Finally, consider an arithmetic scheme X corresponding to $\mathcal{E}(K)$; the latter fibers over $\mathbf{Z}$, see [268, Example 4.2.2] for the details. It can be immediately seen that the RHS of our last equation coincides with the scheme X, where the regular fiber over p corresponds to $\mathcal{E}(\mathbb{F}_p)$ [268]. The argument finishes the proof of Theorem 5.3.5. □

Guide to the literature

The standard reference to complex projective varieties is the monograph [108]. Twisted homogeneous coordinate rings of projective varieties are covered in the excellent survey [265]. The Serre C^*-algebras were introduced in [187, 206].

5.4 Tate curves and UHF-algebras

We study a link between geometry of the Tate curves and the class of the UHF-algebras. Namely, we prove that a stabilization of the norm closure of a self-adjoint representation of the twisted homogeneous coordinate ring of a Tate curve contains a copy of the UHF-algebra.

5.4.1 Elliptic curve over p-adic numbers

A *Tate curve* $\mathcal{E}_q$ is an affine cubic over the field $\mathbf{Q}_p$ of p-adic numbers,

$$y^2 + xy = x^3 - \left(5\sum_{n=1}^{\infty}\frac{n^3q^n}{1-q^n}\right)x - \frac{1}{12}\sum_{n=1}^{\infty}\frac{(5n^3+7n^5)q^n}{1-q^n},$$

where q is a p-adic integer satisfying condition $0 < |q| < 1$. If $q^{\mathbf{Z}} = \{q^n : n \in \mathbf{Z}\}$ is a lattice and $\mathbf{Q}_p^*$ is the group of units of $\mathbf{Q}_p$, then the action $x \mapsto qx$ is discrete; in particular, the quotient $\mathbf{Q}_p^*/q^{\mathbf{Z}}$ is a Hausdorff topological space. It was proved by Tate that there exists an (analytic) isomorphism $\phi : \mathbf{Q}_p^*/q^{\mathbf{Z}} \to \mathcal{E}_q$ [274, p. 190].

On the other hand, it is known that each p-adic integer $0 < |q| < 1$ is the limit of a convergent series of the rational integers $\alpha_k = \sum_{i=1}^k b_i p^i$, where $0 \le b_i \le p-1$ and $q = \sum_{i=1}^\infty b_i p^i$, see, e. g., [99, p. 66] for the details. For each α_k, one can define a supernatural number $\mathbf{n}(\alpha_k)$ of the form $p_1^\infty \cdots p_s^\infty$, where $\mathcal{P}_k := \{p_1, \dots, p_s\}$ the finite set of all primes dividing α_k. By M_{α_k} we shall understand (the stable isomorphism class of) an UHF-algebra such that $K_0(M_{\alpha_k}) \cong \mathbf{Q}(\mathbf{n}(\alpha_k))$. Let $\{\pi_1, \pi_2, \dots\}$ be a (finite or infinite) set of all primes, such that $\pi_j \in \bigcup_{k=1}^\infty \mathcal{P}_k$. By $\mathbf{n}(q)$ we shall understand a supernatural number of the form $\pi_1^\infty \pi_2^\infty \cdots$ and by M_q an UHF-algebra, such that $K_0(M(q)) \cong \mathbf{Q}(\mathbf{n}(q))$; in other words, M_q is the smallest UHF-algebra containing all the UHF-algebras M_{α_k}. Our main result can be stated as follows.

Theorem 5.4.1. *A stabilization of the norm closure of a self-adjoint representation of the (quotient of) twisted homogeneous coordinate ring of the Tate curve $\mathcal{E}_q$ contains a copy the UHF-algebra M_q.*

5.4.2 Proof of Theorem 5.4.1

Let the ground field k be the complex numbers $\mathbf{C}$. We shall split the proof in a series of lemmas starting with the following elementary

Lemma 5.4.1. *The ideal of free algebra $\mathbf{C}\langle x_1, x_2, x_3, x_4\rangle$, generated by equations*

$$\begin{cases} x_1x_2 - x_2x_1 = \alpha(x_3x_4 + x_4x_3), \\ x_1x_2 + x_2x_1 = x_3x_4 - x_4x_3, \\ x_1x_3 - x_3x_1 = \beta(x_4x_2 + x_2x_4), \\ x_1x_3 + x_3x_1 = x_4x_2 - x_2x_4, \\ x_1x_4 - x_4x_1 = \gamma(x_2x_3 + x_3x_2), \\ x_1x_4 + x_4x_1 = x_2x_3 - x_3x_2, \end{cases}$$

is invariant under involution $x_1^* = x_2, x_3^* = x_4$ if and only if $\bar{\alpha} = \alpha, \beta = 1$ and $\gamma = -1$.

Proof. (i) Let us consider the first two equations above; this pair is invariant of the involution. Indeed, by the rules of composition for an involution, we have

$$\begin{cases} (x_1x_2)^* = x_2^*x_1^* = x_1x_2, \\ (x_2x_1)^* = x_1^*x_2^* = x_2x_1, \\ (x_3x_4)^* = x_4^*x_3^* = x_3x_4, \\ (x_4x_3)^* = x_3^*x_4^* = x_4x_3. \end{cases}$$

Since $\alpha^* = \bar{\alpha} = \alpha$, the first two equations remain invariant of the involution.

(ii) Let us consider the middle pair of the above equations; by the rules of composition for an involution, we get

$$\begin{cases} (x_1x_3)^* = x_3^*x_1^* = x_4x_2, \\ (x_3x_1)^* = x_1^*x_3^* = x_2x_4, \\ (x_2x_4)^* = x_4^*x_2^* = x_3x_1, \\ (x_4x_2)^* = x_2^*x_4^* = x_1x_3. \end{cases}$$

One can apply the involution to the first equation $x_1x_3 - x_3x_1 = \beta(x_4x_2 + x_2x_4)$; then one gets $x_4x_2 - x_2x_4 = \bar{\beta}(x_1x_3 + x_3x_1)$. But the second equation says that $x_1x_3 + x_3x_1 = x_4x_2 - x_2x_4$; the last two equations are compatible if and only if $\bar{\beta} = 1$. Thus, $\beta = 1$.

The second equation in involution writes as $x_4x_2 + x_2x_4 = x_1x_3 - x_3x_1$; the last equation coincides with the first equation for $\beta = 1$.

Therefore, $\beta = 1$ is necessary and sufficient for invariance of the middle pair of equations with respect to the involution.

(iii) Let us consider the last pair of equations; by the rules of composition for an involution, we obtain

$$\begin{cases} (x_1x_4)^* = x_4^*x_1^* = x_3x_2, \\ (x_4x_1)^* = x_1^*x_4^* = x_2x_3, \\ (x_2x_3)^* = x_3^*x_2^* = x_4x_1, \\ (x_3x_2)^* = x_2^*x_3^* = x_1x_4. \end{cases}$$

One can apply the involution to the first equation $x_1x_4 - x_4x_1 = \gamma(x_2x_3 + x_3x_2)$; then one gets $x_3x_2 - x_2x_3 = \bar{\gamma}(x_4x_1 + x_1x_4)$. But the second equation says that $x_1x_4 + x_4x_1 = x_2x_3 - x_3x_2$; the last two equations are compatible if and only if $\bar{\gamma} = -1$. Thus, $\gamma = -1$.

The second equation in involution writes as $x_3x_2 + x_2x_3 = x_4x_1 - x_1x_4$; the last equation coincides with the first equation for $\gamma = -1$.

Therefore, $\gamma = -1$ is necessary and sufficient for invariance of the last pair of equations with respect to the involution.

(iv) It remains to verify that condition $\alpha + \beta + \gamma + \alpha\beta\gamma = 0$ is satisfied by $\beta = 1$ and $\gamma = -1$ for any $\alpha \in k$. Lemma 5.4.1 follows. □

Remark 5.4.1. The Sklyanin algebra $\mathfrak{S}_{\alpha,1,-1}(\mathbf{C})$ with $\alpha \in \mathbf{R}$ is a $*$-algebra with the involution $x_1^* = x_2$ and $x_3^* = x_4$.

Lemma 5.4.2. *The first pair of equations of Lemma 5.4.1,*

$$\begin{cases} x_1x_2 - x_2x_1 = \alpha(x_3x_4 + x_4x_3), \\ x_1x_2 + x_2x_1 = x_3x_4 - x_4x_3, \end{cases}$$

is equivalent to the pair

$$\begin{cases} x_2x_1 = \frac{1+\alpha}{1-\alpha}x_1x_2 - \frac{2\alpha}{1-\alpha}x_3x_4, \\ x_4x_3 = -\frac{2}{1-\alpha}x_1x_2 + \frac{1+\alpha}{1-\alpha}x_3x_4. \end{cases}$$

Proof. Let us isolate x_2x_1 and x_4x_3 in the above equations; for that, we shall write them in the form

$$\begin{cases} x_2x_1 + \alpha x_4x_3 = x_1x_2 - \alpha x_3x_4, \\ x_2x_1 + x_4x_3 = -x_1x_2 + x_3x_4. \end{cases}$$

Consider the above as a linear system of equations relative to x_2x_1 and x_4x_3; since $\alpha \neq 1$, it has a unique solution

$$\begin{cases} x_2x_1 = \frac{1}{1-\alpha}\begin{vmatrix} x_1x_2 - \alpha x_3x_4 & \alpha \\ -x_1x_2 + x_3x_4 & 1 \end{vmatrix} = \frac{1+\alpha}{1-\alpha}x_1x_2 - \frac{2\alpha}{1-\alpha}x_3x_4, \\ x_4x_3 = \frac{1}{1-\alpha}\begin{vmatrix} 1 & x_1x_2 - \alpha x_3x_4 \\ 1 & -x_1x_2 + x_3x_4 \end{vmatrix} = -\frac{2}{1-\alpha}x_1x_2 + \frac{1+\alpha}{1-\alpha}x_3x_4. \end{cases}$$

Lemma 5.4.2 follows. □

Remark 5.4.2. In new variables $(x_2x_1)' = (1-\alpha)x_2x_1$ and $(x_4x_3)' = (1-\alpha)x_4x_3$, the above equations can be written in the form

$$\begin{cases} x_2x_1 = (1+\alpha)x_1x_2 - 2\alpha x_3x_4, \\ x_4x_3 = -2x_1x_2 + (1+\alpha)x_3x_4. \end{cases}$$

Lemma 5.4.3 (Main lemma). *Let $\alpha_k = \sum_{i=1}^{k} b_ip^i$ be a rational integer and $A = \begin{pmatrix} 1+\alpha_k & -2\alpha_k \\ -2 & 1+\alpha_k \end{pmatrix}$. Suppose M_{α_k} is the UHF-algebra defined in Section 5.4.1 and $\mathbb{A}$ is a stationary AF-algebra given by the inductive limit $\mathbf{Z}^2 \xrightarrow{A^T} \mathbf{Z}^2 \xrightarrow{A^T} \cdots$, where A^T is the transpose of matrix A. Let I_0 be the (two-sided) ideal of the Sklyanin $*$-algebra $\mathfrak{S}_{\alpha_k,1,-1}(\mathbf{C})$ generated by relation $x_1x_2 + x_3x_4 = 1$ and J_0 the ideal of $\mathcal{O}_A^0$ generated by relations $x_4x_2 - x_1x_3 = x_3x_1 + x_2x_4 = x_4x_1 - x_2x_3 = x_3x_2 + x_1x_4 = 0$. Then there exists a $*$-isomorphism*

$$\mathfrak{S}_{\alpha_k,1,-1}(\mathbf{C})/I_0 \cong \mathcal{O}_A^0/J_0,$$

where

$$\overline{\mathcal{O}_A^0} \cong \mathcal{O}_A \quad and \quad \mathcal{O}_A \otimes \mathcal{K} \supset \mathbb{A} \supset M_{\alpha_k}$$

are inclusions of the C^*-algebras.

Proof. Recall that there exists a dense inclusion $\mathbf{Z} \hookrightarrow \mathbf{Z}_p$ given by formula $\alpha_k \mapsto \sum_{i=1}^k b_i p^i$, where $0 \le b_i \le p-1$ are integer numbers, see, e. g., [99, Proposition 3.3.4(ii)]; we shall use the inclusion to identify p-adic integer $\sum_{i=1}^k b_i p^i$ with the corresponding rational integer α_k. Because $\alpha_k \in \mathbf{R}$, one gets a Sklyanin $*$-algebra $\mathfrak{S}_{\alpha_k,1,-1}(\mathbf{C})$ with the involution $x_1^* = x_2$ and $x_3^* = x_4$, see Remark 5.4.1. Notice that the inclusion $\mathbf{Z} \hookrightarrow \mathbf{Z}_p$ induces a $*$-isomorphism between the following Sklyanin algebras:

$$\mathfrak{S}_{\alpha_k,\,1,\,-1}(\mathbf{C}) \cong \mathfrak{S}_{\sum_{i=1}^k b_i p^i,\,1,\,-1}(\mathbf{Q}_p).$$

However, the norm convergence in the above families of Sklyanin algebras is different; in what follows, we deal with the Sklyanin algebras $\mathfrak{S}_{\alpha_k,1,-1}(\mathbf{C})$ endowed with the usual archimedean norm.

Part I. To prove our formula, one compares relations defining the Sklyanin algebra $\mathfrak{S}_{\alpha,\beta,\gamma}(k)$ with relations defining dense subalgebra $\mathcal{O}_A^0$ of the Cuntz–Krieger algebra $\mathcal{O}_A$. It is easy to see that ideal J_0 is generated by the last four relations of corresponding to the case $\beta = -\gamma = 1$. Likewise, ideal I_0 is generated by the last relation of the system.

As for the first pair of relations, they are identical after a substitution $a_{11} = a_{22} = 1+\alpha_k$, $a_{12} = -2\alpha_k$, and $a_{21} = -2$, see also Lemma 5.4.2 and Remark 5.4.2. Thus, one gets the required isomorphism, where matrix A is given by the formula $A = \begin{pmatrix} 1+\alpha_k & -2\alpha_k \\ -2 & 1+\alpha_k \end{pmatrix}$.

Part II. One can prove inclusions stated in the lemma in the following steps.

(i) The isomorphism $\overline{\mathcal{O}_A^0} \cong \mathcal{O}_A$ follows from definition of the Cuntz–Krieger algebra as the norm closure of algebra $\mathcal{O}_A^0$.

(ii) In view of the isomorphism $\mathcal{O}_A \otimes \mathcal{K} \cong \mathbb{A} \rtimes_\alpha \mathbf{Z}$, there exists a sub-$C^*$-algebra $\mathbb{A} \subset \mathcal{O}_A \otimes \mathcal{K}$; the sub-$C^*$-algebra is the stationary AF-algebra (see [74, Chapter 6]) given by the following inductive limit:

$$\mathbf{Z}^2 \xrightarrow{\begin{pmatrix} 1+\alpha_k & -2 \\ -2\alpha_k & 1+\alpha_k \end{pmatrix}} \mathbf{Z}^2 \xrightarrow{\begin{pmatrix} 1+\alpha_k & -2 \\ -2\alpha_k & 1+\alpha_k \end{pmatrix}} \cdots$$

Notice that, since α_k are positive integers, matrix A^T has two negative off-diagonal entries. However, since $\mathrm{tr}(A^T) > 2$, there exists a matrix in the similarity class of A^T all of whose entries are positive; the above inductive limit is invariant of the similarity class.

(iii) To establish inclusion $M_{\alpha_k} \subset \mathbb{A}$, let us calculate dimension group of the AF-algebra $\mathbb{A}$. It is known that for stationary AF-algebra $\mathbb{A}$ the dimension group is order-isomorphic to $\mathbf{Z}[\frac{1}{\lambda_{A^T}}]$, where λ_{A^T} is the maximal eigenvalue of matrix A^T. We encourage the reader to verify that

$$\mathbf{Z}\left[\frac{1}{\lambda_{A^T}}\right] = \mathbf{Z}\left[\frac{1+\alpha_k+2\sqrt{\alpha_k}}{(\alpha_k-1)^2}\right].$$

It follows from the above that

$$\mathbf{Z}\left[\frac{1}{\alpha_k-1}\right] \subset \mathbf{Z}\left[\frac{1}{\lambda_{A^T}}\right]$$

is an inclusion of dimension groups. It is easy to see that one can replace $\alpha_k - 1$ by $-\alpha_k$ in equation of the elliptic curve defined by the Sklyanin algebra; therefore, one gets the inclusion

$$\mathbf{Z}\left[\frac{1}{\alpha_k}\right] \subset \mathbf{Z}\left[\frac{1}{\lambda_{A^T}}\right].$$

Because $\alpha_k \notin \{0; \pm 1\}$, one concludes that $\mathbf{Z}[\frac{1}{\alpha_k}]$ is a dense abelian subgroup of the rational numbers $\mathbf{Q}$. It remains to notice that the dimension group $\mathbf{Z}[\frac{1}{\alpha_k}]$ is order-isomorphic to such of the UHF-algebra M_{α_k}; see the definition of M_{α_k}. Thus the above inclusion implies the inclusion $M_{\alpha_k} \subset \mathbb{A}$. Lemma 5.4.3 follows. □

Theorem 5.4.1 follows from Lemma 5.4.3.

5.4.3 Example

We shall consider an example illustrating Theorem 5.4.1. Let p be a prime number and consider the p-adic integer of the form $q = p$; notice that in this case $b_1 = 1$ and $b_2 = b_3 = \cdots = 0$. One gets therefore a supernatural number $\mathbf{n}(q)$ of the form p^∞. The $\mathbf{n}(q)$ corresponds to a dense subgroup of $\mathbf{Q}$ of the form

$$\mathbf{Q}(\mathbf{n}) = \mathbf{Z}\left[\frac{1}{p}\right].$$

It is easy to see, that the UHF-algebra corresponding to the Tate curve $\mathcal{E}_p = \mathbf{Q}_p^*/p^{\mathbf{Z}}$ has the form

$$M_{p^\infty} := M_p(\mathbf{C}) \otimes M_p(\mathbf{C}) \otimes \cdots.$$

By Theorem 5.4.1, the UHF-algebra M_{p^∞} is (a sub-C^*-algebra of the stable closure of infinite-dimensional representation of quotient ring of) the twisted homogeneous coordinate ring of the Tate curve $\mathcal{E}_p$. In particular, for $\mathcal{E}_2$ such a coordinate ring is the

UHF-algebra M_{2^∞}; the latter is known as a Canonical Anticommutation Relations C^*-algebra (the CAR or Fermion algebra) [74, p. 13].

Guide to the literature

An excellent introduction to the p-numbers is written in [99]. Elliptic curves over the p-adic numbers are defined in [274]. The UHF-algebras were introduced in [94]. A link between the two was studied in [192]. An independent study of a related problem has been undertaken in [45].

5.5 Mapping class group

Such a group $\mathrm{Mod}(X)$ is defined as the group of isotopy classes of the orientation-preserving diffeomorphisms of a surface X of genus $g \geq 1$. The group is known to be prominent in algebraic geometry [104], topology [275], and dynamics [276]. The mapping class group of torus is isomorphic to the group $\mathrm{SL}_2(\mathbf{Z})$. For $g \geq 2$, a similar representation was conjectured in [110, p. 267]. Below we prove Harvey's conjecture by constructing a faithful representation of the $\mathrm{Mod}(X)$ into the matrix group $\mathrm{GL}_{6g-6}(\mathbf{Z})$ using functor $F : \textbf{Alg-Gen} \to \textbf{AF-Toric}$ of Theorem 5.2.1.

5.5.1 Harvey's conjecture

The group G is called *linear* if there exists a faithful representation of G into the matrix group $\mathrm{GL}(m, R)$, where R is a commutative ring. The braid groups are known to be linear [24]. Using a modification of the argument for the braid groups, it is possible to prove that $\mathrm{Mod}(X)$ is linear in the case $g = 2$ [25].

Definition 5.5.1 (Conjecture [110, p. 267]). Group $\mathrm{Mod}(X)$ is linear for all $g \geq 3$.

Recall that a covariant functor $F : \textbf{Alg-Gen} \to \textbf{AF-Toric}$ from a category of generic Riemann surfaces (i. e., complex algebraic curves) to a category of toric AF-algebras was constructed in Section 5.2; the functor maps any pair of isomorphic Riemann surfaces to a pair of Morita equivalent (stably isomorphic) toric AF-algebras. Since each isomorphism of Riemann surfaces is given by an element of $\mathrm{Mod}(X)$ [104], it is natural to ask about a representation of $\mathrm{Mod}(X)$ by the stable isomorphisms of toric AF-algebras. Recall that the stable isomorphisms of toric AF-algebras are well understood and surprisingly simple; provided the automorphism group of the algebra is trivial (this is true for a generic algebra), its group of stable isomorphism admits a faithful representation into the matrix group $\mathrm{GL}_m(\mathbf{Z})$, see, e. g., [74]. This fact, combined with the properties of functor F, implies a positive solution to the Harvey conjecture.

Theorem 5.5.1. *For every surface X of genus $g \geq 2$, there exists a faithful representation* $\rho : \mathrm{Mod}(X) \to \mathrm{GL}_{6g-6}(\mathbf{Z})$.

5.5.2 Proof of Theorem 5.5.1

Let **AF-Toric** denote the set of all toric AF-algebras of genus $g \geq 2$. Let G be a finitely presented group and

$$G \times \textbf{AF-Toric} \longrightarrow \textbf{AF-Toric}$$

be its action on **AF-Toric** by the stable isomorphisms (Morita equivalences) of toric AF-algebras; in other words, $\gamma(\mathbb{A}_\theta) \otimes \mathcal{K} \cong \mathbb{A}_\theta \otimes \mathcal{K}$ for all $\gamma \in G$ and all $\mathbb{A}_\theta \in$ **AF-Toric.** The following preparatory lemma will be important.

Lemma 5.5.1. *For each $\mathbb{A}_\theta \in$ **AF-Toric**, there exists a representation*

$$\rho_{\mathbb{A}_\theta} : G \to \mathrm{GL}_{6g-6}(\mathbf{Z}).$$

Proof. The proof of lemma is based on the following well-known criterion of the stable isomorphism for the (toric) AF-algebras: a pair of such algebras $\mathbb{A}_\theta, \mathbb{A}_{\theta'}$ are stably isomorphic if and only if their Bratteli diagrams coincide, except (possibly) for a finite part of the diagram, see, e. g., [74, Theorem 2.3].

Remark 5.5.1. Note that the order isomorphism between the dimension groups [74], translates to the language of the Bratteli diagrams as stated.

Let G be a finitely presented group on the generators $\{\gamma_1, \dots, \gamma_m\}$ subject to relations $r_1, \dots, r_n$. Let $\mathbb{A}_\theta \in$ **AF-Toric.** Since G acts on the toric AF-algebra $\mathbb{A}_\theta$ by stable isomorphisms, the toric AF-algebras $\mathbb{A}_{\theta_1} := \gamma_1(\mathbb{A}_\theta), \dots, \mathbb{A}_{\theta_m} := \gamma_m(\mathbb{A}_\theta)$ are stably isomorphic to $\mathbb{A}_\theta$; moreover, by transitivity, they are also pairwise stably isomorphic. Therefore, the Bratteli diagrams of $\mathbb{A}_{\theta_1}, \dots, \mathbb{A}_{\theta_m}$ coincide everywhere except, possibly, for some finite parts. We shall denote by $\mathbb{A}_{\theta_{\max}} \in$ **AF-Toric** a toric AF-algebra, whose Bratteli diagram is the maximal common part of the Bratteli diagrams of $\mathbb{A}_{\theta_i}$ for $1 \leq i \leq m$; such a choice is unique and defined correctly because the set $\{\mathbb{A}_{\theta_i}\}$ is a finite set. By Definition 5.2.1 of a toric AF-algebra, the vectors $\theta_i = (1, \theta_1^{(i)}, \dots, \theta_{6g-7}^{(i)})$ are related to the vector $\theta_{\max} = (1, \theta_1^{(\max)}, \dots, \theta_{6g-7}^{(\max)})$ by the formula

$$\begin{pmatrix} 1 \\ \theta_1^{(i)} \\ \vdots \\ \theta_{6g-7}^{(i)} \end{pmatrix} = \underbrace{\begin{pmatrix} 0 & 0 & \dots & 0 & 1 \\ 1 & 0 & \dots & 0 & b_1^{(1)(i)} \\ \vdots & \vdots & \ddots & \vdots & \vdots \\ 0 & 0 & \dots & 1 & b_{6g-7}^{(1)(i)} \end{pmatrix} \cdots \begin{pmatrix} 0 & 0 & \dots & 0 & 1 \\ 1 & 0 & \dots & 0 & b_1^{(k)(i)} \\ \vdots & \vdots & \ddots & \vdots & \vdots \\ 0 & 0 & \dots & 1 & b_{6g-7}^{(k)(i)} \end{pmatrix}}_{A_i} \begin{pmatrix} 1 \\ \theta_1^{(\max)} \\ \vdots \\ \theta_{6g-7}^{(\max)} \end{pmatrix}.$$

The above expression can be written in the matrix form $\theta_i = A_i\theta_{max}$, where $A_i \in GL_{6g-6}(\mathbf{Z})$. Thus, one gets a matrix representation of the generator γ_i, given by the formula

$$\rho_{\mathbb{A}_\theta}(\gamma_i) := A_i.$$

The map $\rho_{\mathbb{A}_\theta} : G \to GL_{6g-6}(\mathbf{Z})$ extends to the rest of the group G via its values on the generators; namely, for every $g \in G$ one sets $\rho_{\mathbb{A}_\theta}(g) = A_1^{k_1} \cdots A_m^{k_m}$, whenever $g = \gamma_1^{k_1} \cdots \gamma_m^{k_m}$. Let us verify that the map $\rho_{\mathbb{A}_\theta}$ is a well-defined homomorphism of groups G and $GL_{6g-6}(\mathbf{Z})$. Indeed, let us write $g_1 = \gamma_1^{k_1} \cdots \gamma_m^{k_m}$ and $g_2 = \gamma_1^{s_1} \cdots \gamma_m^{s_m}$ for a pair of elements $g_1, g_2 \in G$; then their product $g_1g_2 = \gamma_1^{k_1} \cdots \gamma_m^{k_m}\gamma_1^{s_1} \cdots \gamma_m^{s_m} = \gamma_1^{l_1} \cdots \gamma_m^{l_m}$, where the last equality is obtained by a reduction of words using the relations $r_1, \dots, r_n$. One can write relations r_i in their matrix form $\rho_{\mathbb{A}_\theta}(r_i)$; thus, one gets the matrix equality $A_1^{l_1} \cdots A_m^{l_m} = A_1^{k_1} \cdots A_m^{k_m} A_1^{s_1} \cdots A_m^{s_m}$. It is immediate from the last equation, that $\rho_{\mathbb{A}_\theta}(g_1g_2) = A_1^{l_1} \cdots A_m^{l_m} = A_1^{k_1} \cdots A_m^{k_m} A_1^{s_1} \cdots A_m^{s_m} = \rho_{\mathbb{A}_\theta}(g_1)\rho_{\mathbb{A}_\theta}(g_2)$ for $\forall g_1, g_2 \in G$, i. e., $\rho_{\mathbb{A}_\theta}$ is a homomorphism. Lemma 5.5.1 follows. □

Let **AF-Toric-Aper** $\subset$ **AF-Toric** be a set consisting of the toric AF-algebras whose Bratteli diagrams are *not* periodic; these are known as nonstationary toric AF-algebras (Section 3.5.2) and they are generic in the set **AF-Toric** endowed with the natural topology.

Definition 5.5.2. The action of group G on the toric AF-algebra $\mathbb{A}_\theta \in$ **AF-Toric** will be called free if $\gamma(\mathbb{A}_\theta) = \mathbb{A}_\theta$ implies $\gamma = \mathrm{Id}$.

Lemma 5.5.2. *If $\mathbb{A}_\theta \in$ **AF-Toric-Aper** and the action of group G on the $\mathbb{A}_\theta$ is free, then $\rho_{\mathbb{A}_\theta}$ is a faithful representation.*

Proof. Since the action of G is free, to prove that $\rho_{\mathbb{A}_\theta}$ is faithful, it remains to show that, in the formula $\theta_i = A_i\theta_{max}$, it holds that $A_i = I$ if and only if $\theta_i = \theta_{max}$, where I is the unit matrix. Indeed, it is immediate that $A_i = I$ implies $\theta_i = \theta_{max}$. Suppose now that $\theta_i = \theta_{max}$ and let to the contrary $A_i \neq I$. One gets $\theta_i = A_i\theta_{max} = \theta_{max}$. Such an equation has a nontrivial solution if and only if the vector θ_{max} has a periodic Jacobi–Perron fraction; the period of such a fraction is given by the matrix A_i. This is impossible since it has been assumed that $\mathbb{A}_{\theta_{max}} \in$ **AF-Toric-Aper**. The contradiction proves Lemma 5.5.2. □

Let $G = \mathrm{Mod}(X)$, where X is a surface of genus $g \geq 2$. The group G is finitely presented, see [66]; it acts on the Teichmueller space $T(g)$ by isomorphisms of the Riemann surfaces. Moreover, the action of G is free on a generic set, $U \subset T(g)$, consisting of the Riemann surfaces with the trivial group of automorphisms. On the other hand, there exists a functor

$$F : \textbf{Alg-Gen} \longrightarrow \textbf{AF-Toric}$$

between the Riemann surfaces (complex algebraic curves) and toric AF-algebras, see Theorem 5.2.1.

Lemma 5.5.3. *The preimage* F^{-1}(**AF-Toric-Aper**) *is a generic set in the space* $T(g)$.

Proof. Note that the set of stationary toric AF-algebras is a countable set. The functor F is a surjective map, which is continuous with respect to the natural topology on the sets **Alg-Gen** and **AF-Toric**. Therefore, the preimage of the complement of a countable set is a generic set. Lemma 5.5.3 follows. □

Consider the set $U \cap F^{-1}$(**AF-Toric-Aper**); this set is nonempty since it is the intersection of two generic subsets of $T(g)$, see Lemma 5.5.3. Let

$$S \in U \cap F^{-1}(\mathbf{AF\text{-}Toric\text{-}Aper})$$

be a point (a Riemann surface) in the above set. In view of Lemma 5.5.1, group G acts on the toric AF-algebra $\mathbb{A}_\theta = F(S)$ by the stable isomorphisms. By the construction, the action is free and $\mathbb{A}_\theta \in$ **AF-Toric-Aper**. In view of Lemma 5.5.2, one gets a faithful representation $\rho = \rho_{\mathbb{A}_\theta}$ of the group $G \cong \mathrm{Mod}(X)$ into the matrix group $GL_{6g-6}(\mathbf{Z})$. Theorem 5.5.1 is proved.

Guide to the literature

The mapping class groups were introduced by M. Dehn [66]. For a primer on the mapping class groups, we refer the reader to the textbook [82]. The Harvey conjecture was formulated in [110]. The braid groups are known to be linear [24]. Using a modification of the argument for the braid groups, it is possible to prove that $\mathrm{Mod}(X)$ is linear in the case $g = 2$ [25]. Some infinite-dimensional (asymptotic) faithfulness of the mapping class groups was proved in [2]. A faithful representation of $\mathrm{Mod}(X)$ in the matrix group $GL_{6g-6}(\mathbf{Z})$ was constructed in [197].

Exercises

1. Prove that the skew-symmetric relations

$$\begin{cases} x_3x_1 = q_{13}x_1x_3, \\ x_4x_2 = q_{24}x_2x_4, \\ x_4x_1 = q_{14}x_1x_4, \\ x_3x_2 = q_{23}x_2x_3, \\ x_2x_1 = q_{12}x_1x_2, \\ x_4x_3 = q_{34}x_3x_4 \end{cases}$$

are invariant of the involution $x_1^* = x_2, x_3^* = x_4$ if and only if the following restrictions on the constants q_{ij} hold:

$$\begin{cases} q_{13} = (\bar{q}_{24})^{-1}, \\ q_{24} = (\bar{q}_{13})^{-1}, \\ q_{14} = (\bar{q}_{23})^{-1}, \\ q_{23} = (\bar{q}_{14})^{-1}, \\ q_{12} = \bar{q}_{12}, \\ q_{34} = \bar{q}_{34}, \end{cases}$$

where $\bar{q}_{ij}$ means the complex conjugate of $q_{ij} \in \mathbf{C} \setminus \{0\}$.

2. Prove that a family of free algebras $\mathbf{C}\langle x_1, x_2, x_3, x_4 \rangle$ modulo an ideal generated by six skew-symmetric quadratic relations

$$\begin{cases} x_3 x_1 = \mu e^{2\pi i\theta} x_1 x_3, \\ x_4 x_2 = \dfrac{1}{\mu} e^{2\pi i\theta} x_2 x_4, \\ x_4 x_1 = \mu e^{-2\pi i\theta} x_1 x_4, \\ x_3 x_2 = \dfrac{1}{\mu} e^{-2\pi i\theta} x_2 x_3, \\ x_2 x_1 = x_1 x_2, \\ x_4 x_3 = x_3 x_4 \end{cases}$$

consists of the pairwise nonisomorphic algebras for different values of $\theta \in S^1$ and $\mu \in (0, \infty)$.

3. Prove that the system of relations for noncommutative torus $\mathcal{A}_\theta$, namely

$$\begin{cases} x_3 x_1 = e^{2\pi i\theta} x_1 x_3, \\ x_4 x_2 = e^{2\pi i\theta} x_2 x_4, \\ x_4 x_1 = e^{-2\pi i\theta} x_1 x_4, \\ x_3 x_2 = e^{-2\pi i\theta} x_2 x_3, \\ x_2 x_1 = x_1 x_2 = e, \\ x_4 x_3 = x_3 x_4 = e, \end{cases}$$

is equivalent to the system of relations

$$\begin{cases} x_3x_1x_4 = e^{2\pi i\theta}x_1, \\ x_4 = e^{2\pi i\theta}x_2x_4x_1, \\ x_4x_1x_3 = e^{-2\pi i\theta}x_1, \\ x_2 = e^{-2\pi i\theta}x_4x_2x_3, \\ x_1x_2 = x_2x_1 = e, \\ x_3x_4 = x_4x_3 = e. \end{cases}$$

(Hint: Use the last two relations.)

4. Prove that the system of relations for the Sklyanin $*$-algebra plus the scaled unit relation, i. e.,

$$\begin{cases} x_3x_1 = \mu e^{2\pi i\theta}x_1x_3, \\ x_4x_2 = \frac{1}{\mu}e^{2\pi i\theta}x_2x_4, \\ x_4x_1 = \mu e^{-2\pi i\theta}x_1x_4, \\ x_3x_2 = \frac{1}{\mu}e^{-2\pi i\theta}x_2x_3, \\ x_2x_1 = x_1x_2 = \frac{1}{\mu}e, \\ x_4x_3 = x_3x_4 = \frac{1}{\mu}e, \end{cases}$$

is equivalent to the system

$$\begin{cases} x_3x_1x_4 = e^{2\pi i\theta}x_1, \\ x_4 = e^{2\pi i\theta}x_2x_4x_1, \\ x_4x_1x_3 = e^{-2\pi i\theta}x_1, \\ x_2 = e^{-2\pi i\theta}x_4x_2x_3, \\ x_2x_1 = x_1x_2 = \frac{1}{\mu}e, \\ x_4x_3 = x_3x_4 = \frac{1}{\mu}e. \end{cases}$$

(Hint: Use multiplication and cancellation involving the last two equations.)

5. If **Proj-Alg** is the category of all complex projective varieties X (of dimension n) and **C*-Serre** the category of all Serre C^*-algebras $\mathcal{A}_X$, then the formula $X \mapsto \mathcal{A}_X$ gives rise to a map

$$F : \textbf{Proj-Alg} \longrightarrow \textbf{C*-Serre}.$$

Prove that the map F is a functor which takes isomorphisms between projective varieties to the Morita equivalences (stable isomorphisms) between the corresponding Serre C^*-algebras. (Hint: Repeat the argument for elliptic curves given in Section 5.1.1.)

6. Prove Remark 5.3.3, i. e., that equations $b_k^{\alpha} t = t b_k$ and $t^* = t^{-1}$ imply equation $(b_k^*)^{\alpha} = (b_k^{\alpha})^*$.

6 Number theory

A restriction of functor $F : \mathbf{CRng} \to \mathbf{Grp\text{-}Rng}$ to the arithmetic schemes provides a link to number theory. For instance, if $\mathcal{E}_{CM}$ is an elliptic curve with complex multiplication, then $F(\mathcal{E}_{CM}) = \mathcal{A}_{RM}$, where $\mathcal{A}_{RM}$ is a noncommutative torus with real multiplication; we refer the reader to Section 6.1 for a proof. We use this fact in Section 6.2 to relate the rank of $\mathcal{E}_{CM}$ to an invariant of $\mathcal{A}_{RM}$ called arithmetic complexity. We use it again in Section 6.3 to prove that the complex number $e^{2\pi i\theta+\log\log\varepsilon}$ is algebraic, whenever θ and ε are algebraic numbers in a real quadratic field, and again in Section 6.4 to find generators of the abelian extension of a real quadratic number field. We define an L-function $L(\mathcal{A}_{RM}, s)$ of $\mathcal{A}_{RM}$ in Section 6.5 and prove that it coincides with the Hasse–Weil function $L(\mathcal{E}_{CM}, s)$ of $\mathcal{E}_{CM}$; the obtained localization formula tells us that the crossed products are analogs of the prime ideals used in algebraic geometry. In Section 6.6 we extend the function $L(\mathcal{A}_{RM}, s)$ to the even-dimensional noncommutative tori $\mathcal{A}_{RM}^{2n}$ and sketch an analog of the Langlands Conjecture for such tori; we refer the reader to [91] for an introduction to the Langlands program. The number of points of projective variety $V(\mathbb{F}_q)$ over a finite field $\mathbb{F}_q$ in terms of the invariants of the Serre C^*-algebra $F(V_{\mathbf{C}})$ of the complex projective variety $V_{\mathbf{C}}$ is calculated in Section 6.7; for an introduction to the Weil Conjectures, we refer the reader to [108, Appendix C].

6.1 Isogenies of elliptic curves

An elliptic curve is a subset of the complex projective plane of the form $\mathcal{E}(\mathbf{C}) = \{(x,y,z) \in \mathbf{C}P^2 \mid y^2z = 4x^3 - g_2xz^2 - g_3z^3\}$, where g_2 and g_3 are constants. The j-invariant of $\mathcal{E}(\mathbf{C})$ is a complex number

$$j(\mathcal{E}(\mathbf{C})) = \frac{1728g_2^3}{g_2^3 - 27g_3^2}.$$

If $\mathcal{E}$ and $\mathcal{E}'$ are isomorphic over $\mathbf{C}$, then $j(\mathcal{E}') = j(\mathcal{E})$. The Weierstrass function $\wp(z)$ defines an isomorphism $\mathcal{E}(\mathbf{C}) \cong \mathbf{C}/(\mathbf{Z} + \mathbf{Z}\tau)$ from elliptic curves to complex tori. By $\mathcal{E}_\tau$ we understand an elliptic curve isomorphic to the complex torus $\mathbf{C}/(\mathbf{Z} + \mathbf{Z}\tau)$.

Definition 6.1.1. An isogeny between elliptic curves $\mathcal{E}_\tau$ and $\mathcal{E}_{\tau'}$ is an analytic map $\varphi : \mathcal{E}_\tau \to \mathcal{E}_{\tau'}$ such that $\varphi(0) = 0$. An invertible isogeny corresponds to an isomorphism between elliptic curves.

Remark 6.1.1. The elliptic curves $\mathcal{E}_\tau$ and $\mathcal{E}_{\tau'}$ are isogenous if and only if

$$\tau' = \frac{a\tau + b}{c\tau + d} \quad \text{for a matrix } \begin{pmatrix} a & b \\ c & d \end{pmatrix} \in M_2(\mathbf{Z}) \text{ with } ad - bc > 0.$$

https://doi.org/10.1515/9783110788709-006

Remark 6.1.2. The endomorphism ring $\mathrm{End}(\mathcal{E}_\tau) := \{\alpha \in \mathbf{C} : \alpha L_\tau \subseteq L_\tau\}$ of elliptic curve $\mathcal{E}_\tau = \mathbf{C}/L_\tau$ is given by the formula

$$\mathrm{End}(\mathcal{E}_\tau) \cong \begin{cases} \mathbf{Z} & \text{if } \tau \in \mathbf{C} - \mathcal{Q}, \\ \mathbf{Z} + f O_k & \text{if } \tau \in \mathcal{Q}, \end{cases}$$

where $\mathcal{Q}$ is the set of all imaginary quadratic numbers and integer $f \geq 1$ is conductor of an order in the ring of integers O_k of the imaginary quadratic number field $k = \mathbf{Q}(\sqrt{-D})$.

Definition 6.1.2. Elliptic curve $\mathcal{E}_\tau$ is said to have complex multiplication if the ring $\mathrm{End}(\mathcal{E}_\tau)$ exceeds $\mathbf{Z}$, i. e., τ is a quadratic irrationality. We write $\mathcal{E}_{\mathrm{CM}}^{(-D,f)}$ to denote elliptic curves with complex multiplication by an order of conductor f in the quadratic field $\mathbf{Q}(\sqrt{-D})$.

Remark 6.1.3 (Hilbert class field of k). The class group of a ring $R := \mathrm{End}(\mathcal{E}_\tau)$ is denoted by $\mathrm{Cl}(R)$. There exist $|\mathrm{Cl}(R)|$ nonisomorphic curves $\mathcal{E}_\tau$ with the same ring R. Since $j(\mathcal{E}_{\mathrm{CM}}^{(-D,f)})$ is an algebraic number, one gets an isomorphism $\mathrm{Gal}(K|k) \cong \mathrm{Cl}(R)$, where $K = k(j(\mathcal{E}_{\mathrm{CM}}^{(-D,f)}))$ and $\mathrm{Gal}(K|k)$ is the Galois group of the extension $K|k$. Thus K is the Hilbert class field of the imaginary quadratic field k. Moreover, $\mathcal{E}_{\mathrm{CM}}^{(-D,f)} \cong \mathcal{E}(K)$, where constants $g_2, g_3 \in K$ [250].

6.1.1 Symmetry of complex and real multiplication

Definition 6.1.3. The algebra $\mathcal{A}_\theta$ is stably homomorphic to $\mathcal{A}_{\theta'}$ if there exists a homomorphism $h : \mathcal{A}_\theta \otimes \mathcal{K} \to \mathcal{A}_{\theta'} \otimes \mathcal{K}$, where $\mathcal{K}$ is the C^*-algebra of compact operators.

Remark 6.1.4. It follows from the argument of [238] that $\mathcal{A}_\theta$ and $\mathcal{A}_{\theta'}$ are stably homomorphic if and only if

$$\theta' = \frac{a\theta + b}{c\theta + d} \quad \text{for a matrix } \begin{pmatrix} a & b \\ c & d \end{pmatrix} \in M_2(\mathbf{Z}) \text{ with } ad - bc > 0.$$

We denote by **Ell-Isgn** the category of elliptic curves $\mathcal{E}_\tau$ whose arrows are isogenies of the elliptic curves. Likewise, we shall write **NC-Tor-Homo** to denote the category of noncommutative tori $\mathcal{A}_\theta$ whose the arrows are stable homomorphisms of the noncommutative tori.

Theorem 6.1.1 (Functor F on isogenous elliptic curves). *There exists a covariant functor*

$$F : \textbf{Ell-Isgn} \longrightarrow \textbf{NC-Tor-Homo},$$

which maps isogenous elliptic curves $\mathcal{E}_\tau$ to the stably homomorphic noncommutative tori $\mathcal{A}_\theta$, see Fig. 6.1; functor F is noninjective with $\mathrm{Ker}\, F \cong (0, \infty)$.

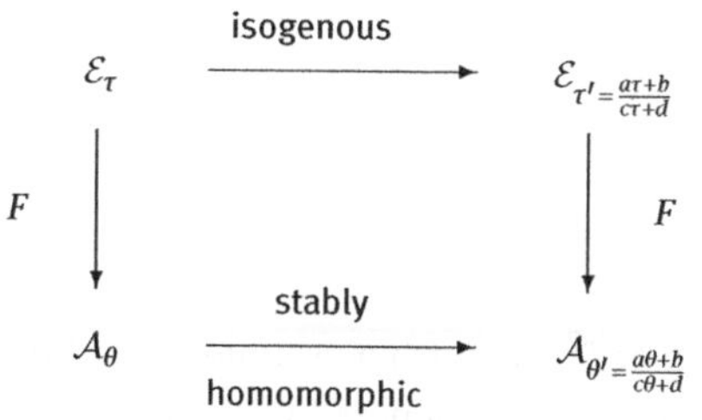

Figure 6.1: Functor on isogenous elliptic curves.

Theorem 6.1.2 (Functor F on $\mathcal{E}_{\mathrm{CM}}$). *Let* $\mathrm{Isom}(\mathcal{E}_{\mathrm{CM}})$ *be an isomorphism class of elliptic curve with complex multiplication and* $\mathfrak{m}_{\mathrm{CM}} := \mu_{\mathrm{CM}}(\mathbf{Z} + \mathbf{Z}\theta_{\mathrm{CM}}) \subset \mathbf{R}$ *be a* $\mathbf{Z}$*-module such that* $\mathcal{A}_{\theta_{\mathrm{CM}}} = F(\mathcal{E}_{\mathrm{CM}})$ *and* $\mu_{\mathrm{CM}} \in \mathrm{Ker}\, F$*. Then:*
(i) $\mathfrak{m}_{\mathrm{CM}}$ *is an invariant of* $\mathrm{Isom}(\mathcal{E}_{\mathrm{CM}})$*;*
(ii) $\mathfrak{m}_{\mathrm{CM}}$ *is a full module in the real quadratic number field.*

In particular, $\mathcal{A}_{\theta_{\mathrm{CM}}}$ *is a noncommutative torus with real multiplication.*

Let $\mathcal{A}_{\mathrm{RM}}^{(D,f)}$ be a noncommutative torus with real multiplication and let $X(\mathcal{A}_{\mathrm{RM}}^{(D,f)})$ be a Riemann surface such that its geodesic spectrum on $\mathbb{H}$ contains the set $\{\tilde{\gamma}(x,\bar{x}) : \forall x \in \mathfrak{m}_{\mathrm{CM}}\}$, where

$$\tilde{\gamma}(x,\bar{x}) = \frac{xe^{\frac{t}{2}} + i\bar{x}e^{-\frac{t}{2}}}{e^{\frac{t}{2}} + ie^{-\frac{t}{2}}}, \qquad -\infty \le t \le \infty,$$

is the geodesic half-circle through the pair of conjugate quadratic irrationalities $x, \bar{x} \in \mathfrak{m}_{\mathrm{CM}} \subset \partial\mathbb{H}$; we refer the reader to Definition 6.1.4 for the details. The theorem below describes functor F of Theorem 6.1.2 in terms of D and f.

Theorem 6.1.3 (Symmetry of complex and real multiplication). *For each square-free integer* $D > 1$ *and integer* $f \ge 1$*, there exists a holomorphic map* $F^{-1} : X(\mathcal{A}_{\mathrm{RM}}^{(D,f')}) \to \mathcal{E}_{\mathrm{CM}}^{(-D,f)}$*, where* $F(\mathcal{E}_{\mathrm{CM}}^{(-D,f)}) = \mathcal{A}_{\mathrm{RM}}^{(D,f')}$ *and* f' *is the least integer satisfying the equation* $|\mathrm{Cl}(\mathbf{Z} + fO_{\mathbf{Q}(\sqrt{-D})})| = |\mathrm{Cl}(\mathbf{Z} + f'O_{\mathbf{Q}(\sqrt{D})})|$.

6.1.2 Proof of Theorem 6.1.1

Let $\phi = \mathrm{Re}\,\omega$ be a 1-form defined by a holomorphic form ω on the complex torus S. Since ω is holomorphic, ϕ is a closed 1-form on topological torus T^2. The $\mathbf{R}$-isomorphism $h_q : H^0(S,\Omega) \to \mathrm{Hom}(H_1(T^2);\mathbf{R})$, as explained, is given by the formulas

$$\begin{cases} \lambda_1 = \int\limits_{\gamma_1} \phi, \\ \lambda_2 = \int\limits_{\gamma_2} \phi, \end{cases}$$

where $\{\gamma_1, \gamma_2\}$ is a basis in the first homology group of T^2. We further assume that, after a proper choice of the basis, λ_1, λ_2 are positive real numbers. Denote by Φ_{T^2} the space of measured foliations on T^2. Each $\mathcal{F} \in \Phi_{T^2}$ is measure equivalent to a foliation by a family of the parallel lines of a slope θ and the invariant transverse measure μ, see Fig. 6.2.

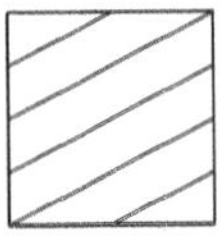

Figure 6.2: Measured foliation $\mathcal{F}$ on $T^2 = \mathbf{R}^2/\mathbf{Z}^2$.

We use the notation $\mathcal{F}_\theta^\mu$ for such a foliation. There exists a simple relationship between the reals (λ_1, λ_2) and (θ, μ). Indeed, the closed 1-form ϕ = const defines a measured foliation $\mathcal{F}_\theta^\mu$, so that

$$\begin{cases} \lambda_1 = \int\limits_{\gamma_1} \phi = \int\limits_0^1 \mu dx, \\ \lambda_2 = \int\limits_{\gamma_2} \phi = \int\limits_0^1 \mu dy, \end{cases} \quad \text{where } \frac{dy}{dx} = \theta.$$

By integration,

$$\begin{cases} \lambda_1 = \int\limits_0^1 \mu dx = \mu, \\ \lambda_2 = \int\limits_0^1 \mu \theta dx = \mu\theta. \end{cases}$$

Thus one gets $\mu = \lambda_1$ and $\theta = \frac{\lambda_2}{\lambda_1}$. Recall that the Hubbard–Masur theory establishes a homeomorphism $h : T_S(1) \to \Phi_{T^2}$, where $T_S(1) \cong \mathbb{H} = \{\tau : \operatorname{Im} \tau > 0\}$ is the Teichmüller space of the torus, see Corollary 5.1.1. Denote by ω_N an invariant (Néron) differential of the complex torus $\mathbf{C}/(\omega_1\mathbf{Z} + \omega_2\mathbf{Z})$. It is well known that $\omega_1 = \int_{\gamma_1} \omega_N$ and $\omega_2 = \int_{\gamma_2} \omega_N$, where γ_1 and γ_2 are the meridians of the torus. Let π be a projection acting by the formula $(\theta,\mu)\mapsto\theta$. An explicit formula for the functor F : **Ell-Isgn** $\to$ **NC-Tor-Homo** is given by the composition $F=\pi\circ h$, where h is the Hubbard–Masur homeomorphism. In other words, one gets the following explicit correspondence between the complex and noncommutative tori:

$$\mathcal{E}_\tau=\mathcal{E}_{(\int_{\gamma_2}\omega_N)/(\int_{\gamma_1}\omega_N)}\overset{h}{\longmapsto}\mathcal{F}^{\int_{\gamma_1}\phi}_{(\int_{\gamma_2}\phi)/(\int_{\gamma_1}\phi)}\overset{\pi}{\longmapsto}\mathcal{A}_{(\int_{\gamma_2}\phi)/(\int_{\gamma_1}\phi)}=\mathcal{A}_\theta,$$

where $\mathcal{E}_\tau=\mathbf{C}/(\mathbf{Z}+\mathbf{Z}\tau)$. Let

$$\varphi:\mathcal{E}_\tau\longrightarrow\mathcal{E}_{\tau'}$$

be an isogeny of the elliptic curves. The action of φ on the homology basis $\{\gamma_1,\gamma_2\}$ of T^2 is given by the formulas

$$\begin{cases}\gamma_1'=a\gamma_1+b\gamma_2\\ \gamma_2'=c\gamma_1+d\gamma_2\end{cases},\quad\text{where }\begin{pmatrix}a&b\\c&d\end{pmatrix}\in M_2(\mathbf{Z}).$$

Recall that the functor F : **Ell-Isgn** $\to$ **NC-Tor-Homo** is given by the formula

$$\tau=\frac{\int_{\gamma_2}\omega_N}{\int_{\gamma_1}\omega_N}\longmapsto\theta=\frac{\int_{\gamma_2}\phi}{\int_{\gamma_1}\phi},$$

where ω_N is an invariant differential on $\mathcal{E}_\tau$ and $\phi=\operatorname{Re}\omega$ is a closed 1-form on T^2.

(i) From the left-hand side of the above equation, one obtains

$$\begin{cases}\omega_1'=\int\limits_{\gamma_1'}\omega_N=\int\limits_{a\gamma_1+b\gamma_2}\omega_N=a\int\limits_{\gamma_1}\omega_N+b\int\limits_{\gamma_2}\omega_N=a\omega_1+b\omega_2,\\ \omega_2'=\int\limits_{\gamma_2'}\omega_N=\int\limits_{c\gamma_1+d\gamma_2}\omega_N=c\int\limits_{\gamma_1}\omega_N+d\int\limits_{\gamma_2}\omega_N=c\omega_1+d\omega_2,\end{cases}$$

and therefore $\tau'=\frac{\int_{\gamma_2'}\omega_N}{\int_{\gamma_1'}\omega_N}=\frac{c+d\tau}{a+b\tau}$.

(ii) From the right-hand side, one obtains

$$\begin{cases}\lambda_1'=\int\limits_{\gamma_1'}\phi=\int\limits_{a\gamma_1+b\gamma_2}\phi=a\int\limits_{\gamma_1}\phi+b\int\limits_{\gamma_2}\phi=a\lambda_1+b\lambda_2,\\ \lambda_2'=\int\limits_{\gamma_2'}\phi=\int\limits_{c\gamma_1+d\gamma_2}\phi=c\int\limits_{\gamma_1}\phi+d\int\limits_{\gamma_2}\phi=c\lambda_1+d\lambda_2,\end{cases}$$

and therefore $\theta' = \frac{\int_{\gamma_2'} \phi}{\int_{\gamma_1'} \phi} = \frac{c+d\theta}{a+b\theta}$. Comparing (i) and (ii), one gets the conclusion of the first part of Theorem 6.1.1. To prove the second part, recall that the invertible isogeny is an isomorphism of the elliptic curves. In this case $\left(\begin{smallmatrix} a & b \\ c & d \end{smallmatrix}\right) \in SL_2(\mathbf{Z})$ and $\theta' = \theta \bmod SL_2(\mathbf{Z})$. Therefore F sends the isomorphic elliptic curves to the stably isomorphic noncommutative tori. The second part of Theorem 6.1.1 is proved. It follows from the proof that F : **Ell-Isgn** → **NC-Tor-Homo** is a covariant functor. Indeed, F preserves the morphisms and does not reverse the arrows, $F(\varphi_1\varphi_2) = \varphi_1\varphi_2 = F(\varphi_1)F(\varphi_2)$, for any pair of the isogenies $\varphi_1, \varphi_2 \in$ Mor(**Ell-Isgn**). Theorem 6.1.1 follows.

6.1.3 Proof of Theorem 6.1.2

Lemma 6.1.1. *Let $\mathfrak{m} \subset \mathbf{R}$ be a module of the rank 2, i. e., $\mathfrak{m} = \mathbf{Z}\lambda_1 + \mathbf{Z}\lambda_2$, where $\theta = \frac{\lambda_2}{\lambda_1} \notin \mathbf{Q}$. If $\mathfrak{m}' \subseteq \mathfrak{m}$ is a submodule of the rank 2, then $\mathfrak{m}' = k\mathfrak{m}$, where either*
(i) $k \in \mathbf{Z} - \{0\}$ and $\theta \in \mathbf{R} - \mathbf{Q}$, or
(ii) k and θ are the irrational numbers of a quadratic number field.

Proof. Any rank 2 submodule of m can be written as $\mathfrak{m}' = \lambda_1'\mathbf{Z} + \lambda_2'\mathbf{Z}$, where

$$\begin{cases} \lambda_1' = a\lambda_1 + b\lambda_2 \\ \lambda_2' = c\lambda_1 + d\lambda_2 \end{cases} \quad \text{and} \quad \begin{pmatrix} a & b \\ c & d \end{pmatrix} \in M_2(\mathbf{Z}).$$

(i) Let us assume that $b \neq 0$. Let $\Delta = (a+d)^2 - 4(ad - bc)$ and $\Delta' = (a+d)^2 - 4bc$. We shall consider the following cases.

Case 1: $\Delta > 0$ and $\Delta \neq m^2$, $m \in \mathbf{Z} - \{0\}$. The real number k can be determined from the equations

$$\begin{cases} \lambda_1' = k\lambda_1 = a\lambda_1 + b\lambda_2, \\ \lambda_2' = k\lambda_2 = c\lambda_1 + d\lambda_2. \end{cases}$$

Since $\theta = \frac{\lambda_2}{\lambda_1}$, one gets the equation $\theta = \frac{c+d\theta}{a+b\theta}$ by taking the ratio of two equations above. A quadratic equation for θ is written as $b\theta^2 + (a-d)\theta - c = 0$. The discriminant of the equation coincides with Δ, and therefore there exist real roots $\theta_{1,2} = \frac{d-a\pm\sqrt{\Delta}}{2b}$. Moreover, $k = a + b\theta = \frac{1}{2}(a + d \pm \sqrt{\Delta})$. Since Δ is not the square of an integer, k and θ are irrationalities of the quadratic number field $\mathbf{Q}(\sqrt{\Delta})$.

Case 2: $\Delta > 0$ and $\Delta = m^2$, $m \in \mathbf{Z} - \{0\}$. Note that $\theta = \frac{a-d\pm|m|}{2c}$ is a rational number. Since θ does not satisfy the rank assumption of the lemma, the case should be omitted.

Case 3: $\Delta = 0$. The quadratic equation has a double root $\theta = \frac{a-d}{2c} \in \mathbf{Q}$. This case leads to a module of the rank 1, which is contrary to the assumption of the lemma.

Case 4: $\Delta < 0$ and $\Delta' \neq m^2, m \in \mathbf{Z} - \{0\}$. Let us define a new basis $\{\lambda_1'', \lambda_2''\}$ in $\mathfrak{m}'$ so that

$$\begin{cases} \lambda_1'' = \lambda_1', \\ \lambda_2'' = -\lambda_2'. \end{cases}$$

Then

$$\begin{cases} \lambda_1'' = a\lambda_1 + b\lambda_2, \\ \lambda_2'' = -c\lambda_1 - d\lambda_2, \end{cases}$$

and $\theta = \frac{\lambda_2''}{\lambda_1''} = \frac{-c-d\theta}{a+b\theta}$. The quadratic equation for θ has the form $b\theta^2 + (a+d)\theta + c = 0$, whose discriminant is $\Delta' = (a+d)^2 - 4bc$. Let us show that $\Delta' > 0$. Indeed, $\Delta = (a+d)^2 - 4(ad-bc) < 0$ and the evident inequality $-(a-d)^2 \leq 0$ have the same sign, and we shall add them up. After an obvious elimination, one gets $bc < 0$. Therefore Δ' is a sum of the two positive integers, which is always a positive integer. Thus, there exist the real roots $\theta_{1,2} = \frac{-a-d\pm\sqrt{\Delta'}}{2b}$. Moreover, $k = a + b\theta = \frac{1}{2}(a - d \pm \sqrt{\Delta'})$. Since Δ' is not the square of an integer, k and θ are irrational numbers in the quadratic field $\mathbf{Q}(\sqrt{\Delta'})$.

Case 5: $\Delta < 0$ and $\Delta' = m^2, m \in \mathbf{Z} - \{0\}$. Note that $\theta = \frac{-a-d\pm|m|}{2b}$ is a rational number. Since θ does not satisfy the rank assumption of the lemma, the case should be omitted.

(ii) Assume that $b = 0$.

Case 1: $a - d \neq 0$. The quadratic equation for θ degenerates to a linear equation $(a - d)\theta + c = 0$. The root $\theta = \frac{c}{d-a} \in \mathbf{Q}$ does not satisfy the rank assumption again, and we omit the case.

Case 2: $a = d$ and $c \neq 0$. It is easy to see that the set of solutions for θ is an empty set.

Case 3: $a = d$ and $c = 0$. Finally, in this case all the coefficients of the quadratic equation vanish, so that any $\theta \in \mathbf{R} - \mathbf{Q}$ is a solution. Note that $k = a = d \in \mathbf{Z}$. Thus, one gets case (i) of the lemma. Since there are no other possibilities left, Lemma 6.1.1 is proved. □

Lemma 6.1.2. *Let $\mathcal{E}_{CM}$ be an elliptic curve with complex multiplication and consider a $\mathbf{Z}$-module $F(\mathrm{Isom}(\mathcal{E}_{CM})) = \mu_{CM}(\mathbf{Z} + \mathbf{Z}\theta_{CM}) := \mathfrak{m}_{CM}$. Then:*
(i) θ_{CM} is a quadratic irrationality,
(ii) $\mu_{CM} \in \mathbf{Q}$ (up to a choice of map F).

Proof. (i) Since $\mathcal{E}_{CM}$ has complex multiplication, one gets $\mathrm{End}(\mathcal{E}_{CM}) > \mathbf{Z}$. In particular, there exists a nontrivial isogeny

$$\varphi : \mathcal{E}_{CM} \longrightarrow \mathcal{E}_{CM},$$

i. e., an endomorphism which is *not* the multiplication by $k \in \mathbf{Z}$. By Theorem 6.1.1 and Remark 6.1.4, each isogeny φ defines a rank 2 submodule $\mathfrak{m}'$ of module $\mathfrak{m}_{CM}$. By

Lemma 6.1.1, $\mathfrak{m}' = k\mathfrak{m}_{CM}$ for a $k \in \mathbf{R}$. Because φ is a nontrivial endomorphism, we get $k \notin \mathbf{Z}$; thus, option (i) of Lemma 6.1.1 is excluded. Therefore, by Lemma 6.1.1(ii), the real number θ_{CM} must be a quadratic irrationality.

(ii) Recall that $E_{\mathcal{F}} \subset Q - \{0\}$ is the space of holomorphic differentials on the complex torus, whose horizontal trajectory structure is equivalent to given measured foliation $\mathcal{F} = \mathcal{F}_\theta^\mu$. We shall vary $\mathcal{F}_\theta^\mu$, thus varying the Hubbard–Masur homeomorphism $h = h(\mathcal{F}_\theta^\mu) : E_{\mathcal{F}} \to T(1)$, see Section 6.1.2. Namely, consider a 1-parameter continuous family of such maps $h = h_\mu$, where $\theta = \text{const}$ and $\mu \in \mathbf{R}$. Recall that $\mu_{CM} = \lambda_1 = \int_{\gamma_1} \phi$, where $\phi = \operatorname{Re} \omega$ and $\omega \in E_{\mathcal{F}}$. The family h_μ generates a family $\omega_\mu = h_\mu^{-1}(C)$, where C is a fixed point in $T(1)$. Denote by ϕ_μ and λ_1^μ the corresponding families of the closed 1-forms and their periods, respectively. By the continuity, λ_1^μ takes on a rational value for a $\mu = \mu'$. (Actually, every neighborhood of μ_0 contains such a μ'.) Thus, $\mu_{CM} \in \mathbf{Q}$ for the Hubbard–Masur homeomorphism $h = h_{\mu'}$. Lemma 6.1.2 follows. □

Claim (ii) of Theorem 6.1.2 follows from Lemma 6.1.2(i) and Theorem 6.1.2(i). To prove claim (i) of Theorem 6.1.2, notice that whenever $\mathcal{E}_1, \mathcal{E}_2 \in \operatorname{Isom}(\mathcal{E}_{CM})$ the respective $\mathbf{Z}$-modules coincide, i. e., $\mathfrak{m}_1 = \mathfrak{m}_2$; this happens because an isomorphism between elliptic curves corresponds to a change of basis in the module $\mathfrak{m}$, see Theorem 6.1.1 and Remark 6.1.4. Theorem 6.1.2 is proved.

6.1.4 Proof of Theorem 6.1.3

Let $N \geq 1$ be an integer; recall that $\Gamma_1(N)$ is a subgroup of the modular group $SL_2(\mathbf{Z})$ consisting of matrices of the form

$$\left\{ \begin{pmatrix} a & b \\ c & d \end{pmatrix} \in SL_2(\mathbf{Z}) \;\middle|\; a, d \equiv 1 \mod N,\; c \equiv 0 \mod N \right\};$$

the corresponding Riemann surface $\mathbb{H}/\Gamma_1(N)$ will be denoted by $X_1(N)$. Consider the geodesic spectrum of $X_1(N)$, i. e., the set $\operatorname{Spec} X_1(N)$ consisting of all closed geodesics of the surface $X_1(N)$; each geodesic $\gamma \in \operatorname{Spec} X_1(N)$ is the image under the covering map $\mathbb{H} \to \mathbb{H}/\Gamma_1(N)$ of a geodesic half-circle $\tilde{\gamma} \in \mathbb{H}$ passing through the points x and $\bar{x}$ fixed by the linear fractional transformation $x \mapsto \frac{ax+b}{cx+d}$, where matrix $(a, b, c, d) \in \Gamma_1(N)$. It is not hard to see that x and $\bar{x}$ are quadratic irrational numbers; the numbers are real when $|a + d| > 2$.

Definition 6.1.4. We shall say that the Riemann surface X is associated to the noncommutative torus $\mathcal{A}_{RM}^{(D,f)}$, if $\{\tilde{\gamma}(x, \bar{x}) : \forall x \in \mathfrak{m}_{RM}^{(D,f)}\} \subset \widetilde{\operatorname{Spec} X}$, where $\widetilde{\operatorname{Spec} X} \subset \mathbb{H}$ is the set of geodesic half-circles covering the geodesic spectrum of X and $\mathfrak{m}_{RM}^{(D,f)}$ is a $\mathbf{Z}$-module (a pseudo-lattice) in $\mathbf{R}$ generated by torus $\mathcal{A}_{RM}^{(D,f)}$; the associated Riemann surface will be denoted by $X(\mathcal{A}_{RM}^{(D,f)})$.

Lemma 6.1.3. $X(\mathcal{A}_{\mathrm{RM}}^{(D,f)}) \cong X_1(fD)$.

Proof. Recall that $\mathfrak{m}_{\mathrm{RM}}^{(D,f)}$ is a $\mathbf{Z}$-module (a pseudo-lattice) with real multiplication by an order R in the real quadratic number field $\mathbf{Q}(\sqrt{D})$; it is known that $\mathfrak{m}_{\mathrm{RM}}^{(D,f)} \subseteq R$ and $R = \mathbf{Z} + (f\omega)\mathbf{Z}$, where $f \geq 1$ is the conductor of R and

$$\omega = \begin{cases} \frac{1+\sqrt{D}}{2} & \text{if } D \equiv 1 \mod 4, \\ \sqrt{D} & \text{if } D \equiv 2,3 \mod 4, \end{cases}$$

see, e. g., [38, pp. 130–131]. Recall that matrix $(a,b,c,d) \in \mathrm{SL}_2(\mathbf{Z})$ has a pair of real fixed points x and $\bar{x}$ if and only if $|a+d| > 2$ (the hyperbolic matrix); the fixed points can be found from the equation $x = (ax+b)(cx+d)^{-1}$ by the formulas

$$x = \frac{a-d}{2c} + \sqrt{\frac{(a+d)^2-4}{4c^2}}, \quad \bar{x} = \frac{a-d}{2c} - \sqrt{\frac{(a+d)^2-4}{4c^2}}.$$

Case I. If $D \equiv 1 \bmod 4$, then the above formulas imply that $R = (1+\frac{f}{2})\mathbf{Z} + \frac{\sqrt{f^2D}}{2}\mathbf{Z}$. If $x \in \mathfrak{m}_{\mathrm{RM}}^{(D,f)}$ is fixed point of a transformation $(a,b,c,d) \in \mathrm{SL}_2(\mathbf{Z})$, then

$$\begin{cases} \dfrac{a-d}{2c} = \left(1+\dfrac{f}{2}\right)z_1, \\ \dfrac{(a+d)^2-4}{4c^2} = \dfrac{f^2D}{4}z_2^2 \end{cases}$$

for some integer numbers z_1 and z_2. The second equation can be written in the form $(a+d)^2-4 = c^2f^2Dz_2^2$; we have therefore $(a+d)^2 \equiv 4 \bmod (fD)$ and $a+d \equiv \pm 2 \bmod (fD)$. Without loss of generality, we assume $a+d \equiv 2 \bmod (fD)$ since matrix $(a,b,c,d) \in \mathrm{SL}_2(\mathbf{Z})$ can be multiplied by -1. Notice that the last equation admits a solution $a = d \equiv 1 \bmod (fD)$. The first equation yields us $\frac{a-d}{c} = (2+f)z_1$, where $c \neq 0$ since the matrix (a,b,c,d) is hyperbolic. Notice that $a-d \equiv 0 \bmod (fD)$; since the ratio $\frac{a-d}{c}$ must be an integer, we conclude that $c \equiv 0 \bmod (fD)$. Summing up, we get

$$a \equiv 1 \mod (fD), \quad d \equiv 1 \mod (fD), \quad c \equiv 0 \mod (fD).$$

Case II. If $D \equiv 2$ or $3 \bmod 4$, then $R = \mathbf{Z} + (\sqrt{f^2D})\,\mathbf{Z}$. If $x \in \mathfrak{m}_{\mathrm{RM}}^{(D,f)}$ is a fixed point of a transformation $(a,b,c,d) \in \mathrm{SL}_2(\mathbf{Z})$, then

$$\begin{cases} \dfrac{a-d}{2c} = z_1, \\ \dfrac{(a+d)^2-4}{4c^2} = f^2Dz_2^2 \end{cases}$$

for some integer numbers z_1 and z_2. The second equation gives $(a+d)^2-4=4c^2f^2Dz_2^2$; therefore $(a+d)^2 \equiv 4 \mod (fD)$ and $a+d \equiv \pm 2 \mod (fD)$. Again without loss of generality, we assume $a+d \equiv 2 \mod (fD)$ since matrix $(a,b,c,d) \in SL_2(\mathbf{Z})$ can be multiplied by -1. The last equation admits a solution $a = d \equiv 1 \mod (fD)$. The first equation is $\frac{a-d}{c} = 2z_1$, where $c \neq 0$. Since $a-d \equiv 0 \mod (fD)$ and the ratio $\frac{a-d}{c}$ must be integer, one concludes that $c \equiv 0 \mod (fD)$. All together, one gets

$$a \equiv 1 \mod (fD), \quad d \equiv 1 \mod (fD), \quad c \equiv 0 \mod (fD).$$

Since all possible cases are exhausted, Lemma 6.1.3 follows. □

Remark 6.1.5. There exist other finite index subgroups of $SL_2(\mathbf{Z})$ whose geodesic spectrum contains the set $\{\tilde{\gamma}(x,\bar{x}) : \forall x \in \mathfrak{m}_{RM}^{(D,f)}\}$; however, $\Gamma_1(fD)$ is a unique group with such a property among subgroups of the principal congruence group.

Remark 6.1.6. Not all geodesics of $X_1(fD)$ have the above form; thus the set $\{\tilde{\gamma}(x,\bar{x}) : \forall x \in \mathfrak{m}_{RM}^{(D,f)}\}$ is strictly included in the geodesic spectrum of modular curve $X_1(fD)$.

Definition 6.1.5. The group

$$\Gamma(N) := \left\{ \begin{pmatrix} a & b \\ c & d \end{pmatrix} \in SL_2(\mathbf{Z}) \;\middle|\; a,d \equiv 1 \mod N,\ b,c \equiv 0 \mod N \right\}$$

is called a principal congruence group of level N; the corresponding compact modular curve will be denoted by $X(N) = \mathbb{H}/\Gamma(N)$.

Lemma 6.1.4 (Hecke). *There exists a holomorphic map $X(fD) \to \mathcal{E}_{CM}^{(-D,f)}$.*

Proof. A detailed proof of this beautiful fact is given in [113]. For the sake of clarity, we shall give an idea of the proof. Let $\mathfrak{R}$ be an order of conductor $f \geq 1$ in the imaginary quadratic number field $\mathbf{Q}(\sqrt{-D})$; consider an L-function attached to $\mathfrak{R}$, namely

$$L(s,\psi) = \prod_{\mathfrak{P} \subset \mathfrak{R}} \frac{1}{1 - \frac{\psi(\mathfrak{P})}{N(\mathfrak{P})^s}}, \quad s \in \mathbf{C},$$

where $\mathfrak{P}$ is a prime ideal in $\mathfrak{R}$, $N(\mathfrak{P})$ its norm and ψ a Grössencharacter. A crucial observation of Hecke says that the series $L(s,\psi)$ converges to a cusp form $w(s)$ of the principal congruence group $\Gamma(fD)$. By Deuring Theorem, $L(\mathcal{E}_{CM}^{(-D,f)}, s) = L(s,\psi)L(s,\bar{\psi})$, where $L(\mathcal{E}_{CM}^{(-D,f)}, s)$ is the Hasse–Weil L-function of the elliptic curve and $\bar{\psi}$ a conjugate of the Grössencharacter, see, e. g., [259, p. 175]; moreover, $L(\mathcal{E}_{CM}^{(-D,f)}, s) = L(w,s)$, where $L(w,s) := \sum_{n=1}^{\infty} \frac{c_n}{n^s}$ and c_n the Fourier coefficients of the cusp form $w(s)$. In other words, $\mathcal{E}_{CM}^{(-D,f)}$ is a modular elliptic curve. One can now apply the modularity principle: if A_w is an abelian variety given by the periods of holomorphic differential $w(s)ds$ (and its conjugates) on $X(fD)$, then the diagram in Fig. 6.3 is commutative. The holomorphic map $X(fD) \to \mathcal{E}_{CM}^{(-D,f)}$ is obtained as a composition of the canonical embedding $X(fD) \to$

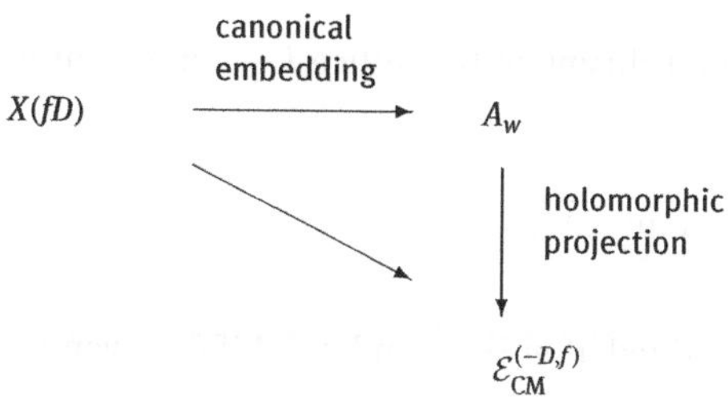

Figure 6.3: Hecke lemma.

A_w with the subsequent holomorphic projection $A_w \to \mathcal{E}_{\mathrm{CM}}^{(-D,f)}$. Lemma 6.1.4 is proved. □

Remark 6.1.7. The diagram in Fig. 6.3 is invariant of the Galois automorphisms if and only if $|\mathrm{Cl}(\mathbf{Z} + fO_{\mathbf{Q}(\sqrt{-D})})| = |\mathrm{Cl}(\mathbf{Z} + f'O_{\mathbf{Q}(\sqrt{D})})|$, where f' is the smallest conductor satisfying the equation. Since f is a divisor of f', one gets a holomorphic map $X(f'D) \to \mathcal{E}_{\mathrm{CM}}^{(-D,f)}$.

Lemma 6.1.5. *The functor F acts by the formula $\mathcal{E}_{\mathrm{CM}}^{(-D,f)} \mapsto \mathcal{A}_{\mathrm{RM}}^{(D,f)}$.*

Proof. Let L_{CM} be a lattice with complex multiplication by an order $\mathfrak{R} = \mathbf{Z} + (f\omega)\mathbf{Z}$ in the imaginary quadratic field $\mathbf{Q}(\sqrt{-D})$; the multiplication by $\alpha \in \mathfrak{R}$ generates an endomorphism $(a,b,c,d) \in M_2(\mathbf{Z})$ of the lattice L_{CM}. It is known from Section 6.1.3, **Case 4**, that the endomorphisms of lattice L_{CM} and endomorphisms of the pseudo-lattice $\mathfrak{m}_{\mathrm{RM}} = F(L_{\mathrm{CM}})$ are related by the following explicit map:

$$\begin{pmatrix} a & b \\ c & d \end{pmatrix} \in \mathrm{End}(L_{\mathrm{CM}}) \longmapsto \begin{pmatrix} a & b \\ -c & -d \end{pmatrix} \in \mathrm{End}(\mathfrak{m}_{\mathrm{RM}}),$$

Moreover, one can always assume $d = 0$ in a proper basis of L_{CM}. We shall consider the following two cases.

Case I. If $D \equiv 1 \bmod 4$ then we have $\mathfrak{R} = \mathbf{Z} + (\frac{f+\sqrt{-f^2D}}{2})\mathbf{Z}$; thus $\alpha = \frac{2m+fn}{2} + \sqrt{\frac{-f^2Dn^2}{4}}$ for some $m, n \in \mathbf{Z}$. Therefore multiplication by α corresponds to an endomorphism $(a,b,c,0) \in M_2(\mathbf{Z})$, where

$$\begin{cases} a = \mathrm{Tr}(\alpha) = \alpha + \bar{\alpha} = 2m + fn, \\ b = -1, \\ c = N(\alpha) = \alpha\bar{\alpha} = \left(\dfrac{2m+fn}{2}\right)^2 + \dfrac{f^2Dn^2}{4}. \end{cases}$$

To calculate a primitive generator of endomorphisms of the lattice L_{CM}, one should find a multiplier $\alpha_0 \neq 0$ such that

$$|\alpha_0| = \min_{m,n\in\mathbf{Z}} |\alpha| = \min_{m,n\in\mathbf{Z}} \sqrt{N(\alpha)}.$$

From the equation for c, the minimum is attained at $m = -\frac{f}{2}$ and $n = 1$ if f is even or $m = -f$ and $n = 2$ if f is odd. Thus

$$\alpha_0 = \begin{cases} \pm\frac{f}{2}\sqrt{-D} & \text{if } f \text{ is even,} \\ \pm f\sqrt{-D} & \text{if } f \text{ is odd.} \end{cases}$$

To find the matrix form of the endomorphism α_0, we shall substitute in the corresponding formula $a = d = 0, b = -1$, and $c = \frac{f^2D}{4}$ if f is even or $c = f^2D$ if f is odd. Thus functor F maps the multiplier α_0 into

$$F(\alpha_0) = \begin{cases} \pm\frac{f}{2}\sqrt{D} & \text{if } f \text{ is even,} \\ \pm f\sqrt{D} & \text{if } f \text{ is odd.} \end{cases}$$

Comparing the above equations, one verifies that formula $F(\mathcal{E}_{CM}^{(-D,f)}) = \mathcal{A}_{RM}^{(D,f)}$ is true in this case.

Case II. If $D \equiv 2$ or $3 \bmod 4$ then $\mathfrak{R} = \mathbf{Z}+(\sqrt{-f^2D})\,\mathbf{Z}$; thus the multiplier $\alpha = m+\sqrt{-f^2Dn^2}$ for some $m, n \in \mathbf{Z}$. A multiplication by α corresponds to an endomorphism $(a, b, c, 0) \in M_2(\mathbf{Z})$, where

$$\begin{cases} a = \mathrm{Tr}(\alpha) = \alpha + \bar{\alpha} = 2m, \\ b = -1, \\ c = N(\alpha) = \alpha\bar{\alpha} = m^2 + f^2Dn^2. \end{cases}$$

We shall repeat the argument of Case I; then from the equation for c, the minimum of $|\alpha|$ is attained at $m = 0$ and $n = \pm 1$. Thus $\alpha_0 = \pm f\sqrt{-D}$. To find the matrix form of the endomorphism α_0, we substitute in the corresponding equation $a = d = 0, b = -1$, and $c = f^2D$. Thus functor F maps the multiplier $\alpha_0 = \pm f\sqrt{-D}$ into $F(\alpha_0) = \pm f\sqrt{D}$. In other words, formula $F(\mathcal{E}_{CM}^{(-D,f)}) = \mathcal{A}_{RM}^{(D,f)}$ is true in this case as well. Since all possible cases are exhausted, Lemma 6.1.5 is proved. ☐

Lemma 6.1.6. *For every $N \geq 1$, there exists a holomorphic map $X_1(N) \to X(N)$.*

Proof. Indeed, $\Gamma(N)$ is a normal subgroup of index N of the group $\Gamma_1(N)$; therefore there exists a degree N holomorphic map $X_1(N) \to X(N)$. Lemma 6.1.6 follows. ☐

Theorem 6.1.3 follows from Remark 6.1.7, Lemmas 6.1.3–6.1.5 and Lemma 6.1.6 for $N = fD$.

Guide to the literature
Hilbert counted complex multiplication as not only the most beautiful part of mathematics but also of entire science; it surely does as it links complex analysis and number theory. One cannot beat [250] for an introduction, but more comprehensive [259, Chapter 2] is a must. Real multiplication has been introduced in [155]. The link between the two was the subject of [181] and the inverse functor F^{-1} was constructed in [189].

6.2 Ranks of elliptic curves

If $\mathcal{E}(K)$ an elliptic curve over the number field K, then Mordell–Néron Theorem says that the K-rational points of $\mathcal{E}(K)$ form a finitely generated abelian group of rank $\operatorname{rk}\mathcal{E}(K)$ [260, Chapter 1]. We prove that $\operatorname{rk}\mathcal{E}(K)$ is one less the so-called arithmetic complexity of the algebra $\mathcal{A}_{\mathrm{RM}} = F(\mathcal{E}(K))$. As a corollary, one gets a simple estimate for the rank in terms of the length of a continued fraction attached to the noncommutative torus. As an illustration, we consider a family of elliptic curves with complex multiplication and a family of rational elliptic curves.

6.2.1 Arithmetic complexity of $\mathcal{A}_{\mathrm{RM}}$

Recall that a quadratic irrationality can be written in the form $\theta_d = \frac{a+b\sqrt{d}}{c}$, where $a, b, c \in \mathbf{Z}$ and $d > 0$ is a square-free integer. It is well known that the continued fraction of θ_d is eventually periodic, i. e., $\theta_d = [g_1, \dots, g_m, \overline{k_1, \dots, k_{n-m}}]$, where $(k_1, \dots, k_{n-m})$ is the minimal period and $n \geq m + 1 \geq 1$. Consider a one-parameter family $\{\theta_x \mid x > 0\}$ of the form

$$\theta_x := \left\{ \frac{a + b\sqrt{x}}{c} \;\middle|\; x \in U_d;\ a, b, c = \mathrm{const} \right\},$$

where U_d is a set of the square-free integers containing d. A system of the polynomial equations in the ring $\mathbf{Z}[g_1, \dots, g_m;\ k_1, \dots, k_{n-m}]$ is called *Euler's* if each $\{\theta_x,\ x \in U_d\}$ can be written as $\theta_x = [g_1(x), \dots, g_m(x), \overline{k_1(x), \dots, k_{n-m}(x)}]$, where $g_i(x)$ and $k_i(x)$ are polynomials in the ring $\mathbf{Z}[x]$.

Remark 6.2.1. For an immediate example of the Euler equations in the case $a = 0$ and $b = c = 1$, we refer the reader to [225, p. 88].

The Euler equations define an affine algebraic set. By the *Euler variety* V_E we understand the projective closure of an irreducible affine variety containing the point $x = d$ of this set.

Definition 6.2.1. The Krull dimension of the Euler variety V_E is called an arithmetic complexity $c(\mathcal{A}_{\mathrm{RM}})$ of the algebra $\mathcal{A}_{\mathrm{RM}}$.

Remark 6.2.2. The $c(\mathcal{A}_{RM})$ counts the number of the algebraically independent entries in the continued fraction of θ_d. The $c(\mathcal{A}_{RM})$ is an isomorphism invariant of the algebra $\mathcal{A}_{RM}$.

Theorem 6.2.1. $\operatorname{rk}\mathcal{E}(K) = c(\mathcal{A}_{RM}) - 1$, *where* $\mathcal{A}_{RM} = F(\mathcal{E}(K))$.

Corollary 6.2.1. $\operatorname{rk}\mathcal{E}(K) \le n - 1$.

6.2.2 Mordell AF-algebra

Denote by S^1 a unit circle in the complex plane; let G be a multiplicative subgroup of S^1 given by a finite set of generators $\{y_j\}_{j=1}^s$.

Lemma 6.2.1. *There exists a bijection between groups* $G \subset S^1$ *and the dimension groups* $(\Lambda_G, \Lambda_G^+, u)$ *given by the formula*

$$\begin{cases} \Lambda_G := \mathbf{Z} + \mathbf{Z}\omega_1 + \cdots + \mathbf{Z}\omega_s \subset \mathbf{R}, & \omega_j = \dfrac{1}{2\pi}\operatorname{Arg}(y_j), \\ \Lambda_G^+ := \Lambda_G \cap \mathbf{R}^+, \\ u := \textit{an order unit}. \end{cases}$$

The rank of $(\Lambda_G, \Lambda_G^+, u)$ *is equal to* $s - t + 1$, *where* t *is the total number of roots of unity among* y_j.

Proof. If y_j is a root of unity, then $\omega_j = \frac{p_j}{q_j}$ is a rational number. We have

$$\begin{aligned} \Lambda_G &\cong \mathbf{Z} + \mathbf{Z}\frac{p_1}{q_1} + \cdots + \mathbf{Z}\frac{p_t}{q_t} + \mathbf{Z}\omega_1 + \cdots + \mathbf{Z}\omega_{s-t} \\ &\cong \mathbf{Z} + \mathbf{Z}\omega_1 + \cdots + \mathbf{Z}\omega_{s-t}, \end{aligned}$$

where ω_i are linearly independent irrational numbers. Since the set Λ_G is dense in $\mathbf{R}$, the triple $(\Lambda_G, \Lambda_G^+, u)$ is a dimension group [74, Corollary 4.7]. Clearly, the rank of such a group is equal to $s - t + 1$. The converse statement is proved similarly. Lemma 6.2.1 follows. □

Let $\mathcal{E}(\mathbf{R})$ (resp. $\mathcal{E}(\mathbf{C})$) be the real (resp. complex) points of $\mathcal{E}(K)$; we have the following inclusions:

$$\mathcal{E}(K) \subset \mathcal{E}(\mathbf{R}) \subset \mathcal{E}(\mathbf{C}).$$

In view of Mordell–Néron Theorem, we shall write $\mathcal{E}(K) \cong \mathbf{Z}^r \oplus \mathcal{E}_{tors}(K)$, where $r = \operatorname{rk}\mathcal{E}(K)$ and $\mathcal{E}_{tors}(K)$ are the torsion points of $\mathcal{E}(K)$. The Weierstrass $\wp$-function maps $\mathcal{E}(\mathbf{C})$ (resp. $\mathcal{E}(\mathbf{R})$) to a complex torus $\mathbf{C}/\Lambda$ (resp. the one-dimensional compact connected Lie group S^1). In view of the above inclusion, the points of $\mathcal{E}(K)$ map to an abelian subgroup G of S^1 such that the $\mathcal{E}_{tors}(K)$ consists of the roots of unity, see, e. g., [260, p. 42]. Denote by t the total number of generators of $\mathcal{E}_{tors}(K)$; then $\mathcal{E}(K)$ has $s = r+t$ generators. Lemma 6.2.1 implies a bijection between elliptic curves $\mathcal{E}(K)$ modulo their torsion points and dimension groups $(\Lambda_G, \Lambda_G^+, u)$ of rank $r+1$; the following definition is natural.

Definition 6.2.2. By the Mordell AF-algebra $\mathbb{A}_{\mathcal{E}(K)}$ of an elliptic curve $\mathcal{E}(K)$ we understand an AF-algebra given by an isomorphism

$$K_0(\mathbb{A}_{\mathcal{E}(K)}) \cong (\Lambda_G, \Lambda_G^+, u),$$

where $(\Lambda_G, \Lambda_G^+, u)$ is a dimension group of the rank $r+1$.

6.2.3 Proof of Theorem 6.2.1

Theorem 6.2.1 follows from an observation that the Mordell AF-algebra $\mathbb{A}_{\mathcal{E}(K)}$ is a noncommutative coordinate ring of the Euler variety V_E. We shall split the proof in a series of lemmas.

Lemma 6.2.2. *The V_E is a flat family of abelian varieties A_E over the projective line $\mathbf{C}P^1$.*

Proof. To prove Lemma 6.2.2, we shall adopt and generalize an argument of [225, Proposition 3.17]. Let us show that there exists a flat morphism $p : V_E \to \mathbf{C}P^1$ of the Euler variety V_E into $\mathbf{C}P^1$ with the fiber $p^{-1}(x)$ an abelian variety A_E.

Indeed, recall that the algebras $\mathcal{A}_\theta$ and $\mathcal{A}_{\theta'}$ are Morita equivalent if and only if $\theta' = \frac{\alpha\theta+\beta}{\gamma\theta+\delta}$, where $\alpha, \beta, \gamma, \delta \in \mathbf{Z}$ satisfy the equality $\alpha\delta - \beta\gamma = \pm 1$ [239]. In other words, the continued fractions of θ and θ' must coincide everywhere but a finite number of entries [225, Section 13]. On the other hand, $\mathcal{A}_{\theta'} \cong \mathcal{A}_\theta$ are isomorphic algebras if and only if $\theta' = \theta$ or $\theta' = 1 - \theta$ [239].

Remark 6.2.3. The K-isomorphism class of elliptic curve $\mathcal{E}(K)$ corresponds to the isomorphism class of the algebra $\mathcal{A}_{RM} = F(\mathcal{E}(K))$, while the $\bar{K}$-isomorphism class of $\mathcal{E}(K)$ corresponds to the Morita equivalence class of the algebra $\mathcal{A}_{RM}$, where $\bar{K} \cong \mathbf{C}$ is the algebraic closure of the field K. (In other words, all twists of the elliptic curve $\mathcal{E}(K)$ are enumerated by the isomorphism classes contained in the Morita equivalence class of the algebra $\mathcal{A}_{RM}$.) This fact follows from a canonical isomorphism $A(K) \otimes M_n \cong A(K^{(n)})$, where $A(K)$ is an algebra over the field K, M_n is the matrix algebra of rank n and $K^{(n)}$ is an extension of degree n of K.

Let $\theta_d = \frac{a+b\sqrt{d}}{c} = [g_1, \ldots, g_m, \overline{k_1, \ldots, k_{n-m}}]$, where $(k_1, \ldots, k_{n-m})$ is the minimal period of the continued fraction of θ_d. To find the Euler equations for θ_d, let $D = b^2 d$ and write θ_D in the form

$$\theta_D = \frac{A_n\theta_D + A_{n-1}}{B_n\theta_D + B_{n-1}},$$

where $\frac{A_i}{B_i}$ is the ith partial fraction of θ_D and $A_nB_{n-1} - A_{n-1}B_n = \pm 1$ [225, Section 19]; in particular,

$$\begin{cases} A_n = k_nA_{n-1} + A_{n-2}, \\ B_n = k_nB_{n-1} + B_{n-2}. \end{cases}$$

From the above equation, one finds that

$$\theta_D = \frac{A_n - B_{n-1} + \sqrt{(A_n - B_{n-1})^2 + 4A_{n-1}B_n}}{2B_n}.$$

Comparing the last equation and with conditions above, we conclude that the Euler equations for θ_D have the form

$$\begin{cases} A_n - B_{n-1} = c_1, \\ 2B_n = c_2, \\ D = c_1^2 + 2c_2A_{n-1}, \end{cases}$$

where $c_1, c_2 \in \mathbf{Z}$ are constants. Using formulas above, one can rewrite c_1 and c_2 in the form

$$\begin{cases} c_1 = k_nA_{n-1} + A_{n-2} - B_{n-1}, \\ c_2 = 2k_nB_{n-1} + 2B_{n-2}. \end{cases}$$

We substitute c_1 and c_2 into the last equation of the above system and get an equation

$$D - k_n^2A_{n-1}^2 - 2(A_{n-1}A_{n-2} - A_{n-1}B_{n-1} + 2B_{n-1})k_n = (A_{n-2} + B_{n-1})^2 \pm 4.$$

But the equation above is a linear diophantine equation in variables $D - k_n^2A_{n-1}$ and k_n; since it has one integer solution, it has infinitely many such solutions. In other words, the projective closure V_E of an affine variety defined by the Euler equations is a fiber bundle over the projective line $\mathbf{C}P^1$.

Let us show that the fiber $p^{-1}(x)$ of a flat morphism $p : V_E \to \mathbf{C}P^1$ is an abelian variety A_E. Indeed, the fraction $[g_1, \dots, g_m, \overline{k_1, \dots, k_{n-m}}]$ corresponding to the quadratic irrationality θ_d can be written in a normal form for which the Euler equations $A_n - B_{n-1} = c_1$ and $2B_n = c_2$ in the above system become a trivial identity. (For instance, if $c_1 = 0$, then the normal fraction has the form $[y_1, \overline{\kappa_1, \kappa_2, \dots, \kappa_2, \kappa_1, 2y_1}]$, see, e. g., [225, Section 24].) Denote by $[y_1, \dots, y_m, \overline{\kappa_1, \dots, \kappa_{n-m}}]$ the normal form of the fraction

$[g_1,\dots,g_m, \overline{k_1,\dots,k_{n-m}}]$. Then the Euler equations reduce to a single equation $D = c_1^2 + c_2A_{n-1}$, which is equivalent to the linear equation above. Thus the quotient of the Euler variety V_E by $\mathbf{C}P^1$ corresponds to the projective closure A_E of an affine variety defined by y_i and κ_j. But the A_E has an obvious translation symmetry, i. e., $\{y_i' = y_i + c_i, \kappa_j' = \kappa_j + c_j \mid c_i, c_j \in \mathbf{Z}\}$ define an isomorphic projective variety. In other words, the fiber $p^{-1}(x) \cong A_E$ is an abelian group, i. e., the A_E is an abelian variety. Lemma 6.2.2 follows. □

Corollary 6.2.2. $\dim_{\mathbf{C}} V_E = 1 + \dim_{\mathbf{C}} A_E$.

Proof. Since the morphism $p : V_E \to \mathbf{C}P^1$ is flat, we conclude that

$$\dim_{\mathbf{C}} V_E = \dim_{\mathbf{C}}(\mathbf{C}P^1) + \dim_{\mathbf{C}} p^{-1}(x).$$

But $\dim_{\mathbf{C}}(\mathbf{C}P^1) = 1$ and $p^{-1}(x) \cong A_E$. Corollary 6.2.2 follows from the above formula. □

Lemma 6.2.3. *The algebra $\mathbb{A}_{\mathcal{E}(K)}$ is a noncommutative coordinate ring of the abelian variety A_E.*

Proof. This fact follows from an equivalence of four categories $\mathbf{E}_0 \cong \mathbf{A}_0 \cong \mathbf{M} \cong \mathbf{V}$, where $\mathbf{E}_0 \subset \mathbf{E}$ is a subcategory of the category $\mathbf{E}$ of all nonsingular elliptic curves consisting of the curves defined over a number field K, $\mathbf{A}_0 \subset \mathbf{A}$ is a subcategory of the category $\mathbf{A}$ of all noncommutative tori consisting of the tori with real multiplication, $\mathbf{M}$ is the category of all Mordell AF-algebras, and $\mathbf{V}$ is the category of all Euler varieties. The morphisms in the categories $\mathbf{E}$ and $\mathbf{V}$ are isomorphisms between the projective varieties and morphisms in the categories $\mathbf{A}$ and $\mathbf{M}$ are Morita equivalences (stable isomorphisms) between the corresponding C^*-algebras [170]. Let us pass to a detailed argument.

The equivalence $\mathbf{E}_0 \cong \mathbf{A}_0$ follows from the results of [183]. It remains to show that $\mathbf{E}_0 \cong \mathbf{M}$ and $\mathbf{A}_0 \cong \mathbf{V}$.

(i) Let us prove that $\mathbf{E}_0 \cong \mathbf{M}$. Let $\mathcal{E}(K) \in \mathbf{E}_0$ be an elliptic curve and $\mathbb{A}_{\mathcal{E}(K)} \in \mathbf{M}$ be the corresponding Mordell AF-algebra. Then $K_0(\mathbb{A}_{\mathcal{E}(K)}) \cong \mathcal{E}(K)$ by an isomorphism of dimension groups. Note that the order unit u depends on the choice of generators of the abelian group $\mathcal{E}(K)$ and, therefore, an isomorphism class of $\mathcal{E}(K)$ corresponds to the Morita equivalence class of the AF-algebra $\mathbb{A}_{\mathcal{E}(K)}$. Thus the above formula defines an equivalence between the categories $\mathbf{E}_0$ and $\mathbf{M}$.

(ii) Let us prove that $\mathbf{A}_0 \cong \mathbf{V}$. Let $\mathcal{A}_{RM} \in \mathbf{A}_0$ be a noncommutative torus with real multiplication and let $(k_1,\dots,k_n)$ be the minimal period of continued fraction corresponding to the quadratic irrational number θ_d. The period is a Morita invariant of the algebra $\mathcal{A}_{RM}$ and a cyclic permutation of k_i gives an algebra $\mathcal{A}_{RM}'$ which is Morita equivalent to the $\mathcal{A}_{RM}$. On the other hand, it is not hard to see that the period $(k_1,\dots,k_n)$ can be uniquely recovered from the coefficients of the Euler equations and a cyclic permu-

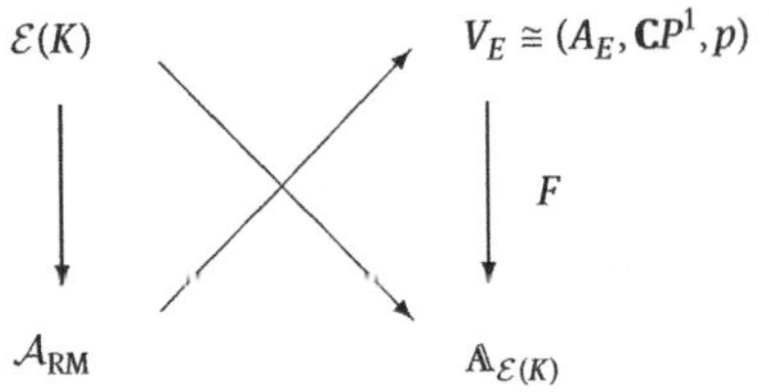

Figure 6.4: Functor $F : \mathbf{V} \to \mathbf{M}$.

tation of k_i defines an Euler variety V'_E which is isomorphic to the V_E. In other words, the categories $\mathbf{A}_0$ and $\mathbf{V}$ are isomorphic.

Comparing (i) and (ii) with the equivalence $\mathbf{E}_0 \cong \mathbf{A}_0$, one concludes that $\mathbf{V} \cong \mathbf{M}$ are equivalent categories, where a functor $F : \mathbf{V} \to \mathbf{M}$ acts by the formula $V_E \mapsto \mathbb{A}_{\mathcal{E}(K)}$ given by the closure of arrows of the commutative diagram in Fig. 6.4. It remains to recall that the V_E is a fiber bundle over the $\mathbf{C}P^1$; therefore all geometric data of V_E is recorded by the fiber A_E alone. We conclude that the algebra $\mathbb{A}_{\mathcal{E}(K)}$ is a coordinate ring of the abelian variety A_E, i. e., an isomorphism class of A_E corresponds to the Morita equivalence class of the algebra $\mathbb{A}_{\mathcal{E}(K)}$. Lemma 6.2.3 follows. □

Remark 6.2.4. The diagram in Fig. 6.4 implies an equivalence of the categories: (i) $\mathbf{E}_0 \cong \mathbf{V}$ and (ii) $\mathbf{A}_0 \cong \mathbf{M}$. The bijection (i) is realized by the formula $\mathcal{E}(K) \mapsto A_E \cong \mathrm{Jac}(X(N))$, where $X(N)$ is an Eichler–Shimura–Taniyama modular curve of level N over the $\mathcal{E}(K)$ and $\mathrm{Jac}(X(N))$ is the Jacobian of $X(N)$. The bijection (ii) is given by the formula $\mathcal{A}_{RM} \mapsto \mathbb{A}_{\mathcal{E}(K)}$ such that the number field $K_0(\mathbb{A}_{\mathcal{E}(K)}) \otimes \mathbf{Q}$ is the maximal abelian extension of a real quadratic number field $K_0(\mathcal{A}_{RM}) \otimes \mathbf{Q}$.

Lemma 6.2.4. $\mathrm{rk}\,\mathcal{E}(K) = \dim_{\mathbf{C}} A_E$.

Proof. Let A_E be an abelian variety over the field $\mathbf{Q}$ and let $n = \dim_{\mathbf{C}} A_E$. In this case it is known that the coordinate ring of A_E is isomorphic to the algebra $\mathcal{A}_{RM}^{2n}$, where $\mathcal{A}_{RM}^{2n}$ is a $2n$-dimensional noncommutative torus with real multiplication. Moreover,

$$K_0(\mathcal{A}_{RM}^{2n}) \cong \mathbf{Z} + \mathbf{Z}\theta_1 + \cdots + \mathbf{Z}\theta_n,$$

see Remark 6.6.2. But we know from Lemma 6.2.3 that a coordinate ring of the abelian variety A_E is isomorphic to the algebra $\mathbb{A}_{\mathcal{E}(K)}$ such that

$$\mathrm{rk}\,K_0(\mathbb{A}_{\mathcal{E}(K)}) = 1 + \mathrm{rk}\,\mathcal{E}(K).$$

Comparing the formulas above, we conclude that $\mathrm{rk}\,\mathcal{E}(K) = n = \dim_{\mathbf{C}} A_E$. Lemma 6.2.4 follows. □

Proof. Theorem 6.2.1 follows from Corollary 6.2.2, Lemma 6.2.4, and the definition of the arithmetic complexity $c(\mathcal{A}_{RM}) = \dim_{\mathbf{C}} V_E$. □

6.2.4 Numerical examples

We shall illustrate Theorem 6.2.1 by two series of examples; the first are the so-called **Q**-curves introduced in [101] and the second is a family of the rational elliptic curves.

6.2.4.1 Q-curves

The **Q**-curves were introduced in Section 1.4.2. The **Q**-ranks of the **Q**-curves and the arithmetic complexity of the corresponding noncommutative tori are shown in Fig. 1.5 for the primes $1 < p < 100$. The reader can see that in the above examples the rank of elliptic curves and the corresponding arithmetic complexity satisfy the equality

$$rk_{\mathbf{Q}}(\mathcal{E}_{CM}^{(-p,1)}) = c(\mathcal{A}_{RM}^{(p,1)}) - 1,$$

where $\mathcal{A}_{RM}^{(p,1)} = F(\mathcal{E}_{CM}^{(-p,1)})$.

6.2.4.2 Rational elliptic curves

For an integer $b \geq 3$, we shall consider a family of the rational elliptic curves given by the equation

$$\mathcal{E}_b(\mathbf{Q}) \cong \left\{(x,y,z) \in \mathbf{C}P^2 \;\middle|\; y^2 z = x(x-z)\left(x - \frac{b-2}{b+2}z\right)\right\}.$$

It was shown in [196] that $F(\mathcal{E}_b(\mathbf{Q})) = \mathcal{A}_{\theta_d}$, where $\theta_d = \frac{1}{2}(b + \sqrt{b^2-4})$. The following evaluation of the rank of elliptic curves in the family $\mathcal{E}_b(\mathbf{Q})$ is true.

Corollary 6.2.3. $\mathrm{rk}\,\mathcal{E}_b(\mathbf{Q}) \leq 2$.

Proof. It is easy to see that

$$\theta_d = \frac{b + \sqrt{b^2-4}}{2} = [b-1, \overline{1, b-2}].$$

Thus $m = 1$ and $n = 3$. From Corollary 6.2.1, one gets $\mathrm{rk}\,\mathcal{E}_b(\mathbf{Q}) \leq 2$. □

Corollary 6.2.4. *The elliptic curve $\mathcal{E}_3(\mathbf{Q})$ has a twist of rank* 0.

Proof. From the above equation, one gets $\theta_5 = \frac{1}{2}(3 + \sqrt{5}) = [2, \overline{1}]$. But $\mathcal{A}_{\theta_5}$ is Morita equivalent to an algebra $\mathcal{A}_{\theta_5'}$, where

$$\theta_5' = \frac{1+\sqrt{5}}{2} = [\overline{1}] = 1 + \cfrac{1}{1 + \cfrac{1}{1 + \cdots}}$$

is a purely periodic continued fraction. In particular, we have $m = 0$ and $n = 1$. From Corollary 6.2.1 and Remark 6.2.3, one concludes that the rank of a twist of $\mathcal{E}_3(\mathbf{Q})$ is equal to 0. □

Guide to the literature

The problem of ranks dates back to [234, p. 493]. It was proved in [165] that the rank of any rational elliptic curve is always a finite number. The result was extended in [172] to the elliptic curves over any number field K. The ranks of individual elliptic curves are calculated by the method of descent, see, e. g., [49, p. 205]. An analytic approach uses the Hasse–Weil L-function $L(\mathcal{E}_\tau, s)$; it was conjectured by B. J. Birch and H. P. F. Swinnerton-Dyer that the order of zero of such a function at $s = 1$ is equal to the rank of $\mathcal{E}_\tau$, see, e. g., [274, p. 198]. A rank conjecture involving invariants of the noncommutative tori was formulated in [181]. Torsion points were studied in [186].

6.3 Transcendental number theory

Whether the irrational value of a transcendental function is algebraic or transcendental for certain algebraic arguments is an old problem of number theory. The algebraic values are particularly remarkable and worthy of thorough investigation [115, p. 456]. Only a few general results are known; we refer the reader to the monograph [20]. For instance, the Gelfond–Schneider Theorem says that $e^{\beta\log\alpha}$ is a transcendental number, whenever $\alpha \notin \{0,1\}$ is an algebraic and β an irrational algebraic number. In contrast, Klein's invariant $j(\tau)$ is known to take algebraic values whenever $\tau \in \mathbb{H} := \{x + iy \in \mathbf{C} \mid y > 0\}$ is an imaginary quadratic number. We use functor $F : \mathbf{Ell} \to \mathbf{NC\text{-}Tor}$ to prove that the value of $\{\mathcal{J}(\theta,\varepsilon) := e^{2\pi i\theta+\log\log\varepsilon} \mid \theta,\varepsilon \in \mathbf{Q}(\sqrt{D})\}$ is an algebraic number generating the Hilbert class field of the imaginary quadratic field $\mathbf{Q}(\sqrt{-D})$.

6.3.1 Algebraic values of $\mathcal{J}(\theta,\varepsilon) = e^{2\pi i\theta+\log\log\varepsilon}$

Let $k = \mathbf{Q}(\sqrt{D})$ be a real quadratic field and $R_f = \mathbf{Z} + fO_k$ be an order of conductor $f \geq 1$ in k. Let $h = |Cl(R_f)|$ be the class number of the ring R_f. Denote by $\{\mathbf{Z} + \mathbf{Z}\theta_i \mid 1 \leq i \leq h\}$ the set of the pairwise nonisomorphic pseudo-lattices in k having the same endomorphism ring R_f [155, Lemma 1.1.1]. Suppose that ε is the fundamental unit of R_f. Finally, let f' be the least integer satisfying equation $|Cl(\mathbf{Z} + f'O_{k'})| = h$, where $k' := \mathbf{Q}(\sqrt{-D})$ is the imaginary quadratic field.

Theorem 6.3.1. *For each square-free positive integer $D \notin \{1,2,3,7,11,19,43,67,163\}$, the values $\{\mathcal{J}(\theta_i,\varepsilon) \mid 1 \leq i \leq h\}$ are algebraically conjugate numbers generating the Hilbert class field of k' modulo conductor f'.*

Remark 6.3.1. The absolute value $|z| = (z\bar{z})^{\frac{1}{2}}$ of an algebraic number z is always an "abstract" algebraic number, i. e., a root of a polynomial with integer coefficients; yet Theorem 6.3.1 implies that $|\mathcal{J}(\theta_i, \varepsilon)| = \log \varepsilon$ is a transcendental number. This observation is not a contradiction since quadratic extensions of the field $\mathbf{Q}(z\bar{z})$ may have no real embeddings in general; in other words, our extension cannot be a subfield of $\mathbf{R}$.

6.3.2 Proof of Theorem 6.3.1

From Lemma 5.1.3, the system of defining relations

$$\begin{cases} x_3x_1 = e^{2\pi i\theta}x_1x_3, \\ x_4x_2 = e^{2\pi i\theta}x_2x_4, \\ x_4x_1 = e^{-2\pi i\theta}x_1x_4, \\ x_3x_2 = e^{-2\pi i\theta}x_2x_3, \\ x_2x_1 = x_1x_2 = e, \\ x_4x_3 = x_3x_4 = e, \end{cases}$$

for the noncommutative torus $\mathcal{A}_\theta$ and defining relations

$$\begin{cases} x_3x_1 = \mu e^{2\pi i\theta}x_1x_3, \\ x_4x_2 = \frac{1}{\mu}e^{2\pi i\theta}x_2x_4, \\ x_4x_1 = \mu e^{-2\pi i\theta}x_1x_4, \\ x_3x_2 = \frac{1}{\mu}e^{-2\pi i\theta}x_2x_3, \\ x_2x_1 = x_1x_2, \\ x_4x_3 = x_3x_4, \end{cases}$$

for the Sklyanin $*$-algebra $S(q_{13})$ with $q_{13} = \mu e^{2\pi i\theta} \in \mathbf{C}$ are identical modulo the scaled unit relation

$$x_1x_2 = x_3x_4 = \frac{1}{\mu}e.$$

On the other hand, we have:

(i) $q_{13} = \mu e^{2\pi i\theta} \in K$, where $K = k(j(\mathcal{E}_{CM}^{(-D,f')}))$ is the Hilbert class field of the imaginary quadratic field $\mathbf{Q}(\sqrt{-D})$ modulo conductor f';

(ii) $\theta \in \mathbf{Q}(\sqrt{D})$, since $F(\mathcal{E}_{CM}^{(-D,f')}) = \mathcal{A}_{RM}^{(D,f)}$ by Theorem 6.1.3.

Thus one gets an inclusion

$$\mu e^{2\pi i\theta} \in K, \quad \text{where } \theta \in \mathbf{Q}(\sqrt{D}).$$

Lemma 6.3.1. *For each noncommutative torus $\mathcal{A}_{\mathrm{RM}}^{(D,f)}$, the constant $\mu = \log \varepsilon$, where $\varepsilon > 1$ is the Perron–Frobenius eigenvalue of the positive integer matrix*

$$A = \begin{pmatrix} a_1 & 1 \\ 1 & 0 \end{pmatrix} \cdots \begin{pmatrix} a_n & 1 \\ 1 & 0 \end{pmatrix}$$

and $\overline{(a_1, \dots, a_n)}$ is the period of continued fraction of the corresponding quadratic irrationality θ.

Proof. (i) Recall that the range of the canonical trace τ on projections of algebra $\mathcal{A}_\theta \otimes \mathcal{K}$ is given by pseudo-lattice $\Lambda = \mathbf{Z} + \mathbf{Z}\theta$ [239, p. 195]. Because $\tau(\frac{1}{\mu}e) = \frac{1}{\mu}\tau(e) = \frac{1}{\mu}$, the pseudo-lattice corresponding to the algebra $\mathcal{A}_\theta$ with a scaled unit can be written as $\Lambda_\mu = \mu(\mathbf{Z} + \mathbf{Z}\theta)$.

(ii) To express μ in terms of the inner invariants of pseudo-lattice Λ, denote by R the ring of endomorphisms of Λ and by $U_R \subset R$ the multiplicative group of automorphisms (units) of Λ. For each $\varepsilon, \varepsilon' \in U_R$, it must hold that

$$\mu(\varepsilon\varepsilon'\Lambda) = \mu(\varepsilon\varepsilon')\Lambda = \mu(\varepsilon)\Lambda + \mu(\varepsilon')\Lambda,$$

since μ is an additive functional on the pseudo-lattice Λ. Canceling Λ in the above equation, one gets

$$\mu(\varepsilon\varepsilon') = \mu(\varepsilon) + \mu(\varepsilon'), \quad \forall \varepsilon, \varepsilon' \in U_R.$$

The only real-valued function on U_R with such a property is the logarithmic function (a *regulator* of U_R); thus $\mu(\varepsilon) = \log \varepsilon$.

(iii) Notice that U_R is generated by a single element ε. To calculate the generator, recall that pseudo-lattice $\Lambda = \mathbf{Z} + \mathbf{Z}\theta$ is isomorphic to a pseudo-lattice $\Lambda' = \mathbf{Z} + \mathbf{Z}\theta'$, where $\theta' = \overline{(a_1, \dots, a_n)}$ is purely periodic continued fraction and $(a_1, \dots, a_n)$ is the period of continued fraction of θ. From the standard facts of the theory of continued fractions, one gets that ε coincides with the Perron–Frobenius eigenvalue of the matrix

$$A = \begin{pmatrix} a_1 & 1 \\ 1 & 0 \end{pmatrix} \cdots \begin{pmatrix} a_n & 1 \\ 1 & 0 \end{pmatrix}.$$

Clearly, $\varepsilon > 1$ and it is an invariant of the stable isomorphism class of algebra $\mathcal{A}_\theta$. Lemma 6.3.1 is proved. □

Remark 6.3.2 (Second proof of Lemma 6.3.1). Lemma 6.3.1 follows from a purely measure-theoretic argument. Indeed, if $h_x : \mathbf{R} \to \mathbf{R}$ is a "stretch-out" automorphism of the real line $\mathbf{R}$ given by the formula $t \mapsto tx$, $\forall t \in \mathbf{R}$, then the only h_x-invariant measure μ on $\mathbf{R}$ is the "scale-back" measure $d\mu = \frac{1}{t}dt$. Taking the antiderivative and integrating between $t_0 = 1$ and $t_1 = x$, one gets

$$\mu = \log x.$$

It remains to notice that for pseudo-lattice $K_0^+(\mathcal{A}_{RM}^{(D,f)}) \cong \mathbf{Z} + \mathbf{Z}\theta \subset \mathbf{R}$, the automorphism h_x corresponds to $x = \varepsilon$, where $\varepsilon > 1$ is the Perron–Frobenius eigenvalue of matrix A. Lemma 6.3.1 follows.

Theorem 6.3.1 is an implication of the following argument:

(i) Let $D \notin \{1, 2, 3, 7, 11, 19, 43, 67, 163\}$ be a positive square-free integer; for the sake of simplicity, let $f' = 1$, i. e., the endomorphism ring of $\mathcal{E}_{CM}^{(-D,f')}$ coincides with the ring of integers O_k of the imaginary quadratic field $k = \mathbf{Q}(\sqrt{-D})$. In this case, $\mathcal{E}_{CM}^{(-D,f')} \cong \mathcal{E}(K)$, where $K = k(j(\mathcal{E}(K)))$ is the Hilbert class field of k. It follows from the well-known facts of complex multiplication that condition $D \notin \{1, 2, 3, 7, 11, 19, 43, 67, 163\}$ is necessary and sufficient for $K \not\cong \mathbf{Q}$, i. e., the field K has a complex embedding.

(ii) Let $\mathcal{E}_1(K), \dots, \mathcal{E}_h(K)$ be pairwise nonisomorphic curves with $\mathrm{End}(\mathcal{E}_i(K)) \cong O_k$, where h is the class number of O_k. Repeating for each $\mathcal{E}_i(K)$ the argument at the beginning of the proof of Theorem 6.3.1, we conclude that $\mu e^{2\pi i\theta_i} \in K$.

(iii) Lemma 6.3.1 says that $\mu = \log \varepsilon$; thus for each $1 \le i \le h$, one gets

$$(\log \varepsilon)e^{2\pi i\theta_i} = e^{2\pi i\theta_i + \log\log \varepsilon} = \mathcal{J}(\theta_i, \varepsilon) \in K.$$

(iv) The transitive action of the ideal class group $\mathrm{Cl}(k) \cong \mathrm{Gal}(K|k)$ on the elliptic curves $\mathcal{E}_i(K)$ extends to the algebraic numbers $\mathcal{J}(\theta_i, \varepsilon)$; thus $\mathcal{J}(\theta_i, \varepsilon) \in K$ are algebraically conjugate.

Theorem 6.3.1 is proved.

6.3.3 Comments on a note by M. Waldschmidt

In [282] it was argued that the value of $\mu = \log \varepsilon$ must be equal to 1. As a consequence, Theorem 6.3.1 was claimed to be false. In this section we explain why $\mu = \log \varepsilon$ cannot be equal to 1 thus upholding Theorem 6.3.1.

Our comments touch [282, second paragraph], since [282, first paragraph] is covered in Remark 6.3.1. Recall that the β-KMS (Kubo–Martin–Schwinger) state on a C^*-algebra $\mathcal{A}$ is a real number β such that

$$\omega(x\sigma_{i\beta}(y)) = \omega(yx), \quad \forall x, y \in \mathcal{A},$$

where σ_t is the 1-parameter group of automorphisms of $\mathcal{A}$ and ω is a weight on $\mathcal{A}$ (Section 11.1.1). Such a state is the fundamental invariant of algebra $\mathcal{A}$ known in the quantum statistical mechanics as an inverse temperature. If $\mathcal{A}$ is an AF (Approximately Finite) C^*-algebra given by a constant incidence matrix $A \in GL_n(\mathbf{Z})$, the inverse temperature was calculated in [42, Chapter 4 and Figs. 1–5] by the remarkable formula

$$\beta = \log \lambda_A,$$

where $\lambda_A > 1$ is the Perron–Frobenius eigenvalue of the matrix A. Moreover, this KMS state is unique [42].

In our case, $n = 2$ and the Pimsner–Voiculescu theorem says that the noncommutative torus $\mathcal{A}_\theta$ embeds into an AF-algebra preserving the respective K_0-groups (Theorem 3.5.3). In particular, the incidence matrix A is nothing but the matrix form of the fundamental unit $\varepsilon > 1$ of the number field $\mathbb{Q}(\theta)$ and therefore the inverse temperature of $\mathcal{A}_\theta$ is

$$\beta = \log \varepsilon.$$

Since μ is known to parametrize the state space of the C^*-algebra $\mathcal{A}_\theta$, we identify it with β, keeping the notation below.

Assuming $\beta = 1$, [282, second paragraph] would simply mean that a representation of the Sklyanin $*$-algebra $\{S(q) \mid q = \beta e^{2\pi i\theta}\}$ by the linear operators on a Hilbert space does not exist since there is no scalar product to apply the standard GNS construction (Section 3.1). The same is true for all other values of β, unless $\beta = \log \varepsilon$, in which case the GNS construction gives a representation of the Sklyanin algebra. The existence of such a representation is critical for the validity of Theorem 6.3.1. This resolves the issue raised in [282].

Guide to the literature

A complex number is algebraic if it is a root of a polynomial with integer coefficients. It is notoriously hard to prove that a given complex number is algebraic. Thanks to Ch. Hermite and C. L. Lindemann the numbers e and π are not algebraic, but even for $e \pm \pi$ the answer is unknown. The Seventh Hilbert Problem deals with such type of questions [115, p. 456]. The famous Gelfond–Schneider Theorem says that $e^{\beta \log \alpha}$ is a transcendental number, whenever $\alpha \notin \{0, 1\}$ is an algebraic and β an irrational algebraic number. We refer the reader to the book [20] for an excellent introduction. The noncommutative invariants were linked to the transcendental number theory in [198].

6.4 Class field theory

Such a theory deals with the extensions $K|k$ of a number field k, where the Galois group $\mathrm{Gal}(K|k)$ is abelian. A general formula for the abelian extensions in terms of the arithmetic of field k is known, see, e. g., [258, p. 115]. Yet generators of such extensions are unknown, except for the cases $k \cong \mathbf{Q}$ or an imaginary quadratic field $k \cong \mathbf{Q}(\sqrt{-D})$. We use functor $F : \mathbf{Ell} \to \mathbf{NC\text{-}Tor}$ to construct generators of the Hilbert class field of a real quadratic field $k \cong \mathbf{Q}(\sqrt{D})$. Such a solution was conjectured in [155].

6.4.1 Hilbert class field of a real quadratic field

The *Kronecker's Jugendtraum* (Hilbert's 12th problem) is a conjecture that the maximal nonramified abelian extension of any algebraic number field is generated by the special values of modular functions attached to an abelian variety. The conjecture is true for the rational field and imaginary quadratic fields with the modular functions being an exponent and the j-invariant, respectively. In the case of an arbitrary number field, a description of the abelian extensions is given by the class field theory, but an explicit formula for the generators of these abelian extensions, in the sense sought by Kronecker, is unknown even for the real quadratic fields.

We construct generators of the Hilbert class field (i. e., the maximal nonramified extension) of a real quadratic field based on a modularity and a symmetry of complex and real multiplication. To give an idea, let

$$\Gamma_1(N) = \left\{ \begin{pmatrix} a & b \\ c & d \end{pmatrix} \in SL_2(\mathbf{Z}) \;\middle|\; a \equiv d \equiv 1 \mod N,\ c \equiv 0 \mod N \right\}$$

be a congruence subgroup of level $N \geq 1$ and $\mathbb{H}$ be the Lobachevsky half-plane; let $X_1(N) := \mathbb{H}/\Gamma_1(N)$ be the corresponding modular curve and $S_2(\Gamma_1(N))$ the space of all cusp forms on $\Gamma_1(N)$ of weight 2. Let $\mathcal{E}_{CM}^{(-D,f)}$ be an elliptic curve with complex multiplication by an order $R_f = \mathbf{Z} + fO_k$ in the field $k = \mathbf{Q}(\sqrt{-D})$ [259, Chapter II]. Denote by $\mathcal{K}^{ab}(k) := k(j(\mathcal{E}_{CM}^{(-D,f)}))$ the Hilbert class field of k modulo conductor $f \geq 1$ and let $N = fD$; let $\mathrm{Jac}(X_1(fD))$ be the Jacobian of modular curve $X_1(fD)$. There exists an abelian subvariety $A_\phi \subset \mathrm{Jac}(X_1(fD))$ such that its points of finite order generate $\mathcal{K}^{ab}(k)$ [113], [253, Theorem 1], and [254, Section 8]. The field $\mathcal{K}^{ab}(k)$ is a *CM-field*, i. e., a totally imaginary quadratic extension of the totally real field $\mathcal{K}_\phi$ generated by the Fourier coefficients of the Hecke eigenform $\phi(z) \in S_2(\Gamma_1(fD))$ [254, p. 137]. In particular, there exists a holomorphic map $X_1^0(fD) \to \mathcal{E}_{CM}^{(-D,f)}$, where $X_1^0(fD)$ is a Riemann surface such that $\mathrm{Jac}(X_1^0(fD)) \cong A_\phi$; we refer to the above as a *modularity* of complex multiplication.

On the other hand, Theorem 6.1.3 says that the noncommutative torus $\mathcal{A}_\theta = F(\mathcal{E}_{CM}^{(-D,f)})$ has real multiplication by an order $\mathfrak{R}_{\mathfrak{f}} = \mathbf{Z} + \mathfrak{f}O_{\mathfrak{k}}$ in the field $\mathfrak{k} = \mathbf{Q}(\sqrt{D})$. We assume that $\mathfrak{f} = f^m$ for the minimal integer m satisfying an isomorphism $\mathrm{Cl}(\mathfrak{R}_{f^m}) \cong$

$Cl(R_f)$, where $Cl(R_f)$ and $Cl(\mathfrak{R}_\mathfrak{f})$ are the ideal class groups of orders R_f and $\mathfrak{R}_\mathfrak{f}$, respectively. (We refer the reader to Remark 6.4.6 for a motivation.) As usual, we denote by $\mathcal{A}_{RM}^{(D,\mathfrak{f})}$ a noncommutative torus with real multiplication by $\mathfrak{R}_\mathfrak{f}$.

Remark 6.4.1. The isomorphism $Cl(\mathfrak{R}_{f^m}) \cong Cl(R_f)$ can be calculated using the well-known formula for the class number of a nonmaximal order $\mathbf{Z} + fO_K$ of a quadratic field $K = \mathbf{Q}(\sqrt{D})$,

$$h_{\mathbf{Z}+fO_K} = \frac{h_{O_K} f}{e_f} \prod_{p|f} \left(1 - \left(\frac{D}{p}\right)\frac{1}{p}\right),$$

where h_{O_K} is the class number of the maximal order O_K, e_f is the index of the group of units of $\mathbf{Z}+fO_K$ in the group of units of O_K, p is a prime number and $(\frac{D}{p})$ is the Legendre symbol, see, e. g., [38, p. 153] and [111, pp. 297 and 351].

The (twisted homogeneous) coordinate ring of the Riemann surface $X_1^0(fD)$ is an AF-algebra $\mathbb{A}_{\phi^0}$ linked to a holomorphic differential $\phi^0(z)dz$ on $X_1^0(fD)$, see Definition 6.4.1 and Remark 6.4.5 for the details. The Grothendieck semigroup $K_0^+(\mathbb{A}_{\phi^0})$ is a pseudo-lattice $\mathbf{Z} + \mathbf{Z}\theta_1 + \cdots + \mathbf{Z}\theta_{n-1}$ in the number field $\mathcal{K}_\phi$, where n equals the genus of $X_1^0(fD)$. Moreover, a holomorphic map $X_1^0(fD) \to \mathcal{E}_{CM}^{(-D,f)}$ induces the C^*-algebra homomorphism $\mathbb{A}_{\phi^0} \to \mathcal{A}_{RM}^{(D,\mathfrak{f})}$ between the corresponding coordinate rings, so that the diagram in Fig. 6.5 is commutative. But $K_0^+(\mathcal{A}_{RM}^{(D,\mathfrak{f})})$ is a pseudo-lattice $\mathbf{Z} + \mathbf{Z}\theta$ in the field $\mathfrak{k}$, such that $\mathrm{End}(\mathbf{Z} + \mathbf{Z}\theta) \cong \mathfrak{R}_\mathfrak{f}$. In other words, one can use the diagram in Fig. 6.5 to control the arithmetic of the field $\mathcal{K}_\phi$ by such of the real quadratic field $\mathfrak{k}$. This observation solves Hilbert's 12th problem for the real quadratic fields.

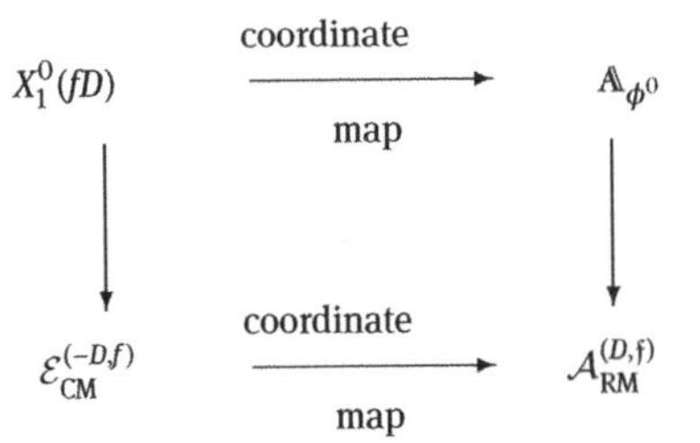

Figure 6.5: Symmetry of complex and real multiplication.

Theorem 6.4.1. *The Hilbert class field of a real quadratic field $\mathfrak{k} = \mathbf{Q}(\sqrt{D})$ modulo conductor f^m is an extension of $\mathfrak{k}$ by the Fourier coefficients of the Hecke eigenform $\phi(z) \in S_2(\Gamma_1(fD))$, where m is the least integer satisfying isomorphism $Cl(\mathfrak{R}_{f^m}) \cong Cl(R_f)$.*

Remark 6.4.2. Theorem 6.4.1 can be used to compute concrete extensions. For instance, Theorem 6.4.1 says that for the quadratic field $\mathbf{Q}(\sqrt{15})$ its Hilbert class field is isomorphic to $\mathbf{Q}(\sqrt{-1+\sqrt{15}})$ and for $\mathbf{Q}(\sqrt{14})$ such a field modulo conductor $\mathfrak{f} = 8$ is isomorphic to $\mathbf{Q}(\sqrt[4]{-27+8\sqrt{14}})$.

6.4.2 AF-algebra of the Hecke eigenform

Let $N \geq 1$ be a natural number and consider a (finite index) subgroup of the modular group given by the formula:

$$\Gamma_1(N) = \left\{ \begin{pmatrix} a & b \\ c & d \end{pmatrix} \in SL_2(\mathbf{Z}) \;\middle|\; a \equiv d \equiv 1 \mod N,\; c \equiv 0 \mod N \right\}.$$

Let $\mathbb{H} = \{z = x + iy \in \mathbf{C} \mid y > 0\}$ be the upper half-plane and let $\Gamma_1(N)$ act on $\mathbb{H}$ by the linear fractional transformations; consider an orbifold $\mathbb{H}/\Gamma_1(N)$. To compactify the orbifold at the cusps, one adds a boundary to $\mathbb{H}$, so that $\mathbb{H}^* = \mathbb{H} \cup \mathbf{Q} \cup \{\infty\}$ and the compact Riemann surface $X_1(N) = \mathbb{H}^*/\Gamma_1(N)$ is called a modular curve. The meromorphic functions $\phi(z)$ on $\mathbb{H}$ that vanish at the cusps and such that

$$\phi\left(\frac{az+b}{cz+d}\right) = (cz+d)^2\phi(z), \quad \forall \begin{pmatrix} a & b \\ c & d \end{pmatrix} \in \Gamma_0(N),$$

are called cusp forms of weight two; the (complex linear) space of such forms will be denoted by $S_2(\Gamma_1(N))$. The formula $\phi(z) \mapsto \omega = \phi(z)dz$ defines an isomorphism $S_2(\Gamma_1(N)) \cong \Omega_{hol}(X_1(N))$, where $\Omega_{hol}(X_1(N))$ is the space of all holomorphic differentials on the Riemann surface $X_1(N)$. Note that $\dim_{\mathbf{C}}(S_2(\Gamma_1(N))) = \dim_{\mathbf{C}}(\Omega_{hol}(X_1(N))) = g$, where $g = g(N)$ is the genus of the surface $X_1(N)$. A Hecke operator, T_n, acts on $S_2(\Gamma_1(N))$ by the formula $T_n\phi = \sum_{m\in\mathbf{Z}} \gamma(m)q^m$, where $\gamma(m) = \sum_{a|\mathbf{GCD}(m,n)} ac_{mn/a^2}$ and $\phi(z) = \sum_{m\in\mathbf{Z}} c(m)q^m$ is the Fourier series of the cusp form ϕ at $q = e^{2\pi iz}$. Further, T_n is a self-adjoint linear operator on the vector space $S_2(\Gamma_1(N))$ endowed with the Petersson inner product; the algebra $\mathbb{T}_N := \mathbf{Z}[T_1, T_2, \dots]$ is a commutative algebra. Any cusp form $\phi \in S_2(\Gamma_1(N))$ that is an eigenvector for one (and hence all) of T_n is referred to as a Hecke eigenform. The Fourier coefficients $c(m)$ of ϕ are algebraic integers, and we denote by $\mathcal{K}_\phi = \mathbf{Q}(c(m))$ an extension of the field $\mathbf{Q}$ by the Fourier coefficients of ϕ. Then $\mathcal{K}_\phi$ is a real algebraic number field of degree $1 \leq \deg(\mathcal{K}_\phi|\mathbf{Q}) \leq g$, where g is the genus of the surface $X_1(N)$, see, e. g., [68, Proposition 6.6.4]. Any embedding $\sigma : \mathcal{K}_\phi \to \mathbf{C}$ conjugates ϕ by acting on its coefficients; we write the corresponding Hecke eigenform $\phi^\sigma(z) := \sum_{m\in\mathbf{Z}} \sigma(c(m))q^m$ and call ϕ^σ a *conjugate* of the Hecke eigenform ϕ.

Let $\omega = \phi(z)dz \in \Omega_{hol}(X)$ be a holomorphic differential on a Riemann surface X. We shall denote by $\mathrm{Re}(\omega)$ a closed form on X (the real part of ω) and consider its periods $\lambda_i = \int_{\gamma_i} \mathrm{Re}(\omega)$ against a basis γ_i in the (relative) homology group $H_1(X, Z(\mathrm{Re}(\omega)); \mathbf{Z})$, where $Z(\mathrm{Re}(\omega))$ is the set of zeros of the form $\mathrm{Re}(\omega)$. Assume $\lambda_i > 0$ and consider the vector $\theta = (\theta_1, \dots, \theta_{n-1})$ with $\theta_i = \lambda_{i+1}/\lambda_1$. The Jacobi–Perron continued fraction of θ is given by the formula

$$\begin{pmatrix}1\\ \theta\end{pmatrix} = \lim_{i\to\infty}\begin{pmatrix}0 & 1\\ I & b_1\end{pmatrix}\cdots\begin{pmatrix}0 & 1\\ I & b_i\end{pmatrix}\begin{pmatrix}0\\ \mathbb{I}\end{pmatrix} = \lim_{i\to\infty} B_i\begin{pmatrix}0\\ \mathbb{I}\end{pmatrix},$$

where $b_i = (b_1^{(i)},\dots,b_{n-1}^{(i)})^T$ is a vector of nonnegative integers, I is the unit matrix, and $\mathbb{I} = (0,\dots,0,1)^T$, see, e. g., [23]. By $\mathbb{A}_\phi$ we shall understand the AF-algebra given the Bratteli diagram with partial multiplicity matrices B_i. If $\phi(z) \in S_2(\Gamma_1(N))$ is a Hecke eigenform, then the corresponding AF-algebra $\mathbb{A}_\phi$ is a stationary AF-algebra with the partial multiplicity matrices $B_i = \text{const} = B$; moreover, each conjugate eigenform ϕ^σ defines a *companion* AF-algebra $\mathbb{A}_{\phi^\sigma}$, i. e., the AF-algebras given by nonsimilar matrices B and B' having the same characteristic polynomial. It is known that $K_0^+(\mathbb{A}_\phi) \cong \mathbf{Z} + \mathbf{Z}\theta_1 + \cdots + \mathbf{Z}\theta_{n-1} \subset \mathcal{K}_\phi$, where $\mathcal{K}_\phi$ is an algebraic number field generated by the Fourier coefficients of ϕ. We refer the reader to [184] for the details.

6.4.3 Proof of Theorem 6.4.1

Definition 6.4.1. Let $A_\phi \subset \text{Jac}(X_1(fD))$ be an abelian variety associated to the Hecke eigenform $\phi(z) \in S_2(\Gamma_1(fD))$, see, e. g., [68, Definition 6.6.3]. By $X_1^0(fD)$ we shall understand the Riemann surface of genus g such that

$$\text{Jac}(X_1^0(fD)) \cong A_\phi.$$

By $\phi^0(z)dz \in \Omega_{\text{hol}}(X_1^0(fD))$ we denote the image of the Hecke eigenform $\phi(z)dz \in \Omega_{\text{hol}}(X_1(fD))$ under the holomorphic map $X_1(fD) \to X_1^0(fD)$.

Remark 6.4.3. The surface $X_1^0(fD)$ is correctly defined. Indeed, since the abelian variety A_ϕ is the product of g copies of an elliptic curve with the complex multiplication, there exists a holomorphic map from A_ϕ to the elliptic curve. For a Riemann surface X of genus g covering the elliptic curve $\mathcal{E}_{CM}$ by a holomorphic map (such a surface and a map always exist), one gets a period map $X \to A_\phi$ by closing the arrows of a commutative diagram $A_\phi \to \mathcal{E}_{CM} \leftarrow X$. It is easy to see that the Jacobian of X coincides with A_ϕ and we set $X_1^0(fD) := X$.

Lemma 6.4.1. $g(X_1^0(fD)) = \deg(\mathcal{K}^{ab}(k)|k)$.

Proof. By definition, abelian variety A_ϕ is the quotient of $\mathbf{C}^n$ by a lattice of periods of the Hecke eigenform $\phi(z) \in S_2(\Gamma_1(fD))$ and all its conjugates $\phi^\sigma(z)$ on the Riemann surface $X_1(fD)$. These periods are complex algebraic numbers generating the Hilbert class field $\mathcal{K}^{ab}(k)$ over imaginary quadratic field $k = \mathbf{Q}(\sqrt{-D})$ modulo conductor f [113, 253] and [254, Section 8]. The number of linearly independent periods is equal to the total number of the conjugate eigenforms $\phi^\sigma(z)$, i. e., $|\sigma| = n = \dim_{\mathbf{C}}(A_\phi)$. Since real dimension $\dim_{\mathbf{R}}(A_\phi) = 2n$, we conclude that $\deg(\mathcal{K}^{ab}(k)|\mathbf{Q}) = 2n$ and, therefore,

$\deg(\mathcal{K}^{ab}(k)|k) = n$. But $\dim_{\mathbf{C}}(A_\phi) = g(X_1^0(fD))$, and one gets $g(X_1^0(fD)) = \deg(\mathcal{K}^{ab}(k)|k)$. Lemma 6.4.1 follows. □

Corollary 6.4.1. $g(X_1^0(fD)) = |Cl(R_f)|$.

Proof. Because $\mathcal{K}^{ab}(k)$ is the Hilbert class field over k modulo conductor f, we must have

$$\mathrm{Gal}(\mathcal{K}^{ab}(k)|k) \cong Cl(R_f),$$

where Gal is the Galois group of the extension $\mathcal{K}^{ab}(k)|k$ and $Cl(R_f)$ is the class group of ring R_f, see, e. g., [259, p. 112]. But $|\mathrm{Gal}(\mathcal{K}^{ab}(k)|k)| = \deg(\mathcal{K}^{ab}(k)|k)$ and, by Lemma 6.4.1, we have $\deg(\mathcal{K}^{ab}(k)|k) = g(X_1^0(fD))$. In view of this and isomorphism $\mathrm{Gal}(\mathcal{K}^{ab}(k)|k) \cong Cl(R_f)$, one gets $|Cl(R_f)| = |\mathrm{Gal}(\mathcal{K}^{ab}|k)| = g(X_1^0(fD))$. Corollary 6.4.1 follows. □

Lemma 6.4.2. $g(X_1^0(fD)) = \deg(\mathcal{K}_\phi|\mathbf{Q})$.

Proof. It is known that $\dim_{\mathbf{C}}(A_\phi) = \deg(\mathcal{K}_\phi|\mathbf{Q})$, see, e. g., [68, Proposition 6.6.4]. But abelian variety $A_\phi \cong \mathrm{Jac}(X_1^0(fD))$ and, therefore, $\dim_{\mathbf{C}}(A_\phi) = \dim_{\mathbf{C}}(\mathrm{Jac}(X_1^0(fD))) = g(X_1^0(fD))$, hence the lemma. □

Corollary 6.4.2. $\deg(\mathcal{K}_\phi|\mathbf{Q}) = |Cl(\mathfrak{R}_{\mathfrak{f}})|$.

Proof. From Lemma 6.4.2 and Corollary 6.4.1, one gets $\deg(\mathcal{K}_\phi|\mathbf{Q}) = |Cl(R_f)|$. In view of this and isomorphism $Cl(\mathfrak{R}_{f^m}) \cong Cl(R_f)$, one gets the conclusion of Corollary 6.4.2. □

Lemma 6.4.3 (Basic lemma). $\mathrm{Gal}(\mathcal{K}_\phi|\mathbf{Q}) \cong Cl(\mathfrak{R}_{\mathfrak{f}})$.

Proof. Let us outline the proof. In view of Lemma 6.4.2 and Corollaries 6.4.1–6.4.2, we denote by h the single integer $g(X_1^0(fD)) = |Cl(R_f)| = |Cl(\mathfrak{R}_{\mathfrak{f}})| = \deg(\mathcal{K}_\phi|\mathbf{Q})$. Since $\deg(\mathcal{K}_\phi|\mathbf{Q}) = h$, there exist $\{\phi_1, \dots, \phi_h\}$ conjugate Hecke eigenforms $\phi_i(z) \in S_2(\Gamma_1(fD))$ [68, Theorem 6.5.4]. Thus one gets h holomorphic forms $\{\phi_1^0, \dots, \phi_h^0\}$ on the Riemann surface $X_1^0(fD)$. Let $\{\mathbb{A}_{\phi_1^0}, \dots, \mathbb{A}_{\phi_h^0}\}$ be the corresponding stationary AF-algebras; the $\mathbb{A}_{\phi_i^0}$ are *companion* AF-algebras. Recall that the characteristic polynomial for the partial multiplicity matrices $B_{\phi_i^0}$ of companion AF-algebras $\mathbb{A}_{\phi_i^0}$ is the same; it is a minimal polynomial of degree h and let $\{\lambda_1, \dots, \lambda_h\}$ be the roots of such a polynomial, compare with [74, Corollary 6.3]. Since $\det(B_{\phi_i^0}) = 1$, the numbers λ_i are algebraic units of the field $\mathcal{K}_\phi$. Moreover, λ_i are algebraically conjugate and can be taken for generators of the extension $\mathcal{K}_\phi|\mathbf{Q}$; since $\deg(\mathcal{K}_\phi|\mathbf{Q}) = h = |Cl(\mathfrak{R}_{\mathfrak{f}})|$, there exists a natural action of group $Cl(\mathfrak{R}_{\mathfrak{f}})$ on these generators. The action extends to automorphisms of the entire field $\mathcal{K}_\phi$ preserving $\mathbf{Q}$; thus one gets the Galois group of extension $\mathcal{K}_\phi|\mathbf{Q}$ and an isomorphism $\mathrm{Gal}(\mathcal{K}_\phi|\mathbf{Q}) \cong Cl(\mathfrak{R}_{\mathfrak{f}})$. Let us pass to a step-by-step argument.

(i) Let $h := g(X_1^0(fD)) = |Cl(R_f)| = |Cl(\mathfrak{R}_{\mathfrak{f}})|$ and let $\phi(z) \in S_2(\Gamma_1(fD))$ be the Hecke eigenform. It is known that there exist $\{\phi_1, \dots, \phi_h\}$ conjugate Hecke eigenforms, so that $\phi(z)$ is one of them, see [68, Theorem 6.5.4]. Let $\{\phi_1^0, \dots, \phi_h^0\}$ be the corresponding forms on the Riemann surface $X_1^0(fD)$.

Remark 6.4.4. The forms $\{\phi_1^0, \dots, \phi_h^0\}$ can be taken for a basis in the space $\Omega_{\text{hol}}(X_1^0(fD))$; we leave it to the reader to verify that abelian variety A_ϕ is isomorphic to the quotient of $\mathbf{C}^h$ by the lattice of periods of holomorphic differentials $\phi_i^0(z)dz$ on $X_1^0(fD)$.

(ii) Let $\mathbb{A}_{\phi_i^0}$ be the AF-algebra corresponding to holomorphic differential $\phi_i^0(z)dz$ on $X_1^0(fD)$. The set $\{\mathbb{A}_{\phi_1^0}, \dots, \mathbb{A}_{\phi_h^0}\}$ consists of the companion AF-algebras. It is known that each $\mathbb{A}_{\phi_i^0}$ is a stationary AF-algebra, i. e., its partial multiplicity matrix is a constant; we shall denote such a matrix by $B_{\phi_i^0}$.

(iii) By definition, the matrices $B_{\phi_i^0}$ of companion AF-algebras $\mathbb{A}_{\phi_i}$ have the same characteristic polynomial $p(x) \in \mathbf{Z}[x]$; the matrices $B_{\phi_i^0}$ itself are not pairwise similar and, therefore, the AF-algebras $\mathbb{A}_{\phi_i^0}$ are not pairwise isomorphic. The total number h of such matrices is equal to the class number of the endomorphism ring of pseudo-lattice $K_0^+(\mathbb{A}_{\phi_i^0}) \cong \mathbf{Z} + \mathbf{Z}\theta_1^i + \dots + \mathbf{Z}\theta_{h-1}^i \subset \mathcal{K}_\phi$ [74, Corollary 6.3].

Remark 6.4.5. Notice that there are $\{X_1, \dots, X_h\}$ pairwise nonisomorphic Riemann surfaces $X := X_1^0(fD)$ endowed with a holomorphic map $X_i \to \mathcal{E}_i$, where $\{\mathcal{E}_1, \dots, \mathcal{E}_h\}$ are pairwise nonisomorphic elliptic curves $\mathcal{E}_{\text{CM}}^{(-D,f)}$ corresponding to elements of the group $\text{Cl}(R_f)$. Thus the companion AF-algebras $\{\mathbb{A}_{\phi_1^0}, \dots, \mathbb{A}_{\phi_h^0}\}$ can be viewed as coordinate rings of $\{X_1, \dots, X_h\}$; the latter means that $\mathbb{A}_{\phi_i^0}$ discern nonisomorphic Riemann surfaces and $K_0^+(\mathbb{A}_{\phi_i^0}) \cong \mathbf{Z} + \mathbf{Z}\theta_1^i + \dots + \mathbf{Z}\theta_{h-1}^i$ represents the moduli space of $X_1^0(fD)$.

(iv) The polynomial $p(x)$ is minimal and splits in the totally real field $\mathcal{K}_\phi$. Indeed, matrices $B_{\phi_i^0}$ generate the Hecke algebra $\mathbb{T}_N$ on $S_2(\Gamma_1(N))$; thus each $B_{\phi_i^0}$ is self-adjoint and, therefore, all eigenvalues are real of multiplicity one; since $B_{\phi_i^0}$ is integer, all roots of characteristic polynomial $p(x)$ of $B_{\phi_i^0}$ belong to the field $\mathcal{K}_\phi$.

(v) Let $p(x) = (x-\lambda_1)\cdots(x-\lambda_h)$. It is easy to see that λ_i are algebraic units of the field $\mathcal{K}_\phi$ because $\det(B_{\phi_i^0}) = 1$; note that numbers $\{\lambda_1, \dots, \lambda_h\}$ are algebraically conjugate. Since $\deg(\mathcal{K}_\phi|\mathbf{Q}) = h$, the numbers λ_i can be taken for generators of the field $\mathcal{K}_\phi$, i. e., $\mathcal{K}_\phi = \mathbf{Q}(\lambda_1, \dots, \lambda_h)$.

(vi) Finally, let us establish an explicit formula for the isomorphism

$$\text{Cl}(\mathfrak{R}_\mathfrak{f}) \to \text{Gal}(\mathcal{K}_\phi|\mathbf{Q}).$$

Since $\text{Gal}(\mathcal{K}_\phi|\mathbf{Q})$ is an automorphism group of the field $\mathcal{K}_\phi$ preserving $\mathbf{Q}$, it will suffice to define the action $*$ of an element $a \in \text{Cl}(\mathfrak{R}_\mathfrak{f})$ on the generators λ_i of $\mathcal{K}_\phi$. Let $\{a_1, \dots, a_h\}$ be the set of all elements of the group $\text{Cl}(\mathfrak{R}_\mathfrak{f})$. For an element $a \in \text{Cl}(\mathfrak{R}_\mathfrak{f})$, define an index function α by the formula $a_i a = a_{\alpha(i)}$. Then the action $*$ of an element $a \in \text{Cl}(\mathfrak{R}_\mathfrak{f})$ on the generators λ_i of the field $\mathcal{K}_\phi$ is given by the formula

$$a * \lambda_i := \lambda_{\alpha(i)}, \quad \forall a \in \text{Cl}(\mathfrak{R}_\mathfrak{f}).$$

It is easy to verify that above formula gives an isomorphism $Cl(\mathfrak{R}_{\mathfrak{f}}) \to Gal(\mathcal{K}_\phi|\mathbf{Q})$, which is independent of the choice of $\{a_i\}$ and $\{\lambda_i\}$. This argument completes the proof of Lemma 6.4.3. □

Remark 6.4.6. The class field theory says that $\mathfrak{f} = f^m$, i. e., the extensions of fields k and $\mathfrak{k}$ must ramify over the same set of prime ideals. Indeed, consider the commutative diagram in Fig. 6.6, where I_f and $I_{\mathfrak{f}}$ are groups of all ideals of k and $\mathfrak{k}$, which are relatively prime to the principal ideals (f) and $(\mathfrak{f})$, respectively. Since $Gal(\mathcal{K}^{ab}(k)|\mathbf{Q}) \cong Gal(\mathcal{K}_\phi|\mathbf{Q})$, one gets an isomorphism $I_f \cong I_{\mathfrak{f}}$, i. e., $\mathfrak{f} = f^m$ for some positive integer m.

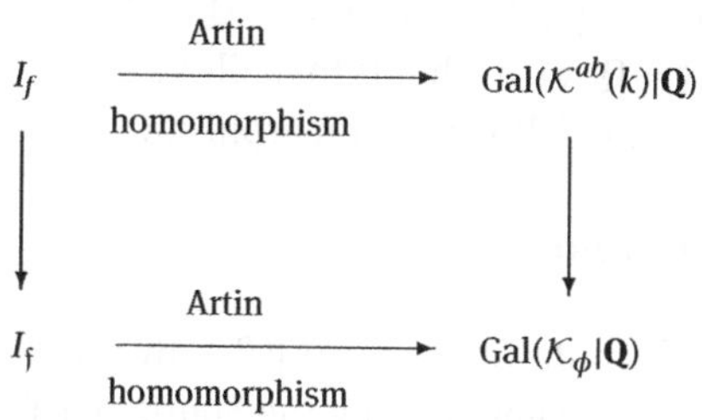

Figure 6.6: Artin homomorphism.

Corollary 6.4.3. *The Hilbert class field of real quadratic field $\mathfrak{k} = \mathbf{Q}(\sqrt{D})$ modulo conductor $\mathfrak{f} \geq 1$ is isomorphic to the field $\mathfrak{k}(\mathcal{K}_\phi)$ generated by the Fourier coefficients of the Hecke eigenform $\phi(z) \in S_2(\Gamma_1(fD))$.*

Proof. As in the classical case of imaginary quadratic fields, notice that $\deg(\mathcal{K}_\phi|\mathbf{Q}) = \deg(\mathfrak{k}(\mathcal{K}_\phi)|\mathfrak{k}) = Cl(\mathfrak{R}_{\mathfrak{f}})$; therefore Corollary 6.4.3 is an implication of Lemma 6.4.3 and isomorphism $Gal(\mathcal{K}_\phi|\mathbf{Q}) \cong Gal(\mathfrak{k}(\mathcal{K}_\phi)|\mathfrak{k}) \cong Cl(\mathfrak{R}_{\mathfrak{f}})$. □

Theorem 6.4.1 follows from Corollary 6.4.3.

6.4.4 Examples

Along with the method of Stark's units [52], Theorem 6.4.1 can be used in the computational number theory. For the sake of clarity, we shall consider the simplest examples; the rest can be found in Table 6.1.

Example 6.4.1. Let $D = 15$. The class number of quadratic field $k = \mathbf{Q}(\sqrt{-15})$ is known to be 2; such a number for quadratic field $\mathfrak{k} = \mathbf{Q}(\sqrt{15})$ is also equal to 2. Thus,

$$Cl(\mathfrak{R}_{\mathfrak{f}=1}) \cong Cl(R_{f=1}) \cong \mathbf{Z}/2\mathbf{Z},$$

and isomorphism $Cl(\mathfrak{R}_{f^m}) \cong Cl(R_f)$ is trivially satisfied for each power m, i. e., one obtains an unramified extension. By Theorem 6.4.1, the Hilbert class field of $\mathfrak{k}$ is generated by the Fourier coefficients of the Hecke eigenform $\phi(z) \in S_2(\Gamma_1(15))$. Using the computer program *SAGE* created by William A. Stein, one finds an irreducible factor $p(x) = x^2 - 4x + 5$ of the characteristic polynomial of the Hecke operator $T_{p=2}$ acting on the space $S_2(\Gamma_1(15))$. Therefore, the Fourier coefficient $c(2)$ coincides with a root of equation $p(x) = 0$; in other words, we arrive at an extension of $\mathfrak{k}$ by the polynomial $p(x)$. The generator x of the field $\mathcal{K}_\phi = \mathbf{Q}(c(2))$ is a root of the biquadratic equation $[(x-2)^2+1]^2 - 15 = 0$; it is easy to see that $x = 2 + \sqrt{-1+\sqrt{15}}$. One concludes that the field $\mathcal{K}_\phi \cong \mathbf{Q}(\sqrt{-1+\sqrt{15}})$ is the Hilbert class field of quadratic field $\mathfrak{k} = \mathbf{Q}(\sqrt{15})$.

Example 6.4.2. Let $D = 14$. It is known that for the quadratic field $k = \mathbf{Q}(\sqrt{-14})$ we have $|\mathrm{Cl}(R_{f=1})| = 4$, while for the quadratic field $\mathfrak{k} = \mathbf{Q}(\sqrt{14})$ it holds that $|\mathrm{Cl}(\mathfrak{R}_{\mathfrak{f}=1})| = 1$. However, for the ramified extensions, one obtains the following isomorphism:

$$\mathrm{Cl}(\mathfrak{R}_{\mathfrak{f}=2^3}) \cong \mathrm{Cl}(R_{f=2}) \cong \mathbf{Z}/4\mathbf{Z},$$

where $m = 3$ is the smallest integer satisfying formula $\mathrm{Cl}(\mathfrak{R}_{f^m}) \cong \mathrm{Cl}(R_f)$. By Theorem 6.4.1, the Hilbert class field of $\mathfrak{k}$ modulo $\mathfrak{f} = 8$ is generated by the Fourier coefficients of the Hecke eigenform $\phi(z) \in S_2(\Gamma_1(2\times 14))$. Using *SAGE*, one finds that the characteristic polynomial of the Hecke operator $T_{p=3}$ on $S_2(\Gamma_1(2\times 14))$ has an irreducible factor $p(x) = x^4 + 3x^2 + 9$. Thus the Fourier coefficient $c(3)$ is a root of the polynomial $p(x)$ and one gets an extension of $\mathfrak{k}$ by the polynomial $p(x)$. In other words, generator x of the field $\mathcal{K}_\phi = \mathbf{Q}(c(3))$ is a root of the polynomial equation $(x^4+3x^2+9)^2 - 4\times 14 = 0$. The biquadratic equation $x^4 + 3x^2 + 9 - 2\sqrt{14} = 0$ has discriminant $-27 + 8\sqrt{14}$ and one finds a generator of $\mathcal{K}_\phi$ to be $\sqrt[4]{-27+8\sqrt{14}}$. Thus the field $\mathbf{Q}(\sqrt[4]{-27+8\sqrt{14}})$ is the Hilbert class field over $\mathbf{Q}(\sqrt{14})$ modulo conductor $\mathfrak{f} = 8$. Clearly, the extension is ramified over the prime ideal $\mathfrak{P} = (2)$.

Remark 6.4.7. Table 6.1 lists quadratic fields for some square-free discriminants $2 \le D \le 101$. The conductors f and $\mathfrak{f}$ satisfying isomorphism $\mathrm{Cl}(\mathfrak{R}_{f^m}) \cong \mathrm{Cl}(R_f)$ were calculated using tables for the class number of nonmaximal orders in quadratic fields posted at www.numbertheory.org; the site is maintained by Keith Matthews. We focused on small conductors; the interested reader can compute the higher conductors using a pocket calculator. In contrast, computation of generator x of the Hilbert class field requires the online program *SAGE* created by William A. Stein. We write an explicit formula for x or its minimal polynomial $p(x)$ over $\mathfrak{k}$.

Guide to the literature

The abelian extensions of a real quadratic field were first studied in [112]. A description of such extensions in terms of coordinates of points of finite order on abelian varieties associated with certain modular curves was obtained in [254]. Stark formulated a number of conjectures on abelian extension of arbitrary number fields, which in the real

Table 6.1: Square-free discriminants $2 \le D \le 101$.

D	f	$Cl(R_f)$	$\mathfrak{f}$	Hilbert class field of $\mathbf{Q}(\sqrt{D})$ modulo conductor $\mathfrak{f}$
2	1	trivial	1	$\mathbf{Q}(\sqrt{2})$
3	1	trivial	1	$\mathbf{Q}(\sqrt{3})$
7	1	trivial	1	$\mathbf{Q}(\sqrt{7})$
11	1	trivial	1	$\mathbf{Q}(\sqrt{11})$
14	2	$\mathbf{Z}/4\mathbf{Z}$	8	$\mathbf{Q}(\sqrt[4]{-27+8\sqrt{14}})$
15	1	$\mathbf{Z}/2\mathbf{Z}$	1	$\mathbf{Q}(\sqrt{-1+\sqrt{15}})$
19	1	trivial	1	$\mathbf{Q}(\sqrt{19})$
21	2	$\mathbf{Z}/4\mathbf{Z}$	8	$\mathbf{Q}(\sqrt[4]{-3+2\sqrt{21}})$
35	1	$\mathbf{Z}/2\mathbf{Z}$	1	$\mathbf{Q}(\sqrt{17+\sqrt{35}})$
43	1	trivial	1	$\mathbf{Q}(\sqrt{43})$
51	1	$\mathbf{Z}/2\mathbf{Z}$	1	$\mathbf{Q}(\sqrt{17+\sqrt{51}})$
58	1	$\mathbf{Z}/2\mathbf{Z}$	1	$\mathbf{Q}(\sqrt{-1+\sqrt{58}})$
67	1	trivial	1	$\mathbf{Q}(\sqrt{67})$
82	1	$\mathbf{Z}/4\mathbf{Z}$	1	$x^4-2x^3+4x^2-8x+16$
91	1	$\mathbf{Z}/2\mathbf{Z}$	1	$\mathbf{Q}(\sqrt{-3+\sqrt{91}})$

quadratic case amount to specifying generators of these extensions using special values of Artin L-functions, see [266]. Based on an analogy with complex multiplication, Manin suggested to use pseudo-lattices $\mathbf{Z}+\mathbf{Z}\theta$ in $\mathbf{R}$ having nontrivial real multiplications to produce abelian extensions of real quadratic fields [155]. Similar to the case of complex multiplication, the endomorphism ring $\mathfrak{R}_{\mathfrak{f}} = \mathbf{Z}+\mathfrak{f}O_{\mathfrak{k}}$ of pseudo-lattice $\mathbf{Z}+\mathbf{Z}\theta$ is an order in the real quadratic field $\mathfrak{k} = \mathbf{Q}(\theta)$, where $O_{\mathfrak{k}}$ is the ring of integers of $\mathfrak{k}$ and $\mathfrak{f}$ is the conductor of $\mathfrak{R}_{\mathfrak{f}}$. The AF-algebra of the Hecke eigenform was studied in [184]. The real multiplication problem was solved in [191].

6.5 Noncommutative reciprocity

We define an L-function of the noncommutative torus $\mathcal{A}_{RM}$ and prove that it coincides with the Hasse–Weil L-function of an elliptic curve $\mathcal{E}_{CM}$ such that $\mathcal{A}_{RM} = F(\mathcal{E}_{CM})$. As a corollary, it is proved that the crossed product of the $\mathcal{A}_{RM}$ by an endomorphism corresponds to localization of the $\mathcal{E}_{CM}$ at a prime p.

6.5.1 *L*-function of noncommutative tori

Let p be a prime number. For an elliptic curve $\mathcal{E}_{CM}$, denote by $\mathcal{E}_{CM}(\mathbb{F}_p)$ a localization of the $\mathcal{E}_{CM}$ at the prime ideal $\mathfrak{P}$ over p [259, p. 171]. We look for an analog of the

$\mathcal{E}_{CM}(\mathbb{F}_p)$ in terms of the algebra $\mathcal{A}_{RM} = F(\mathcal{E}_{CM})$. Recall that the integers $|\mathcal{E}_{CM}(\mathbb{F}_p)|$ generate the Hasse–Weil function $L(\mathcal{E}_{CM}, s)$ of the curve $\mathcal{E}_{CM}$ [259, p. 172]. Thus to solve the localization problem for $\mathcal{A}_{RM}$, we define an L-function of the $\mathcal{A}_{RM}$ equal to the Hasse–Weil function of the $\mathcal{E}_{CM}$. Namely, let θ be a quadratic irrationality whose continued fraction has the minimal period $(a_1, \dots, a_k)$. Consider a two-by-two integer matrix $A = \prod_{i=1}^{k}(a_i, 1, 1, 0)$. Let tr be trace of the matrix and $\pi(n)$ an integer-valued function defined in Supplement 6.5.3. Consider a matrix $L_p = \begin{pmatrix} \mathrm{tr}(A^{\pi(p)}) & p \\ -1 & 0 \end{pmatrix}$ and define an endomorphism of the $\mathcal{A}_{RM}$ by the action of L_p on its generators u and v. The crossed product $\mathcal{A}_{RM} \rtimes_{L_p} \mathbf{Z}$ is stably isomorphic to the Cuntz–Krieger algebra $\mathcal{O}_{L_p}$ [32, § 10.11.9]. For $z \in \mathbf{C}$ and $\alpha \in \{-1, 0, 1\}$, define

$$\zeta_p(\mathcal{A}_{RM}, z) := \exp\left(\sum_{n=1}^{\infty} \frac{|K_0(\mathcal{O}_{\varepsilon_n})|}{n} z^n\right), \quad \varepsilon_n = \begin{cases} L_p^n & \text{if } p \nmid \mathrm{tr}^2(A) - 4, \\ 1 - \alpha^n & \text{if } p \mid \mathrm{tr}^2(A) - 4, \end{cases}$$

a local zeta function of the $\mathcal{A}_{RM}$. An *L-function* of the $\mathcal{A}_{RM}$ is a product of the local zetas over all primes, i. e., $L(\mathcal{A}_{RM}, s) = \prod_p \zeta_p(\mathcal{A}_{RM}, p^{-s})$, where $s \in \mathbf{C}$.

Theorem 6.5.1. *If $\mathcal{A}_{RM} = F(\mathcal{E}_{CM})$, then $L(\mathcal{A}_{RM}, s) = L(\mathcal{E}_{CM}, s)$ and $K_0(\mathcal{O}_{\varepsilon_n}) \cong \mathcal{E}_{CM}(\mathbb{F}_{p^n})$.*

Remark 6.5.1 (Localization formula). The isomorphism $K_0(\mathcal{O}_{\varepsilon_n}) \cong \mathcal{E}_{CM}(\mathbb{F}_{p^n})$ is a localization formula for the algebra $\mathcal{A}_{RM}$.

6.5.2 Proof of Theorem 6.5.1

Let p be such that $\mathcal{E}_{CM}$ has a good reduction at $\mathfrak{P}$; the corresponding local zeta function $\zeta_p(\mathcal{E}_{CM}, z) = (1 - \mathrm{tr}(\psi_{\mathcal{E}(K)}(\mathfrak{P}))z + pz^2)^{-1}$, where $\psi_{\mathcal{E}(K)}$ is the Grössencharacter on K and tr is the trace of algebraic number. We have to prove that $\zeta_p(\mathcal{E}_{CM}, z) = \zeta_p(\mathcal{A}_{RM}, z) := (1 - \mathrm{tr}(A^{\pi(p)})z + pz^2)^{-1}$; the last equality is a consequence of definition of $\zeta_p(\mathcal{A}_{RM}, z)$. Let $\mathcal{E}_{CM} \cong \mathbf{C}/L_{CM}$, where $L_{CM} = \mathbf{Z} + \mathbf{Z}\tau$ is a lattice in the complex plane [259, pp. 95–96]; let $K_0(\mathcal{A}_{RM}) \cong \mathfrak{m}_{RM}$, where $\mathfrak{m}_{RM} = \mathbf{Z} + \mathbf{Z}\theta$ is a pseudo-lattice in $\mathbf{R}$ [155]. Roughly speaking, we construct an invertible element (a unit) u of the ring $\mathrm{End}(\mathfrak{m}_{RM})$ attached to pseudo-lattice $\mathfrak{m}_{RM} = F(L_{CM})$ such that

$$\mathrm{tr}(\psi_{\mathcal{E}(K)}(\mathfrak{P})) = \mathrm{tr}(u) = \mathrm{tr}(A^{\pi(p)}).$$

The latter will be achieved with the help of an explicit formula connecting endomorphisms of lattice L_{CM} with such of the pseudo-lattice $\mathfrak{m}_{RM}$

$$\begin{pmatrix} a & b \\ c & d \end{pmatrix} \in \mathrm{End}(L_{CM}) \longmapsto \begin{pmatrix} a & b \\ -c & -d \end{pmatrix} \in \mathrm{End}(\mathfrak{m}_{RM}),$$

see proof of Lemma 6.1.5 for the details. We shall split the proof into a series of lemmas, starting with the following elementary lemma.

Lemma 6.5.1. *Let $A = (a, b, c, d)$ be an integer matrix with $ad - bc \neq 0$ and $b = 1$. Then A is similar to the matrix $(a + d, 1, c - ad, 0)$.*

Proof. Indeed, take a matrix $(1, 0, d, 1) \in SL_2(\mathbf{Z})$. The matrix realizes the similarity, i. e.,

$$\begin{pmatrix} 1 & 0 \\ -d & 1 \end{pmatrix}\begin{pmatrix} a & 1 \\ c & d \end{pmatrix}\begin{pmatrix} 1 & 0 \\ d & 1 \end{pmatrix} = \begin{pmatrix} a+d & 1 \\ c-ad & 0 \end{pmatrix}.$$

Lemma 6.5.1 follows. □

Lemma 6.5.2. *The matrix $A = (a + d, 1, c - ad, 0)$ is similar to its transpose $A^t = (a + d, c - ad, 1, 0)$.*

Proof. We shall use the following criterion: the integer matrices A and B are similar if and only if the characteristic matrices $xI - A$ and $xI - B$ have the same Smith normal form. The calculation for the matrix $xI - A$ gives

$$\begin{aligned}\begin{pmatrix} x-a-d & -1 \\ ad-c & x \end{pmatrix} &\sim \begin{pmatrix} x-a-d & -1 \\ x^2-(a+d)x+ad-c & 0 \end{pmatrix} \\ &\sim \begin{pmatrix} 1 & 0 \\ 0 & x^2-(a+d)x+ad-c \end{pmatrix},\end{aligned}$$

where $\sim$ are the elementary operations between the rows (columns) of the matrix. Similarly, a calculation for the matrix $xI - A^t$ gives

$$\begin{aligned}\begin{pmatrix} x-a-d & ad-c \\ -1 & x \end{pmatrix} &\sim \begin{pmatrix} x-a-d & x^2-(a+d)x+ad-c \\ -1 & 0 \end{pmatrix} \\ &\sim \begin{pmatrix} 1 & 0 \\ 0 & x^2-(a+d)x+ad-c \end{pmatrix}.\end{aligned}$$

Thus, $(xI - A) \sim (xI - A^t)$ and Lemma 6.5.2 follows. □

Corollary 6.5.1. *The matrices $(a, 1, c, d)$ and $(a + d, c - ad, 1, 0)$ are similar.*

Proof. The Corollary 6.5.1 follows from Lemmas 6.5.1–6.5.2. □

Recall that if $\mathcal{E}_{CM}^{(-D,f)}$ is an elliptic curve with complex multiplication by order $R = \mathbf{Z} + fO_k$ in imaginary quadratic field $k = \mathbf{Q}(\sqrt{-D})$, then $\mathcal{A}_{RM}^{(D,f)} = F(\mathcal{E}_{CM}^{(-D,f)})$ is the noncommutative torus with real multiplication by the order $\mathfrak{R} = \mathbf{Z} + fO_{\mathfrak{k}}$ in real quadratic field $\mathfrak{k} = \mathbf{Q}(\sqrt{D})$.

Lemma 6.5.3. *Each $\alpha \in R$ goes under F into an $\omega \in \mathfrak{R}$ such that $\mathrm{tr}(\alpha) = \mathrm{tr}(\omega)$, where $\mathrm{tr}(x) = x + \bar{x}$ is the trace of an algebraic number x.*

Proof. Recall that each $\alpha \in R$ can be written in a matrix form for a given basis $\{\omega_1, \omega_2\}$ of the lattice L_{CM}. Namely,

$$\begin{cases} \alpha\omega_1 = a\omega_1 + b\omega_2, \\ \alpha\omega_2 = c\omega_1 + d\omega_2, \end{cases}$$

where (a, b, c, d) is an integer matrix with $ad - bc \neq 0$ and $\mathrm{tr}(\alpha) = a + d$. The first equation implies $\alpha = a + b\tau$; since both α and τ are algebraic integers, one concludes that $b = 1$. In view of Corollary 6.5.1, in a basis $\{\omega_1', \omega_2'\}$, α has a matrix form $(a + d, c - ad, 1, 0)$. To calculate a real quadratic $\omega \in \mathfrak{R}$ corresponding to α, recall an explicit formula obtained in the proof of Lemma 6.1.5; namely, each endomorphism (a, b, c, d) of the lattice L_{CM} gives rise to the endomorphism $(a, b, -c, -d)$ of pseudo-lattice $\mathfrak{m}_{RM} = F(L_{CM})$. Thus, one gets a map

$$F : \begin{pmatrix} a + d & c - ad \\ 1 & 0 \end{pmatrix} \mapsto \begin{pmatrix} a + d & c - ad \\ -1 & 0 \end{pmatrix}.$$

In other words, for a given basis $\{\lambda_1, \lambda_2\}$ of the pseudo-lattice $\mathbf{Z} + \mathbf{Z}\theta$, one can write

$$\begin{cases} \omega\lambda_1 = (a + d)\lambda_1 + (c - ad)\lambda_2, \\ \omega\lambda_2 = -\lambda_1. \end{cases}$$

It is easy to verify that ω is a real quadratic integer with $\mathrm{tr}(\omega) = a + d$. The latter coincides with the $\mathrm{tr}(\alpha)$. Lemma 6.5.3 follows. □

Let $\omega \in \mathfrak{R}$ be an endomorphism of the pseudo-lattice $\mathfrak{m}_{RM} = \mathbf{Z} + \mathbf{Z}\theta$ of degree $\deg(\omega) := \omega\bar{\omega} = n$. The endomorphism maps $\mathfrak{m}_{RM}$ to a sublattice $\mathfrak{m}_0 \subset \mathfrak{m}_{RM}$ of index n; any such has the form $\mathfrak{m}_0 = \mathbf{Z} + (n\theta)\mathbf{Z}$, see, e. g., [38, p. 131]. Moreover, ω generates an automorphism u of the pseudo-lattice $\mathfrak{m}_0$; the traces of ω and u are related.

Lemma 6.5.4. $\mathrm{tr}(u) = \mathrm{tr}(\omega)$.

Proof. Let us calculate the action of endomorphism $\omega = (a + d, c - ad, -1, 0)$ on the pseudo-lattice $\mathfrak{m}_0 = \mathbf{Z} + (n\theta)\mathbf{Z}$. Since $\deg(\omega) = c - ad = n$, one gets

$$\begin{pmatrix} a + d & n \\ -1 & 0 \end{pmatrix} \begin{pmatrix} 1 \\ \theta \end{pmatrix} = \begin{pmatrix} a + d & 1 \\ -1 & 0 \end{pmatrix} \begin{pmatrix} 1 \\ n\theta \end{pmatrix},$$

where $\{1, \theta\}$ and $\{1, n\theta\}$ are bases of the pseudo-lattices $\mathfrak{m}_{RM}$ and $\mathfrak{m}_0$, respectively, and $u = (a + d, 1, -1, 0)$ is an automorphism of $\mathfrak{m}_0$. It is easy to see that $\mathrm{tr}(u) = a + d = \mathrm{tr}(\omega)$. Lemma 6.5.4 follows. □

Remark 6.5.2 (Second proof of Lemma 6.5.4). There exists a canonical proof of Lemma 6.5.4 based on the notion of a subshift of finite type [281]; we shall give such a proof below, since it generalizes to pseudo-lattices of any rank. Consider a dimension group ([32, p. 55]) corresponding to the endomorphism ω of lattice $\mathbf{Z}^2$, i. e., the limit $G(\omega)$,

$$\mathbf{Z}^2 \stackrel{\omega}{\to} \mathbf{Z}^2 \stackrel{\omega}{\to} \mathbf{Z}^2 \stackrel{\omega}{\to} \cdots .$$

It is known that $G(\omega) \cong \mathbf{Z}[\frac{1}{\lambda}]$, where $\lambda > 1$ is the Perron–Frobenius eigenvalue of ω. We shall write $\hat{\omega}$ to denote the shift automorphism of dimension group $G(\omega)$ [74, p. 37] and $\zeta_\omega(t) = \exp(\sum_{k=1}^{\infty} \frac{\mathrm{tr}(\omega^k)}{k} t^k)$ and $\zeta_{\hat{\omega}}(t) = \exp(\sum_{k=1}^{\infty} \frac{\mathrm{tr}(\hat{\omega}^k)}{k} t^k)$ the corresponding Artin–Mazur zeta functions [281, p. 273]. Since the Artin–Mazur zeta function of the subshift of finite type is an invariant of shift equivalence, we conclude that $\zeta_\omega(t) \equiv \zeta_{\hat{\omega}}(t)$; in particular, $\mathrm{tr}(\omega) = \mathrm{tr}(\hat{\omega})$. Hence the matrix form of $\hat{\omega} = (a + d, 1, -1, 0) = u$ and, therefore, $\mathrm{tr}(u) = \mathrm{tr}(\omega)$. Lemma 6.5.4 is proved by a different method.

Lemma 6.5.5. *The automorphism u is a unit of the ring $\mathfrak{R}_0 := \mathrm{End}(\mathfrak{m}_0)$; it is the fundamental unit of $\mathfrak{R}_0$, whenever $n = p$ is a prime number and $\mathrm{tr}(u) = \mathrm{tr}(\psi_{\mathcal{E}(K)}(\mathfrak{P}))$, where $(\psi_{\mathcal{E}(K)}(\mathfrak{P}))$ is the Grössencharacter associated to prime p, see Supplement* 6.5.3.

Proof. (i) Since $\deg(u) = 1$, the element u is invertible and, therefore, a unit of the ring $\mathfrak{R}_0$; in general, unit u is not the fundamental unit of $\mathfrak{R}_0$, since it is possible that $u = \varepsilon^a$, where ε is another unit of $\mathfrak{R}_0$ and $a \geq 1$.

(ii) When $n = p$ is a prime number, then we let $\psi_{\mathcal{E}(K)}(\mathfrak{P})$ be the corresponding Grössencharacter on K attached to an elliptic curve $\mathcal{E}_{CM} \cong \mathcal{E}(K)$, see Supplement 6.3.3 for the notation. The Grössencharacter can be identified with a complex number $\alpha \in k$ of the imaginary quadratic field k associated to the complex multiplication. Let $\mathrm{tr}(u) = \mathrm{tr}(\psi_{\mathcal{E}(K)}(\mathfrak{P}))$ and suppose to the contrary that u is not the fundamental unit of $\mathfrak{R}_0$, i. e., $u = \varepsilon^a$ for a unit $\varepsilon \in \mathfrak{R}_0$ and an integer $a \geq 1$. Then there exists a Grössencharacter $\psi'_{\mathcal{E}(K)}(\mathfrak{P})$, such that

$$\mathrm{tr}(\psi'_{\mathcal{E}(K)}(\mathfrak{P})) < \mathrm{tr}(\psi_{\mathcal{E}(K)}(\mathfrak{P})).$$

Since $\mathrm{tr}(\psi_{\mathcal{E}(K)}(\mathfrak{P})) = q_{\mathfrak{P}} + 1 - \#\tilde{E}(\mathbb{F}_{\mathfrak{P}})$, one concludes that $\#\tilde{E}(\mathbb{F}'_{\mathfrak{P}}) > \#\tilde{E}(\mathbb{F}_{\mathfrak{P}})$; in other words, there exists a nontrivial extension $\mathbb{F}'_{\mathfrak{P}} \supset \mathbb{F}_{\mathfrak{P}}$ of the finite field $\mathbb{F}_{\mathfrak{P}}$. The latter is impossible since any extension of $\mathbb{F}_{\mathfrak{P}}$ has the form $\mathbb{F}_{\mathfrak{P}^n}$ for some $n \geq 1$; thus $a = 1$, i. e., unit u is the fundamental unit of the ring $\mathfrak{R}_0$. Lemma 6.5.5 is proved. □

Lemma 6.5.6. $\mathrm{tr}(\psi_{\mathcal{E}(K)}(\mathfrak{P})) = \mathrm{tr}(A^{\pi(p)})$.

Proof. Recall that the fundamental unit of the order $\mathfrak{R}_0$ is given by the formula $\varepsilon_p = \varepsilon^{\pi(p)}$, where ε is the fundamental unit of the ring $O_{\mathfrak{k}}$ and $\pi(p)$ an integer number, see Hasse's Lemma 6.5.10 of Supplement 6.5.3. On the other hand, matrix $A = \prod_{i=1}^{n}(a_i, 1, 1, 0)$, where $\theta = \overline{(a_1, \ldots, a_n)}$ is a purely periodic continued fraction. Therefore

$$A \begin{pmatrix} 1 \\ \theta \end{pmatrix} = \varepsilon \begin{pmatrix} 1 \\ \theta \end{pmatrix},$$

where $\varepsilon > 1$ is the fundamental unit of the real quadratic field $\mathfrak{k} = \mathbf{Q}(\theta)$. In other words, A is the matrix form of the fundamental unit ε. Therefore the matrix form of the fundamental unit $\varepsilon_p = \varepsilon^{\pi(p)}$ of $\mathfrak{R}_0$ is given by matrix $A^{\pi(p)}$. One can apply Lemma 6.5.5 and get

$$\mathrm{tr}(\psi_{\mathcal{E}(K)}(\mathfrak{P})) = \mathrm{tr}(\varepsilon_p) = \mathrm{tr}(A^{\pi(p)}).$$

Lemma 6.5.6 follows. □

One can finish the proof of Theorem 6.5.1 by comparing the local L-series of the Hasse–Weil L-function for the $\mathcal{E}_{\mathrm{CM}}$ with that of the local zeta for the $\mathcal{A}_{\mathrm{RM}}$. The local L-series for $\mathcal{E}_{\mathrm{CM}}$ are $L_{\mathfrak{P}}(\mathcal{E}(K), T) = 1 - a_{\mathfrak{P}}T + q_{\mathfrak{P}}T^2$ if the $\mathcal{E}_{\mathrm{CM}}$ has a good reduction at $\mathfrak{P}$ and $L_{\mathfrak{P}}(\mathcal{E}(K), T) = 1 - \alpha T$ otherwise; here

$$\begin{cases} q_{\mathfrak{P}} = N_{\mathbf{Q}}^{K}\mathfrak{P} = \#\mathbb{F}_{\mathfrak{P}} = p, \\ a_{\mathfrak{P}} = q_{\mathfrak{P}} + 1 - \#\tilde{E}(\mathbb{F}_{\mathfrak{P}}) = \mathrm{tr}(\psi_{\mathcal{E}(K)}(\mathfrak{P})), \\ \alpha \in \{-1, 0, 1\}. \end{cases}$$

Therefore,

$$L_{\mathfrak{P}}(\mathcal{E}_{\mathrm{CM}}, T) = \begin{cases} 1 - \mathrm{tr}(\psi_{\mathcal{E}(K)}(\mathfrak{P}))T + pT^2, & \text{for good reduction,} \\ 1 - \alpha T, & \text{for bad reduction.} \end{cases}$$

Lemma 6.5.7. *For $\mathcal{A}_{\mathrm{RM}} = F(\mathcal{E}_{\mathrm{CM}})$, it holds that*

$$\zeta_p^{-1}(\mathcal{A}_{\mathrm{RM}}, T) = 1 - \mathrm{tr}(A^{\pi(p)})T + pT^2,$$

whenever $p \nmid \mathrm{tr}^2(A) - 4$.

Proof. By the formula $K_0(\mathcal{O}_B) = \mathbf{Z}^2/(I - B^t)\mathbf{Z}^2$, one gets

$$|K_0(\mathcal{O}_{L_p^n})| = \left|\frac{\mathbf{Z}^2}{(I - (L_p^n)^t)\mathbf{Z}^2}\right| = |\det(I - (L_p^n)^t)| = |\mathrm{Fix}(L_p^n)|,$$

where $\mathrm{Fix}(L_p^n)$ is the set of (geometric) fixed points of the endomorphism $L_p^n : \mathbf{Z}^2 \to \mathbf{Z}^2$. Thus,

$$\zeta_p(\mathcal{A}_{\mathrm{RM}}, z) = \exp\left(\sum_{n=1}^{\infty} \frac{|\mathrm{Fix}(L_p^n)|}{n} z^n\right), \quad z \in \mathbf{C}.$$

But the latter series is an Artin–Mazur zeta function of the endomorphism L_p; it converges to a rational function $\det^{-1}(I - zL_p)$, see, e. g., [108, p. 455]. Thus, $\zeta_p(\mathcal{A}_{\mathrm{RM}}, z) =$

$\det^{-1}(I - zL_p)$. The substitution $L_p = (\mathrm{tr}(A^{\pi(p)}), p, -1, 0)$ gives us

$$\det(I - zL_p) = \det\begin{pmatrix} 1 - \mathrm{tr}(A^{\pi(p)})z & -pz \\ z & 1 \end{pmatrix} = 1 - \mathrm{tr}(A^{\pi(p)})z + pz^2.$$

Put $z = T$ and get $\zeta_p(\mathcal{A}_{RM}, T) = (1 - \mathrm{tr}(A^{\pi(p)})T + pT^2)^{-1}$, which is a conclusion of Lemma 6.5.7. □

Lemma 6.5.8. *For $\mathcal{A}_{RM} = F(\mathcal{E}_{CM})$, it holds that*

$$\zeta_p^{-1}(\mathcal{A}_{RM}, T) = 1 - \alpha T,$$

whenever $p \mid \mathrm{tr}^2(A) - 4$.

Proof. Indeed, $K_0(\mathcal{O}_{1-\alpha^n}) = \mathbf{Z}/(1 - 1 + \alpha^n)\mathbf{Z} = \mathbf{Z}/\alpha^n\mathbf{Z}$. Thus, $|K_0(\mathcal{O}_{1-\alpha^n})| = \det(\alpha^n) = \alpha^n$. By definition,

$$\zeta_p(\mathcal{A}_{RM}, z) = \exp\left(\sum_{n=1}^{\infty} \frac{\alpha^n}{n} z^n\right) = \exp\left(\sum_{n=1}^{\infty} \frac{(\alpha z)^n}{n}\right) = \frac{1}{1 - \alpha z}.$$

The substitution $z = T$ gives the conclusion of Lemma 6.5.8. □

Lemma 6.5.9. *Let $\mathfrak{P} \subset K$ be a prime ideal over p; then $\mathcal{E}_{CM} = \mathcal{E}(K)$ has a bad reduction at $\mathfrak{P}$ if and only if $p \mid \mathrm{tr}^2(A) - 4$.*

Proof. Let k be a field of complex multiplication of $\mathcal{E}_{CM}$; its discriminant we shall write as $\Delta_k < 0$. It is known that whenever $p \mid \Delta_k$, $\mathcal{E}_{CM}$ has a bad reduction at the prime ideal $\mathfrak{P}$ over p. On the other hand, the explicit formula for functor F applied to the matrix L_p gives us $F : (\mathrm{tr}(A^{\pi(p)}), p, -1, 0) \mapsto (\mathrm{tr}(A^{\pi(p)}), p, 1, 0)$, see the proof of Lemma 6.1.5. The characteristic polynomials of the above matrices are $x^2 - \mathrm{tr}(A^{\pi(p)})x + p$ and $x^2 - \mathrm{tr}(A^{\pi(p)})x - p$, respectively. They generate an imaginary (resp. real) quadratic field k (resp. $\mathfrak{k}$) with the discriminant $\Delta_k = \mathrm{tr}^2(A^{\pi(p)}) - 4p < 0$ (resp. $\Delta_{\mathfrak{k}} = \mathrm{tr}^2(A^{\pi(p)}) + 4p > 0$). Thus, $\Delta_{\mathfrak{k}} - \Delta_k = 8p$. It is easy to see that $p \mid \Delta_{\mathfrak{k}}$ if and only if $p \mid \Delta_k$. It remains to express the discriminant $\Delta_{\mathfrak{k}}$ in terms of the matrix A. Since the characteristic polynomial for A is $x^2 - \mathrm{tr}(A)x + 1$, it follows that $\Delta_{\mathfrak{k}} = \mathrm{tr}^2(A) - 4$. Lemma 6.5.9 follows. □

Let us prove that the first part of Theorem 6.5.1 implies the first claim of its second part; notice that the critical piece of information is provided by Lemma 6.5.6, which says that $\mathrm{tr}(\psi_{\mathcal{E}(K)}(\mathfrak{P})) = \mathrm{tr}(A^{\pi(p)})$. Thus, Lemmas 6.5.7–6.5.9 imply that $L_{\mathfrak{P}}(\mathcal{E}_{CM}, T) \equiv \zeta_p^{-1}(\mathcal{A}_{RM}, T)$. The first claim of part (ii) of Theorem 6.5.1 follows.

A. Let *p* be a good prime. Let us prove the second claim of part (ii) of Theorem 6.5.1 in the case $n = 1$. From the left-hand side, $K_0(\mathcal{A}_{RM} \rtimes_{L_p} \mathbf{Z}) \cong K_0(\mathcal{O}_{L_p}) \cong \mathbf{Z}^2/(I - L_p^t)\mathbf{Z}^2$, where $L_p = (\mathrm{tr}(A^{\pi(p)}), p, -1, 0)$. To calculate the abelian group $\mathbf{Z}^2/(I - L_p^t)\mathbf{Z}^2$, we shall use a reduction of the matrix $I - L_p^t$ to the Smith normal form:

$$I - L_p^t = \begin{pmatrix} 1 - \operatorname{tr}(A^{\pi(p)}) & 1 \\ -p & 1 \end{pmatrix} \sim \begin{pmatrix} 1 + p - \operatorname{tr}(A^{\pi(p)}) & 0 \\ -p & 1 \end{pmatrix}$$
$$\sim \begin{pmatrix} 1 & 0 \\ 0 & 1 + p - \operatorname{tr}(A^{\pi(p)}) \end{pmatrix}.$$

Therefore, $K_0(\mathcal{O}_{L_p}) \cong \mathbf{Z}_{1+p-\operatorname{tr}(A^{\pi(p)})}$. From the right-hand side, $\mathcal{E}_{CM}(\mathbb{F}_{\mathfrak{P}})$ is an elliptic curve over the field of characteristic p. Recall, that the chord and tangent law turns the $\mathcal{E}_{CM}(\mathbb{F}_{\mathfrak{P}})$ into a finite abelian group. The group is cyclic and has the order $1 + q_{\mathfrak{P}} - a_{\mathfrak{P}}$. But $q_{\mathfrak{P}} = p$ and $a_{\mathfrak{P}} = \operatorname{tr}(\psi_{\mathcal{E}(K)}(\mathfrak{P})) = \operatorname{tr}(A^{\pi(p)})$, see Lemma 6.5.6. Thus, $\mathcal{E}_{CM}(\mathbb{F}_{\mathfrak{P}}) \cong \mathbf{Z}_{1+p-\operatorname{tr}(A^{\pi(p)})}$; therefore $K_0(\mathcal{O}_{L_p}) \cong \mathcal{E}_{CM}(\mathbb{F}_p)$. The general case $n \geq 1$ is treated likewise. Repeating the argument of Lemmas 6.5.1–6.5.2, it follows that

$$L_p^n = \begin{pmatrix} \operatorname{tr}(A^{n\pi(p)}) & p^n \\ -1 & 0 \end{pmatrix}.$$

Then one gets $K_0(\mathcal{O}_{L_p^n}) \cong \mathbf{Z}_{1+p^n-\operatorname{tr}(A^{n\pi(p)})}$ on the left-hand side. From the right-hand side, $|\mathcal{E}_{CM}(\mathbb{F}_{p^n})| = 1 + p^n - \operatorname{tr}(\psi^n_{\mathcal{E}(K)}(\mathfrak{P}))$; but a repetition of the argument of Lemma 6.5.6 yields us $\operatorname{tr}(\psi^n_{\mathcal{E}(K)}(\mathfrak{P})) = \operatorname{tr}(A^{n\pi(p)})$. Comparing the left- and right-hand sides, one gets that $K_0(\mathcal{O}_{L_p^n}) \cong \mathcal{E}_{CM}(\mathbb{F}_{p^n})$. This argument finishes the proof of the second claim of part (ii) of Theorem 6.5.1 for the good primes.

B. Let p be a bad prime. From the proof of Lemma 6.5.8, one gets for the left-hand side $K_0(\mathcal{O}_{\varepsilon_n}) \cong \mathbf{Z}_{\alpha^n}$. From the right-hand side, it holds that $|\mathcal{E}_{CM}(\mathbb{F}_{p^n})| = 1 + q_{\mathfrak{P}} - a_{\mathfrak{P}}$, where $q_{\mathfrak{P}} = 0$ and $a_{\mathfrak{P}} = \operatorname{tr}(\varepsilon_n) = \varepsilon_n$. Thus, $|\mathcal{E}_{CM}(\mathbb{F}_{p^n})| = 1 - \varepsilon_n = 1 - (1 - \alpha^n) = \alpha^n$. Comparing the left- and right-hand sides, we conclude that $K_0(\mathcal{O}_{\varepsilon_n}) \cong \mathcal{E}_{CM}(\mathbb{F}_{p^n})$ at the bad primes.

All cases are exhausted; thus part (i) of Theorem 6.5.1 implies its part (ii). The proof of converse consists in a step by step claims similar to just proved and is left to the reader. Theorem 6.5.1 is proved.

6.5.3 Supplement: Grössencharacters, units, and $\pi(n)$

We shall briefly review the well-known facts about complex multiplication and units in subrings of the ring of integers in algebraic number fields; for the detailed account, we refer the reader to [259] and [111], respectively.

6.5.3.1 Grössencharacters

Let $\mathcal{E}_{CM} \cong \mathcal{E}(K)$ be an elliptic curve with complex multiplication and $K \cong k(j(\mathcal{E}_{CM}))$ the Hilbert class field attached to $\mathcal{E}_{CM}$. For each prime ideal $\mathfrak{P}$ of K, let $\mathbb{F}_{\mathfrak{P}}$ be a residue field of K at $\mathfrak{P}$ and $q_{\mathfrak{P}} = N^K_{\mathbf{Q}}\mathfrak{P} = \#\mathbb{F}_{\mathfrak{P}}$, where $N^K_{\mathbf{Q}}$ is the norm of the ideal $\mathfrak{P}$. If $\mathcal{E}(K)$ has a good reduction at $\mathfrak{P}$, one defines $a_{\mathfrak{P}} = q_{\mathfrak{P}} + 1 - \#\tilde{\mathcal{E}}(\mathbb{F}_{\mathfrak{P}})$, where $\tilde{\mathcal{E}}(\mathbb{F}_{\mathfrak{P}})$ is a reduction of $\mathcal{E}(K)$ modulo the prime ideal $\mathfrak{P}$. If $\mathcal{E}(K)$ has good reduction at $\mathfrak{P}$, the polynomial

$$L_{\mathfrak{P}}(\mathcal{E}(K), T) = 1 - a_{\mathfrak{P}}T + q_{\mathfrak{P}}T^2,$$

is called the *local L-series* of $\mathcal{E}(K)$ at $\mathfrak{P}$. If $\mathcal{E}(K)$ has bad reduction at $\mathfrak{P}$, the local L-series are $L_{\mathfrak{P}}(\mathcal{E}(K), T) = 1 - T$ (resp. $L_{\mathfrak{P}}(\mathcal{E}(K), T) = 1 + T$; $L_{\mathfrak{P}}(\mathcal{E}(K), T) = 1$) if $\mathcal{E}(K)$ has split multiplicative reduction at $\mathfrak{P}$ (if $\mathcal{E}(K)$ has nonsplit multiplicative reduction at $\mathfrak{P}$; if $\mathcal{E}(K)$ has additive reduction at $\mathfrak{P}$).

Definition 6.5.1. By the Hasse–Weil L-function of an elliptic curve $\mathcal{E}(K)$ one understands the global L-series defined by the Euler product

$$L(\mathcal{E}(K), s) = \prod_{\mathfrak{P}} [L_{\mathfrak{P}}(\mathcal{E}(K), q_{\mathfrak{P}}^{-s})]^{-1}.$$

Definition 6.5.2. If A_K^* be the idele group of the number field K, then by a Grössencharacter on K one understands a continuous homomorphism

$$\psi : A_K^* \longrightarrow \mathbf{C}^*$$

with the property $\psi(K^*) = 1$; the asterisk denotes the group of invertible elements of the corresponding ring. The Hecke L-series attached to the Grössencharacter $\psi : A_K^* \to \mathbf{C}^*$ is defined by the Euler product

$$L(s, \psi) = \prod_{\mathfrak{P}} (1 - \psi(\mathfrak{P}) q_{\mathfrak{P}}^{-s})^{-1},$$

where the product is taken over all prime ideals of K.

Remark 6.5.3. For a prime ideal $\mathfrak{P}$ of field K at which $\mathcal{E}(K)$ has good reduction and $\tilde{\mathcal{E}}(\mathbb{F}_{\mathfrak{P}})$ is the reduction of $\mathcal{E}(K)$ at $\mathfrak{P}$, we let

$$\phi_{\mathfrak{P}} : \tilde{\mathcal{E}}(\mathbb{F}_{\mathfrak{P}}) \longrightarrow \tilde{\mathcal{E}}(\mathbb{F}_{\mathfrak{P}})$$

denote the associated Frobenius map; if $\psi_{\mathcal{E}(K)} : A_K^* \to k^*$ is the Grössencharacter attached to $\mathcal{E}_{CM}$, then the diagram in Fig. 6.7 is known to be commutative, see [259, p. 174]. In particular, $\psi_{\mathcal{E}(K)}(\mathfrak{P})$ is an endomorphism of the $\mathcal{E}(K)$ given by the complex number $\alpha_{\mathcal{E}(K)}(\mathfrak{P}) \in R$, where $R = \mathbf{Z} + fO_k$ is an order in imaginary quadratic field k. If $\overline{\psi}_{\mathcal{E}(K)}(\mathfrak{P})$ is the conjugate Grössencharacter viewed as a complex number, then Deuring Theorem says that the Hasse–Weil L-function of $\mathcal{E}(K)$ is related to the Hecke L-series of the $\psi_{\mathcal{E}(K)}$ by the formula

$$L(\mathcal{E}(K), s) \equiv L(s, \psi_{\mathcal{E}(K)}) L(s, \overline{\psi}_{\mathcal{E}(K)}).$$

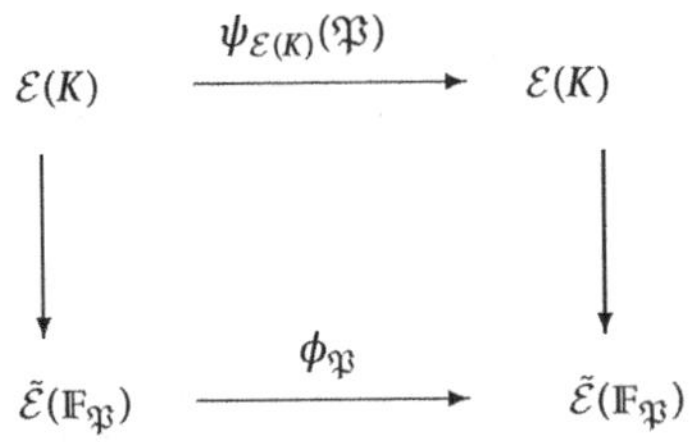

Figure 6.7: The Grössencharacter $\psi_{\mathcal{E}(K)}(\mathfrak{P})$.

6.5.3.2 Units and function $\pi(n)$

Let $\mathfrak{k} = \mathbf{Q}(\sqrt{D})$ be a real quadratic number field and $O_{\mathfrak{k}}$ its ring of integers. For a rational integer $n \geq 1$, we shall write $\mathfrak{R}_n \subseteq O_{\mathfrak{k}}$ to denote an order (i. e., a subring containing 1) of $O_{\mathfrak{k}}$. The order $\mathfrak{R}_n$ has a basis $\{1, n\omega\}$, where

$$\omega = \begin{cases} \frac{\sqrt{D}+1}{2} & \text{if } D \equiv 1 \mod 4, \\ \sqrt{D} & \text{if } D \equiv 2,3 \mod 4. \end{cases}$$

In other words, $\mathfrak{R}_n = \mathbf{Z} + (n\omega)\mathbf{Z}$. It is clear that $\mathfrak{R}_1 = O_{\mathfrak{k}}$, and the fundamental unit of $O_{\mathfrak{k}}$ we shall denote by ε. Each $\mathfrak{R}_n$ has its own fundamental unit, which we shall write as ε_n; notice that $\varepsilon_n \neq \varepsilon$ unless $n = 1$. There exists a well-known formula, which relates ε_n to the fundamental unit ε, see, e. g., [111, p. 297]. Denote by $\mathfrak{G}_n := U(O_{\mathfrak{k}}/nO_{\mathfrak{k}})$ the multiplicative group of invertible elements (units) of the residue ring $O_{\mathfrak{k}}/nO_{\mathfrak{k}}$; clearly, all units of $O_{\mathfrak{k}}$ map (under the natural mod$n$ homomorphism) to $\mathfrak{G}_n$. Likewise, let $\mathfrak{G}_n := U(\mathfrak{R}_n/n\mathfrak{R}_n)$ be the group of units of the residue ring $\mathfrak{R}_n/n\mathfrak{R}_n$; it is not hard to prove [111, p. 296] that $\mathfrak{G}_n \cong U(\mathbf{Z}/n\mathbf{Z})$ the rational unit group of the residue ring $\mathbf{Z}/n\mathbf{Z}$. Similarly, all units of the order $\mathfrak{R}_n$ map to $\mathfrak{G}_n$. Since units of $\mathfrak{R}_n$ are also units of $O_{\mathfrak{k}}$ (but not vice versa), $\mathfrak{R}_n$ is a subgroup of $\mathfrak{G}_n$; in particular, $|\mathfrak{G}_n|/|\mathfrak{R}_n|$ is an integer number and $|\mathfrak{G}_n| = \varphi(n)$, where $\varphi(n)$ is the Euler totient function. In general, the following formula is true:

$$\frac{|\mathfrak{G}_n|}{|\mathfrak{R}_n|} = n \prod_{p_i | n} \left(1 - \left(\frac{D}{p_i}\right)\frac{1}{p_i}\right),$$

where $(\frac{D}{p_i})$ is the Legendre symbol, see [111, p. 351].

Definition 6.5.3. By the function $\pi(n)$ one understands the least integer number dividing $|\mathfrak{G}_n|/|\mathfrak{R}_n|$ and such that $\varepsilon^{\pi(n)}$ is a unit of $\mathfrak{R}_n$, i. e., belongs to $\mathfrak{G}_n$.

Lemma 6.5.10 ([111, p. 298]). $\varepsilon_n = \varepsilon^{\pi(n)}$.

Remark 6.5.4. Lemma 6.5.10 asserts the existence of the number $\pi(n)$ as one of the divisors of $|\mathfrak{G}_n|/|\mathfrak{R}_n|$, yet no analytic formula for $\pi(n)$ is known; it would be rather interesting to have such a formula.

Remark 6.5.5. In the special case when $n = p$ is a prime number, the following formula is true:

$$\frac{|\mathfrak{G}_p|}{|\mathfrak{R}_p|} = p - \left(\frac{D}{p}\right).$$

Guide to the literature

The Hasse–Weil L-functions $L(\mathcal{E}(K), s)$ of the K-rational elliptic curves are covered in the textbooks [122, 136, 259]; see also the survey [274]. The reciprocity of $L(\mathcal{E}(K), s)$ with an L-function obtained from certain cusp form of weight two is subject of the Eichler–Shimura theory, see, e. g., [136, Chapter XI]; such a reciprocity coupled with the Shimura–Taniyama Conjecture was critical to solution of the Fermat Last Theorem by A. Wiles. The noncommutative reciprocity was proved in [204]. Such a reciprocity can be viewed as an analog of the Eichler–Shimura theory.

6.6 Langlands Conjecture for $\mathcal{A}_{\mathrm{RM}}^{2n}$

We construct an L-function $L(\mathcal{A}_{\mathrm{RM}}^{2n}, s)$, where $\mathcal{A}_{\mathrm{RM}}^{2n}$ is an even-dimensional noncommutative torus with real multiplication. We formulate an analog of the Langlands Conjecture for $\mathcal{A}_{\mathrm{RM}}^{2n}$, i. e., that for each $n \geq 1$ and each irreducible representation

$$\sigma : \mathrm{Gal}(E|\mathbf{Q}) \longrightarrow \mathrm{GL}_n(\mathbf{C}),$$

the corresponding Artin L-function $L(\sigma, s)$ coincides with $L(\mathcal{A}_{\mathrm{RM}}^{2n}, s)$. Our main result, Theorem 6.6.1, says that the conjecture is true for $n = 1$ (resp. $n = 0$) and E being the Hilbert class field of an imaginary quadratic field k (resp. the rational field $\mathbf{Q}$). We refer the reader to [91] for an excellent introduction the Langlands Program.

6.6.1 $L(\mathcal{A}_{\mathrm{RM}}^{2n}, s)$

The $2n$-dimensional noncommutative tori were introduced in Section 3.4.1; we remind the notation below. Let $\Theta = (\theta_{ij})$ be a real skew-symmetric matrix of even dimension $2n$. By $\mathcal{A}_{\Theta}^{2n}$ we shall mean the even-dimensional noncommutative torus defined by matrix Θ, i. e., a universal C^*-algebra on the unitary generators $u_1, \ldots, u_{2n}$ and relations

$$u_j u_i = e^{2\pi i \theta_{ij}} u_i u_j, \quad 1 \leq i, j \leq 2n.$$

Every generic real even-dimensional skew-symmetric matrix can be brought by the orthogonal linear transformations to the normal form

$$\Theta_0 = \begin{pmatrix} 0 & \theta_1 & & & \\ -\theta_1 & 0 & & & \\ & & \ddots & & \\ & & & 0 & \theta_n \\ & & & -\theta_n & 0 \end{pmatrix},$$

where $\theta_i > 0$ are linearly independent over $\mathbf{Q}$. We shall consider the noncommutative tori $\mathcal{A}^{2n}_{\Theta_0}$, given by matrix in the above normal form; we refer to the family as a normal family. Recall that $K_0(\mathcal{A}^{2n}_{\Theta_0}) \cong \mathbf{Z}^{2^{2n-1}}$ and the positive cone $K_0^+(\mathcal{A}^{2n}_{\Theta_0})$ is given by the pseudo-lattice

$$\mathbf{Z} + \theta_1\mathbf{Z} + \cdots + \theta_n\mathbf{Z} + \sum_{i=n+1}^{2^{2n-1}} p_i(\theta)\mathbf{Z} \subset \mathbf{R},$$

where $p_i(\theta) \in \mathbf{Z}[1, \theta_1, \dots, \theta_n]$ [77].

Definition 6.6.1. The noncommutative torus $\mathcal{A}^{2n}_{\Theta_0}$ is said to have real multiplication if the endomorphism ring $\mathrm{End}(K_0^+(\mathcal{A}^{2n}_{\Theta_0}))$ is nontrivial, i. e., exceeds the ring $\mathbf{Z}$; we shall denote such a torus by $\mathcal{A}^{2n}_{\mathrm{RM}}$.

Remark 6.6.1. It is easy to see that if $\mathcal{A}^{2n}_{\Theta_0}$ has real multiplication, then θ_i are algebraic integers; we leave the proof to the reader. (Hint: Each endomorphism of $K_0^+(\mathcal{A}^{2n}_{\Theta_0}) \cong \mathbf{Z} + \theta_1\mathbf{Z} + \cdots + \theta_n\mathbf{Z} + \sum_{i=n+1}^{2^{2n-1}} p_i(\theta)\mathbf{Z}$ is a multiplication by a real number; thus the endomorphism is described by an integer matrix, which defines a polynomial equation involving θ_i.)

Remark 6.6.2. Remark 6.6.1 says that θ_i are algebraic integers whenever $\mathcal{A}^{2n}_{\Theta_0}$ has real multiplication; so will be the values of polynomials $p_i(\theta)$ in this case. Since such values belong to the number field $\mathbf{Q}(\theta_1, \dots, \theta_n)$, one concludes that

$$K_0^+(\mathcal{A}^{2n}_{\mathrm{RM}}) \cong \mathbf{Z} + \theta_1\mathbf{Z} + \cdots + \theta_n\mathbf{Z} \subset \mathbf{R}.$$

Let $A \in GL_{n+1}(\mathbf{Z})$ be a positive matrix such that

$$A \begin{pmatrix} 1 \\ \theta_1 \\ \vdots \\ \theta_n \end{pmatrix} = \lambda_A \begin{pmatrix} 1 \\ \theta_1 \\ \vdots \\ \theta_n \end{pmatrix},$$

where λ_A is the Perron–Frobenius eigenvalue of A. In other words, A is a matrix corresponding to the shift automorphism σ_A of $K_0^+(\mathcal{A}^{2n}_{\mathrm{RM}})$ regarded as a stationary dimension group; we refer the reader to Definition 3.5.4. For each prime number p, consider the characteristic polynomial of matrix $A^{\pi(p)}$, where $\pi(n)$ is the integer-valued function introduced in Section 6.5.3; in other words,

$$\mathbf{Char}(A^{\pi(p)}) := \det(xI - A^{\pi(p)}) = x^{n+1} - a_1x^n - \cdots - a_nx - 1 \in \mathbf{Z}[x].$$

Definition 6.6.2. By a local zeta function of the noncommutative torus $\mathcal{A}_{RM}^{2n}$ we understand the function

$$\zeta_p(\mathcal{A}_{RM}^{2n}, z) := \frac{1}{1 - a_1z + a_2z^2 - \cdots - a_nz^n + pz^{n+1}}, \quad z \in \mathbf{C}.$$

Remark 6.6.3. To explain the structure of $\zeta_p(\mathcal{A}_{RM}^{2n}, z)$, consider the companion matrix

$$J = \begin{pmatrix} a_1 & 1 & 0 & \dots & 0 & 0 \\ a_2 & 0 & 1 & \dots & 0 & 0 \\ \vdots & \vdots & \vdots & \ddots & \vdots & \vdots \\ a_n & 0 & 0 & \dots & 0 & 1 \\ 1 & 0 & 0 & \dots & 0 & 0 \end{pmatrix}$$

of polynomial $\mathbf{Char}(A^{\pi(p)}) = x^{n+1} - a_1x^n - \cdots - a_nx - 1$, i. e., the matrix J such that $\det(xI - J) = x^{n+1} - a_1x^n - \cdots - a_nx - 1$. It is not hard to see that the nonnegative integer matrix J corresponds to the shift automorphism of a stationary dimension group

$$\mathbf{Z}^{n+1} \xrightarrow{J_p} \mathbf{Z}^{n+1} \xrightarrow{J_p} \mathbf{Z}^{n+1} \xrightarrow{J_p} \cdots,$$

where

$$J_p = \begin{pmatrix} a_1 & 1 & 0 & \dots & 0 & 0 \\ a_2 & 0 & 1 & \dots & 0 & 0 \\ \vdots & \vdots & \vdots & \ddots & \vdots & \vdots \\ a_n & 0 & 0 & \dots & 0 & 1 \\ p & 0 & 0 & \dots & 0 & 0 \end{pmatrix}.$$

On the other hand, the companion matrix of polynomial $\mathbf{Char}(\sigma(\mathrm{Fr}_p)) = \det(xI - \sigma(\mathrm{Fr}_p)) = x^{n+1} - a_1x^n + \cdots - a_nx + p$ has the form

$$W_p = \begin{pmatrix} a_1 & 1 & 0 & \dots & 0 & 0 \\ -a_2 & 0 & 1 & \dots & 0 & 0 \\ \vdots & \vdots & \vdots & \ddots & \vdots & \vdots \\ a_n & 0 & 0 & \dots & 0 & 1 \\ -p & 0 & 0 & \dots & 0 & 0 \end{pmatrix}.$$

We refer the reader to Section 6.6.3 for the definition of $\sigma(\mathrm{Fr}_p)$. The action of functor F : **Alg-Num** → **NC-Tor** on the corresponding companion matrices W_p and J_p is given by the formula

$$F: \begin{pmatrix} a_1 & 1 & 0 & \dots & 0 & 0 \\ -a_2 & 0 & 1 & \dots & 0 & 0 \\ \vdots & \vdots & \vdots & \ddots & \vdots & \vdots \\ a_n & 0 & 0 & \dots & 0 & 1 \\ -p & 0 & 0 & \dots & 0 & 0 \end{pmatrix} \mapsto \begin{pmatrix} a_1 & 1 & 0 & \dots & 0 & 0 \\ a_2 & 0 & 1 & \dots & 0 & 0 \\ \vdots & \vdots & \vdots & \ddots & \vdots & \vdots \\ a_n & 0 & 0 & \dots & 0 & 1 \\ p & 0 & 0 & \dots & 0 & 0 \end{pmatrix}.$$

It remains to compare our formula for $\zeta_p(\mathcal{A}_{\mathrm{RM}}^{2n}, z)$ with the well-known formula for the Artin zeta function

$$\zeta_p(\sigma_n, z) = \frac{1}{\det(I_n - \sigma_n(\mathrm{Fr}_p)z)},$$

where $z = x^{-1}$ [91, p. 181].

Definition 6.6.3. By an L-function of the noncommutative torus $\mathcal{A}_{\mathrm{RM}}^{2n}$ one understand the product

$$L(\mathcal{A}_{\mathrm{RM}}^{2n}, s) := \prod_p \zeta_p(\mathcal{A}_{\mathrm{RM}}^{2n}, p^{-s}), \quad s \in \mathbf{C},$$

over all but a finite number of primes p.

Conjecture 6.6.1 (Langlands conjecture for noncommutative tori). *For each finite extension E of the field of rational numbers $\mathbf{Q}$ with the Galois group* $\mathrm{Gal}(E|\mathbf{Q})$ *and each irreducible representation*

$$\sigma_{n+1} : \mathrm{Gal}(E|\mathbf{Q}) \to \mathrm{GL}_{n+1}(\mathbf{C}),$$

there exists a 2n-dimensional noncommutative torus with real multiplication $\mathcal{A}_{\mathrm{RM}}^{2n}$ such that

$$L(\sigma_{n+1}, s) \equiv L(\mathcal{A}_{\mathrm{RM}}^{2n}, s),$$

where $L(\sigma_{n+1}, s)$ is the Artin L-function attached to representation σ_{n+1} and $L(\mathcal{A}_{\mathrm{RM}}^{2n}, s)$ is the L-function of the noncommutative torus $\mathcal{A}_{\mathrm{RM}}^{2n}$.

Remark 6.6.4. Conjecture 6.6.1 says that the Galois extensions (abelian or not) of the field $\mathbf{Q}$ are in a one-to-one correspondence with the even-dimensional noncommutative tori with real multiplication. In the context of the Langlands program, the noncommutative torus $\mathcal{A}_{\mathrm{RM}}^{2n}$ can be regarded as an analog of the automorphic cuspidal representation $\pi_{\sigma_{n+1}}$ of the group $\mathrm{GL}(n+1)$. The tori $\mathcal{A}_{\mathrm{RM}}^{2n}$ are natural in the context of the Langlands program, because they classify the irreducible infinite-dimensional representations of the Lie group $GL(n+1)$ [233].

Theorem 6.6.1. *Conjecture* 6.6.1 *is true for* $n = 1$ *(resp.* $n = 0$*) and* E *abelian extension of an imaginary quadratic field* k *(resp. the rational field* $\mathbf{Q}$*).*

6.6.2 Proof of Theorem 6.6.1

6.6.2.1 Case $n = 1$

This case is equivalent to Theorem 6.5.1; it was a model example for Conjecture 6.6.1. Using the Grössencharacters, one can identify the Artin L-function for abelian extensions of the imaginary quadratic fields k with the Hasse–Weil L-function $L(\mathcal{E}_{CM}, s)$, where $\mathcal{E}_{CM}$ is an elliptic curve with complex multiplication by k; but Theorem 6.5.1 says that $L(\mathcal{E}_{CM}, s) \equiv L(\mathcal{A}_{RM}, s)$, where $L(\mathcal{A}_{RM}, s)$ is the special case $n = 1$ of our function $L(\mathcal{A}_{RM}^{2n}, s)$.

To give the details, let k be an imaginary quadratic field and let $\mathcal{E}_{CM}$ be an elliptic curve with complex multiplication by (an order) in k. By the theory of complex multiplication, the Hilbert class field K of k is given by the j-invariant of $\mathcal{E}_{CM}$, i. e.,

$$K \cong k(j(\mathcal{E}_{CM})),$$

and $\mathrm{Gal}(K|k) \cong \mathrm{Cl}(k)$, where $\mathrm{Cl}(k)$ is the ideal class group of k; moreover,

$$\mathcal{E}_{CM} \cong \mathcal{E}(K),$$

see, e. g., [259]. Recall that functor $F : \mathbf{Ell} \to \mathbf{NC\text{-}Tor}$ maps $\mathcal{E}_{CM}$ to a two-dimensional noncommutative torus with real multiplication $\mathcal{A}_{RM}^2$. To calculate $L(\mathcal{A}_{RM}^2, s)$, let $A \in GL_2(\mathbf{Z})$ be positive matrix corresponding the *shift automorphism* of $\mathcal{A}_{RM}^2$, i. e.,

$$A\begin{pmatrix}1\\ \theta\end{pmatrix} = \lambda_A\begin{pmatrix}1\\ \theta\end{pmatrix},$$

where θ is a quadratic irrationality and λ_A the Perron–Frobenius eigenvalue of A. If p is a prime, then the characteristic polynomial of matrix $A^{\pi(p)}$ can be written as $\mathbf{Char}\, A^{\pi(p)} = x^2 - \mathrm{tr}(A^{\pi(p)})x - 1$; therefore, the local zeta function of torus $\mathcal{A}_{RM}^2$ has the form

$$\zeta_p(\mathcal{A}_{RM}^2, z) = \frac{1}{1 - \mathrm{tr}(A^{\pi(p)})z + pz^2}.$$

On the other hand, Lemma 6.5.6 says that

$$\mathrm{tr}(A^{\pi(p)}) = \mathrm{tr}(\psi_{\mathcal{E}(K)}(\mathfrak{P})),$$

where $\psi_{\mathcal{E}(K)}$ is the Grössencharacter on K and $\mathfrak{P}$ the prime ideal of K over p. But we know that the local zeta function of $\mathcal{E}_{CM}$ has the form

$$\zeta_p(\mathcal{E}_{CM}, z) = \frac{1}{1 - \operatorname{tr}(\psi_{\mathcal{E}(K)}(\mathfrak{P}))z + pz^2};$$

thus for each prime p it holds that $\zeta_p(\mathcal{A}^2_{RM}, z) = \zeta_p(\mathcal{E}_{CM}, z)$. Leaving aside the bad primes, one derives the following important equality of the L-functions

$$L(\mathcal{A}^2_{RM}, s) \equiv L(\mathcal{E}_{CM}, s),$$

where $L(\mathcal{E}_{CM}, s)$ is the Hasse–Weil L-function of elliptic curve $\mathcal{E}_{CM}$. Case $n = 1$ of Theorem 6.6.1 becomes an implication of the following lemma.

Lemma 6.6.1. *$L(\mathcal{E}_{CM}, s) \equiv L(\sigma_2, s)$, where $L(\sigma_2, s)$ the Artin L-function for an irreducible representation $\sigma_2 : \mathrm{Gal}(K|k) \to \mathrm{GL}_2(\mathbf{C})$.*

Proof. Deuring Theorem says that

$$L(\mathcal{E}_{CM}, s) = L(\psi_K, s)L(\overline{\psi}_K, s),$$

where $L(\psi_K, s)$ is the Hecke L-series attached to the Grössencharacter $\psi : \mathbb{A}^*_K \to \mathbf{C}^*$; here $\mathbb{A}^*_K$ denotes the adele ring of the field K and the bar means the complex conjugation, see, e. g., [259, p. 175]. Because our elliptic curve has complex multiplication, the group $\mathrm{Gal}(K|k)$ is abelian; one can apply the result of [137, Theorem 5.1], which says that the Hecke L-series $L(\sigma_1 \circ \theta_{K|k}, s)$ equals the Artin L-function $L(\sigma_1, s)$, where $\psi_K = \sigma \circ \theta_{K|k}$ is the Grössencharacter and $\theta_{K|k} : \mathbb{A}^*_K \to \mathrm{Gal}(K|k)$ the canonical homomorphism. Thus one gets

$$L(\mathcal{E}_{CM}, s) \equiv L(\sigma_1, s)L(\overline{\sigma}_1, s),$$

where $\overline{\sigma}_1 : \mathrm{Gal}(K|k) \to \mathbf{C}$ means a (complex) conjugate representation of the Galois group. Consider the local factors of the Artin L-functions $L(\sigma_1, s)$ and $L(\overline{\sigma}_1, s)$; it is immediate that they are $(1 - \sigma_1(\mathrm{Fr}_p)p^{-s})^{-1}$ and $(1 - \overline{\sigma}_1(\mathrm{Fr}_p)p^{-s})^{-1}$, respectively. Let us consider a representation $\sigma_2 : \mathrm{Gal}(K|k) \to \mathrm{GL}_2(\mathbf{C})$ such that

$$\sigma_2(\mathrm{Fr}_p) = \begin{pmatrix} \sigma_1(\mathrm{Fr}_p) & 0 \\ 0 & \overline{\sigma}_1(\mathrm{Fr}_p) \end{pmatrix}.$$

It can be verified that $\det^{-1}(I_2 - \sigma_2(\mathrm{Fr}_p)p^{-s}) = (1 - \sigma_1(\mathrm{Fr}_p)p^{-s})^{-1}(1 - \overline{\sigma}_1(\mathrm{Fr}_p)p^{-s})^{-1}$, i. e., $L(\sigma_2, s) = L(\sigma_1, s)L(\overline{\sigma}_1, s)$. Lemma 6.6.1 follows. □

From Lemma 6.6.1 and $L(\mathcal{A}^2_{RM}, s) \equiv L(\mathcal{E}_{CM}, s)$, one gets

$$L(\mathcal{A}^2_{RM}, s) \equiv L(\sigma_2, s)$$

for an irreducible representation $\sigma_2 : \mathrm{Gal}(K|k) \to \mathrm{GL}_2(\mathbf{C})$. It remains to notice that $L(\sigma_2, s) = L(\sigma_2', s)$, where $\sigma_2' : \mathrm{Gal}(K|\mathbf{Q}) \to \mathrm{GL}_2(\mathbf{C})$, see, e. g., [4, Section 3]. Case $n = 1$ of Theorem 6.6.1 is proved.

6.6.2.2 Case $n = 0$

When $n = 0$, one gets a one-dimensional (degenerate) noncommutative torus; such an object, $\mathcal{A}_{\mathbf{Q}}$, can be obtained from the 2-dimensional torus $\mathcal{A}_\theta^2$ by forcing $\theta = p/q \in \mathbf{Q}$ be a rational number (hence our notation). One can always assume $\theta = 0$ and, thus,

$$K_0^+(\mathcal{A}_{\mathbf{Q}}) \cong \mathbf{Z}.$$

The group of automorphisms of $\mathbf{Z}$-module $K_0^+(\mathcal{A}_{\mathbf{Q}}) \cong \mathbf{Z}$ is trivial, i. e., the multiplication by ± 1; hence matrix A corresponding to the shift automorphisms is either 1 or -1. Since A must be positive, one gets $A = 1$. However, $A = 1$ is not a primitive; indeed, for any $N > 1$, matrix $A' = \zeta_N$ gives us $A = (A')^N$, where $\zeta_N = e^{\frac{2\pi i}{N}}$ is the Nth root of unity. Therefore, one gets

$$A = \zeta_N.$$

Since for the field $\mathbf{Q}$ it holds that $\pi(n) = n$, one obtains $\mathrm{tr}(A^{\pi(p)}) = \mathrm{tr}(A^p) = \zeta_N^p$. A degenerate noncommutative torus, corresponding to the matrix $A = \zeta_N$, we shall write as $\mathcal{A}_{\mathbf{Q}}^N$.

Suppose that $\mathrm{Gal}(K|\mathbf{Q})$ is abelian and let $\sigma : \mathrm{Gal}(K|\mathbf{Q}) \to \mathbf{C}^\times$ be a homomorphism. By the *Artin reciprocity*, there exist an integer N_σ and the Dirichlet character

$$\chi_\sigma : (\mathbf{Z}/N_\sigma\mathbf{Z})^\times \to \mathbf{C}^\times$$

such that $\sigma(\mathrm{Fr}_p) = \chi_\sigma(p)$, see, e. g., [91]. On the other hand, it is verified directly that $\zeta_{N_\sigma}^p = e^{\frac{2\pi i}{N_\sigma}p} = \chi_\sigma(p)$. Therefore $\mathbf{Char}(A^p) = \chi_\sigma(p)x - 1$, and one gets

$$\zeta_p(\mathcal{A}_{\mathbf{Q}}^{N_\sigma}, z) = \frac{1}{1 - \chi_\sigma(p)z},$$

where $\chi_\sigma(p)$ is the Dirichlet character. Therefore, $L(\mathcal{A}_{\mathbf{Q}}^{N_\sigma}, s) \equiv L(s, \chi_\sigma)$ is the Dirichlet L-series; such a series, by construction, coincides with the Artin L-series of the representation $\sigma : \mathrm{Gal}(K|\mathbf{Q}) \to \mathbf{C}^\times$. Case $n = 0$ of Theorem 6.6.1 is proved.

6.6.3 Supplement: Artin L-function

The *Class Field Theory* (CFT) studies algebraic extensions of the number fields; the objective of CFT is a description of arithmetic of the extension E in terms of arithmetic of the ground field k and the Galois group $\mathrm{Gal}(E|k)$ of the extension. Unless $\mathrm{Gal}(E|k)$ is abelian, the CFT is out of reach so far; yet a series of conjectures called the *Langlands program* (LP) are designed to attain the goals of CFT. We refer the interested reader to [91] for an introduction to the CFT and LP; roughly speaking, the LP consists in an n-dimensional generalization of the Artin reciprocity based on the ideas and methods of representation theory of the locally compact Lie groups. The centerpiece of LP is the *Artin L-function* attached to representation $\sigma : \mathrm{Gal}(E|k) \to GL_n(\mathbf{C})$ of the Galois group of E; we shall give a brief account of this L-function following the survey [91].

The fundamental problem in algebraic number theory is to describe how an ordinary prime p factors into prime ideals $\mathfrak{P}$ in the ring of integers of an arbitrary finite extensions E of the rational field $\mathbf{Q}$. Let O_E be the ring of integers of the extension E and pO_E a principal ideal; it is known that

$$pO_E = \prod \mathfrak{P}_i,$$

where $\mathfrak{P}_i$ are prime ideals of O_E. If E is the Galois extension of $\mathbf{Q}$ and $\mathrm{Gal}(E|\mathbf{Q})$ is the corresponding Galois group, then each automorphism $g \in \mathrm{Gal}(E|\mathbf{Q})$ "moves around" the ideals $\mathfrak{P}_i$ in the prime decomposition of p over E. An *isotropy subgroup* of $\mathrm{Gal}(E|\mathbf{Q})$ (for given p) consists of the elements of $\mathrm{Gal}(E|\mathbf{Q})$ which fix all the ideals $\mathfrak{P}_i$. For simplicity, we shall assume that p is *unramified* in E, i. e., all $\mathfrak{P}_i$ are distinct; in this case the isotropy subgroup are *cyclic*. The (conjugacy class of) generator in the cyclic isotropy subgroup of $\mathrm{Gal}(E|\mathbf{Q})$ corresponding to p is called the *Frobenius element* and denoted by Fr_p. The element $\mathrm{Fr}_p \in \mathrm{Gal}(E|\mathbf{Q})$ describes completely the factorization of p over E and the major goal of the CFT is to express Fr_p in terms of arithmetic of the ground field $\mathbf{Q}$. To handle this hard problem, it was suggested by E. Artin to consider the n-dimensional irreducible representations

$$\sigma_n : \mathrm{Gal}(E|\mathbf{Q}) \longrightarrow GL_n(\mathbf{C}),$$

of the Galois group $\mathrm{Gal}(E|\mathbf{Q})$, see [4]. The idea was to use the characteristic polynomial $\mathbf{Char}(\sigma_n(\mathrm{Fr}_p)) := \det(I_n - \sigma_n(\mathrm{Fr}_p)z)$ of the matrix $\sigma_n(\mathrm{Fr}_p)$; the polynomial is independent of the similarity class of $\sigma_n(\mathrm{Fr}_p)$ in the group $GL_n(\mathbf{C})$ and provides an *intrinsic* description of the Frobenius element Fr_p.

Definition 6.6.4. By an Artin zeta function of representation σ_n one understands the function

$$\zeta_p(\sigma_n, z) := \frac{1}{\det(I_n - \sigma_n(\mathrm{Fr}_p)z)}, \quad z \in \mathbf{C}.$$

By an Artin L-function of representation σ_n one understands the product

$$L(\sigma_n, s) := \prod_p \zeta_p(\sigma_n, p^{-s}), \quad s \in \mathbf{C},$$

over all but a finite set of primes p.

Remark 6.6.5 (Artin reciprocity). If $n = 1$ and $\mathrm{Gal}(E|\mathbf{Q}) \cong \mathbf{Z}/N\mathbf{Z}$ is abelian, then the calculation of the Artin L-function gives the equality

$$L(\chi, s) = \prod_p \frac{1}{1 - \chi(p)p^{-s}},$$

where $\chi : (\mathbf{Z}/N\mathbf{Z}) \to \mathbf{C}^\times$ is the Dirichlet character; the RHS of the equality is known as the Dirichlet L-series for χ. Thus one gets a formula

$$\sigma(\mathrm{Fr}_p) = \chi(p)$$

called the Artin reciprocity law; the formula generalizes many classical reciprocity results known for the particular values of N.

Guide to the literature

The Artin L-function first appeared in [4]. The origins of the Langlands program (and philosophy) can be found in his letter to André Weil, see [145]. An excellent introduction to the Langlands program has been written in [91]. The Langlands program for the even-dimensional noncommutative tori was the subject of [183].

6.7 Projective varieties over finite fields

We calculate the number of points of variety $V(\mathbb{F}_q)$ defined over a finite field $\mathbb{F}_q$ in terms of the Serre C^*-algebra $\mathcal{A}_V$, where V is a complex projective variety such that the $V(\mathbb{F}_q)$ is a reduction of V modulo q. We illustrate our formula for the elliptic curves $\mathcal{E}_{\mathrm{CM}}$ and $\mathcal{E}(\mathbf{Q})$.

6.7.1 Traces of Frobenius endomorphisms

The number of solutions of a system of polynomial equations over a finite field is an important invariant of the system and an old problem dating back to Gauss. Recall that if $\mathbb{F}_q$ is a field with $q = p^r$ elements and $V(\mathbb{F}_q)$ a smooth n-dimensional projective variety over $\mathbb{F}_q$, then one can define a zeta function $Z(V; t) := \exp(\sum_{r=1}^{\infty} |V(\mathbb{F}_{q^r})| \frac{t^r}{r})$; the function is rational, i. e.,

$$Z(V; t) = \frac{P_1(t)P_3(t)\cdots P_{2n-1}(t)}{P_0(t)P_2(t)\cdots P_{2n}(t)},$$

where $P_0(t) = 1 - t$, $P_{2n}(t) = 1 - q^n t$, and for each $1 \le i \le 2n - 1$ the polynomial $P_i(t) \in \mathbf{Z}[t]$ can be written as $P_i(t) = \prod_{j=1}^{\deg P_i(t)} (1 - \alpha_{ij} t)$ so that α_{ij} are algebraic integers with $|\alpha_{ij}| = q^{\frac{i}{2}}$, see, e. g., [108, pp. 454–457]. The polynomial $P_i(t)$ can be viewed as the characteristic polynomial of the Frobenius endomorphism Fr_q^i of the ith ℓ-adic cohomology group $H^i(V)$; such an endomorphism is induced by the map acting on points of variety $V(\mathbb{F}_q)$ according to the formula $(a_1,\dots,a_n)\mapsto(a_1^q,\dots,a_n^q)$; we assume throughout the Standard Conjectures [102]. If $V(\mathbb{F}_q)$ is defined by a system of polynomial equations, then the number of solutions of the system is given by the formula

$$|V(\mathbb{F}_q)| = \sum_{i=0}^{2n}(-1)^i \mathrm{tr}(Fr_q^i),$$

where tr is the trace of Frobenius endomorphism, see [108].

Let $V(K)$ be a complex projective variety defined over an algebraic number field $K\subset\mathbf{C}$; suppose that projective variety $V(\mathbb{F}_q)$ is the reduction of $V(K)$ modulo the prime ideal $\mathfrak{P}\subset K$ corresponding to $q=p^r$. Denote by $\mathcal{A}_V$ the *Serre C^*-algebra* of projective variety $V(K)$, see Section 5.3.1. Consider the stable C^*-algebra of $\mathcal{A}_V$, i. e., the C^*-algebra $\mathcal{A}_V\otimes\mathcal{K}$, where $\mathcal{K}$ is the C^*-algebra of compact operators on $\mathcal{H}$. Let $\tau:\mathcal{A}_V\otimes\mathcal{K}\to\mathbf{R}$ be the unique normalized trace (tracial state) on $\mathcal{A}_V\otimes\mathcal{K}$, i. e., a positive linear functional of norm 1 such that $\tau(yx)=\tau(xy)$ for all $x,y\in\mathcal{A}_V\otimes\mathcal{K}$, see [32, p. 31]. Recall that $\mathcal{A}_V$ is the crossed product C^*-algebra of the form $\mathcal{A}_V\cong C(V)\rtimes\mathbf{Z}$, where $C(V)$ is the commutative C^*-algebra of complex valued functions on V and the product is taken by an automorphism of algebra $C(V)$ induced by the map $\sigma:V\to V$; we refer the reader to Lemma 5.3.2. From the Pimsner–Voiculescu six-term exact sequence for crossed products, one gets the short exact sequence of algebraic K-groups

$$0\to K_0(C(V))\xrightarrow{i_*}K_0(\mathcal{A}_V)\to K_1(C(V))\to 0,$$

where map i_* is induced by an embedding of $C(V)$ into $\mathcal{A}_V$, see [32, p. 83] for the details. We have $K_0(C(V))\cong K^0(V)$ and $K_1(C(V))\cong K^{-1}(V)$, where K^0 and K^{-1} are the topological K-groups of variety V, see [32, p. 80]. By the Chern character formula, one gets

$$\begin{cases} K^0(V)\otimes\mathbf{Q}\cong H^{\mathrm{even}}(V;\mathbf{Q}),\\ K^{-1}(V)\otimes\mathbf{Q}\cong H^{\mathrm{odd}}(V;\mathbf{Q}),\end{cases}$$

where H^{even} (resp. H^{odd}) is the direct sum of even (resp. odd) cohomology groups of V.

Remark 6.7.1. It is known that $K_0(\mathcal{A}_V\otimes\mathcal{K})\cong K_0(\mathcal{A}_V)$ because of stability of the K_0-group with respect to tensor products by the algebra $\mathcal{K}$, see, e. g., [32, p. 32].

Thus one gets the commutative diagram shown in Fig. 6.8, where τ_* denotes a homomorphism induced on K_0 by the canonical trace τ on the C^*-algebra $\mathcal{A}_V\otimes\mathcal{K}$.

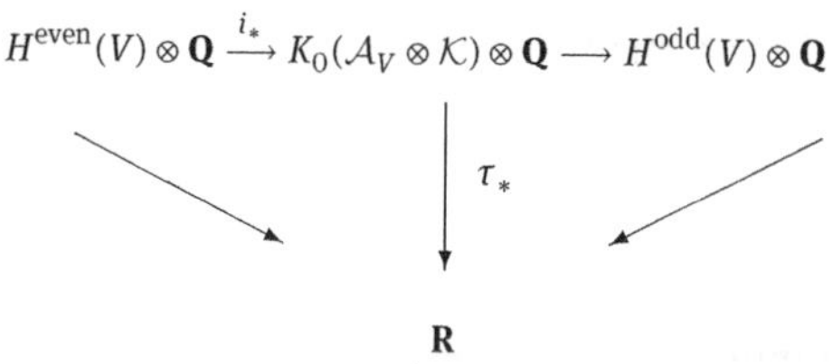

Figure 6.8: K-theory of the Serre C^*-algebra $\mathcal{A}_V$.

Because $H^{\mathrm{even}}(V) := \bigoplus_{i=0}^{n} H^{2i}(V)$ and $H^{\mathrm{odd}}(V) := \bigoplus_{i=1}^{n} H^{2i-1}(V)$, one gets for each $0 \le i \le 2n$ an injective homomorphism

$$H^i(V) \to \mathbf{R},$$

and we shall denote by Λ_i an additive abelian subgroup of real numbers defined by the homomorphism. The subgroup Λ_i is called a pseudo-lattice [155, Section 1].

Recall that endomorphisms of a pseudo-lattice are given as multiplication of points of Λ_i by the real numbers α such that $\alpha\Lambda_i \subseteq \Lambda_i$. It is known that $\mathrm{End}(\Lambda_i) \cong \mathbf{Z}$ or $\mathrm{End}(\Lambda_i) \otimes \mathbf{Q}$ is a real algebraic number field such that $\Lambda_i \subset \mathrm{End}(\Lambda_i) \otimes \mathbf{Q}$, see, e. g., [155, Lemma 1.1.1] for the case of quadratic fields. We shall write ε_i to denote the unit of the order in the field $K_i := \mathrm{End}(\Lambda_i) \otimes \mathbf{Q}$, which induces the shift automorphism of Λ_i, see [74, p. 38] for the details and terminology. Let p be a good prime in the reduction $V(\mathbb{F}_q)$ of complex projective variety $V(K)$ modulo a prime ideal over $q = p^r$. Consider a sublattice Λ_i^q of Λ_i of the index q; by an index of the sublattice we understand its index as an abelian subgroup of Λ_i. We shall write $\pi_i(q)$ to denote an integer such that multiplication by $\varepsilon_i^{\pi_i(q)}$ induces the shift automorphism of Λ_i^q. The trace of an algebraic number will be written as $\mathrm{tr}(\bullet)$. The following result relates invariants ε_i and $\pi_i(q)$ of the C^*-algebra $\mathcal{A}_V$ to the cardinality of the set $V(\mathbb{F}_q)$.

Theorem 6.7.1. $|V(\mathbb{F}_q)| = \sum_{i=0}^{2n} (-1)^i \mathrm{tr}\,(\varepsilon_i^{\pi_i(q)})$.

6.7.2 Proof of Theorem 6.7.1

Lemma 6.7.1. *There exists a symplectic unitary matrix $\Theta_q^i \in \mathrm{Sp}(\deg P_i; \mathbf{R})$ such that*

$$Fr_q^i = q^{\frac{i}{2}} \Theta_q^i.$$

Proof. Recall that the eigenvalues of Fr_q^i have absolute value $q^{\frac{i}{2}}$; they come in the complex conjugate pairs. On the other hand, symplectic unitary matrices in group $\mathrm{Sp}(\deg P_i; \mathbf{R})$ are known to have eigenvalues of absolute value 1 coming in complex conjugate pairs. Since the spectrum of a matrix defines the similarity class of matrix, one can write the characteristic polynomial of Fr_q^i in the form

$$P_i(t) = \det(I - q^{\frac{i}{2}}\Theta_q^i t),$$

where matrix $\Theta_q^i \in \mathrm{Sp}(\deg P_i; \mathbf{Z})$ and its eigenvalues have absolute value 1. It remains to compare the above equation with the formula

$$P_i(t) = \det(I - \mathrm{Fr}_q^i t),$$

i. e., $\mathrm{Fr}_q^i = q^{\frac{i}{2}}\Theta_q^i$. Lemma 6.7.1 follows. □

Lemma 6.7.2. *Using a symplectic transformation, one can bring matrix Θ_q^i to the block form*

$$\Theta_q^i = \begin{pmatrix} A & I \\ -I & 0 \end{pmatrix},$$

where A is a positive symmetric and I the identity matrix.

Proof. Let us write Θ_q^i in the block form

$$\Theta_q^i = \begin{pmatrix} A & B \\ C & D \end{pmatrix},$$

where matrices A, B, C, D are invertible and their transposes A^T, B^T, C^T, D^T satisfy the symplectic equations

$$\begin{cases} A^T D - C^T B = I, \\ A^T C - C^T A = 0, \\ B^T D - D^T B = 0. \end{cases}$$

Recall that symplectic matrices correspond to the linear fractional transformations $\tau \mapsto \frac{A\tau+B}{C\tau+D}$ of the Siegel half-space $\mathbb{H}_n = \{\tau = (\tau_j) \in \mathbf{C}^{\frac{n(n+1)}{2}} \mid \mathfrak{I}(\tau_j) > 0\}$ consisting of symmetric $n \times n$ matrices, see, e. g., [168, p. 173]. One can always multiply the nominator and denominator of such a transformation by B^{-1} without affecting the transformation; thus with no loss of generality, we can assume that $B = I$. We shall consider the symplectic matrix T and its inverse T^{-1} given by the formulas

$$T = \begin{pmatrix} I & 0 \\ D & I \end{pmatrix} \quad \text{and} \quad T^{-1} = \begin{pmatrix} I & 0 \\ -D & I \end{pmatrix}.$$

It is verified directly that

$$T^{-1}\Theta_q^i T = \begin{pmatrix} I & 0 \\ -D & I \end{pmatrix}\begin{pmatrix} A & I \\ C & D \end{pmatrix}\begin{pmatrix} I & 0 \\ D & I \end{pmatrix} = \begin{pmatrix} A+D & I \\ C-DA & 0 \end{pmatrix}.$$

The system of symplectic equations with $B = I$ implies the following two equations:

$$A^T D - C^T = I \quad \text{and} \quad D = D^T.$$

Applying transposition to both parts of the first equation of the above equations, one gets $(A^T D - C^T)^T = I^T$ and, therefore, $D^T A - C = I$. But the second equation says that $D^T = D$; thus one arrives at the equation $DA - C = I$. The latter gives us $C - DA = -I$, which we substitute into the above equations and get (in a new notation) the conclusion of Lemma 6.7.2. Finally, the middle of the symplectic equations with $C = -I$ implies $A = A^T$, i. e., A is a symmetric matrix. Since the eigenvalues of symmetric matrix are always real and in view of $\mathrm{tr}(A) > 0$ (because $\mathrm{tr}(Fr_q^i) > 0$), one concludes that A is similar to a positive matrix, see, e. g., [105, Theorem 1]. Lemma 6.7.2 follows. □

Lemma 6.7.3. *The symplectic unitary transformation Θ_q^i of $H^i(V;\mathbf{Z})$ descends to an automorphism of Λ_i given by the matrix*

$$M_q^i = \begin{pmatrix} A & I \\ I & 0 \end{pmatrix}.$$

Remark 6.7.2. In other words, Lemma 6.7.3 says that functor F : **Proj-Alg** → **C*-Serre** acts between matrices Θ_q^i and M_q^i according to the formula

$$F : \begin{pmatrix} A & I \\ -I & 0 \end{pmatrix} \mapsto \begin{pmatrix} A & I \\ I & 0 \end{pmatrix}.$$

Proof. Since $\Lambda_i \subset K_i$, there exists a basis of Λ_i consisting of algebraic numbers; denote by $(\mu_1, \ldots, \mu_k; \nu_1, \ldots, \nu_k)$ a basis of Λ_i consisting of positive algebraic numbers $\mu_i > 0$ and $\nu_i > 0$. Using the injective homomorphism τ_*, one can descend Θ_q^i to an automorphism of Λ_i so that

$$\begin{pmatrix} \mu' \\ \nu' \end{pmatrix} = \begin{pmatrix} A & I \\ -I & 0 \end{pmatrix} \begin{pmatrix} \mu \\ \nu \end{pmatrix} = \begin{pmatrix} A\mu + \nu \\ -\mu \end{pmatrix},$$

where $\mu = (\mu_1, \ldots, \mu_k)$ and $\nu = (\nu_1, \ldots, \nu_k)$. Because vectors μ and ν consist of positive entries and A is a positive matrix, it is immediate that $\mu' = A\mu + \nu > 0$ while $\nu' = -\mu < 0$.

Remark 6.7.3. All automorphisms in the (Markov) category of pseudo-lattices come from multiplication of the basis vector $(\mu_1, \ldots, \mu_k; \nu_1, \ldots, \nu_k)$ of Λ_i by an algebraic unit $\lambda > 0$ of field K_i; in particular, any such an automorphism must be given by a non-negative matrix, whose Perron–Frobenius eigenvalue coincides with λ. Thus for any automorphism of Λ_i it must hold that $\mu' > 0$ and $\nu' > 0$.

In view of the above, we shall consider an automorphism of Λ_i given by matrix $M_q^i = (A, I, I, 0)$; clearly, for M_q^i it holds that $\mu' = A\mu + \nu > 0$ and $\nu' = \mu > 0$. Therefore

M_q^i is a nonnegative matrix satisfying the necessary condition to belong to the Markov category. It is also a sufficient one because the similarity class of M_q^i contains a representative whose Perron–Frobenius eigenvector can be taken for a basis (μ, ν) of Λ_i. This argument finishes the proof of Lemma 6.7.3. □

Corollary 6.7.1. $\mathrm{tr}(M_q^i) = \mathrm{tr}(\Theta_q^i)$.

Proof. This fact is an implication of the above formulas and a direct computation $\mathrm{tr}(M_q^i) = \mathrm{tr}(A) = \mathrm{tr}(\Theta_q^i)$. □

Definition 6.7.1. We shall call $q^{\frac{i}{2}} M_q^i$ a Markov endomorphism of Λ_i and denote it by Mk_q^i.

Lemma 6.7.4. $\mathrm{tr}(\mathrm{Mk}_q^i) = \mathrm{tr}(\mathrm{Fr}_q^i)$.

Proof. Corollary 6.7.1 says that $\mathrm{tr}(M_q^i) = \mathrm{tr}(\Theta_q^i)$, and therefore

$$\mathrm{tr}(\mathrm{Mk}_q^i) = \mathrm{tr}(q^{\frac{i}{2}} M_q^i) = q^{\frac{i}{2}} \mathrm{tr}(M_q^i) = q^{\frac{i}{2}} \mathrm{tr}(\Theta_q^i) = \mathrm{tr}(q^{\frac{i}{2}} \Theta_q^i) = \mathrm{tr}(\mathrm{Fr}_q^i).$$

In words, Frobenius and Markov endomorphisms have the same trace, i. e., $\mathrm{tr}(\mathrm{Mk}_q^i) = \mathrm{tr}(\mathrm{Fr}_q^i)$. Lemma 6.7.4 follows. □

Remark 6.7.4. Notice that, unless i or r are even, neither Θ_q^i nor M_q^i are integer matrices; yet Fr_q^i and Mk_q^i are always integer matrices.

Lemma 6.7.5. *There exists an algebraic unit $\omega_i \in K_i$ such that:*

(i) *ω_i corresponds to the shift automorphism of an index q sublattice of pseudo-lattice Λ_i;*
(ii) $\mathrm{tr}(\omega_i) = \mathrm{tr}(\mathrm{Mk}_q^i)$.

Proof. (i) To prove Lemma 6.7.5, we shall use the notion of a stationary dimension group and the corresponding shift automorphism; we refer the reader to [74, p. 37] and [105, p. 57] for the notation and details on stationary dimension groups and a survey of [281] for the general theory of subshifts of finite type. Consider a stationary dimension group, $G(\mathrm{Mk}_q^i)$, generated by the Markov endomorphism Mk_q^i,

$$\mathbf{Z}^{b_i} \xrightarrow{\mathrm{Mk}_q^i} \mathbf{Z}^{b_i} \xrightarrow{\mathrm{Mk}_q^i} \mathbf{Z}^{b_i} \xrightarrow{\mathrm{Mk}_q^i} \cdots,$$

where $b_i = \deg P_i(t)$. Let λ_M be the Perron–Frobenius eigenvalue of matrix M_q^i. It is known that $G(\mathrm{Mk}_q^i)$ is order-isomorphic to a dense additive abelian subgroup $\mathbf{Z}[\frac{1}{\lambda_M}]$ of $\mathbf{R}$; here $\mathbf{Z}[x]$ is the set of all polynomials in one variable with the integer coefficients. Let $\widehat{\mathrm{Mk}_q^i}$ be a shift automorphism of $G(\mathrm{Mk}_q^i)$ [74, p. 37]. To calculate the automorphism, notice that multiplication of $\mathbf{Z}[\frac{1}{\lambda_M}]$ by λ_M induces an automorphism of dimension group $\mathbf{Z}[\frac{1}{\lambda_M}]$. Since the determinant of matrix M_q^i (i. e., the degree of Markov endomorphism) is equal to q^n, one concludes that such an automorphism corresponds to a unit of the endomorphism ring of a sublattice of Λ_i of index q^n. We shall denote such a unit by ω_i. Clearly, ω_i generates the required shift automorphism $\widehat{\mathrm{Mk}_q^i}$ through multiplication of dimension group $\mathbf{Z}[\frac{1}{\lambda_M}]$ by the algebraic number ω_i. Item (i) of Lemma 6.7.5 follows.

(ii) Consider the Artin–Mazur zeta function of Mk_q^i,

$$\zeta_{\mathrm{Mk}_q^i}(t) = \exp\left(\sum_{k=1}^{\infty} \frac{\mathrm{tr}\,[(\mathrm{Mk}_q^i)^k]}{k} t^k\right),$$

and that of $\widehat{\mathrm{Mk}_q^i}$, namely

$$\zeta_{\widehat{\mathrm{Mk}_q^i}}(t) = \exp\left(\sum_{k=1}^{\infty} \frac{\mathrm{tr}\,[(\widehat{\mathrm{Mk}_q^i})^k]}{k} t^k\right).$$

Since Mk_q^i and $\widehat{\mathrm{Mk}_q^i}$ are shift-equivalent matrices, one concludes that $\zeta_{\mathrm{Mk}_q^i}(t) \equiv \zeta_{\widehat{\mathrm{Mk}_q^i}}(t)$, see [281, p. 273]. In particular,

$$\mathrm{tr}(\mathrm{Mk}_q^i) = \mathrm{tr}(\widehat{\mathrm{Mk}_q^i}).$$

But $\mathrm{tr}(\widehat{\mathrm{Mk}_q^i}) = \mathrm{tr}(\omega_i)$, where on the right-hand side is the trace of an algebraic number. In view of the above, one gets the conclusion of Lemma 6.7.5(ii). □

Lemma 6.7.6. *There exists a positive integer $\pi_i(q)$ such that*

$$\omega_i = \varepsilon_i^{\pi_i(q)},$$

where $\varepsilon_i \in \mathrm{End}(\Lambda_i)$ is the fundamental unit corresponding to the shift automorphism of pseudo-lattice Λ_i.

Proof. Given an automorphism ω_i of a finite-index sublattice of Λ_i, one can extend ω_i to an automorphism of entire Λ_i since $\omega_i\Lambda_i = \Lambda_i$. Therefore each unit of endomorphism ring of a sublattice is also a unit of the host pseudo-lattice. Notice that the converse statement is false, in general. On the other hand, by virtue of the Dirichlet Unit Theorem, each unit of $\mathrm{End}(\Lambda_i)$ is a product of a finite number of (powers of) fundamental units of $\mathrm{End}(\Lambda_i)$. We shall denote by $\pi_i(q)$ the least positive integer such that $\varepsilon_i^{\pi_i(q)}$ is the shift automorphism of a sublattice of index q of pseudo-lattice Λ_i. The number $\pi_i(q)$ exists and uniquely defined, albeit no general formula for its calculation is known, see Remark 6.7.5. It is clear from construction that $\pi_i(q)$ satisfies the claim of Lemma 6.7.6. □

Remark 6.7.5. No general formula for the number $\pi_i(q)$ as a function of q is known; however, if the rank of Λ_i is two (i. e., $n = 1$), then there are classical results recorded in [111, p. 298]; see also Section 6.5.3.2.

Theorem 6.7.1 follows from Lemmas 6.7.4–6.7.6 and the known formula $|V(\mathbb{F}_q)| = \sum_{i=0}^{2n}(-1)^i \mathrm{tr}(Fr_q^i)$.

6.7.3 Examples

Let $V(\mathbf{C}) \cong \mathcal{E}_\tau$ be an elliptic curve; it is well known that its Serre C^*-algebra $\mathcal{A}_{\mathcal{E}_\tau}$ is isomorphic to the noncommutative torus $\mathcal{A}_\theta$ with the unit scaled by a constant $0 < \log\mu < \infty$. Furthermore, $K_0(\mathcal{A}_\theta) \cong K_1(\mathcal{A}_\theta) \cong \mathbf{Z}^2$ and the canonical trace τ on $\mathcal{A}_\theta$ gives us the following formula:

$$\tau_*(K_0(\mathcal{A}_{\mathcal{E}_\tau} \otimes \mathcal{K})) = \mu(\mathbf{Z} + \mathbf{Z}\theta).$$

Because $H^0(\mathcal{E}_\tau; \mathbf{Z}) = H^2(\mathcal{E}_\tau; \mathbf{Z}) \cong \mathbf{Z}$ while $H^1(\mathcal{E}_\tau; \mathbf{Z}) \cong \mathbf{Z}^2$, one gets the following pseudo-lattices:

$$\Lambda_0 = \Lambda_2 \cong \mathbf{Z} \quad \text{and} \quad \Lambda_1 \cong \mu(\mathbf{Z} + \mathbf{Z}\theta).$$

For the sake of simplicity, we shall focus on the following families of elliptic curves.

6.7.3.1 Complex multiplication

Suppose that $\mathcal{E}_\tau$ has complex multiplication; recall that such a curve was denoted by $\mathcal{E}_{CM}^{(-D,f)}$, i. e., the endomorphism ring of $\mathcal{E}_\tau$ is an order of conductor $f \geq 1$ in the imaginary quadratic field $\mathbf{Q}(\sqrt{-D})$. By the results of Section 5.1 on elliptic curve $\mathcal{E}_{CM}^{(-D,f)}$, the formulas for Λ_i are as follows:

$$\Lambda_0 = \Lambda_2 \cong \mathbf{Z} \quad \text{and} \quad \Lambda_1 = \varepsilon[\mathbf{Z} + (f\omega)\mathbf{Z}],$$

where $\omega = \frac{1}{2}(1 + \sqrt{D})$ if $D \equiv 1 \bmod 4$ and $D \neq 1$ or $\omega = \sqrt{D}$ if $D \equiv 2, 3 \bmod 4$ and $\varepsilon > 1$ is the fundamental unit of order $\mathbf{Z} + (f\omega)\mathbf{Z}$.

Remark 6.7.6. The reader can verify, that $\Lambda_1 \subset K_1$, where $K_1 \cong \mathbf{Q}(\sqrt{D})$.

Let p be a good prime. Consider a localization $\mathcal{E}(\mathbb{F}_p)$ of curve $\mathcal{E}_{CM}^{(-D,f)} \cong \mathcal{E}(K)$ at the prime ideal $\mathfrak{P}$ over p. It is well known, that the Frobenius endomorphism of elliptic curve with complex multiplication is defined by the Grössencharacter; the latter is a complex number $\alpha_{\mathfrak{P}} \in \mathbf{Q}(\sqrt{-D})$ of absolute value $\sqrt{p}$. Moreover, multiplication of the lattice $L_{CM} = \mathbf{Z} + \mathbf{Z}\tau$ by $\alpha_{\mathfrak{P}}$ induces the Frobenius endomorphism Fr_p^1 on $H^1(\mathcal{E}(K); \mathbf{Z})$, see, e. g., [259, p. 174]. Thus one arrives at the following matrix form for the Frobenius and Markov endomorphisms and the shift automorphism, respectively:

$$\begin{cases} \mathrm{Fr}_p^1 = \begin{pmatrix} \mathrm{tr}(\alpha_{\mathfrak{P}}) & p \\ -1 & 0 \end{pmatrix}, \\ \mathrm{Mk}_p^1 = \begin{pmatrix} \mathrm{tr}(\alpha_{\mathfrak{P}}) & p \\ 1 & 0 \end{pmatrix}, \\ \widehat{\mathrm{Mk}_p^1} = \begin{pmatrix} \mathrm{tr}(\alpha_{\mathfrak{P}}) & 1 \\ 1 & 0 \end{pmatrix}. \end{cases}$$

To calculate positive integer $\pi_1(p)$ appearing in Theorem 6.7.1, denote by $(\frac{D}{p})$ the Legendre symbol of D and p. A classical result of the theory of real quadratic fields asserts that $\pi_1(p)$ must be one of the divisors of the integer number

$$p - \left(\frac{D}{p}\right),$$

see [111, p. 298]. Thus the trace of Frobenius endomorphism on $H^1(\mathcal{E}(K); \mathbf{Z})$ is given by the formula

$$\mathrm{tr}(\alpha_{\mathfrak{P}}) = \mathrm{tr}(\varepsilon^{\pi_1(p)}).$$

The right-hand side of the above equation can be further simplified, since

$$\mathrm{tr}(\varepsilon^{\pi_1(p)}) = 2T_{\pi_1(p)}\left[\frac{1}{2}\mathrm{tr}(\varepsilon)\right],$$

where $T_{\pi_1(p)}(x)$ is the Chebyshev polynomial (of the first kind) of degree $\pi_1(p)$. Thus one obtains a formula for the number of (projective) solutions of a cubic equation over field $\mathbb{F}_p$ in terms of invariants of pseudo-lattice Λ_1,

$$|\mathcal{E}(\mathbb{F}_p)| = 1 + p - 2T_{\pi_1(p)}\left[\frac{1}{2}\mathrm{tr}(\varepsilon)\right].$$

6.7.3.2 Rational elliptic curve

Let $b \geq 3$ be an integer and consider a rational elliptic curve $\mathcal{E}(\mathbf{Q}) \subset \mathbf{C}P^2$ given by the homogeneous Legendre equation

$$y^2 z = x(x - z)\left(x - \frac{b-2}{b+2}z\right).$$

The Serre C^*-algebra of projective variety $V \cong \mathcal{E}(\mathbf{Q})$ is isomorphic (modulo an ideal) to the Cuntz–Krieger algebra $\mathcal{O}_B$, where

$$B = \begin{pmatrix} b-1 & 1 \\ b-2 & 1 \end{pmatrix},$$

see [196]. Recall that $\mathcal{O}_B \otimes \mathcal{K}$ is the crossed product C^*-algebra of a stationary AF C^*-algebra by its shift automorphism, see [32, p. 104]; the AF C^*-algebra has the following dimension group:

$$\mathbf{Z}^2 \xrightarrow{B^T} \mathbf{Z}^2 \xrightarrow{B^T} \mathbf{Z}^2 \xrightarrow{B^T} \cdots,$$

where B^T is the transpose of matrix B. Because μ must be a positive eigenvalue of matrix B^T, one gets

$$\mu = \frac{2-b+\sqrt{b^2-4}}{2}.$$

Likewise, since θ must be the corresponding positive eigenvector $(1,\theta)$ of the same matrix, one gets

$$\theta = \frac{1}{2}\left(\sqrt{\frac{b+2}{b-2}} - 1\right).$$

Therefore, pseudo-lattices Λ_i are $\Lambda_0 = \Lambda_2 \cong \mathbf{Z}$ and

$$\Lambda_1 \cong \frac{2-b+\sqrt{b^2-4}}{2}\left[\mathbf{Z} + \frac{1}{2}\left(\sqrt{\frac{b+2}{b-2}} - 1\right)\mathbf{Z}\right].$$

Remark 6.7.7. The pseudo-lattice $\Lambda_1 \subset K_1$, where $K_1 = \mathbf{Q}(\sqrt{b^2-4})$.

Let p be a good prime and let $\mathcal{E}(\mathbb{F}_p)$ be the reduction of our rational elliptic curve modulo p. It follows from Section 6.3.3 that $\pi_1(p)$ as one of the divisors of integer number

$$p - \left(\frac{b^2-4}{p}\right).$$

Unlike the case of complex multiplication, the Grössencharacter is no longer available for $\mathcal{E}(\mathbf{Q})$; yet the trace of Frobenius endomorphism can be computed using Theorem 6.7.1, i. e.,

$$\operatorname{tr}(\mathrm{Fr}_p^1) = \operatorname{tr}[(B^T)^{\pi_1(p)}].$$

Using the Chebyshev polynomials, one can write the last equation in the form

$$\mathrm{tr}(\mathrm{Fr}_p^1) = 2T_{\pi_1(p)}\left[\frac{1}{2}\mathrm{tr}(B^T)\right].$$

Since $\mathrm{tr}(B^T) = b$, one gets

$$\mathrm{tr}(\mathrm{Fr}_p^1) = 2T_{\pi_1(p)}\left(\frac{b}{2}\right).$$

Thus one obtains a formula for the number of solutions of equation $y^2z = x(x-z)(x-\frac{b-2}{b+2}z)$ over field $\mathbb{F}_p$ in terms of the noncommutative invariants of pseudo-lattice Λ_1 of the form

$$|\mathcal{E}(\mathbb{F}_p)| = 1 + p - 2T_{\pi_1(p)}\left(\frac{b}{2}\right).$$

We shall conclude by a concrete example comparing the obtained formula with the known results for rational elliptic curves in the Legendre form, see [108, p. 333] and [135, pp. 49–50].

Example 6.7.1 (Comparison to classical invariants). Suppose that $b \equiv 2 \bmod 4$. Recall that the j-invariant takes the same value on λ, $1-\lambda$, and $\frac{1}{\lambda}$, see [108, p. 320]. Therefore, one can bring equation $y^2z = x(x-z)(x-\frac{b-2}{b+2}z)$ to the form

$$y^2z = x(x-z)(x-\lambda z),$$

where $\lambda = \frac{1}{4}(b+2) \in \{2, 3, 4, \dots\}$. Notice that for the above curve

$$\mathrm{tr}(B^T) = b = 2(2\lambda - 1).$$

To calculate $\mathrm{tr}(\mathrm{Fr}_p^1)$ for our elliptic curve, recall that in view of last equality, one gets

$$\mathrm{tr}(\mathrm{Fr}_p^1) = 2T_{\pi_1(p)}(2\lambda - 1).$$

It will be useful to express Chebyshev polynomial $T_{\pi_1(p)}(2\lambda - 1)$ in terms of the hypergeometric function ${}_2F_1(a, b; c; z)$; the standard formula brings our last equation to the form

$$\mathrm{tr}(\mathrm{Fr}_p^1) = 2\,{}_2F_1\left(-\pi_1(p),\ \pi_1(p);\ \frac{1}{2};\ 1-\lambda\right).$$

We leave to the reader to prove the identity

$$2\,{}_2F_1\left(-\pi_1(p),\ \pi_1(p);\ \frac{1}{2};\ 1-\lambda\right) = (-1)^{\pi_1(p)}\ {}_2F_1(\pi_1(p)+1,\ \pi_1(p)+1;\ 1;\ \lambda).$$

In the latter formula,

$$ {}_2F_1(\pi_1(p)+1,\ \pi_1(p)+1;\ 1;\ \lambda) = \sum_{r=0}^{\pi_1(p)} \binom{\pi_1(p)}{r}^2 \lambda^r, $$

see [46, p. 328]. Recall that $\pi_1(p)$ is a divisor of $p - (\frac{b^2-4}{p})$, which in our case takes the value $\frac{p-1}{2}$. Bringing together the above formulas, one gets

$$ |\mathcal{E}(\mathbb{F}_p)| = 1 + p + (-1)^{\frac{p-1}{2}} \sum_{r=0}^{\frac{p-1}{2}} \binom{\frac{p-1}{2}}{r}^2 \lambda^r. $$

The reader is encouraged to compare the obtained formula with the classical result in [108, p. 333] and [135, pp. 49–50]; notice also a link to the Hasse invariant.

Guide to the literature

The Weil Conjectures (WC) were formulated in [285]; along with the Langalands Program, the WC shaped the modern look of number theory. The theory of motives was elaborated in [102] to solve the WC. An excellent introduction to the WC can be found in [108, Appendix C]. The related noncommutative invariants were calculated in [187] and a proof of the WC using such invariants can be found in [195].

Exercises, problems, and conjectures

1. Prove that elliptic curves $\mathcal{E}_\tau$ and $\mathcal{E}_{\tau'}$ are isogenous if and only if

$$ \tau' = \frac{a\tau + b}{c\tau + d} \quad \text{for some matrix} \begin{pmatrix} a & b \\ c & d \end{pmatrix} \in M_2(\mathbf{Z}) \quad \text{with } ad - bc > 0. $$

   (Hint: Notice that $z \mapsto \alpha z$ is an invertible holomorphic map for each $\alpha \in \mathbf{C} - \{0\}$.)
2. Prove that typically $\mathrm{End}(\mathcal{E}_\tau) \cong \mathbf{Z}$, i. e., the only endomorphisms of $\mathcal{E}_\tau$ are the multiplication-by-m endomorphisms.
3. Prove that for a countable set of τ,

$$ \mathrm{End}(\mathcal{E}_\tau) \cong \mathbf{Z} + fO_k, $$

   where $k = \mathbf{Q}(\sqrt{-D})$ is an imaginary quadratic field, O_k its ring of integers, and $f \geq 1$ is the conductor of a finite index subring of O_k; prove that in such a case $\tau \in \mathrm{End}(\mathcal{E}_\tau)$, i. e., complex modulus itself is an imaginary quadratic number.

4. Show that the noncommutative tori $\mathcal{A}_\theta$ and $\mathcal{A}_{\theta'}$ are stably homomorphic if and only if

$$\theta' = \frac{a\theta + b}{c\theta + d} \quad \text{for some matrix} \begin{pmatrix} a & b \\ c & d \end{pmatrix} \in M_2(\mathbf{Z}) \text{ with } ad - bc > 0.$$

(Hint: Follow and modify the argument of [238].)

5. Prove that part (ii) of Theorem 6.5.1 implies its part (i). (Hint: Repeat the step by step argument of Section 6.3.2.)
6. Prove Remark 6.6.1, i. e., that if noncommutative torus $\mathcal{A}^{2n}_{\Theta_0}$ has real multiplication, then θ_i are algebraic integers. (Hint: Each endomorphism of $K_0^+(\mathcal{A}^{2n}_{\Theta_0}) \cong \mathbf{Z} + \theta_1\mathbf{Z} + \cdots + \theta_n\mathbf{Z} + \sum_{i=n+1}^{2^{2n-1}} p_i(\theta)\mathbf{Z}$ is multiplication by a real number; thus the endomorphism is described by an integer matrix, which defines a polynomial equation involving θ_i.)
7. (Langlands conjecture for noncommutative tori) Prove Conjecture 6.6.1, i. e., that for each finite extension E of the field of rational numbers $\mathbf{Q}$ with the Galois group $\mathrm{Gal}(E|\mathbf{Q})$ and each irreducible representation

$$\sigma_{n+1} : \mathrm{Gal}(E|\mathbf{Q}) \to \mathrm{GL}_{n+1}(\mathbf{C}),$$

there exists a $2n$-dimensional noncommutative torus with real multiplication $\mathcal{A}^{2n}_{\mathrm{RM}}$ such that

$$L(\sigma_{n+1}, s) \equiv L(\mathcal{A}^{2n}_{\mathrm{RM}}, s),$$

where $L(\sigma_{n+1}, s)$ is the Artin L-function attached to representation σ_{n+1} and $L(\mathcal{A}^{2n}_{\mathrm{RM}}, s)$ is the L-function of the noncommutative torus $\mathcal{A}^{2n}_{\mathrm{RM}}$.
8. Prove the identity

$$2\,{}_2F_1\left(-\pi_1(p), \pi_1(p); \frac{1}{2}; 1 - \lambda\right) = (-1)^{\pi_1(p)}\,{}_2F_1(\pi_1(p) + 1, \pi_1(p) + 1; 1; \lambda),$$

where ${}_2F_1(a, b; c; z)$ is the hypergeometric function.

7 Arithmetic topology

The arithmetic topology studies an interplay between the 3-dimensional manifolds and the fields of algebraic numbers [166]. The idea dates back to C. F. Gauss. Namely, an analogy (a map) between the 3-dimensional manifolds M^3 and the algebraic number fields K is conjectured. Such a map transforms the links $L \subset M^3$ (knots $K \subset M^3$, resp.) into the ideals (prime ideals, resp.) of the ring of integers O_K of K. In Section 7.1, using the cluster algebras (Section 4.4.1), we construct an explicit functor F on the category of 3-dimensional manifolds M^3 with values in the category of algebraic number fields K realizing the above axioms of the arithmetic topology. In Section 7.2 the domain of F is extended to the 4-dimensional manifolds M^4 and the hyper-algebraic number fields $\mathbb{K}$, i. e., fields with noncommutative multiplication. In Section 7.3 we use the functor F and Wedderburn's Theorem on finite division rings to give an algebraic proof to the topological fact that all knots and links in M^4 are trivial. In Section 7.4 an analog of the ideals for simple noncommutative rings is introduced and applied to a classification of the knotted surfaces in the 4-dimensional manifolds. Finally, in Section 7.5 we use a canonical C^*-algebra introduced by Gábor Etesi to classify all smooth structures on the topological manifold M^4.

7.1 Arithmetic topology of 3-manifolds

The arithmetic topology studies an analogy between knots and primes; we refer the reader to the book of Morishita [166] for an excellent introduction. To give an idea of the analogy, let us quote Mazur's notes [159]: "Guided by the results of Artin and Tate applied to the calculation of the Grothendieck Cohomology Groups of the schemes:

$$\operatorname{Spec}(\mathbf{Z}/p\mathbf{Z}) \subset \operatorname{Spec}\mathbf{Z}$$

Mumford has suggested a most elegant model as a geometric interpretation of the above situation: $\operatorname{Spec}(\mathbf{Z}/p\mathbf{Z})$ is like a one-dimensional knot in $\operatorname{Spec}\mathbf{Z}$ which is like a simply connected three-manifold."

Roughly speaking, the idea is this. The 3-dimensional sphere S^3 corresponds to the field of rational numbers $\mathbf{Q}$. The prime ideals $p\mathbf{Z}$ in the ring of integers $\mathbf{Z}$ correspond to the knots $K \subset S^3$ and the ideals in $\mathbf{Z}$ correspond to the links $L \subset S^3$. In general, a 3-dimensional manifold M^3 corresponds to an algebraic number field K. The prime ideals in the ring of integers O_K of the field K correspond to the knots $K \subset M^3$ and the ideals in O_K correspond to the links $L \subset M^3$.

In this section we construct a functor from the category of closed 3-dimensional manifolds to a category of algebraic number fields realizing all axioms of the arithmetic topology. The construction of such a functor is based on a representation of the braid group into a cluster C^*-algebra, see Section 4.4.3.

https://doi.org/10.1515/9783110788709-007

Namely, denote by $S_{g,n}$ a Riemann surface of genus g with n cusps. Recall that a cluster algebra $\mathbb{A}(\mathbf{x}, S_{g,n})$ of the $S_{g,n}$ is a subring of the ring of the Laurent polynomials $\mathbf{Z}[\mathbf{x}^{\pm 1}]$ with integer coefficients and variables $\mathbf{x} := (x_1, \dots, x_{6g-6+2n})$. The algebra $\mathbb{A}(\mathbf{x}, S_{g,n})$ is a coordinate ring of the Teichmüller space $T_{g,n}$ of surface $S_{g,n}$ [288, Section 3]. The algebra $\mathbb{A}(\mathbf{x}, S_{g,n})$ is a commutative algebra with an additive abelian semigroup consisting of the Laurent polynomials with positive coefficients. In particular, $\mathbb{A}(\mathbf{x}, S_{g,n})$ is an abelian group with order satisfying the Riesz interpolation property, i. e., a dimension group, see Definition 3.5.2. By a cluster C^*-algebra $\mathbb{A}(\mathbf{x}, S_{g,n})$ we understand an AF-algebra such that $K_0(\mathbb{A}(\mathbf{x}, S_{g,n})) \cong \mathbb{A}(\mathbf{x}, S_{g,n})$, where $\cong$ is an isomorphism of the dimension groups. We refer the reader to Section 4.4.3 for the details and examples. The algebra $\mathbb{A}(\mathbf{x}, S_{g,n})$ is a noncommutative coordinate ring of the Teichmüller space $T_{g,n}$. This observation and Birman–Hilden Theorem imply a representation

$$\rho : B_{2g+n} \to \mathbb{A}(\mathbf{x}, S_{g,n}), \quad n \in \{0; 1\},$$

where $B_{2g+n} := \{\sigma_1, \dots, \sigma_{2g+n-1} \mid \sigma_i\sigma_{i+1}\sigma_i = \sigma_{i+1}\sigma_i\sigma_{i+1}, \ \sigma_i\sigma_j = \sigma_j\sigma_i \text{ if } |i-j| \geq 2\}$ is the braid group, the map ρ acts by the formula $\sigma_i \mapsto e_i + 1$ and e_i are projections of the algebra $\mathbb{A}(\mathbf{x}, S_{g,n})$, see Chapter 4.4.

Let $b \in B_{2g+n}$ be a braid. Denote by L_b a link obtained by the closure of b and let $\pi_1(L_b)$ be the fundamental group of L_b. Recall that

$$\pi_1(L_b) \cong \langle x_1, \dots, x_{2g+n} \mid x_i = r(b)x_i, \ 1 \leq i \leq 2g+n \rangle,$$

where x_i are generators of the free group $\mathbf{F}^{2g+n}$ and $r : B_{2g+n} \to \mathrm{Aut}(\mathbf{F}^{2g+n})$ is the Artin representation of B_{2g+n} [5, Theorem 6]. Let $\mathcal{I}_b$ be a two-sided ideal in the algebra $\mathbb{A}(\mathbf{x}, S_{g,n})$ generated by relations $x_i = r(b)x_i$. In particular, the ideal $\mathcal{I}_b$ is self-adjoint and one gets a representation

$$R : \pi_1(L_b) \to \mathbb{A}(\mathbf{x}, S_{g,n})/\mathcal{I}_b := \mathbb{A}_b.$$

The above $\mathbb{A}_b$ is a stationary AF-algebra of rank $6g-6+2n$ [74, Chapter 5]. It is known that the group $K_0(\mathbb{A}_b) \cong O_K$, where K is a number field of degree $6g-6+2n$ over $\mathbf{Q}$ [74].

Thus one obtains a map $F : \mathcal{L} \to \mathcal{O}$, where $\mathcal{L}$ is a category of all links L modulo a homotopy equivalence and $\mathcal{O}$ is a category of rings of the algebraic integers O_K modulo an isomorphism. The map F acts by the formula

$$L_b \overset{\pi_1}{\longmapsto} \pi_1(L_b) \overset{R}{\longmapsto} \mathbb{A}_b \overset{K_0}{\longmapsto} O_K.$$

Remark 7.1.1. Using Lickorish–Wallace Theorem [150], one can extend map F to a category of 3-dimensional manifolds. Indeed, recall that if M^3 is a closed, orientable, connected 3-dimensional manifold, then there exists a link $L \subset S^3$ such that the Dehn surgery of L with the ± 1 coefficients is homeomorphic to M^3. (Notice that such a link is not unique, but, using Kirby calculus, one can define a canonical link L attached to M^3.) Thus we get a map $M^3 \mapsto L$.

Theorem 7.1.1. *The map F is a functor, such that:*
(i) $F(S^3) = \mathbf{Z}$;
(ii) *each ideal* $I \subseteq O_K = F(M^3)$ *corresponds to a link* $L \subset M^3$;
(iii) *each prime ideal* $I \subseteq O_K = F(M^3)$ *corresponds to a knot* $K \subset M^3$.

7.1.1 Braids, links, and Galois covering

By an n-string braid b_n one understands two parallel copies of the plane $\mathbf{R}^2$ in $\mathbf{R}^3$ with n distinguished points taken together with n disjoint smooth paths ("strings") joining pairwise the distinguished points of the planes; the tangent vector to each string is never parallel to the planes. The braids b are endowed with a natural equivalence relation: two braids b and b' are equivalent if b can be deformed into b' without intersection of the strings and so that at each moment of the deformation b remains a braid. By an n-string braid group B_n one understands the set of all n-string braids b endowed with a multiplication operation of the concatenation of $b \in B_n$ and $b' \in B_n$, i. e., the identification of the bottom of b with the top of b'. The group is noncommutative and the identity is given by the trivial braid. The B_n is isomorphic to a group on generators $\sigma_1, \sigma_2, \dots, \sigma_{n-1}$ satisfying the relations $\sigma_i\sigma_{i+1}\sigma_i = \sigma_{i+1}\sigma_i\sigma_{i+1}$ and $\sigma_i\sigma_j = \sigma_j\sigma_i$ if $|i-j| \geq 2$. The Artin representation is an injective homomorphism $r : B_n \to \mathrm{Aut}(\mathbf{F}^n)$ into the group of automorphisms of the free group on generators $x_1, \dots, x_n$ given by the formula $\sigma_i : x_i \mapsto x_i x_{i+1} x_i^{-1}$, $\sigma_{i+1} : x_{i+1} \mapsto x_i$ and $\sigma_k = \mathrm{Id}$ if $k \neq i$ or $k \neq i+1$.

A closure of the braid b is a link $L_b \subset \mathbf{R}^3$ obtained by gluing the endpoints of strings at the top of the braid with such at the bottom of the braid. The closure of two braids $b \in B_n$ and $b' \in B_m$ give the same link $L_b \subset \mathbf{R}^3$ if and only if b and b' can be connected by a sequence of the Markov moves of type I, $b \mapsto aba^{-1}$ for a braid $a \in B_n$, and type II, $b \mapsto b\sigma^{\pm 1} \in B_{n+1}$, where $\sigma \in B_{n+1}$.

Theorem 7.1.2 ([5, Theorem 6]). *The fundamental group of link* L_b *is given by the formula* $\pi_1(L_b) \cong \langle x_1, \dots, x_n \mid x_1 = r(b)x_1, \dots, x_n = r(b)x_n \rangle$, *where* x_i *are generators of the free group* $\mathbf{F}^n$ *and* $r : B_n \to \mathrm{Aut}(\mathbf{F}^n)$ *is the Artin representation of group* B_n.

Let X be a topological space. A covering space of X is a topological space X' and a continuous surjective map $p : X' \to X$ such that for an open neighborhood U of every point $x \in X$ the set $p^{-1}(U)$ is a union of disjoint open sets in X'. A deck transformation of the covering space X' is a homeomorphism $f : X' \to X'$ such that $p \circ f = p$. The set of all deck transformations is a group under composition denoted by $\mathrm{Aut}(X')$. The covering $p : X' \to X$ is called Galois (or regular) if the group $\mathrm{Aut}(X')$ acts transitively on each fiber $p^{-1}(x)$, i. e., for all points $y_1, y_2 \in p^{-1}(x)$ there exists $g \in \mathrm{Aut}(X')$ such that $y_2 = g(y_1)$. The covering $p : X' \to X$ is Galois if and only if the group $G := p_*(\pi_1(X'))$ is a normal subgroup of the fundamental group $\pi_1(X)$. In what follows we consider the

Galois coverings of the space X such that $|\pi_1(X)/G| < \infty$, i. e., the quotient $\pi_1(X)/G$ is a finite group.

7.1.2 Proof of Theorem 7.1.1

We shall split the proof in a series of lemmas.

Lemma 7.1.1. *There exists a faithful representation $R : \pi_1(L_b) \to \mathbb{A}_b$, where $\mathbb{A}_b$ is a stationary AF-algebra of rank $6g-6+2n$.*

Proof. (i) Let us construct a representation

$$R : \pi_1(L_b) \to \mathbb{A}(\mathbf{x}, S_{g,n})/\mathcal{I}_b := \mathbb{A}_b.$$

Suppose that $r : B_{2g+n} \to \mathrm{Aut}(\mathbf{F}^{2g+n})$ is the Artin representation of the braid group B_{2g+n}, see Section 2.1. If x_i is a generator of the free group $\mathbf{F}^{2g+n}$, one can think of x_i as an element of the group $\mathrm{Aut}(\mathbf{F}^{2g+n})$ representing an automorphism of the left multiplication $\mathbf{F}^{2g+n} \to x_i\mathbf{F}^{2g+n}$. Thus we get an embedding $\pi_1(L_b) \hookrightarrow \mathrm{Aut}(\mathbf{F}^{2g+n})$, where $r(b) = \mathrm{Id}$ is a trivial automorphism.

The braid relations

$$\Sigma_i = \{x_i x_{i+1} x_i x_{i+1}^{-1} x_i^{-1} x_{i+1}^{-1} = \mathrm{Id},\ x_i x_j \sigma_i^{-1} x_j^{-1} = \mathrm{Id} \text{ if } |i-j| \geq 2\}$$

correspond to the trivial automorphisms of the group $\mathbf{F}^{2g+n}$. Thus

$$\pi_1(L_b) \cong \langle x_1, \dots, x_{2g+n} \mid r(b) = \Sigma_i = \mathrm{Id},\ 1 \leq i \leq 2g+n-1 \rangle.$$

Recall that the formula $\sigma_i \mapsto e_i + 1$ defines a representation $\rho : B_{2g+n} \to \mathbb{A}(\mathbf{x}, S_{g,n})$, where e_i are projections in the algebra $\mathbb{A}(\mathbf{x}, S_{g,n})$ and $n \in \{0; 1\}$, see Chapter 4.4. Because $B_{2g+n} \cong \langle x_i \mid \Sigma_i = \mathrm{Id},\ 1 \leq i \leq 2g+n-1 \rangle$, so that the ideal $\mathcal{I}_b \subset \mathbb{A}(\mathbf{x}, S_{g,n})$ is generated by the relation $r(b) = \mathrm{Id}$, one gets a faithful representation

$$R : \pi_1(L_b) \to \mathbb{A}(\mathbf{x}, S_{g,n})/\mathcal{I}_b.$$

(ii) Let us show that the quotient $\mathbb{A}_b = \mathbb{A}(\mathbf{x}, S_{g,n})/\mathcal{I}_b$ is a stationary AF-algebra of rank $6g-6+2n$.

First, let us show that the ideal $\mathcal{I}_b \subset \mathbb{A}(\mathbf{x}, S_{g,n})$ is self-adjoint, i. e., $\mathcal{I}_b^* \cong \mathcal{I}_b$. The generating relation $r(b) = \mathrm{Id}$ for such an ideal is invariant under $*$-involution. Indeed, the relation $r(b) = \mathrm{Id}$ has the form $(e_1+1)^{k_1} \cdots (e_{n-1}+1)^{k_{n-1}} = 1$, where $k_i \in \mathbf{Z}$ and e_i are projections. Using the braid relations, one can write the product on the LHS in the form $\sum_{i=1}^{|\mathcal{E}|} a_i \varepsilon_i$, where $a_i \in \mathbf{Z}$ and ε_i are elements of a finite multiplicatively closed set $\mathcal{E}$. Moreover, each ε_i is (Murray–von Neumann equivalent to) a projection. In other words, $\varepsilon_i^* = \varepsilon_i$ and $(\sum_{i=1}^{|\mathcal{E}|} a_i \varepsilon_i)^* = \sum_{i=1}^{|\mathcal{E}|} a_i \varepsilon_i$. We conclude that the relation $r(b) = \mathrm{Id}$ is invariant under $*$-involution. Therefore, we have $\mathcal{I}_b^* \cong \mathcal{I}_b$.

Recall that the quotient of the cluster C^*-algebra $\mathbb{A}(\mathbf{x}, S_{g,n})$ by a self-adjoint (primitive) ideal is a simple AF-algebra of rank $6g-6+2n$. Let us show that the quotient $\mathbb{A}(\mathbf{x}, S_{g,n})/\mathcal{I}_b$ is a stationary AF-algebra.

Consider an inner automorphism φ_b of the group B_{2g+n} given by the formula $x \mapsto b^{-1}xb$. Using the representation $\rho : B_{2g+n} \to \mathbb{A}(\mathbf{x}, S_{g,n})$, one can extend φ_b to an automorphism of the algebra $\mathbb{A}(\mathbf{x}, S_{g,n})$. Since braid b is a fixed point of φ_b, we conclude that φ_b induces a nontrivial automorphism of the AF-algebra $\mathbb{A}_b := \mathbb{A}(\mathbf{x}, S_{g,n})/\mathcal{I}_b$. But each simple AF-algebra with a nontrivial group of automorphisms must be a stationary AF-algebra [74, Chapter 5]. Lemma 7.1.1 follows. □

Lemma 7.1.2. *There is a one-to-one correspondence between normal subgroups of the group $\pi_1(L_b)$ and the AF-subalgebras of the algebra $\mathbb{A}_b$.*

Proof. Consider a representation $R : \pi_1(L_b) \to \mathbb{A}_b$ constructed in Lemma 7.1.1. The algebra $\mathbb{A}_b$ is a closure in the norm topology of a self-adjoint representation of the group ring $\mathbf{C}[\pi_1(L_b)]$ by bounded linear operators acting on a Hilbert space. Namely, such a representation is given by the formula $x_i \mapsto e_i + 1$, where x_i is a generator of the group $\pi_1(L_b)$ and e_i is a projection in the algebra $\mathbb{A}_b$.

Let G be a subgroup of $\pi_1(L_b)$. The $\mathbf{C}[G]$ is a subring of the group ring $\mathbf{C}[\pi_1(L_b)]$. Taking the closure of a self-adjoint representation of $\mathbf{C}[G]$, one gets a C^*-subalgebra $\mathbb{A}_G$ of the algebra $\mathbb{A}_b$.

Let us show that if G is a normal subgroup, then the $\mathbb{A}_G$ is an AF-algebra. Indeed, if G is a normal subgroup one gets an exact sequence

$$0 \to \mathbb{A}_G \to \mathbb{A}_b \to \pi_1(L_b)/G \to 0,$$

where $\pi_1(L_b)/G$ is a finite group. Let $\{M_k(\mathbf{C})\}_{k=1}^{\infty}$ be an ascending sequence of the finite-dimensional C^*-algebras, such that $\mathbb{A}_b = \lim_{k\to\infty} M_k(\mathbf{C})$. Let $M_k'(\mathbf{C}) = M_k(\mathbf{C}) \cap \mathbb{A}_G$. From $0 \to \mathbb{A}_G \to \mathbb{A}_b \to \pi_1(L_b)/G \to 0$, we obtain an exact sequence

$$0 \to M_k'(\mathbf{C}) \to M_k(\mathbf{C}) \to \pi_1(L_b)/G \to 0.$$

Since $|\pi_1(L_b)/G| < \infty$, $M_k'(\mathbf{C})$ is a finite-dimensional C^*-algebra. Thus

$$\mathbb{A}_G = \lim_{k\to\infty} M_k'(\mathbf{C}),$$

i. e., $\mathbb{A}_G$ is an AF-algebra. Lemma 7.1.2 follows. □

Remark 7.1.2. The algebra $\mathbb{A}_G$ is a stationary AF-algebra since it is an AF-subalgebra of a stationary AF-algebra [74, Chapter 5].

Corollary 7.1.1. *There is a one-to-one correspondence between normal subgroups of the group $\pi_1(L_b)$ and ideals in the ring of integers $O_K \cong K_0(\mathbb{A}_b)$ of a number field K, where* $\deg(K|\mathbf{Q}) = 6g-6+2n$.

Proof. A one-to-one correspondence between stationary AF-algebras and the rings of integers in number fields has been established in [105]. Namely, the dimension groups of stationary AF-algebras are in a one-to-one relation with the triples $(\Lambda, [I], i)$, where $\Lambda \subset O_K$ is an order (i. e., a ring with the unit) in the number field K, $[I]$ is the equivalence class of ideals corresponding to Λ, and i is the embedding class of the field K. The degree of K over $\mathbf{Q}$ is equal to the rank of stationary AF-algebra, i. e., $\deg(K|\mathbf{Q}) = 6g - 6 + 2n$.

Assume for simplicity that $\Lambda \cong O_K$, i. e., that Λ is the maximal order in the field K. Recall that we have an inclusion $K_0(\mathbb{A}_G) \subset K_0(\mathbb{A}_b) \cong O_K$. By Remark 7.1.2, $K_0(\mathbb{A}_G)$ is an order in O_K. Since $K_0(\mathbb{A}_G)$ is the kernel of a homomorphism, we conclude that it is an ideal in O_K. The rest of the proof follows from Lemma 7.1.2. Corollary 7.1.1 is proved. □

Lemma 7.1.3. *Let M^3 be a 3-dimensional manifold such that $\pi_1(M^3) \cong \pi_1(L_b)$ and let $O_K \cong K_0(\mathbb{A}_b)$. There is a one-to-one correspondence between the Galois coverings of M^3 ramified over a link $L \subset M^3$ (a knot $K \subset M^3$, resp.) and the ideals (the prime ideals, resp.) of the ring O_K. In other words, each link $L \hookrightarrow M^3$ (each knot $K \hookrightarrow M^3$, resp.) corresponds to an ideal (a prime ideal, resp.) of the ring O_K.*

Proof. Let

$$L \cong \underbrace{S^1 \cup S^1 \cup \cdots \cup S^1}_{k}$$

and let $L \hookrightarrow M^3$ be an embedding of link L into a 3-dimensional manifold M^3. Let M_1^3 be the Galois covering of M^3 ramified over the first component S^1 of the link L and such that the deck transformations fix the remaining components of L. Let

$$L_1 \cong \underbrace{S^1 \cup S^1 \cup \cdots \cup S^1}_{k-1}$$

and let $L_1 \hookrightarrow M_1^3$ be an embedding of link L_1 into M_1^3. Denote by G_1 a normal subgroup of $\pi_1(M^3)$ corresponding to the Galois covering M_1^3.

Let M_2^3 be the Galois covering of M_1^3 ramified over the first component of the link L_1 such that the remaining components are fixed by the corresponding deck transformations. We denote by $G_2 \trianglelefteq G_1$ a normal subgroup of G_1 corresponding to the Galois covering M_2^3.

Proceeding by the induction, one gets the following inclusion of the normal subgroups:

$$G_k \trianglelefteq G_{k-1} \trianglelefteq \cdots \trianglelefteq G_1 \trianglelefteq \pi_1(M^3).$$

By Corollary 7.1.1, the above normal subgroups correspond to a chain of ideals of the ring O_K,

$$I_k \subset I_{k-1} \subset \cdots \subset I_1 \subset O_K.$$

(i) Suppose that $k \geq 2$. In this case, the link L has at least two components, and L is distinct from a knot. Notice that the ideal I_k cannot be a maximal ideal of the ring O_K, since $I_k \subset I_{k-1} \subset O_K$.

(ii) Suppose that $k = 1$. In this case $L \cong K$ is a knot. Clearly, the ideal I_k is the maximal ideal of the ring O_K. Since O_K is the Dedekind domain, we conclude that I_k is a prime ideal. This argument finishes the proof of Lemma 7.1.3. □

Items (ii) and (iii) of Theorem 7.1.1 follow from Lemma 7.1.3.

To prove item (i) of Theorem 7.1.1, one needs to show that $M^3 \cong S^3$ implies $O_K \cong \mathbf{Z}$. Indeed, consider the Riemann surface $S_{0,1}$, i. e., sphere with a cusp. Since the fundamental group $\pi_1(S_{0,1})$ is trivial, the 3-dimensional sphere S^3 is homeomorphic to the mapping torus M_ϕ of surface $S_{0,1}$ by an automorphism $\phi : S_{0,1} \to S_{0,1}$, i. e., $S^3 \cong M_\phi$.

On the other hand, $S_{0,1}$ is homeomorphic to the interior of a planar d-gon. Thus the cluster algebra $\mathbb{A}(\mathbf{x}, S_{0,1})$ is isomorphic to the algebra $\mathbb{A}_{d-3}$ of the triangulated d-gon for $d \geq 3$, see [288, Example 2.2]. It is known that $\mathbb{A}_{d-3}$ is a cluster algebra of finite type, i. e., has finitely many seeds [288]. Therefore the corresponding cluster C^*-algebra must be finite-dimensional, i. e., $\mathbb{A}(\mathbf{x}, S_{0,1}) \cong M_n(\mathbf{C})$. But $K_0(M_n(\mathbf{C})) \cong \mathbf{Z}$. We conclude that the homeomorphism $M^3 \cong S^3$ implies an isomorphism $O_K \cong \mathbf{Z}$. This argument finishes the proof of Theorem 7.1.1(i).

Theorem 7.1.1 follows.

7.1.3 Punctured torus

Let $g = n = 1$, i. e., surface $S_{g,n}$ is homeomorphic to the torus with a cusp (punctured torus). The matrix B associated to an ideal triangulation of surface $S_{1,1}$ has the form

$$B = \begin{pmatrix} 0 & 2 & -2 \\ -2 & 0 & 2 \\ 2 & -2 & 0 \end{pmatrix}.$$

The cluster C^*-algebra $\mathbb{A}(\mathbf{x}, S_{1,1})$ is given by the Bratteli diagram shown in Fig. 4.11. The representation $\rho : B_{2g+n} \to \mathbb{A}(\mathbf{x}, S_{g,n})$ with $g = n = 1$ takes the form

$$\rho : B_3 \to \mathbb{A}(\mathbf{x}, S_{1,1}).$$

For every $b \in B_3$, the inner automorphism $\varphi_b : x \mapsto b^{-1}xb$ of B_3 defines a unique automorphism of the algebra $\mathbb{A}(\mathbf{x}, S_{1,1})$. Since the algebra $\mathbb{A}(\mathbf{x}, S_{1,1})$ is a coordinate ring of the Teichmüller space $T_{1,1}$ and $\mathrm{Aut}(T_{1,1}) \cong \mathrm{SL}_2(\mathbf{Z})$, we conclude that φ_b corresponds to an element of the modular group $\mathrm{SL}_2(\mathbf{Z})$.

An explicit formula for the correspondence $b \mapsto \varphi_b$ is well known. Namely, if σ_1 and σ_2 are the standard generators of the braid group $B_3 \cong \langle \sigma_1, \sigma_2 \mid \sigma_1\sigma_2\sigma_1 = \sigma_2\sigma_1\sigma_2 \rangle$

then the formula

$$\sigma_1 \mapsto \begin{pmatrix} 1 & 1 \\ 0 & 1 \end{pmatrix}, \quad \sigma_2 \mapsto \begin{pmatrix} 1 & 0 \\ -1 & 1 \end{pmatrix}$$

defines a surjective homomorphism $B_3 \to SL_2(\mathbf{Z})$.

Example 7.1.1. Let L_b be a link given by the closure of a braid of the form $b = \sigma_1^p \sigma_2^{-q}$, where $p \geq 1$ and $q \geq 1$. In this case

$$\varphi_b = \begin{pmatrix} pq+1 & p \\ q & 1 \end{pmatrix}.$$

The algebra $\mathbb{A}_b = \mathbb{A}(\mathbf{x}, S_{1,1})/\mathcal{I}_b$ is a stationary AF-algebra of rank 2 given by the Bratteli diagram shown in Fig. 7.1, where the integers correspond to the multiplicity of edges of the graph. The Perron–Frobenius eigenvalue of the matrix φ_b is equal to

$$\lambda_{\varphi_b} = \frac{pq + 2 + \sqrt{pq(pq+4)}}{2}.$$

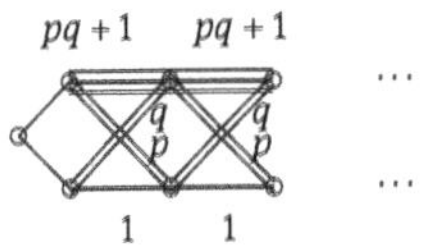

Figure 7.1: Bratteli diagram of the algebra $\mathbb{A}(\mathbf{x}, S_{1,1})/\mathcal{I}_b$.

Therefore $K_0(\mathbb{A}_b) \cong \mathbf{Z} + \mathbf{Z}\sqrt{D}$, where $D = pq(pq+4)$. The number field K corresponding to the link L_b is a real quadratic field of the form

$$K \cong \mathbf{Q}(\sqrt{pq(pq+4)}).$$

Denote by $M^3_{p,q}$ a 3-dimensional manifold such that $\pi_1(M^3_{p,q}) \cong \pi_1(L_{\sigma_1^p \sigma_2^{-q}})$. The manifolds $M^3_{p,q}$ corresponding to the quadratic fields with a small square-free discriminant D are recorded below.

Manifold $M_{p,q}$	Number field $K = F(M_{p,q})$
$M^3_{1,1}$	$\mathbf{Q}(\sqrt{5})$
$M^3_{1,3}$	$\mathbf{Q}(\sqrt{21})$
$M^3_{1,7}$	$\mathbf{Q}(\sqrt{77})$
$M^3_{1,11}$	$\mathbf{Q}(\sqrt{165})$
$M^3_{1,13}$	$\mathbf{Q}(\sqrt{221})$
$M^3_{3,5}$	$\mathbf{Q}(\sqrt{285})$
$M^3_{3,7}$	$\mathbf{Q}(\sqrt{525})$
$M^3_{3,11}$	$\mathbf{Q}(\sqrt{1221})$

Remark 7.1.3. The manifold $M^3_{p,q}$ can be realized as a torus bundle over the circle with the monodromy given by matrix φ_b. The arithmetic invariants of surface bundles over the circle were studied in Section 1.5.

Guide to the literature

The arithmetic topology dates back to C. F. Gauss who introduced the concepts of an algebraic number and a link in the 3-dimensional space. We refer the reader to the excellent monograph by M. Morishita [166]. Topological interpretation of the class field theory based on the Galois cohomology were given in the 1960s by J. T. Tate [273]. M. Artin and J.-L. Verdier obtained a similar interpretation based on the étale cohomology [10]. An analogy between prime ideals and knots was explored by Yu. Manin, B. Mazur, and D. Mumford, see, e. g., [159, 160]. The term "arithmetic topology" was introduced by M. Kapranov [129] and A. Reznikov [237]. New developments in the area belong to A. S. Sikora [257]. Our exposition of the arithmetic topology is based on [207].

7.2 Arithmetic topology of 4-manifolds

In this section we extend the functor F (Theorem 7.1.1) to the smooth 4-dimensional manifolds M^4. The range of F is no longer the fields of algebraic numbers, but the fields $\mathbb{K}$ with noncommutative multiplication, e. g., the quaternions or the cyclic division algebras. We refer to $\mathbb{K}$ as a hyper-algebraic number field. The Galois theory of such fields was elaborated in [47, 123, 134].

To introduce the map F, let N be a finite index subgroup of the mapping class group $\operatorname{Mod} M^3$, where M^3 is a 3-dimensional manifold. The subgroup N gives rise to a smooth branched cover M^4_N of the 4-sphere S^4 [227]. On the other hand, we have $\widehat{N}/\widehat{\mathbf{Z}} \cong G_{\mathbb{K}}$, where $\widehat{N}$ (resp. $\widehat{\mathbf{Z}}$) is a profinite completion of N (resp. $\mathbf{Z}$) and $G_{\mathbb{K}}$ is the absolute Galois group of $\mathbb{K}$ [210, Theorem 1.2]. Recall that the field $\mathbb{K}$ can be recovered up to an isomorphism from the group $G_{\mathbb{K}}$ [279, Corollary 2]. We define F as a composition of maps

$$M^4_N \mapsto \widehat{N}/\widehat{\mathbf{Z}} \cong G_{\mathbb{K}} \mapsto \mathbb{K}.$$

Remark 7.2.1. The map F can be equivalently defined via a C^*-algebra $\mathbb{E}_{M^4}$ generated by the diffeomorphisms of M^4, see Section 7.5. Namely, the K-theory of $\mathbb{E}_{M^4}$ gives rise to a number field K whose central simple algebra is isomorphic to $\mathbb{K}$. In particular, the exotic smoothings of M^4 are classified by the Brauer group of K, see Section 7.5 for the details.

Let $O_{\mathbb{K}}$ be the ring of integers of the field $\mathbb{K}$. Unlike O_K, the ring $O_{\mathbb{K}}$ is usually simple, i. e., all two-sided ideals of $O_{\mathbb{K}}$ are trivial. Hence we must use the dynamical ideals (dynamials), i. e., the crossed products $O_{\mathbb{K}} \rtimes m\mathbf{Z}$ by an endomorphism of $O_{\mathbb{K}}$ of degree $m \geq 1$, see Section 7.4 for the details and motivation. The dynamial $O_{\mathbb{K}} \rtimes m\mathbf{Z}$ is minimal if and only if $m = p$ is a prime number; see Section 7.4.

To formalize our results, recall that the groups $N \not\cong N'$ are a Grothendieck pair if $\widehat{N} \cong \widehat{N'}$. We shall denote by $\mathfrak{M}^4$ the category of all smooth 4-dimensional manifolds M_N^4 such that: (i) the set $\mathfrak{M}^4$ contains no M_N^4 and $M_{N'}^4$ such that N and N' are a Grothendieck pair and (ii) the arrows of $\mathfrak{M}^4$ are differentiable maps between such manifolds. Denote by $\mathfrak{K}$ the category of the hyper-algebraic number fields such that the arrows of $\mathfrak{K}$ are injective homomorphisms between these fields. The knotted surface is a transverse immersion $\iota : X_{g_1} \cup \cdots \cup X_{g_n} \hookrightarrow M^4$ of a collection of the 2-dimensional orientable surfaces X_{g_i} of the genera $g_i \geq 0$. The $\iota(X_{g_1} \cup \cdots \cup X_{g_n})$ is called a surface knot K if $n = 1$ and a surface link L otherwise. Our main result can be formulated as follows.

Theorem 7.2.1. *The map $F : \mathfrak{M}^4 \to \mathfrak{K}$ is a covariant functor such that:*
(i) *$F(S^4) \cong \mathbb{H}$ is the field of quaternions;*
(ii) *the dynamials $O_{\mathbb{K}} \rtimes m\mathbf{Z}$ correspond to the surface links $L \subset M^4$;*
(iii) *the minimal dynamials $O_{\mathbb{K}} \rtimes p\mathbf{Z}$ correspond to the surface knots $K \subset M^4$.*

Remark 7.2.2. The functor F is independent of the choice of manifold M^3 used in its construction. Indeed, if $\{W_i^4 \mid \pi_1(W_i^4) \cong \mathrm{Mod}\, M_i^3\}$ are the manifolds described in Section 7.2.1, one can always take a connected sum $\# W_i^4$, so that $\mathrm{Mod}\, M^3 \cong \pi_1(\# W_i^4)$. This observation is corroborated by the C^*-algebra approach, see Remark 7.2.1.

Let $\mathbb{K}$ be a Galois extension of $\mathbb{H}$, i. e., an extension having the Galois group $\mathrm{Gal}\, \mathbb{K}$ and obeying the fundamental correspondence of the Galois theory [46, 123, 134]. Recall that $\mathbb{K}$ is said to be an abelian extension, if $\mathrm{Gal}\, \mathbb{K}$ is an abelian group. Theorem 7.2.1 implies the following result.

Corollary 7.2.1. *Let $M^4 \in \mathfrak{M}^4$ be a simply connected 4-manifold. Then $\mathbb{K} = F(M^4)$ is an abelian extension.*

Remark 7.2.3. The converse of Corollary 7.2.1 is false.

7.2.1 4-dimensional manifolds

Let M^3 be a closed 3-dimensional manifold. Denote by $\mathrm{Mod}\, M^3$ the mapping class group of M^3, i. e., a group of isotopy classes of the orientation-preserving diffeomorphisms of M^3. The $\mathrm{Mod}\, M^3$ is a finitely presented group [109]. Since each finitely presented group can be realized as the fundamental group of a smooth 4-manifold, we denote by $W^4 \in \mathfrak{M}^4$ a 4-manifold with $\pi_1(W^4) \cong \mathrm{Mod}\, M^3$. It will be useful for us to represent W^4 as a PL (and, therefore, smooth) 4-fold (or 5-fold, if necessary) branched cover of the sphere S^4. Let X_g be a closed surface, which we assume for the sake of brevity to be orientable of genus $g \geq 0$. Recall that there exists a transverse immersion

$\iota : X_g \hookrightarrow S^4$ such that W^4 is the 4-fold PL cover of S^4 branched at the points of X_g [227]. In other words, $\operatorname{Mod} M^3$ is a normal subgroup of index 4 of the fundamental group $\pi_1(S^4 - X_g)$.

Remark 7.2.4. The immersion ι and surface X_g need not be unique for given W^4. Yet one can always restrict to a canonical choice of ι and X_g, which we always assume to be the case.

Definition 7.2.1. By M_N^4 we understand a cover of the manifold W^4 corresponding to the finite index subgroup N of the fundamental group $\pi_1(W^4)$.

Remark 7.2.5. In view of the continuous map $W^4 \to S^4$, the manifold M_N^4 is a smooth $4d$-fold branched cover of the sphere S^4, where d is the index of N in $\operatorname{Mod} M^3$. The manifold M_N^4 is a regular (Galois) cover of S^4 if and only if N is a normal subgroup of $\operatorname{Mod} M^3$.

7.2.2 Galois theory for noncommutative fields

Denote by $\mathbb{K}$ a division ring, i. e., a ring such that the set $\mathbb{K}^\times := \mathbb{K} - \{0\}$ is a group under multiplication. If the group $\mathbb{K}^\times$ is commutative, then $\mathbb{K}$ is a field. Otherwise, we refer to $\mathbb{K}$ as a noncommutative field.

Roughly speaking, the Galois theory for $\mathbb{K}$ is a correspondence between the subfields of $\mathbb{K}$ and the subgroups of a Galois group $\operatorname{Gal}\mathbb{K}$ of the field $\mathbb{K}$. Namely, let G be a group of automorphisms of the field $\mathbb{K}$. It is easy to see that the set

$$\mathbb{I}_G = \{x \in \mathbb{K} \mid g(x) = x \text{ for all } g \in G\}$$

is a subfield of the field $\mathbb{K}$.

Definition 7.2.2. An extension $\mathbb{K}$ of the field $\mathbb{L}$ is called Galois if there exists a group G of automorphisms of the field $\mathbb{K}$ such that $\mathbb{L} \cong \mathbb{I}_G$. The Galois group of $\mathbb{K}$ with respect to $\mathbb{L}$ is defined as $\operatorname{Gal}\mathbb{K} \cong G$. The absolute Galois group $G_{\mathbb{K}}$ is defined as a Galois group of the algebraic closure of $\mathbb{K}$. The $G_{\mathbb{K}}$ is a profinite group.

Remark 7.2.6. By an adaption of the argument for the case of fields [279, Corollary 2], one can show that the group $G_{\mathbb{K}}$ defines the underlying field $\mathbb{K}$ up to an isomorphism.

Unlike in the case of fields, the inner automorphisms of $\mathbb{K}$ are a nontrivial group. Indeed, consider an automorphism $h_g : \mathbb{K} \to \mathbb{K}$ given by the formula

$$x \mapsto g^{-1}xg, \quad x \in \mathbb{K},\ g \in \mathbb{K}^\times.$$

By $\operatorname{Inn}(\mathbb{K})$ we denote a group of such automorphisms under the composition $h_{g_1} \circ h_{g_2} = h_{g_1 g_2}$, where $g_1, g_2 \in \mathbb{K}^\times$. It follows from the above formula that $\operatorname{Inn}(\mathbb{K}) \cong \mathbb{K}^\times / C$, where

C is the center of $\mathbb{K}^\times$. Let G be a finite group of automorphisms of the field $\mathbb{K}$. A normal subgroup $\Gamma := G \cap \mathrm{Inn}(\mathbb{K})$ of G consists of the inner automorphisms of the field $\mathbb{K}$. Consider a group ring

$$\mathbb{B}(\Gamma) = \sum_{g_i \in \Gamma} \sum_{h_j \in C} g_i h_j.$$

It is easy to see that $\mathbb{B}(\Gamma)$ is a finite-dimensional algebra over its center C. The following result has been established in [46, Théorème 1] and [134, p. 139].

Theorem 7.2.2. *If $\mathbb{K}$ is a Galois extension, then the corresponding Galois group G satisfies the short exact sequence of groups*

$$1 \to \Gamma \xrightarrow{\iota} G \to G/\Gamma \to 1,$$

where ι is an inclusion and $|G/\Gamma| = \dim_C \mathbb{B}(\Gamma)$.

Recall that $\mathbb{B}(\Gamma)$ is a Frobenius algebra [134, Theorem 3.5.1]. In other words, there exists a nondegenerate bilinear form

$$Q : \mathbb{B}(\Gamma) \times \mathbb{B}(\Gamma) \to C$$

such that $Q(xy, z) = Q(x, yz)$ for all $x, y, z \in \mathbb{B}(\Gamma)$.

Remark 7.2.7. The above form is symmetric if and only if $\mathbb{B}(\Gamma)$ is a commutative ring.

Proof. $Q(xy, z) = Q(x, yz) = Q(x, zy) = Q(xz, y) = Q(zx, y) = Q(z, xy)$. □

7.2.3 Uchida map

The following construction of F is based on [279]. Let $X_{g,n}$ be an orientable surface of genus $g \geq 0$ with $n \geq 0$ boundary components. Denote by $\mathrm{Mod}\, X_{g,n}$ the mapping class group of $X_{g,n}$, i. e., a group of isotopy classes of the orientation and boundary-preserving diffeomorphisms of the surface $X_{g,n}$. Let N be a finite index subgroup of $\mathrm{Mod}\, X_{g,n}$. We omit the construction of a 3-manifold $M_N^3 \in \mathfrak{M}^3$ from N referring the reader to Section 7.1. As usual, let G_K be the absolute Galois group of $K \in \mathbf{K}$, while $\widehat{N}$ and $\widehat{\mathbf{Z}}$ denote the profinite completion of the groups N and $\mathbf{Z}$, respectively. Let $G_K \mapsto K$ be an injective map defined in [279, Corollary 2].

Theorem 7.2.3 ([210, Theorem 1.2]). *The map*

$$M_N^3 \mapsto \widehat{N}/\widehat{\mathbf{Z}} \cong G_K \mapsto K$$

coincides with the functor F of Theorem 7.1.1. Such a functor is injective, unless N and N' are a Grothendieck pair. Moreover, for every normal finite index subgroup $N' \subseteq N$, there *exist a regular cover $M^3_{N'}$ of M^3_N and an intermediate field $K' = F(M^3_{N'})$ such that $K \subseteq K'$ and* $\mathrm{Gal}(K'|K) \cong N/N'$.

Remark 7.2.8. The construction of map $M^3_N \mapsto K$ extends to the 4-manifolds M^4 and noncommutative fields $\mathbb{K}$.

7.2.4 Proofs

7.2.4.1 Proof of Theorem 7.2.1

We shall focus on the first part by proving that $F : \mathfrak{M}^4 \to \mathfrak{K}$ is a covariant functor, while referring the reader to Section 7.4.4 for the proof of items (i)–(iii) of Theorem 7.2.1. We shall prove F to be a functor with respect to homeomorphisms first, and then generalize to arbitrary differentiable mappings, see Remark 7.2.9.

Part I. Let us show that $F : \mathfrak{M}^4 \to \mathfrak{K}$ is a covariant functor. For clarity, we split the proof in a series of lemmas.

Lemma 7.2.1. *The manifolds $M^4_N, M^4_{N'} \in \mathfrak{M}^4$ are homeomorphic if and only if the subgroups $N, N' \subseteq \mathrm{Mod}\, M^3$ are isomorphic.*

Proof. (i) Let M^4_N be homeomorphic to $M^4_{N'}$ by a homeomorphism $h : M^4_N \to M^4_{N'}$. Since both M^4_N and $M^4_{N'}$ cover the 4-sphere S^4 branched over a surface $X_g \hookrightarrow S^4$, one gets a commutative diagram in Fig. 7.2. By definition, $N \cong p_*(\pi_1(M^4_N)) \subset \pi_1(S^4 - X_g)$. Likewise, $N' \cong p'_*(\pi_1(M^4_{N'})) \subset \pi_1(S^4 - X_g)$. Since M^4_N is homeomorphic to $M^4_{N'}$, one gets an isomorphism of the fundamental groups $\pi_1(M^4_N) \cong \pi_1(M^4_{N'})$. Because the maps p_* and p'_* are injective, we conclude that N and N' are isomorphic subgroups of $\pi_1(S^4 - X_g)$ and, therefore, of the group $\mathrm{Mod}\, M^3$. The necessary condition of Lemma 7.2.1 is proved.

(ii) Let $N \cong N'$ be isomorphic subgroups of $\mathrm{Mod}\, M^3$. As explained in Section 2.1, the groups N and N' define a pair of branched covers M^4_N and $M^4_{N'}$ of S^4. In view of the diagram in Fig. 7.2, one obtains an inclusion of groups $N, N' \subset \pi_1(S^4 - X_g)$. Since $N \cong N'$, the subgroups N and N' are conjugate in the group $\pi_1(S^4 - X_g)$. In other words, there exists an element $\mathbf{g} \in \pi_1(S^4 - X_g)$ such that $N' = \mathbf{g}^{-1} N \mathbf{g}$. It is well known that the conjugate subgroups of $\pi_1(S^4 - X_g)$ correspond to the homeomorphic branched covers M^4_N and $M^4_{N'}$ of S^4. Moreover, the homeomorphism $h : M^4_N \to M^4_{N'}$ is realized by

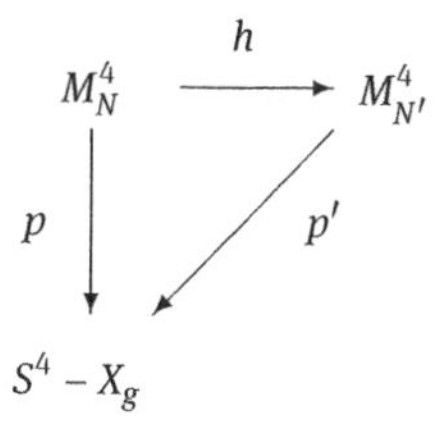

Figure 7.2: Branched covers of S^4.

a deck transformation of the branched cover. The sufficient condition of Lemma 7.2.1 is proved. □

Lemma 7.2.2. *Up to the Grothendieck pairs, the subgroups N and N' of* $\mathrm{Mod}\, M^3$ *are isomorphic if and only if the groups $\widehat{N}/\widehat{\mathbf{Z}} \cong G_{\mathbb{K}}$ and $\widehat{N'}/\widehat{\mathbf{Z}} \cong G_{\mathbb{K}'}$ are isomorphic.*

Proof. (i) Let $N \cong N'$ be a pair of isomorphic subgroups of $\mathrm{Mod}\, M^3$. Recall that a profinite group $\widehat{N}$ is a topological group defined by the inverse limit

$$\widehat{N} := \varprojlim N/N_k,$$

where N_k runs through all open normal finite index subgroups of N. It follows from the above that if $N \cong N'$, then $\widehat{N} \cong \widehat{N'}$. Since $G_{\mathbb{K}} \cong \widehat{N}/\widehat{\mathbf{Z}}$ and $G_{\mathbb{K}'} \cong \widehat{N'}/\widehat{\mathbf{Z}}$, we conclude that $G_{\mathbb{K}} \cong G_{\mathbb{K}'}$. The necessary condition of Lemma 7.2.2 is proved.

(ii) Suppose that $G_{\mathbb{K}} \cong \widehat{N}/\widehat{\mathbf{Z}}$ and $G_{\mathbb{K}'} \cong \widehat{N'}/\widehat{\mathbf{Z}}$ are isomorphic groups. Let us show that $N \cong N'$ are isomorphic subgroups of $\mathrm{Mod}\, M^3$. To the contrary, let $N \not\cong N'$. Since N and N' cannot be a Grothendieck pair, we conclude that $\widehat{N} \not\cong \widehat{N'}$. But $G_{\mathbb{K}} \cong \widehat{N}/\widehat{\mathbf{Z}}$ and $G_{\mathbb{K}'} \cong \widehat{N'}/\widehat{\mathbf{Z}}$ and, therefore, $G_{\mathbb{K}} \not\cong G_{\mathbb{K}'}$. One gets a contradiction proving the sufficient condition of Lemma 7.2.2. □

Lemma 7.2.3. *The absolute Galois groups $G_{\mathbb{K}}$ and $G_{\mathbb{K}'}$ are isomorphic if and only if the underlying noncommutative fields $\mathbb{K}$ and $\mathbb{K}'$ are isomorphic.*

Proof. The proof is an adaption of the argument of [279, Corollary 2] to the case of noncommutative fields. Namely, we introduce a topology on $G_{\mathbb{K}}$ compatible with the formula $M_N^4 \mapsto \widehat{N}/\widehat{\mathbf{Z}} \cong G_{\mathbb{K}} \mapsto \mathbb{K}$. Let $G_{\mathbb{K}}$ and $G_{\mathbb{K}'}$ be open subgroups of an absolute Galois group G, and let $\sigma : G_{\mathbb{K}} \to G_{\mathbb{K}'}$ be a topological isomorphism. To prove our lemma, it is enough to show that σ can be extended to an inner automorphism of G, which corresponds to an isomorphism $\mathbb{K} \cong \mathbb{K}'$. Indeed, one takes an open normal subgroup N of G contained in $G_{\mathbb{K}}$ and $G_{\mathbb{K}'}$. The group N induces an isomorphism $\sigma_N : G_{\mathbb{K}}/N \to G_{\mathbb{K}'}/N$. It can be shown that σ_N extends to an inner automorphism of the group G/N. We repeat the construction over all open normal subgroups of G and obtain an explicit formula for the required inner automorphism of G. Lemma 7.2.3 follows. □

Remark 7.2.9. Lemmas 7.2.1–7.2.3 are true for the differentiable mappings. In this case one obtains an inclusion of the corresponding noncommutative fields.

Lemmas 7.2.1–7.2.3 and formula $M_N^4 \mapsto \widehat{N}/\widehat{\mathbf{Z}} \cong G_{\mathbb{K}} \mapsto \mathbb{K}$ imply that $F : \mathfrak{M}^4 \to \mathfrak{K}$ is a covariant functor.

Part II. The detailed proof of items (i)–(iii) of Theorem 7.2.1 is given in Section 7.4.4, where the corresponding examples of the sphere knots are considered.

Theorem 7.2.1 is proved.

7.2.4.2 Proof of Corollary 7.2.1

If $M^4 \in \mathfrak{M}^4$ is a simply connected manifold, then $\pi_1(M^4)$ is a trivial group. Since M^4 is a cover of S^4 branched over a surface $X_g \hookrightarrow S^4$, we conclude that $\pi_1(S^4 - X_g)$ is a finite group. By Hurewicz Theorem, the homology group

$$H_1(S^4 - X_g; \mathbf{Z}) \cong \pi_1(S^4 - X_g)/[\pi_1(S^4 - X_g), \pi_1(S^4 - X_g)]$$

is also a finite abelian group.

Consider a commutative diagram in Fig. 7.3. An automorphism of $H_1(S^4 - X_g; \mathbf{Z})$ comes from a homeomorphism $h : M^4 \to M^4$ of the manifold M^4. Theorem 7.2.1 says that h corresponds to an automorphism σ_h of the field $\mathbb{K}$ such that $\sigma_h(\mathbb{L}) = \mathbb{L}$ for a subfield $\mathbb{L} \subseteq \mathbb{K}$. (For simplicity, the reader can think $\mathbb{L} \cong \mathbb{H}$ is the field of quaternions.) Since the automorphisms σ_h generate the Galois group of the extension $\mathbb{K} \mid \mathbb{L}$, we conclude that

$$\operatorname{Gal} \mathbb{K} \cong \operatorname{Aut}(H_1(S^4 - X_g; \mathbf{Z})).$$

Recall that the group of automorphisms of a finite abelian group is always an abelian group. Indeed, a finite abelian groups can be written in the form $\mathbf{Z}/p_1\mathbf{Z} \oplus \cdots \oplus \mathbf{Z}/p_k\mathbf{Z}$ for some distinct primes p_i. On the other hand, $\operatorname{Aut}(G \oplus H) \cong \operatorname{Aut}(G) \oplus \operatorname{Aut}(H)$, where G and H are finite abelian groups of the coprime order. Thus

$$\operatorname{Aut}(H_1(S^4 - X_g; \mathbf{Z})) \cong \operatorname{Aut}(\mathbf{Z}/p_1\mathbf{Z}) \oplus \cdots \oplus \operatorname{Aut}(\mathbf{Z}/p_k\mathbf{Z}).$$

Since the group of automorphism of a cyclic group is a cyclic group, we conclude that the group $\operatorname{Aut}(H_1(S^4 - X_g; \mathbf{Z}))$ is abelian. In view of the above, the group $\operatorname{Gal} \mathbb{K}$ is also abelian. Corollary 7.2.1 is proved.

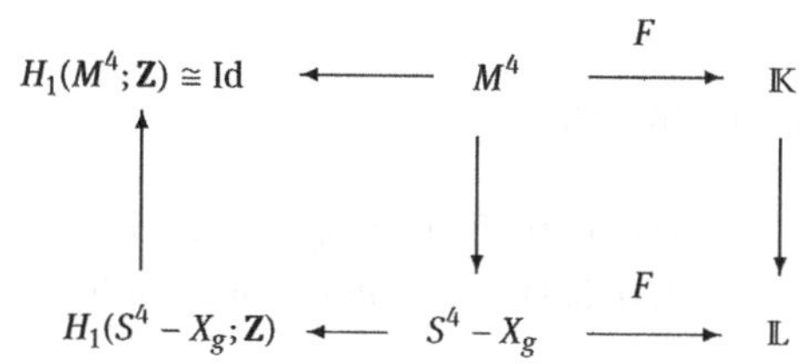

Figure 7.3: Simply connected manifold M^4.

7.2.4.3 Proof of Remark 7.2.3

Let $\mathbb{K}$ be an abelian extension. Then $\operatorname{Gal} \mathbb{K}$ is a finite abelian group. Let us show that the group $H_1(S^4 - X_g; \mathbf{Z})$ can be infinite. Indeed, one can always write $H_1(S^4 - X_g; \mathbf{Z}) \cong \mathbf{Z}^k \oplus \mathrm{Tors}$, where $k \geq 0$ and Tors is a finite abelian group. As explained, we have $\operatorname{Aut}(H_1(S^4 - X_g; \mathbf{Z})) \cong \operatorname{Aut}(\mathbf{Z}^k) \oplus \operatorname{Aut}(\mathrm{Tors})$. Recall that $\operatorname{Aut}(\mathbf{Z}^k) \cong GL_k(\mathbf{Z})$. If $k = 1$, then the group

$\mathrm{Aut}(\mathbf{Z}) \cong \mathbf{Z}/2\mathbf{Z}$ is abelian and finite. Yet the group $H_1(S^4 - X_g; \mathbf{Z}) \cong \mathbf{Z} \oplus \mathrm{Tors}$ is infinite. In other words, the corresponding 4-manifold $M^4 \in \mathfrak{M}^4$ cannot be simply connected.

7.2.5 Rokhlin and Donaldson's Theorems revisited

In this section we give an alternative proof of Rokhlin and Donaldson's Theorems based on the Galois theory of noncommutative fields. To outline the proof, let $M^4 \in \mathfrak{M}^4$ be a simply connected smooth 4-manifold. Corollary 7.2.1 says that the noncommutative field $\mathbb{K} = F(M^4)$ is an abelian extension of the quaternions. Consider a subgroup of the inner automorphisms Γ of the abelian group Gal $\mathbb{K}$. Let $\mathbb{B}(\Gamma)$ be the corresponding group ring. Denote by Q a symmetric bilinear form on $\mathbb{B}(\Gamma)$. For Γ a finite abelian group, we calculate both $\mathbb{B}(\Gamma)$ and Q. On the other hand, Theorem 7.2.1 implies an isomorphism of the $\mathbf{Z}$-modules

$$\mathbb{B}(\Gamma) \cong H_2(M^4; \mathbf{Z}).$$

From the map $Q : \mathbb{B}(\Gamma)\times\mathbb{B}(\Gamma) \to \mathbf{Z}$, one gets a symmetric bilinear form on the homology group $H_2(M^4; \mathbf{Z})$. As a corollary, we recover Rokhlin and Donaldson's Theorems for the simply connected smooth 4-manifolds from the arithmetic of Q. Let us pass to a detailed argument.

Lemma 7.2.4. *The group ring of a finite abelian group $\Gamma \cong \mathbf{Z}/p_1\mathbf{Z}\oplus\cdots\oplus\mathbf{Z}/p_k\mathbf{Z}$ is isomorphic to a direct sum of the cyclotomic fields, i. e.,*

$$\mathbb{B}(\Gamma) \cong \mathbf{Z}(\zeta_{p_1}) \oplus \cdots \oplus \mathbf{Z}(\zeta_{p_k}),$$

where ζ_{p_i} is the p_ith root of unity.

Proof. An elegant proof of this fact can be found in [18]. □

Thus to calculate Q, we can restrict to the cyclotomic fields $\mathbf{Q}(\zeta_{p_i})$ and take the tensor product over p_i. Recall that the trace form on $\mathbf{Q}(\zeta_{p_i})$ is a symmetric bilinear form

$$\mathrm{Tr}_{\mathbf{Q}(\zeta_{p_i})} : \mathbf{Q}(\zeta_{p_i}) \times \mathbf{Q}(\zeta_{p_i}) \to \mathbf{Q} \quad \text{such that } (x, y) \mapsto \mathrm{tr}(xy),$$

where tr is the trace of an algebraic number. The above trace form is equivalent to the form $\mathrm{Tr}_{\mathbf{Q}(\zeta_{p_i})}(x, x) := \mathrm{Tr}_{\mathbf{Q}(\zeta_{p_i})}(x^2)$ via the formula

$$\mathrm{Tr}_{\mathbf{Q}(\zeta_{p_i})}(x, y) = \frac{1}{2}[\mathrm{Tr}_{\mathbf{Q}(\zeta_{p_i})}(x + y)^2 - \mathrm{Tr}_{\mathbf{Q}(\zeta_{p_i})}(x^2) - \mathrm{Tr}_{\mathbf{Q}(\zeta_{p_i})}(y^2)].$$

Lemma 7.2.5. *The following classification is true:*

$$\mathrm{Tr}_{\mathbf{Q}(\zeta_p)}(x^2) = \begin{cases} p\langle 1\rangle, & \text{if } p \text{ is odd prime} \\ 2^n\langle 1\rangle(\langle 1\rangle \oplus \langle -1\rangle \oplus (2^{n-1}-1)\times H), & \text{if } p = 2^n,\ n \geq 4, \end{cases}$$

where we use the Witt ring notation $p\langle 1\rangle$ for the diagonal quadratic form of dimension p, $2^n\langle 1\rangle$ for the Pfister form of dimension 2^n and H for the hyperbolic (split) plane, see [143, Chapter X] for definitions.

Proof. We refer the reader to [218, Theorem 2.1] for the proof. □

Remark 7.2.10. The case $n = 1$ ($n = 2$; $n = 3$) corresponds to the complex numbers (quaternions; octonions), respectively [143, Chapter X]. We omit these values of n, since none of the fields is contained in $\mathbb{K}$.

Lemma 7.2.6. *If $H_2(M^4;\mathbf{Z})$ is the second homology of a simply connected manifold M^4, then $\mathbb{B}(\Gamma) \cong H_2(M^4;\mathbf{Z})$, where $\cong$ is an isomorphism of the corresponding $\mathbf{Z}$-modules.*

Proof. Recall that $\mathrm{Gal}\,\mathbb{K}$ is acting on the field $\mathbb{K}$ by the automorphisms of $\mathbb{K}$. Theorem 7.2.1 says that such automorphisms correspond to the homeomorphisms of the manifold M^4. Recall that each homeomorphism $h : M^4 \to M^4$ defines an linear map $h_* : H_2(M^4;\mathbf{Z}) \to H_2(M^4;\mathbf{Z})$. Since $\Gamma \subseteq \mathrm{Gal}\,\mathbb{K}$, one gets a linear representation

$$\rho : \Gamma \to \mathrm{Aut}(H_2(M^4;\mathbf{Z})).$$

Consider a regular representation of Γ by the automorphisms of $H_2(M^4;\mathbf{Z}) \cong \mathbf{Z}^k$. It is easy to see that such a representation coincides with ρ and, therefore, we conclude that $k = |\Gamma|$. Moreover, since $\mathbb{B}(\Gamma)$ is the group ring of Γ, one gets an isomorphism $\mathbb{B}(\Gamma) \cong H_2(M^4;\mathbf{Z})$ between the corresponding $\mathbf{Z}$-modules. Lemma 7.2.6 is proved. □

Corollary 7.2.2 (Rokhlin and Donaldson).
(i) *Definite intersection form of a simply connected smooth 4-manifold is diagonalizable;*
(ii) *Signature of the intersection form of a simply connected smooth 4-manifold is divisible by* 16.

Proof. As explained, we identify the intersection form

$$H_2(M^4;\mathbf{Z}) \times H_2(M^4;\mathbf{Z}) \to \mathbf{Z}$$

with the trace form $\mathrm{Tr}_{\mathbf{Q}(\zeta_p)}(x, y)$ given by Lemma 7.2.5. Accordingly, we have to consider the following two cases:

(i) Let us consider the case $\mathrm{Tr}_{\mathbf{Q}(\zeta_p)}(x^2) = p\langle 1\rangle$, where p is an odd prime. Let $\{1, \zeta_p, \ldots, \zeta_p^{p-1}\}$ be the standard basis in the ring of integers $\mathbf{Z}[\zeta_p]$ of the cyclotomic field $\mathbf{Q}(\zeta_p)$. The corresponding formula can be written as

$$\mathrm{Tr}_{\mathbf{Q}(\zeta_p)}(x^2) = \sum_{i=0}^{p-1} x_i^2, \quad x_i \in \mathbf{Z},$$

where $x = x_0 + x_1\zeta_p + \cdots + x_{p-1}\zeta_p^{p-1}$ for an $x \in \mathbf{Z}[\zeta_p]$. Thus one gets a symmetric bilinear form on $H_2(M^4;\mathbf{Z})$,

$$\mathrm{Tr}_{\mathbf{Q}(\zeta_p)}(x,y) = \sum_{i=0}^{p-1} x_i y_i, \quad x_i, y_i \in \mathbf{Z}.$$

It remains to notice that $M^4 \in \mathfrak{M}^4$ is a smooth manifold and the trace is a positive definite diagonalizable intersection form. Item (i) of Corollary 7.2.2 follows.

(ii) Let us consider the case

$$\mathrm{Tr}_{\mathbf{Q}(\zeta_p)}(x^2) = 2^n\langle 1\rangle(\langle 1\rangle \oplus \langle -1\rangle \oplus (2^{n-1}-1) \times H),$$

where $p = 2^n$ and $n \geq 4$; see Remark 7.2.10 explaining the restriction $n \geq 4$. Let $\{1, \zeta_{2^{n+1}}, \dots, \zeta_{2^{n+1}}^{2^{n+1}-1}\}$ be the standard basis in the $\mathbf{Z}[\zeta_{2^{n+1}}]$. The corresponding trace formula can be written as

$$\mathrm{Tr}_{\mathbf{Q}(\zeta_{2^{n+1}})}(x^2) = \sum_{i=0}^{2^n-1} x_i^2 + (x_{2^n}^2 - x_{2^n+1}^2) + \left(\sum_{i=2^n+2}^{3\times 2^{n-1}-1} x_i^2 - \sum_{i=3\times 2^{n-1}}^{2^{n+1}-1} x_i^2\right),$$

where $x = x_0 + \cdots + x_{2^{n+1}-1}\zeta_{2^{n+1}}^{2^{n+1}-1} \in \mathbf{Z}[\zeta_{2^{n+1}}]$. Thus one obtains a symmetric bilinear form

$$\begin{aligned}\mathrm{Tr}_{\mathbf{Q}(\zeta_{2^{n+1}})}(x,y) &= \sum_{i=0}^{2^n-1} x_i y_i + (x_{2^n}y_{2^n} - x_{2^n+1}y_{2^n+1}) \\ &\quad + \left(\sum_{i=2^n+2}^{3\cdot 2^{n-1}-1} x_i y_i - \sum_{i=3\cdot 2^{n-1}}^{2^{n+1}-1} x_i y_i\right).\end{aligned}$$

It is easy to see that $\mathrm{Tr}_{\mathbf{Q}(\zeta_{2^{n+1}})}(x,y)$ is a diagonal bilinear form on $H_2(M^4;\mathbf{Z})$. The signature of the above trace form is equal to 2^n, since the number of positive and negative 1s for the terms in brackets is the same, while the signature of the first sum is 2^n. It remains to notice that $n \geq 4$ and, therefore, the signature of $\mathrm{Tr}_{\mathbf{Q}(\zeta_{2^{n+1}})}(x,y)$ is divisible by 16. Since M^4 is a simply connected smooth 4-manifold, one gets item (ii) of Corollary 7.2.2. □

Guide to the literature

Our exposition of the arithmetic topology of the 4-dimensional manifolds is based on [211] and [203].

7.3 Untying knots in 4D and Wedderburn's Theorem

This section is focused on an algebraic proof of the otherwise known topological fact that all knots and links in the smooth 4-dimensional manifolds can be untied, i. e., are trivial. The novelty is a surprising rôle of Wedderburn's Theorem on the finite division rings [284] in the 4-dimensional topology.

Let us recapitulate the content of Sections 7.1 and 7.2. We constructed a functor F between the 3-dimensional manifolds and the fields of algebraic numbers. Such a functor maps 3-dimensional manifolds M^3 to the algebraic number fields K, so that the knots (links, resp.) in M^3 correspond to the prime ideals (ideals, resp.) in the ring of integers O_K. The map F extends to the smooth 4-dimensional manifolds M^4 and the fields of hyper-algebraic numbers $\mathbb{K}$, i. e., fields with a noncommutative multiplication. Denote by $O_{\mathbb{K}}$ the ring of integers of the field $\mathbb{K}$. A ring R is called a domain if R has no zero divisors. The R is called simple if it has only trivial two-sided ideals. We prove the following result.

Theorem 7.3.1. *$O_{\mathbb{K}}$ is a simple domain.*

Remark 7.3.1. Theorem 7.3.1 is false for the algebraic integers, since the domain O_K is never simple.

Corollary 7.3.1. *Any knot or link in M^4 is trivial.*

Proof. If $K \subset M^4$ (resp. $L \subset M^4$) is a nontrivial knot (resp. link), then $F(K)$ (resp. $F(L)$) is a nontrivial two-sided prime ideal (reso, two-sided ideal) in $O_{\mathbb{K}}$. The latter contradicts Theorem 7.3.1 saying that $O_{\mathbb{K}}$ is a simple ring. □

7.3.1 Wedderburn's Theorem

Roughly speaking, Wedderburn's Theorem says that finite noncommutative fields cannot exist [284]. Namely, denote by D a division ring. Let $\mathbb{F}_q$ be a finite field for some $q = p^r$, where p is a prime and $r \geq 1$ is an integer number.

Theorem 7.3.2 (Wedderburn's Theorem). *If $|D| < \infty$ and D is finite dimensional over a division ring, then $D \cong \mathbb{F}_q$ for some $q = p^r$.*

We shall use Theorem 7.3.2 along with a classification of simple rings due to Artin and Wedderburn. Recall that a ring R is called simple if R has only trivial two-sided ideals. By $M_n(D)$ we understand the ring of n by n matrices over D.

Theorem 7.3.3 (Artin–Wedderburn). *If R is a simple ring, then $R \cong M_n(D)$ for a division ring D and an integer $n \geq 1$.*

Remark 7.3.2. The ring $M_n(D)$ is a domain if and only if $n = 1$. For instance, if $n = 2$, then the matrices $\left(\begin{smallmatrix} 1 & 0 \\ 0 & 0 \end{smallmatrix}\right)$ and $\left(\begin{smallmatrix} 0 & 0 \\ 0 & 1 \end{smallmatrix}\right)$ are zero divisors in the ring $M_2(D)$.

7.3.2 Proof of Theorem 7.3.1

Theorem 7.3.1 will be proved by contradiction. Namely, we show that existence of a nontrivial two-sided ideal in $O_{\mathbb{K}}$ contradicts Theorem 7.3.3. To begin, let us prove the following lemma.

Lemma 7.3.1. *$O_{\mathbb{K}}$ is a noncommutative Noetherian domain.*

Proof. Recall that $O_{\mathbb{K}}$ is generated by the zeroes of a noncommutative polynomial $P(x) := \sum_i a_i x b_i x c_i x \cdots e_i x l_i$, where $a_i, b_i, c_i \dots, e_i, l_i \in O_{\mathbb{L}}$ and $\mathbb{K}$ is a finite dimensional extension of $\mathbb{L}$. By Hilbert Basis Theorem for noncommutative rings [1], if $O_{\mathbb{L}}$ is Noetherian, i. e., any ascending chain of the two-sided ideals of $O_{\mathbb{L}}$ stabilizes, then the ring $O_{\mathbb{K}}$ is also Noetherian. Repeating the construction, one arrives at a finite dimensional extension $\mathbb{H} \subset \mathbb{K}$, where $\mathbb{H}$ is the field of quaternions. The ring of the Hurwitz quaternions $O_{\mathbb{H}}$ is known to be Noetherian. Thus $O_{\mathbb{K}}$ is a Noetherian ring. Lemma 7.3.1 is proved. □

Returning to the proof of Theorem 7.3.1, let us assume to the contrary that $\mathbf{I}$ is a nontrivial two-sided ideal of $O_{\mathbb{K}}$. By Lemma 7.3.1, there exists the maximal two-sided ideal $\mathbf{I}_{max}$ such that

$$\mathbf{I} \subseteq \mathbf{I}_{max} \subset O_{\mathbb{K}}.$$

Lemma 7.3.2. *The ring $R := O_{\mathbb{K}}/\mathbf{I}_{max}$ is a simple domain.*

Proof. The ring R is simple, since $\mathbf{I}_{max}$ is the maximal two-sided ideal of $O_{\mathbb{K}}$. The ring R is a domain, since $O_{\mathbb{K}}$ is a domain and the homomorphism

$$h : O_{\mathbb{K}} \to R$$

is surjective. □

Remark 7.3.3. It follows from $R \cong O_{\mathbb{K}}/\mathbf{I}_{max}$ that $|R| < \infty$. Indeed, any non-trivial subgroup of the abelian group $(O_{\mathbb{K}}, +)$ has finite index by the Margulis normal subgroup theorem. In particular, the subgroup $(\mathbf{I}_{max}, +)$ has finite index in $(O_{\mathbb{K}}, +)$.

To finish the proof of Theorem 7.3.1, we write

$$R \cong M_n(D),$$

where D is a division ring (Theorem 7.3.3). Since R is a domain, we conclude that $n = 1$, see Remark 7.3.2. Thus

$$R \cong D.$$

On the other hand, Remark 7.3.3 says that $|R| < \infty$ and by Wedderburn's Theorem one gets $R \cong \mathbb{F}_q$ for some $q = p^r$. In particular, the homomorphism $h : O_{\mathbb{K}} \to R$ implies that the ring $O_{\mathbb{K}}$ is commutative. Indeed, since R is a commutative ring, one gets $h(xy - yx) = h(x)h(y) - h(y)h(x) = h(x)h(y) - h(x)h(y) = 0$, where 0 is the neutral element of R. In other words, the element $xy - yx$ belongs to the kernel of h, which is a two-sided ideal $I_h \subset O_{\mathbb{K}}$. If h is not injective, then I_h is nontrivial and, taking the multiplicative identity $1 \in I_h$, we obtain a contradiction $h(1) = 0$. Thus h is injective and $xy = yx$ for all $x, y \in O_{\mathbb{K}}$, i. e., $O_{\mathbb{K}}$ is a commutative ring. On the other hand, the ring $O_{\mathbb{K}}$ cannot be commutative by an assumption of Theorem 7.3.1. The obtained contradiction completes the proof of Theorem 7.3.1.

Guide to the literature

Our exposition follows [201].

7.4 Dynamical ideals of noncommutative rings

The concept of an ideal is fundamental in commutative algebra. Recall that the prime factorization in the ring of integers O_K of a number field K fails to be unique. To fix the problem, one needs to complete the set of the prime numbers of O_K by the "ideal numbers" lying in an abelian extension of K [140]. A description of an "ideal number" in terms of the ring O_K leads to the notion of an ideal [65].

Formally, an ideal of a commutative ring R is defined as a subset $I \subseteq R$ such that I is additively closed and $IR \subseteq I$. However, the true power of the ideals comes from their geometry. For instance, if R is the coordinate ring of an affine variety V, then the prime ideals of R make up a topological space homeomorphic to V. This link between algebra and geometry is critical, e. g., for the Weil Conjectures [285].

Although the notion of an ideal adapts to the noncommutative rings using the left, right, and two-sided ideals such an approach seems devoid of meaningful geometry. The drawback is that the "coordinate rings" in noncommutative geometry are usually simple, i. e., have only trivial ideals; see, e. g., Section 1.3. Moreover, such rings contain the idempotent elements (projections) and therefore the Ore localization of a domain fails, in general.

In this section we introduce an analog of the ideals for simple noncommutative rings R arising in geometry of the elliptic curves (Section 6.5.1) and topology of the 4-dimensional manifolds (Section 7.2). Our construction is similar to Kummer's construction of the ideals [140]. Roughly, the idea is this. Instead of a subset $I \subseteq R$, we take a partition D_α of R by the orbits $\{\alpha^{\mathbf{Z}}(x) \mid x \in R\}$ of an outer automorphism $\alpha : R \to R$. Let α be given by the formula

$$\alpha(x) = \mathbf{u}x\mathbf{u}^{-1}, \quad \forall x \in R,$$

where **u** is an "ideal number" lying outside R. We want to extend R by **u** so that α becomes an inner automorphism. Such an extension is known to coincide with the crossed product $R\rtimes_\alpha\mathbf{Z}$, where **u** is the generator of **Z**. The dynamical system $D_\alpha := R\rtimes_\alpha\mathbf{Z}$ will be called a dynamical ideal (or dynamial, for short) of R.

The dynamical system D_α is called minimal if the D_α does not split into a union of simpler dynamical subsystems. The D_α is minimal if and only if the crossed product $R\rtimes_\alpha\mathbf{Z}$ is a simple algebra. The following model example shows that the minimal dynamials are a proper generalization of the prime ideals to the case of simple noncommutative rings.

Example 7.4.1 (Section 6.5.1). Let $\mathcal{E}(K)$ be a nonsingular elliptic curve over the number field K having the coordinate ring $\mathfrak{V}(\mathcal{E})$. Let $\mathcal{E}(\mathbf{F}_p) := \mathfrak{V}(\mathcal{E})/\mathcal{P}$ be the localization of $\mathcal{E}(K)$ at the prime ideal $\mathcal{P}\subset\mathfrak{V}(\mathcal{E})$ over a prime number p. Denote by A_θ the noncommutative torus, i. e., a simple C^*-algebra A_θ generated by the unitary operators U and V satisfying the relation

$$VU = e^{2\pi i\theta}UV, \quad \text{where } \theta\in\mathbf{R}-\mathbf{Q}.$$

It is verified directly that the substitution

$$\begin{cases} U' = e^{\pi iac}U^aV^c \\ V' = e^{\pi ibd}U^bV^d \end{cases} \quad \text{with } \begin{pmatrix} a & b \\ c & d\end{pmatrix}\in M_2(\mathbf{Z})$$

brings relations to the form

$$V'U' = e^{2\pi i\theta(ad-bc)}U'V'.$$

If θ is a quadratic irrationality, then A_θ is said to have real multiplication and is denoted by A_{RM}. In particular, A_{RM} is a noncommutative coordinate ring of the elliptic curve $\mathcal{E}(K)$. Let $\alpha : A_{\mathrm{RM}}\to A_{\mathrm{RM}}$ be the shift automorphism of the A_{RM}. Consider a minimal dynamial

$$D_p := A_{\mathrm{RM}}\rtimes_\alpha p\mathbf{Z}\cong A_{\mathrm{RM}}\rtimes_{L_p}\mathbf{Z},$$

where L_p an endomorphism of the A_{RM} corresponding to the matrix

$$\begin{pmatrix} \mathrm{tr}(\varepsilon^{\pi(p)}) & p \\ -1 & 0\end{pmatrix}\in M_2(\mathbf{Z}),$$

see Section 6.5.3.2 for the notation. Let $K_0(D_p)$ be the K_0-group of the C^*-algebra D_p. Then for all but a finite set of primes p, there exists a canonical isomorphism of the finite abelian groups

$$K_0(D_p)\cong\mathfrak{V}(\mathcal{E})/\mathcal{P}.$$

Remark 7.4.1. Letting p run through the set of all primes, one gets a bijective map between the minimal dynamials D_p of the noncommutative ring A_{RM} and the prime ideals $\mathcal{P}$ of the commutative ring $\mathfrak{V}(\mathcal{E})$. In fact, such a map is a functor between the respective categories (Section 6.5.1). A relation of $K_0(D_p)$ to the Brauer group is discussed in Remark 7.4.6. Formula $K_0(D_p) \cong \mathfrak{V}(\mathcal{E})/\mathcal{P}$ is a motivation of the following definition.

Definition 7.4.1. By a Dedekind–Hecke ring R we understand a simple noncommutative topological ring such that $\mathbf{Z} \subseteq \operatorname{Out} R$, where $\operatorname{Out} R$ are the outer automorphisms of R.

Example 7.4.2. Ring A_θ is a Dedekind–Hecke ring in the norm topology. Indeed, it is easy to see that $\operatorname{Out} A_\theta \cong SL_2(\mathbf{Z})$. The upper triangular matrix $\left(\begin{smallmatrix}1 & 1\\ 0 & 1\end{smallmatrix}\right)$ is the generator of a group $\mathbf{Z} \subset \operatorname{Out} A_\theta$.

Remark 7.4.2. If R is a Dedekind–Hecke ring, then

$$R \rtimes_\alpha m\mathbf{Z} \cong R_m \rtimes_\alpha \mathbf{Z} := R \rtimes_{\alpha_m} \mathbf{Z},$$

where $\alpha_m : R \to R_m \subseteq R$ is an endomorphism of degree $m \geq 1$. Indeed, the $R \rtimes_\alpha m\mathbf{Z}$ gives rise to an automorphism of R acting by the formula $x \mapsto \mathbf{u}^m x \mathbf{u}^{-m}$. Denote by $\hat{\alpha}_m$ an extension of α_m to the $R \rtimes_\alpha \mathbf{Z}$. Since $\mathbf{u}^m \in \operatorname{Ker} \hat{\alpha}_m$, one gets an isomorphism $R \rtimes_\alpha m\mathbf{Z} \cong R_m \rtimes_\alpha \mathbf{Z}$. Notice that the crossed product $R \rtimes_{\alpha_m} \mathbf{Z}$ is undefined since α_m is not an automorphism of R, if $m \neq 1$. Hence the $R \rtimes_{\alpha_m} \mathbf{Z}$ is a symbolic notation for either $R \rtimes_\alpha m\mathbf{Z}$ or $R_m \rtimes_\alpha \mathbf{Z}$.

Remark 7.4.3. The endomorphism L_p of the ring A_{RM} is known to commute with the Hecke operator T_p on a lattice $\Lambda \subset \mathbf{C}$ such that $\mathcal{E}(K) \cong \mathbf{C}/\Lambda$ (Section 6.5.1); hence our terminology.

Our main result can be formulated as follows. Fix a generator $\alpha : R \to R$ of the cyclic group $\mathbf{Z} \subseteq \operatorname{Out} R$. For brevity, we write $D_m := R \rtimes_\alpha m\mathbf{Z}$, where $m \geq 1$ is an integer. The product of the dynamials $D_{m_1} \cong (R, G_1, \pi)$ and $D_{m_2} \cong (R, G_2, \pi)$ will be defined as the direct sum of the corresponding transformation groups, i. e.,

$$D_{m_1} D_{m_2} := (R, G_1 \oplus G_2, \pi).$$

Theorem 7.4.1. *The dynamials D_m of the Dedekind–Hecke ring R satisfy the fundamental theorem of arithmetic, i. e.,*

$$D_m = D_{p_1^{k_1}} D_{p_2^{k_2}} \cdots D_{p_n^{k_n}},$$

where $\prod_{i=1}^n p_i^{k_i}$ is the prime factorization of m. The factorization is unique up to the order of the factors. In particular, the dynamial D_m is minimal if and only if $m = p$ is a prime number.

7.4.1 Topological dynamics

Let X be a topological space and let T be a topological group. Consider a continuous map $\pi : X \times T \to X$ such that
(i) $\pi(x, e) = x$ for the identity $e \in T$ and all $x \in X$;
(ii) $\pi(\pi(x, t), s) = \pi(x, ts)$ for all $s, t \in T$ and all $x \in X$.

The triple (X, T, π) is called a transformation group (topological dynamical system). A topological isomorphism of (X, T, π) onto (Y, S, ρ) is a couple (h, φ) consisting of a homeomorphism $h : X \to Y$ and a homeomorphic group-isomorphism $\varphi : T \to S$ such that $(xh, t\varphi)\rho = (x, t)\pi h$ for all $x \in X$ and $t \in T$. The transformation groups (X, T, π) and (Y, S, ρ) are said to be equivalent if $(X, T, \pi) \cong (Y, S, \rho)$ are topologically isomorphic.

Let $S \subseteq T$ be a subgroup and $x \in X$. The S-orbit of x is a subset $O_S(x) = \{xS \mid S \subseteq T\}$ of X. If $S = T$, we omit the subscript and write an orbit as $O(x)$. The orbit $O_S(x)$ is the least S-invariant subset of X. The closure of an S-orbit in X is denoted by $\bar{O}_S(x)$. The set $A \subseteq X$ is called S-minimal provided $\bar{O}_S(x) = A$ for all $x \in A$ and A does not contain a smaller S-orbit closure. We call A a minimal set when $S = T$.

A subset $S \subset T$ is said to be left (right, resp.) syndetic in T, if $T = SK$ ($T = KS$, resp.) for some compact subset $K \subset T$. In particular, if T is discrete and if S is a subgroup of T, then S is syndetic in T if and only if S is a finite index subgroup of T [97, Remark 2.03 (7)]. We shall use the following fact.

Theorem 7.4.2 ([97, Theorem 2.32]). *Let X be compact and minimal under T. Let S be a syndetic invariant subgroup of T. Then the S-orbit closures define a (star-closed) decomposition of X.*

7.4.2 Cyclic division algebras

The real quaternions have been the single example of a division ring (hyper-algebraic number field) until the discovery in 1906 of the cyclic division algebras by Leonard E. Dickson. Roughly speaking, such algebras are an infinite family of the division rings generalizing the quaternions and represented by matrices over an algebraic number field. Let us review the main ideas.

Let K be a number field and let E be a finite Galois extension of K. Denote by $G = \mathrm{Gal}(E|K)$ the Galois group of E over K. Let $n = \dim_K(E)$ be the dimension of E as a vector space over K. Consider the ring $\mathrm{End}_K(E)$ of all K-linear transformations of E. Fixing a basis of E over K, one gets an isomorphism

$$\mathrm{End}_K(E) \cong M_n(K).$$

Denote by $C \subset \mathrm{End}_K(E)$ a subring generated by multiplications by the elements $\alpha \in E$ and the automorphisms $\theta \in G$. It can be verified directly that the commutation relation between the two is given by the formula

$$\theta\alpha = \theta(\alpha)\theta.$$

Further we restrict to the case when $G \cong (\mathbf{Z}/n\mathbf{Z})^{\times}$ is a cyclic group of order n generated by θ. Thus the relation $\theta\alpha = \theta(\alpha)\theta$ is complemented by the relation $\theta^n = 1$. On the other hand, it is easy to see that θ is an invertible element of $\mathcal{C}$ along with any element of the form $y\theta$, where $y \in E$. Notice that

$$(y\theta)^n = N(y)\theta^n = N(y),$$

where $N(y) \in K^{\times}$ is the K-norm of the algebraic number y.

Definition 7.4.2. The cyclic algebra $\mathcal{C}(a)$ is a subring of the ring $M_n(K)$ generated by the elements $\alpha \in E$ and the element $u := y\theta$ satisfying the relations

$$u\alpha = \theta(\alpha)u, \quad u^n = a \in K^{\times}.$$

Example 7.4.3. Let $K \cong \mathbf{R}$ and $E \cong \mathbf{C}$. Then $G \cong (\mathbf{Z}/2\mathbf{Z})^{\times}$ and θ is the complex conjugation. In this case, $\mathcal{C}(1) \cong M_2(\mathbf{R})$ and $\mathcal{C}(-1) \cong \mathbb{H}$, where $\mathbb{H}$ is the algebra of real quaternions.

The $\mathcal{C}(a)$ is a simple algebra of dimension n^2 over K. The field K is the center of $\mathcal{C}(a)$ and E is the maximal subfield of $\mathcal{C}(a)$. The following theorem gives the necessary and sufficient condition for the $\mathcal{C}(a)$ to be a division algebra.

Theorem 7.4.3 (Wedderburn's Norm Criterion). *Algebra $\mathcal{C}(a)$ is a division algebra if and only if a^n is the least power of a which is the norm of an element in E.*

Lemma 7.4.1. *Algebra $\mathcal{C}(a)$ is a Dedekind–Hecke ring.*

Proof. Recall that all automorphisms of the matrix algebra $M_n(K)$ are inner. Since $\mathcal{C}(a) \subset M_n(K)$, we conclude that

$$\operatorname{Out}\mathcal{C}(a) \subset M_n(K)$$

and the group $\operatorname{Out}\mathcal{C}(a)$ consists of the elements $g \in M_n(K)$, such that $g\mathcal{C}(a)g^{-1} = \mathcal{C}(a)$. It is clear that $\mathbf{Z} \subset \operatorname{Out}\mathcal{C}(a)$ if g has infinite order, e. g., is given by an upper triangular matrix. On the other hand, $\mathcal{C}(a)$ is a topological ring endowed, e. g., with the discrete topology. Thus $\mathcal{C}(a)$ is a Dedekind–Hecke ring. □

7.4.3 Piergallini covering

By M^4 we understand a smooth 4-dimensional manifold. Let S^4 be the 4-dimensional sphere and X_g be a closed 2-dimensional orientable surface of genus $g \geq 0$.

Definition 7.4.3. By the knotted surface $X := X_{g_1} \cup \cdots \cup X_{g_n}$ in M^4 one understands a transverse immersion of a collection of $n \geq 1$ surfaces X_{g_i} into M^4, i. e.,

$$\iota : X_{g_1} \cup \cdots \cup X_{g_n} \hookrightarrow M^4.$$

We refer to X a surface knot if $n = 1$ and a surface link if $n \geq 2$.

The following result extends the well-known theorem on the covering of the 3-dimensional sphere S^3 branched over a link in the S^3.

Theorem 7.4.4 ([227]). *Each smooth 4-dimensional manifold M^4 is the 4-fold PL cover of the sphere S^4 branched at the points of a knotted surface $X \subset S^4$.*

7.4.4 Proof of Theorem 7.4.1

We shall split the proof in a series of lemmas.

Lemma 7.4.2. $D_{m_1}D_{m_2} = D_{m_2}D_{m_1}$ *for any integers* $m_1, m_2 \geq 1$.

Proof. Consider an exact sequence of the subgroups $G_1 \cong m_1\mathbf{Z}$ and $G_2 \cong m_2\mathbf{Z}$ of the (additive) abelian group $(m_1m_2)\mathbf{Z}$,

$$0 \to m_1\mathbf{Z} \to (m_1m_2)\mathbf{Z} \to m_2\mathbf{Z} \to 0.$$

It can be verified directly that the above exact sequence splits. We conclude therefore that

$$(m_1m_2)\mathbf{Z} \cong G_1 \oplus G_2.$$

On the other hand, one gets the following sequence of isomorphisms:

$$D_{m_1}D_{m_2} \cong (R, (m_1m_2)\mathbf{Z}, \pi) \cong (R, (m_2m_1)\mathbf{Z}, \pi) \cong D_{m_2}D_{m_1}. \tag{7.1}$$

The conclusion of Lemma 7.4.2 follows from formula (7.1). □

Lemma 7.4.3. $D_{m_1}D_{m_2} = D_{m_1m_2}$ *for any integers* $m_1, m_2 \geq 1$.

Proof. We preserve the notation used in the proof of Lemma 7.4.2. From the definition of a product of the dynamials, one gets the following sequence of the isomorphisms:

$$D_{m_1}D_{m_2} \cong (R, G_1 \oplus G_2, \pi) \cong (R, (m_1m_2)\mathbf{Z}, \pi) \cong D_{m_1m_2}.$$

The conclusion of Lemma 7.4.3 follows from the above formula. □

Lemma 7.4.4. *The index map $i(D_m) = m$ is an isomorphism between the multiplicative semigroup $(\mathfrak{D}_R, \times)$ of all dynamials $\mathfrak{D}_R$ of the ring R and the multiplicative semigroup $(\mathbf{N}, \times)$ of all positive integers $\mathbf{N}$.*

Proof. (i) The unit of the semigroup $(\mathfrak{D}_R, \times)$ is the dynamial $D_1 \cong R \rtimes_\alpha \mathbf{Z}$. Also $i(D_1) = 1$ is the unit of the semigroup $(\mathbf{N}, \times)$ and D_1 is the unique dynamial having index 1. Therefore, the index map is injective.

(ii) It is easy to see, that the index map is surjective. Indeed, for every $m \geq 1$, there exists a dynamial D_m such that $i(D_m) = m$.

(iii) Let us verify that the index map preserves the products, i. e., $i(D_{m_1} D_{m_2}) = i(D_{m_1}) i(D_{m_2})$. Using Lemma 7.4.3, we calculate

$$i(D_{m_1} D_{m_2}) = i(D_{m_1 m_2}) = m_1 m_2 = i(D_{m_1}) i(D_{m_2}).$$

Since the index map is (i) injective, (ii) surjective, and (iii) preserves the products, we conclude that such a map is an isomorphism of the underlying semigroups. Lemma 7.4.4 follows. □

Lemma 7.4.5. *If $m = p_1^{k_1} p_2^{k_2} \cdots p_n^{k_n}$ is the prime factorization of an integer $m \geq 1$, then $D_m = D_{p_1^{k_1}} D_{p_2^{k_2}} \cdots D_{p_n^{k_n}}$. The latter product is unique up to the order of the factors.*

Proof. Let $m \geq 1$ be an integer. According to the fundamental theorem of arithmetic, there exists a prime factorization

$$m = p_1^{k_1} p_2^{k_2} \cdots p_n^{k_n},$$

where p_i are the prime numbers and k_i are positive integers. Moreover, the factorization is unique up to the order of the factors.

(i) Using Lemma 7.4.3, we calculate

$$D_m = D_{p_1^{k_1} p_2^{k_2} \cdots p_n^{k_n}} = D_{p_1^{k_1}} D_{p_2^{k_2}} \cdots D_{p_n^{k_n}}.$$

(ii) Since the index map is an isomorphism, we conclude that the above product is unique up to the order of the factors.

The conclusion of Lemma 7.4.5 follows from items (i) and (ii). □

Lemma 7.4.6. *The dynamial D_m is minimal if and only if $m = p$ is a prime number.*

Proof. (i) Let us assume that $m = p$ is a prime number. In view of Lemma 7.4.5, the dynamical system D_p cannot be a product of two or more dynamical subsystems. In other words, D_p is a minimal dynamial.

(ii) Conversely, let D_m be a minimal dynamical system. Let $(R, m\mathbf{Z}, \pi)$ be the corresponding transformation group. Notice that the abelian group $m\mathbf{Z}$ is discrete. Therefore every finite index subgroup S of $m\mathbf{Z}$ must be syndetic (Section 7.4.1). Let us now assume to the contrary that $m \neq p$. Then there exists a nontrivial syndetic subgroup S of the group $m\mathbf{Z}$. By Theorem 7.4.2, there exists a (star-closed) decomposition of the dynamical system D_m by the S-orbit closures, i. e., by smaller dynamical subsystems. In other words, the dynamical system D_m is not minimal. This contradiction proves the necessary condition of Lemma 7.4.6.

Lemma 7.4.6 is equivalent to items (i) and (ii). □

Theorem 7.4.1 follows from Lemmas 7.4.5 and 7.4.6.

7.4.5 Knotted surfaces in 4-manifolds

In this section we use Theorem 7.4.1 to prove an analog of Theorem 7.1.1 for the smooth 4-dimensional manifolds M^4. Roughly speaking, we replace the ideals $I \subseteq O_K$ by the dynamical ideals $D_m(\mathcal{C}(a))$, where $\mathcal{C}(a)$ is a cyclic division algebra associated to M^4 (Section 7.4.2). It is well known that any knot or link in M^4 is trivial. This fact can be derived from the simplicity of algebra $\mathcal{C}(a)$ and Wedderburn's Theorem (Section 7.3). On the other hand, there exist nontrivially knotted surfaces X_g in M^4 (E. Artin). We show that the dynamials $D_m(\mathcal{C}(a))$ classify the embeddings $X_g \hookrightarrow M^4$. Let us pass to a detailed argument.

Let $\mathfrak{M}^4$ be a category of smooth 4-dimensional manifolds M^4 such that the arrows of $\mathfrak{M}^4$ are homeomorphisms between the manifolds. Denote by $\mathfrak{C}$ a category of the cyclic division algebras $\mathcal{C}(a)$ such that the arrows of $\mathfrak{C}$ are isomorphisms between these algebras. The following result is an extension of Theorem 7.1.1 to the category $\mathfrak{M}^4$.

Theorem 7.4.5. *The exists a covariant functor $F : \mathfrak{M}^4 \to \mathfrak{C}$ such that:*
(i) *$F(S^4) = \mathbb{H}(\mathbf{Q})$, where $\mathbb{H}(\mathbf{Q})$ are the rational quaternions;*
(ii) *the dynamial $D_m(F(M^4))$ corresponds to a surface link $X \subset M^4$;*
(iii) *the minimal dynamial $D_p(F(M^4))$ corresponds to a surface knot $X \subset M^4$.*

Proof. The proof of existence and a detailed construction of functor F can be found in Section 7.2. To prove items (i)–(iii) of Theorem 7.4.5, we shall use Theorem 7.4.4 and the spinning of a link construction dating back to E. Artin, see, e. g., [127, Chapter 10]. Let us prove the following lemmas.

Lemma 7.4.7. *Let E be a cyclic extension of the number field K with the Galois group $G \cong (\mathbf{Z}/n\mathbf{Z})^\times$ of order n generated by an element θ, see Section 2.2. Denote by O_E the ring of integers of E. Then*

$$\mathcal{C}(a) \cong O_E \rtimes_\theta G.$$

Proof. Recall that the algebra $\mathcal{C}(a)$ is generated by multiplications by the elements $\alpha \in E$ and the automorphism $\theta \in G$ (Section 7.4.2). Since $E \cong O_E \otimes \mathbf{Q}$, we can always assume that $\alpha \in O_E$.

On the other hand, the commutation relation can be written as

$$\theta(\alpha) = \theta\alpha\theta^{-1}, \quad \forall \alpha \in O_E.$$

Thus the algebra $\mathcal{C}(a)$ is an extension of the ring O_E by an element θ such that each automorphism of O_E becomes an inner automorphism. In other words, the $\mathcal{C}(a)$ coincided with the crossed product $O_E \rtimes_\theta G$. Lemma 7.4.7 is proved. □

Lemma 7.4.8. *The dynamial $D_m(\mathcal{C}(a))$ defines an ideal $I_E \subseteq O_E$ such that:*
(i) $|O_E/I_E| = m$;
(ii) $D_m(\mathcal{C}(a)) = D(\mathcal{C}_m(a))$, *where* $\mathcal{C}_m(a) = I_E \rtimes_\theta G$.

Proof. From the definition of D_m, one gets

$$D_m(\mathcal{C}(a)) \cong \mathcal{C}(a) \rtimes m\mathbf{Z} \cong (O_E \rtimes_\theta G) \rtimes m\mathbf{Z}.$$

On the other hand, by the commutativity of crossed products by the abelian groups, we have

$$(O_E \rtimes_\theta G) \rtimes m\mathbf{Z} \cong (O_E \rtimes m\mathbf{Z}) \rtimes_\theta G.$$

Finally, repeating the argument of Remark 7.4.2, one gets an isomorphism

$$O_E \rtimes m\mathbf{Z} \cong I_E \rtimes \mathbf{Z},$$

where $I_E \subseteq O_E$ is an ideal such that $|O_E/I_E| = m$. Altogether, we have

$$D_m(\mathcal{C}(a)) \cong (I_E \rtimes \mathbf{Z}) \rtimes_\theta G \cong (I_E \rtimes_\theta G) \rtimes \mathbf{Z} \cong D(\mathcal{C}_m(a)),$$

where $\mathcal{C}_m(a) := I_E \rtimes_\theta G$. Lemma 7.4.8 is proved. □

Remark 7.4.4. It is not hard to see that the crossed product

$$\{R \rtimes_\theta G \mid R \text{ is a commutative ring}\}$$

corresponds to a "spinning" of the underlying topological space around a 2-dimensional plane [134, Chapter 10]. Indeed, if $R \cong O_E$, then the underlying topological space is a 3-dimensional manifold M^3 such that $O_E = F(M^3)$. Therefore the spinning $O_E \rtimes_\theta G \cong \mathcal{C}(a)$ corresponds to a 4-dimensional manifold M^4 such that $\mathcal{C}(a) = F(M^4)$. Likewise, if $R \cong I_E \subseteq O_E$ is an ideal, then the underlying topological space is a link $L \subset M^3$ such that $I_E = F(L)$. Thus the spinning $I_E \rtimes_\theta G \cong \mathcal{C}_m(a)$ gives rise to a 2-dimensional knotted surface $X \subset M^4$ such that $\mathcal{C}_m(a) = F(X)$.

Returning to the proof of Theorem 7.4.5, we shall proceed in the following steps:

(i) Consider a cyclic division algebra $\mathcal{C}(a)$ over the number field K. It is not hard to see that the smallest cyclic division subalgebra of $\mathcal{C}(a)$ consists of the rational quaternions $\mathbb{H}(\mathbf{Q})$. In other words, each $\mathcal{C}(a)$ can be seen as an extension of the rational quaternions,

$$\mathbb{H}(\mathbf{Q}) \subseteq \mathcal{C}(a).$$

Since the $\mathbb{H}(\mathbf{Q})$ is an algebra over $\mathbf{Q}$ and

$$\mathbf{Q} \subseteq K,$$

we conclude from Theorem 7.1.1(i) and Remark 7.4.4 that

$$F(S^4) = \mathbb{H}(\mathbf{Q}),$$

where S^4 is the 4-dimensional sphere. Item (i) of Theorem 7.4.5 is proved.

(ii) Let $m = p_1^{k_1} p_2^{k_2} \cdots p_n^{k_n}$ be the prime factorization of an integer $m \geq 1$. By Lemma 7.4.1, $\mathcal{C}(a)$ is a Dedekind–Hecke ring. Thus we can apply the prime factorization Theorem 7.4.1 to the dynamial $D_m(\mathcal{C}(a))$, i. e.,

$$D_m(\mathcal{C}(a)) = D_{p_1^{k_1}}(\mathcal{C}(a)) D_{p_2^{k_2}}(\mathcal{C}(a)) \cdots D_{p_n^{k_n}}(\mathcal{C}(a)).$$

On the other hand, each $D_i(\mathcal{C}(a))$ defines an ideal $I_i \subseteq O_E$, where E is the maximal number field of the cyclic division algebra $\mathcal{C}(a)$. Moreover, the prime factorization defines a prime factorization of the ideal $I_m \subset O_E$ corresponding to the $D_m(\mathcal{C}(a))$, i. e.,

$$I_m = \mathcal{P}_1^{k_1} \mathcal{P}_2^{k_2} \cdots \mathcal{P}_n^{k_n},$$

where $\mathcal{P}_i$ are the prime ideals of E. We can now use Remark 7.4.4 to conclude that the dynamial $D_m(\mathcal{C}(a))$ corresponds to a surface link $X \subset M^4$ obtained by the spinning of a link L defined by its prime factorization. Item (ii) of Theorem 7.4.5 is proved.

(iii) Let $m = p$ be a prime number. We repeat the argument of item (ii) and we apply Theorem 7.1.1(iii). In this case, one gets a surface knot $X \subset M^4$, since there is only one component in its prime factorization. Item (iii) of Theorem 7.4.5 is proved.

Theorem 7.4.5 is proved. □

Example 7.4.4 (Sphere knots in simply connected 4-manifolds). (i) Let W^4 be a simply connected 4-dimensional manifold. Denote by K^{ab} an abelian extension of the field $\mathbf{Q}$ and by $\mathcal{C}(a, K^{ab})$ a cyclic division algebra over the K^{ab}. It follows from Corollary 7.2.1 that

$$F(W^4) = \mathcal{C}(a, K^{ab}).$$

Indeed, since the Galois group of $\mathcal{C}(a) = F(W^4)$ is abelian, so is any subfield of the $\mathcal{C}(a)$. In particular, the center of $\mathcal{C}(a)$ must be an abelian extension of $\mathbf{Q}$.

(ii) Consider a sphere knot

$$S^2 \hookrightarrow W^4.$$

Theorem 7.4.5(iii) says that such knots are classified by the minimal dynamials:

$$\mathcal{D}_p(\mathcal{C}(a, K^{ab})) = \mathcal{C}(a, K^{ab}) \rtimes p\mathbf{Z} \cong \mathcal{C}(a, K_p^{ab}),$$

where K_p^{ab} is an abelian extension of the field $\mathbf{Q}_p$ of the p-adic numbers. The local class field theory says that the fields K_p^{ab} are classified by the open subgroups of the group $\mathbf{Q}_p^\times$ of index $n = \deg(K^{ab}|\mathbf{Q})$. Thus the classification of the sphere knots depends on the finite index open subgroups of the $\mathbf{Q}_p^\times$.

(iii) Let us restrict to the case $W^4 \cong S^4$ is the 4-dimensional sphere. By Theorem 7.2.1(i), we have $F(S^4) = \mathbb{H}(\mathbf{Q})$ and, therefore,

$$\mathcal{D}_p(\mathbb{H}(\mathbf{Q})) \cong \mathbb{H}(\mathbf{Q}_p).$$

Since the index $n = 1$, we conclude that the sphere knots

$$S^2 \hookrightarrow S^4$$

are classified by the primes p. Since $\mathcal{P} \cong p\mathbf{Z}$, one recovers the Artin classification of sphere knots $\pi_1(S^4 - S^2) \cong \pi_1(S^3 - K)$ by knots in S^3 [134, Chapter 10].

Remark 7.4.5. It is known that all division algebras over a local field are cyclic. The Brauer group of such algebras has the form

$$\mathrm{Br}\,(\mathcal{C}(a, K_p^{ab})) \cong \mathbf{Q}/\mathbf{Z},$$

see, e. g., [226, Section 17.10]. The Brauer group classifies the Morita equivalence classes of the algebras $\mathcal{C}(a, K_p^{ab})$ and, therefore, the corresponding classes of the sphere knots.

Remark 7.4.6. It follows from Remark 7.4.5 that

$$\mathrm{Br}(\mathcal{D}_p) \cong \mathbf{Q}/\mathbf{Z}.$$

Notice a formal resemblance of the above formula to the formula $K_0(\mathcal{D}_p) \cong \mathfrak{V}(\mathcal{E})/\mathcal{P}$ obtained in Example 7.4.1. This observation is part of the Merkurjev–Suslin theory linking the Brauer groups with the algebraic K-theory.

Guide to the literature
Our exposition follows [203].

7.5 Etesi C^*-algebras

Algebraic topology of the 4-dimensional manifolds $\mathcal{M}$ is a vast uncharted area of mathematics. Unlike dimensions 2 and 3, the smooth structures are detached from the topology of $\mathcal{M}$. Due to the works of Rokhlin, Freedman, and Donaldson, it is known that $\mathcal{M}$ can be nonsmooth and, if there exists a smooth structure, it need not be unique. The classification of all smoothings of $\mathcal{M}$ is an open problem.

Let $\mathcal{M}$ be a smooth 4-dimensional manifold. Denote by $\mathrm{Diff}(\mathcal{M})$ a group of the orientation-preserving diffeomorphisms of $\mathcal{M}$ and let $\mathrm{Diff}_0(\mathcal{M})$ be a connected component of $\mathrm{Diff}(\mathcal{M})$ containing the identity. The group $\mathrm{Diff}(\mathcal{M})/\mathrm{Diff}_0(\mathcal{M})$ is discrete and therefore locally compact.

Definition 7.5.1. By the Etesi C^*-algebra $\mathbb{E}_{\mathcal{M}}$ we understand a group C^*-algebra of the locally compact group $\mathrm{Diff}(\mathcal{M})/\mathrm{Diff}_0(\mathcal{M})$.

In this section we classify the smooth structures on $\mathcal{M}$ based on the K-theory of the Etesi C^*-algebra $\mathbb{E}_{\mathcal{M}}$. Recall that a C^*-algebra is called Approximately Finite-dimensional (AF-) if it is an inductive limit of the multi-matrix C^*-algebras $M_{n_1}(\mathbf{C}) \oplus \cdots \oplus M_{n_k}(\mathbf{C})$, see Section 3.5. An AF-algebra is called stationary if the inductive limit depends on a single positive integer matrix $A \in GL(n, \mathbf{Z})$, see Section 3.5.2. Our main result can be formulated as follows.

Theorem 7.5.1. *The Etesi C^*-algebra $\mathbb{E}_{\mathcal{M}}$ is a stationary AF-algebra.*

Let $\lambda_A > 1$ be the Perron–Frobenius eigenvalue of the positive matrix A defined by $\mathbb{E}_{\mathcal{M}}$. Consider a number field $K = \mathbf{Q}(\lambda_A)$. The eigenvector $(v_1, \dots, v_n)$ corresponding to λ_A can always be scaled so that $v_i \in K$. By $\mathfrak{m} := \mathbf{Z}v_1 + \cdots + \mathbf{Z}v_n$ we understand a $\mathbf{Z}$-module in the field K and by Λ the ring of endomorphisms of $\mathfrak{m}$. Let $[\mathfrak{m}]$ be an ideal class of $\mathfrak{m}$ in the ring Λ. The K-theory of stationary AF-algebras says that the triples $(\Lambda, [\mathfrak{m}], K)$ are in a one-to-one correspondence with the Morita equivalence classes of $\mathbb{E}_{\mathcal{M}}$ (Theorem 3.5.4). Let $\mathrm{Br}(K)$ be the Brauer group of the number field K, i. e., a torsion abelian group of the Morita equivalence classes of the central simple algebras over K. By $S(k)$ we understand the connected sum of k copies of $S^2 \times S^2$. We assign an index $k \geq 0$ to each smoothing $\mathcal{M}_k$ of a topological 4-manifold $\mathcal{M}_{top}$ using Gompf's Stable Diffeomorphism Theorem, i. e., a diffeomorphism $\mathcal{M}_k \# S(k) \to \mathcal{M}_0 \# S(k)$, where $\mathcal{M}_0$ is the standard smoothing of $\mathcal{M}_{top}$ [96, Theorem 1]. The sum of $\mathcal{M}_{k_1}$ and $\mathcal{M}_{k_2}$ is defined by the formula $\mathcal{M}_{k_1} \oplus \mathcal{M}_{k_2} := \mathcal{M}_{k_1+k_2}$. An application of Theorem 7.5.1 is as follows.

Corollary 7.5.1. *Let $\mathcal{M}$ be a smooth 4-manifold such that $\mathbb{E}_{\mathcal{M}}$ is an infinite-dimensional C^*-algebra. The following is true:*

(i) *Handelman triple* $(\Lambda, [\mathfrak{m}], K)$ *is an invariant of the homeomorphisms of* $\mathcal{M}$;
(ii) *Elements of the Brauer group* $\mathrm{Br}(K)$ *parametrize smooth structures on* $\mathcal{M}_{\mathrm{top}}$.

In particular, all smoothings of $\mathcal{M}$ *form a torsion abelian group under the group operation* $\oplus$ *with the neutral element* $\mathcal{M}_0$.

Remark 7.5.1. The Brauer group $\mathrm{Br}(K)$ is known to classify the division algebras over K. In other words, Corollary 7.5.1(ii) defines a functor from the smooth 4-manifolds to the division algebras. Such a functor was constructed independently in Section 7.2 using the Galois theory for noncommutative fields.

7.5.1 Minkowski group

Recall from Section 3.3 that $M_\infty(A)$ is the algebraic direct limit of the C^*-algebras $M_n(A)$ under the embeddings $a \mapsto \mathbf{diag}(a, 0)$. The direct limit $M_\infty(A)$ can be thought of as the C^*-algebra of infinite-dimensional matrices whose entries are all zero except for a finite number of the nonzero entries taken from the C^*-algebra A. Two projections $p, q \in M_\infty(A)$ are equivalent if there exists an element $v \in M_\infty(A)$ such that $p = v^*v$ and $q = vv^*$. The equivalence class of projection p is denoted by $[p]$. We write $V(A)$ to denote all equivalence classes of projections in the C^*-algebra $M_\infty(A)$, i. e., $V(A) := \{[p] : p = p^* = p^2 \in M_\infty(A)\}$. The set $V(A)$ has the natural structure of an abelian semigroup with the addition operation defined by the formula $[p] + [q] := \mathbf{diag}(p, q) = [p' \oplus q']$, where $p' \sim p$, $q' \sim q$, and $p' \perp q'$. The identity of the semigroup $V(A)$ is given by $[0]$, where 0 is the zero projection. By the K_0-group $K_0(A)$ of the unital C^*-algebra A one understands the Grothendieck group of the abelian semigroup $V(A)$, i. e., the completion of $V(A)$ by the formal elements $[p] - [q]$. The image of $V(A)$ in $K_0(A)$ is a positive cone $K_0^+(A)$ defining the order structure $\le$ on the abelian group $K_0(A)$. The pair $(K_0(A), K_0^+(A))$ is known as a dimension group of the C^*-algebra A. The scale $\Sigma(A)$ is the image in $K_0^+(A)$ of the equivalence classes of projections in the C^*-algebra A. The $\Sigma(A)$ is a generating, hereditary and directed subset of $K_0^+(A)$, i. e., (i) for each $a \in K_0^+(A)$ there exist $a_1, \dots, a_r \in \Sigma(A)$ such that $a = a_1 + \cdots + a_r$; (ii) if $0 \le a \le b \in \Sigma(A)$, then $a \in \Sigma(A)$, and (iii) given $a, b \in \Sigma(A)$, there exists $c \in \Sigma(A)$ such that $a, b \le c$. Each scale can always be written as $\Sigma(A) = \{a \in K_0^+(A) \mid 0 \le a \le u\}$, where u is an order unit of $K_0^+(A)$. The pair $(K_0(A), K_0^+(A))$ and the triple $(K_0(A), K_0^+(A), \Sigma(A))$ are invariants of the Morita equivalence and isomorphism class of the C^*-algebra A, respectively. If $\mathbb{A}$ is an AF-algebra, then its scaled dimension group (dimension group, resp.) is a complete invariant of the isomorphism (Morita equivalence, resp.) class of $\mathbb{A}$, see Theorem 3.5.2.

Let τ be the canonical trace on the AF-algebra $\mathbb{A}$. Such a trace induces a homomorphism $\tau_* : K_0(\mathbb{A}) \to \mathbf{R}$ and we let $\mathfrak{m} := \tau_*(K_0(\mathbb{A})) \subset \mathbf{R}$. If $\mathbb{A}$ is the stationary AF-algebra given by a matrix $A \in \mathrm{GL}(n, \mathbf{Z})$, then $\mathfrak{m}$ is a $\mathbf{Z}$-module in the number field

$K = \mathbf{Q}(\lambda_A)$ generated by the Perron–Frobenius eigenvalue λ_A of the matrix A. The endomorphism ring of $\mathfrak{m}$ is denoted by Λ and the ideal class of $\mathfrak{m}$ is denoted by $[\mathfrak{m}]$. The triple $(\Lambda, [\mathfrak{m}], K)$ is an invariant of the Morita equivalence class of $\mathbb{A}$ (Theorem 3.5.4).

Remark 7.5.2. Each stationary AF-algebra defines a torsion abelian group. Indeed, let $?_n(x)$ be the n-dimensional Minkowski question-mark function, see [161, p. 172] for $n = 2$ and [221, Theorem 3.5] for $n \geq 2$. The function $?_n(x) : [0,1]^{n-1} \to [0,1]^{n-1}$ is a continuous function with the following properties: (i) $?_n(0) = 0$ and $?_n(1) = 1$; (ii) $?_n(\mathbf{Q}^{n-1}) = (\mathbf{Z}[\frac{1}{2}])^{n-1}$ are dyadic rationals; and (iii) $?_n(\mathcal{K}^{n-1}) = (\mathbf{Q} - \mathbf{Z}[\frac{1}{2}])^{n-1}$, where $\mathcal{K}$ are algebraic numbers of degree n over $\mathbf{Q}$. It is not hard to see that (iv) $?_n(\Delta) = \Delta$ is a monotone function, where $\Delta = [0,1]$ is the normalized diagonal of the simplex $[0,1]^{n-1}$. Recall that $\tau_*(K_0(\mathbb{A})) = \mathfrak{m}$ and $\tau_*(\Sigma(\mathbb{A})) = \mathfrak{m} \cap [0,1]$, where τ is the canonical trace on the AF-algebra $\mathbb{A}$ and $\mathfrak{m}$ is a $\mathbf{Z}$-module in the number field K. We assume that $\tau_*(K_0(\mathbb{A})) \subset \Delta$. By properties (iii) and (iv) of the Minkowski question-mark function, one gets the following inclusion:

$$\mathcal{Y} := ?_n(\tau_*(\Sigma(\mathbb{A}))) \subset \mathbf{Q}/\mathbf{Z}.$$

Definition 7.5.2. By the Minkowski group $\mathrm{Mi}(K)$ of a stationary AF-algebra we understand a torsion abelian group generated by the elements of the set $\mathcal{Y}$.

7.5.2 Gompf's Theorem

We denote by $\mathcal{M}$ a smooth 4-dimensional manifold and always assume $\mathcal{M}$ to be compact. Let S^4 be the 4-dimensional sphere and X_g be a closed 2-dimensional orientable surface of genus $g \geq 0$. By the knotted surface $\mathcal{X} := X_{g_1} \cup \cdots \cup X_{g_n}$ in $\mathcal{M}$ one understands a transverse immersion of a collection of $n \geq 1$ surfaces X_{g_i} into $\mathcal{M}$. We refer to $\mathcal{X}$ a surface knot if $n = 1$ and a surface link if $n \geq 2$. Each smooth 4-dimensional manifold $\mathcal{M}$ is the 4-fold PL cover of the sphere S^4 branched at the points of a knotted surface $\mathcal{X} \subset S^4$ [227].

Let S^2 be the 2-dimensional sphere. By $S(k)$ we understand a smooth 4-dimensional manifold corresponding to a connected sum

$$S(k) := \underbrace{(S^2 \times S^2) \# \cdots \# (S^2 \times S^2)}_{k \text{ times}}.$$

Theorem 7.5.2 (Gompf [96]). *Let $\mathcal{M}$ and $\mathcal{M}'$ be two different smoothings of a topological 4-manifold $\mathcal{M}_{top}$. Then for sufficiently large k there exists an orientation-preserving diffeomorphism*

$$\mathcal{M} \# S(k) \longrightarrow \mathcal{M}' \# S(k).$$

Definition 7.5.3. Let $\mathcal{M}_0$ be the standard smoothing of a topological 4-manifold $\mathcal{M}_{top}$. We denote by $\mathcal{M}_k$ a smoothing of $\mathcal{M}_{top}$, such that $\mathcal{M}_k\#S(k) \to \mathcal{M}_0\#S(k)$ is an orientation-preserving diffeomorphism defined by Gompf's Theorem.

Lemma 7.5.1. *The smoothing $\mathcal{M}_k$ exists for each value of $k \geq 0$.*

Proof. Fix an integer $k \geq 0$ and consider the smooth connected sum $\mathcal{M}_0\#S(k)$. The manifold $\mathcal{M}_k$ can be obtained from $\mathcal{M}_0\#S(k)$ by cutting off $S(k)$ along the sphere S^3 and gluing in a copy of S^4 endowed with the standard smooth structure. (Notice that $\mathcal{M}_k$ is different from $\mathcal{M}_0$.) The reader can verify that $\mathcal{M}_k$ is a smoothing of $\mathcal{M}_{top}$ and there exists Gompf's diffeomorphism $\mathcal{M}_k\#S(k) \to \mathcal{M}_0\#S(k)$. □

Corollary 7.5.2. *The smoothings of $\mathcal{M}_{top}$ have the structure of an abelian monoid under the operation*

$$\mathcal{M}_{k_1} \oplus \mathcal{M}_{k_2} := \mathcal{M}_{k_1+k_2}$$

with the neutral element $\mathcal{M}_0$.

Proof. (i) Let us show that operation ⊕ defines a semigroup. It follows from the above that the sum $\mathcal{M}_{k_1} \oplus \mathcal{M}_{k_2}$ is a smoothing of $\mathcal{M}_{top}$. In other words, the set of all smoothing of $\mathcal{M}_{top}$ is closed under the operation ⊕, i. e., such a set is a semigroup.

(ii) The semigroup of item (i) is abelian since $\mathcal{M}_{k_1} \oplus \mathcal{M}_{k_2} = \mathcal{M}_{k_2} \oplus \mathcal{M}_{k_1}$.

(iii) It follows from Lemma 7.5.1 that $\mathcal{M}_k \oplus \mathcal{M}_0 = \mathcal{M}_k$. Thus $\mathcal{M}_0$ is the neutral element of the semigroup. In other words, our semigroup is a monoid.

Corollary 7.5.2 is proved. □

7.5.3 Proofs

7.5.3.1 Proof of Theorem 7.5.1

We shall split the proof in two lemmas.

Lemma 7.5.2. *The C^*-algebra $\mathbb{E}_{\mathcal{M}}$ is an AF-algebra.*

Proof. The lemma follows from an observation that $\mathbb{E}_{\mathcal{M}}$ is a group C^*-algebra of a locally compact group. Then the C^*-algebra $\mathbb{E}_{\mathcal{M}}$ is an AF-algebra, see, e. g., [32, Corollary 11.1.2]. This fact was proved independently by Gábor Etesi in terms of a von Neumann algebra related to $\mathbb{E}_{\mathcal{M}}$. Namely, such an algebra was shown to be hyperfinite [79, Lemma 2.3]. For the sake of clarity, we adapt the proof to the case of the C^*-algebra $\mathbb{E}_{\mathcal{M}}$.

Let $\mathcal{G} := \mathrm{Diff}(\mathcal{M})/\mathrm{Diff}_0(\mathcal{M})$. Consider a profinite completion of the discrete group $\mathcal{G}$, i. e.,

$$\widehat{\mathcal{G}} := \varprojlim \mathcal{G}/N,$$

where N ranges through the open normal finite index subgroups of $\mathcal{G}$. Recall that if G is a finite group, then the group algebra $\mathbf{C}[G]$ has the form

$$\mathbf{C}[G] \simeq M_{n_1}(\mathbf{C}) \oplus \cdots \oplus M_{n_h}(\mathbf{C}),$$

where n_i are degrees of the irreducible representations of G and h is the total number of such representations [251, Proposition 10]. Thus we have

$$\widehat{\mathcal{G}} \cong \varprojlim G_i,$$

where G_i is a finite group. Consider a group algebra

$$\mathbf{C}[G_i] \cong M_{n_1}^{(i)}(\mathbf{C}) \oplus \cdots \oplus M_{n_h}^{(i)}(\mathbf{C})$$

corresponding to G_i. Notice that the $\mathbf{C}[G_i]$ is a finite-dimensional C^*-algebra. The inverse limit defines an ascending sequence of the finite-dimensional C^*-algebras of the form

$$\varprojlim M_{n_1}^{(i)}(\mathbf{C}) \oplus \cdots \oplus M_{n_h}^{(i)}(\mathbf{C}).$$

The group C^*-algebra $C^*(\widehat{\mathcal{G}})$ of the profinite group $\widehat{\mathcal{G}}$ is the norm closure of the group algebra $\mathbf{C}[\widehat{\mathcal{G}}]$ [69, Section 13.9]. One concludes that $C^*(\widehat{\mathcal{G}})$ is an AF-algebra.

On the other hand, the canonical homomorphism $\mathcal{G} \to \widehat{\mathcal{G}}$ gives rise to an extension of the C^*-algebras

$$C^*(\mathcal{G}) \to C^*(\widehat{\mathcal{G}}) \to \mathcal{B}.$$

Since $C^*(\widehat{\mathcal{G}})$ is an AF-algebra, both $C^*(\mathcal{G})$ and $\mathcal{B}$ must be AF-algebras [106, Lemma I.5(a)]. It remains to recall that $C^*(\mathcal{G}) := \mathbb{E}_{\mathcal{M}}$. Thus $\mathbb{E}_{\mathcal{M}}$ is an AF-algebra. Lemma 7.5.2 is proved. □

Lemma 7.5.3. *The AF-algebra $\mathbb{E}_{\mathcal{M}}$ is stationary.*

Proof. (i) Let S^4 be the 4-dimensional sphere. By Theorem 7.4.4, there exists a 4-fold covering map $\mathcal{M} \to S^4$ branched at the points of a knotted surface $\mathcal{X}$ defined by an embedding

$$\mathcal{X} \hookrightarrow S^4.$$

In view of the inclusion $\mathrm{Diff}(S^4) \subset \mathrm{Diff}(S^4 - \mathcal{X})$, one gets an injective homomorphism of the C^*-algebras

$$\mathbb{E}_{S^4} \hookrightarrow \mathbb{E}_{\mathcal{M}}.$$

(ii) Let us show that $\mathbb{E}_{S^4} \cong \mathbf{C}$. Indeed, since $\mathrm{Diff}(S^4) \cong \mathrm{Diff}_0(S^4)$, the group $\mathrm{Diff}(S^4)/\mathrm{Diff}_0(S^4)$ is trivial. In particular, the group C^*-algebra $\mathbb{E}_{S^4}$ is commutative. Gelfand Theorem says that $\mathbb{E}_{S^4} \cong C_0(X)$, where $C_0(X)$ is the C^*-algebra of continuous complex-valued functions on a locally compact Hausdorff space X. But $X \cong \mathbf{pt}$ is a singleton and therefore $C_0(\mathbf{pt}) \cong \mathbf{C}$. Thus one gets $\mathbb{E}_{S^4} \cong \mathbf{C}$.

(iii) The AF-algebra $\mathbb{E}_{S^4}$ is given by an ascending sequence of the form:

$$\mathbf{C} \xrightarrow{\mathbf{1}} \mathbf{C} \xrightarrow{\mathbf{1}} \cdots,$$

where $\mathbf{1}$ is the identity homomorphism. It follows from the above that the $\mathbb{E}_{S^4}$ is a stationary AF-algebra.

(iv) Since $K_0(\mathbf{C}) \cong \mathbf{Z}$, we conclude that the dimension group of the AF-algebra $\mathbb{E}_{S^4}$ is isomorphic to $(\mathbf{Z}, \mathbf{Z}^+)$, where $\mathbf{Z}^+$ is the semigroup of positive integers. It is easy to see that the Handelman triple corresponding to the stationary AF-algebra $\mathbb{E}_{S^4}$ has the form $(\mathbf{Z}, [\mathbf{Z}], \mathbf{Q})$.

(v) On the other hand, one gets an inclusion of the abelian groups

$$K_0(\mathbb{E}_{S^4}) \subset K_0(\mathbb{E}_{\mathcal{M}}).$$

Moreover, if τ is the canonical trace on the AF-algebra $\mathbb{E}_{\mathcal{M}}$, one obtains the following inclusion of the additive groups of the real line:

$$\mathbf{Z} \subset \tau_*(K_0(\mathbb{E}_{\mathcal{M}})).$$

(vi) Since $\mathbf{Z}$ is a ring, the above group inclusion can be extended to such of the rings. But the only finite degree extension of the ring $\mathbf{Z}$ coincides (up to a scaling constant) with an order Λ in the number field $K = \Lambda \otimes \mathbf{Q}$. We conclude that

$$\tau_*(K_0(\mathbb{E}_{\mathcal{M}})) \subset K,$$

where K is a real number field.

(vii) To finish the proof of Lemma 7.5.3, it remains to apply the result [105, Theorem II(iii)] saying that the AF-algebra $\mathbb{E}_{\mathcal{M}}$ is of a stationary type.

Lemma 7.5.3 is proved. □

Remark 7.5.3. The Etesi C^*-algebra $\mathbb{E}_{\mathcal{M}}$ is simple, i. e., has only trivial two-sided ideals. This fact follows from the strict positivity of the matrix A corresponding to the stationary AF-algebra.

Returning to the proof of Theorem 7.5.1, we apply Lemmas 7.5.2 and 7.5.3. Theorem 7.5.1 follows.

7.5.3.2 Proof of Corollary 7.5.1

We split the proof in a series of lemmas.

Lemma 7.5.4. *The Etesi C^*-algebras satisfy an isomorphism*

$$\mathbb{E}_{\mathcal{M}_1\#\mathcal{M}_2} \cong \mathbb{E}_{\mathcal{M}_1} \otimes \mathbb{E}_{\mathcal{M}_2},$$

where # is the connected sum of manifolds and ⊗ is the tensor product of C^*-algebras.

Proof. We let $\mathcal{G} := \mathrm{Diff}(\mathcal{M}_1\#\mathcal{M}_2)/\mathrm{Diff}_0(\mathcal{M}_1\#\mathcal{M}_2)$. It is not hard to see that

$$\mathcal{G} = \mathcal{G}_1 \times \mathcal{G}_2,$$

where $\mathcal{G}_1 = \mathrm{Diff}(\mathcal{M}_1)/\mathrm{Diff}_0(\mathcal{M}_1)$ and $\mathcal{G}_2 = \mathrm{Diff}(\mathcal{M}_2)/\mathrm{Diff}_0(\mathcal{M}_2)$. It is well known that the group ring $\mathbf{C}[\mathcal{G}]$ of the above product is given by the formula

$$\mathbf{C}[\mathcal{G}] \cong \mathbf{C}[\mathcal{G}_1] \otimes \mathbf{C}[\mathcal{G}_2].$$

Since the $\mathbb{E}_{\mathcal{M}}$ is a nuclear C^*-algebra, the norm closure of the self-adjoint representation of $\mathbf{C}[\mathcal{G}]$ defines an isomorphism $\mathbb{E}_{\mathcal{M}_1\#\mathcal{M}_2} \cong \mathbb{E}_{\mathcal{M}_1} \otimes \mathbb{E}_{\mathcal{M}_2}$. Lemma 7.5.4 is proved. □

Lemma 7.5.5. *The Etesi C^*-algebra of the 4-manifold $S^2 \times S^2$ is given by the formula*

$$\mathbb{E}_{S^2\times S^2} \cong M_4(\mathbf{C}).$$

Proof. (i) It follows from Theorem 7.4.4 that the 4-manifold $S^2 \times S^2$ is a 4-fold covering of the 4-sphere S^4 which one can see by factoring this covering geometrically as

$$S^2 \times S^2 \cong \mathbf{C}P^1 \times \mathbf{C}P^1 \to \mathbf{C}P^2 \to S^4.$$

The covering map induces a homomorphism of the C^*-algebras $\mathbb{E}_{S^2\times S^2} \to \mathbb{E}_{S^4}$ and a homomorphism of the corresponding abelian groups

$$K_0(\mathbb{E}_{S^2\times S^2}) \to K_0(\mathbb{E}_{S^4}).$$

We have $K_0(\mathbb{E}_{S^4}) \cong \mathbf{Z}$, see item (iii) of proof of Lemma 7.5.3. Since the kernel of the map $K_0(\mathbb{E}_{S^2\times S^2}) \to K_0(\mathbb{E}_{S^4})$ is $\mathbf{Z}/4\mathbf{Z}$, one gets an isomorphism

$$K_0(\mathbb{E}_{S^2\times S^2}) \cong \mathbf{Z}.$$

(ii) Recall that $\mathbb{E}_{S^2\times S^2}$ is a stationary AF-algebra. If A is the corresponding matrix, then the eigenvalues of A must be rational and equal to each other. In other words, the AF-algebra $\mathbb{E}_{S^2\times S^2}$ corresponds to an ascending sequence of matrix algebras of the form

$$M_2(\mathbf{C})\otimes M_2(\mathbf{C}) \xrightarrow{\mathbf{1}} M_2(\mathbf{C})\otimes M_2(\mathbf{C}) \xrightarrow{\mathbf{1}} \cdots,$$

where $\mathbf{1} = \left(\begin{smallmatrix}1&0\\0&1\end{smallmatrix}\right)\otimes\left(\begin{smallmatrix}1&0\\0&1\end{smallmatrix}\right)$. Clearly, the inductive limit corresponds to the C^*-algebra $M_2(\mathbf{C})\otimes M_2(\mathbf{C})$. We conclude that $\mathbb{E}_{S^2\times S^2} \cong M_2(\mathbf{C})\otimes M_2(\mathbf{C}) \cong M_4(\mathbf{C})$. Lemma 7.5.5 is proved. □

Corollary 7.5.3. *The Etesi C^*-algebra $\mathbb{E}_{\mathcal{M}\#S(k)}$ is Morita equivalent to $\mathbb{E}_{\mathcal{M}}$.*

Proof. Recall that the 4-manifold $S(k) = (S^2\times S^2)\#\cdots\#(S^2\times S^2)$ is a connected sum of the k copies of $S^2\times S^2$. Thus one gets an isomorphism:

$$\mathbb{E}_{S(k)} \cong \underbrace{M_4(\mathbf{C})\otimes\cdots\otimes M_4(\mathbf{C})}_{k\text{ times}} \cong M_{4^k}(\mathbf{C}).$$

If $\mathcal{M}$ is a smooth 4-manifold, then from Lemma 7.5.4 one obtains an isomorphism

$$\mathbb{E}_{\mathcal{M}\#S(k)} \cong \mathbb{E}_{\mathcal{M}}\otimes M_{4^k}(\mathbf{C}).$$

In other words, the C^*-algebras $\mathbb{E}_{\mathcal{M}}$ and $\mathbb{E}_{\mathcal{M}\#S(k)}$ are Morita equivalent. Corollary 7.5.3 is proved. □

Lemma 7.5.6. *The topological type of manifold $\mathcal{M}$ is determined by the Handelman triple* $(\Lambda,[\mathfrak{m}],K)$.

Proof. Our proof is based on Gompf's Theorem 7.5.2 and Corollary 7.5.3. We shall proceed in two steps.

(i) Recall that by Theorem 7.5.2, for any two smoothings $\mathcal{M}$ and $\mathcal{M}'$ of a topological manifold $\mathcal{M}_{top}$, one can find an integer $k\ge 0$ such that $\mathcal{M}\#S(k)$ and $\mathcal{M}'\#S(k)$ are diffeomorphic. The corresponding C^*-algebras are isomorphic, i. e.,

$$\mathbb{E}_{\mathcal{M}\#S(k)} \cong \mathbb{E}_{\mathcal{M}'\#S(k)}.$$

In view of Corollary 7.5.3, the C^*-algebra $\mathbb{E}_{\mathcal{M}}$ is Morita equivalent to $\mathbb{E}_{\mathcal{M}\#S(k)}$ and the C^*-algebra $\mathbb{E}_{\mathcal{M}\#S(k)}$ is Morita equivalent to $\mathbb{E}_{\mathcal{M}'}$. Therefore the C^*-algebras $\mathbb{E}_{\mathcal{M}}$ and $\mathbb{E}_{\mathcal{M}'}$ are Morita equivalent by the transitivity property.

Thus the Morita equivalence class of the Etesi C^*-algebra $\mathbb{E}_{\mathcal{M}}$ consists of all 4-dimensional manifolds which are homeomorphic but not diffeomorphic to each other.

(ii) Recall that the $\mathbb{E}_{\mathcal{M}}$ is a stationary AF-algebra. The Morita equivalence classes of such AF-algebras are described by the Handelman triples $(\Lambda,[\mathfrak{m}],K)$ (Theorem 3.5.4).

Comparing this fact with the result of item (i), we conclude that the $(\Lambda, [\mathfrak{m}], K)$ is an invariant of the topological type of the manifold $\mathcal{M}$.

Lemma 7.5.6 is proved. □

Remark 7.5.4. Lemma 7.5.6 says that the topological type $\mathcal{M}_{top}$ of manifold $\mathcal{M}$ is defined by the Morita equivalence class of the Etesi C^*-algebra $\mathbb{E}_{\mathcal{M}}$. In contrast, different smoothings of $\mathcal{M}_{top}$ are defined by the isomorphism classes contained in the Morita equivalence class of $\mathbb{E}_{\mathcal{M}}$.

Lemma 7.5.7. *Let K be a number field and let* $\mathrm{Mi}(K)$ *be its Minkowski group. Then:*
(i) *the map $K \to$* $\mathrm{Mi}(K)$ *is a covariant functor which maps isomorphic number fields K to the isomorphic torsion abelian groups* $\mathrm{Mi}(K)$*;*
(ii) $\mathrm{Mi}(K) \cong \mathrm{Br}(K)$.

Proof. (i) Let K be a number field corresponding to the Handelman triple $(\Lambda, [\mathfrak{m}], K)$. Denote by $[\mathbb{A}]$ the Morita equivalence class of stationary AF-algebras defined by the triple $(\Lambda, [\mathfrak{m}], K)$. Since τ is the canonical trace, we conclude that $\tau_*(\Sigma(\mathbb{A})) \subset [0,1]$ does not depend on $\mathbb{A} \in [\mathbb{A}]$. Thus one gets a correctly defined map $K \to \mathcal{Y} := \mathrm{Mi}(K)$. It can be verified directly that if K' is a real embedding of K, then $\mathrm{Mi}(K') \cong \mathrm{Mi}(K)$. Item (i) is proved.

(ii) Let $\mathrm{Br}(K)$ be the Brauer group of a number field K. It is well known that the map $K \to \mathrm{Br}(K)$ is a covariant functor which maps isomorphic number fields K to the isomorphic torsion abelian groups $\mathrm{Br}(K)$. Comparing with item (i), we conclude that there exists a natural transformation between these two functors (Section 2.3). In particular, such a transformation implies an isomorphism of the abelian groups $\mathrm{Br}(K)$ and $\mathrm{Mi}(K)$. Item (ii) and Lemma 7.5.7 are proved. □

Remark 7.5.5. Lemma 7.5.7 can be viewed as part of a correspondence between the Milnor–Voevodsky K-theory and the Galois cohomology.

Corollary 7.5.4. *Distinct smoothings of $\mathcal{M}$ are classified by the elements of the Brauer group* $\mathrm{Br}(K)$*. In particular, all smoothings of $\mathcal{M}$ form a torsion abelian group with the summation operation $\oplus$ and the neutral element $\mathcal{M}_0$, see Corollary 7.5.2.*

Proof. (i) Let us recall the formula

$$\mathcal{Y} = ?_n(\tau_*(\Sigma(\mathbb{E}_{\mathcal{M}}))).$$

Since the Minkowski function $?_n(x)$ is monotone, the above expression defines a bijective correspondence between generators of the torsion abelian groups $\mathrm{Mi}(K) \cong \mathrm{Br}(K)$ and the scale $\Sigma(\mathbb{E}_{\mathcal{M}})$ of the Etesi C^*-algebra $\mathbb{E}_{\mathcal{M}}$.

(ii) Recall that the scale $\Sigma(\mathbb{E}_{\mathcal{M}})$ is a generating subset of $K_0^+(\mathbb{E}_{\mathcal{M}})$, i. e., for each $a \in K_0^+(\mathbb{E}_{\mathcal{M}})$ there exist $a_1, \dots, a_r \in \Sigma(\mathbb{E}_{\mathcal{M}})$ such that $a = a_1 + \cdots + a_r$. We extend the correspondence of item (i) to a bijective map between the elements of the Brauer group $Br(K)$ and the elements of positive cone $K_0^+(\mathbb{E}_{\mathcal{M}})$.

(iii) It is known that the pair $(K_0(\mathbb{E}_{\mathcal{M}}), K_0^+(\mathbb{E}_{\mathcal{M}}))$ is invariant of the Morita equivalence class of the AF-algebra $\mathbb{E}_{\mathcal{M}}$, while the triple

$$(K_0(\mathbb{E}_{\mathcal{M}}), K_0^+(\mathbb{E}_{\mathcal{M}}), \Sigma(\mathbb{E}_{\mathcal{M}}))$$

is invariant of the isomorphism class of $\mathbb{E}_{\mathcal{M}}$, see Section 2.2. Moreover, the scale can be written as $\Sigma(\mathbb{E}_{\mathcal{M}}) = \{a \in K_0^+(\mathbb{E}_{\mathcal{M}}) \mid 0 \leq a \leq u\}$, where $u \in K_0^+(\mathbb{E}_{\mathcal{M}})$ is fixed. Thus running through all $u \in K_0^+(\mathbb{E}_{\mathcal{M}})$, one gets all possible scales $\Sigma(\mathbb{E}_{\mathcal{M}})$, and vice versa. In other words, the elements $u \in K_0^+(\mathbb{E}_{\mathcal{M}})$ parametrize isomorphism classes of $\mathbb{E}_{\mathcal{M}}$ within its Morita equivalence class.

(iv) To finish the proof of Corollary 7.5.4, we compare Remark 7.5.4 with item (iii) and conclude that different smooth structures on $\mathcal{M}$ are in bijection with the elements of $K_0^+(\mathbb{E}_{\mathcal{M}})$. Moreover, $K_0^+(\mathbb{E}_{\mathcal{M}})$ has structure of a torsion abelian group isomorphic to the Brauer group $Br(K)$, see item (ii). Corollary 7.5.4 is proved. □

Corollary 7.5.1 follows from Lemma 7.5.6 and Corollary 7.5.4.

Guide to the literature

The C^*-algebra $\mathbb{E}_{\mathcal{M}}$ has been introduced by Gábor Etesi in the remarkable papers [78, 79], hence the name. In particular, a new topological invariant of the 4-manifolds given by the Murray–von Neumann coupling constant of $\mathbb{E}_{\mathcal{M}}$ has been studied in [79]. Our exposition follows [213].

Exercises

1. Prove the formula $Br(D_p) \cong \mathbf{Q}/\mathbf{Z}$ in Remark 7.4.6 using the Merkurjev–Suslin approach to the Brauer groups via the algebraic K-theory.
2. Prove that Lemma 7.5.7 follows from the Milnor–Voevodsky K-theory and the Galois cohomology, see also Remark 7.5.5.
3. Algebra $\mathbb{E}_{\mathcal{M}}$ can be defined via unitary representation of the group $Diff(\mathcal{M})$ by the bounded linear operators on a Hilbert space [78, Section 2]. Since $Diff(\mathcal{M})$ is a Fréchet manifold, the standard construction of the group C^*-algebra fails in general, see, e. g., [32, Section 11.1]; hence Definition 7.5.1. Prove that the two definitions are equivalent from the standpoint of representation theory.
4. Let $\mathcal{M}$ be the 4-dimensional sphere S^4. Since $\mathbb{E}_{S^4} \cong \mathbf{C}$ is a finite-dimensional C^*-algebra, Corollary 7.5.1 fails. On the other hand, it is known that $K_0(\mathbb{E}_{S^4}) \cong \mathbf{Z}$ and $K \cong \mathbf{Q}$. Denote by $\mathcal{S}$ a subgroup of the Brauer group $Br(\mathbf{Q})$ consisting of all smoothings of S^4. Assuming an analog of Corollary 7.5.1, one can recast the smooth Poincaré Conjecture as follows: "Is the group $\mathcal{S}$ trivial?"

8 Quantum arithmetic

The quantum arithmetic studies an interplay between noncommutative geometry and number theory. Such a theory dates back to the work of Bost and Connes [39] on a characterization of the operators on the Hilbert space $\mathcal{H}$ whose spectrum coincides with the set of all prime numbers (Section 11.4). Other model examples are considered in Chapter 6. In Section 8.1 we recast the Langlands Conjecture on the automorphy of the zeta functions in terms of the C^*-algebras. Using the K-theory of such C^*-algebras, we prove the Langlands Conjecture for the Shimura varieties. In Section 8.2 we calculate the genus of a rational quadratic form using the K-theory of C^*-algebra associated to a noncompact reductive algebraic group. The quantum dynamics of rational elliptic curves is studied in Section 8.3. In Section 8.4 we calculate the Shafarevich–Tate groups of abelian varieties based on a localization of the higher-dimensional noncommutative tori. In Section 8.5 we use noncommutative geometry to study elliptic surfaces and calculate their Picard numbers. In Section 8.6 we apply C^*-algebras to study the class field towers over the real fields.

8.1 Langlands reciprocity for C*-algebras

The Langlands conjectures say that all zeta functions are automorphic [146]. In this section we study (one of) the conjectures in terms of the C^*-algebras. Namely, denote by $G(\mathbf{A}_K)$ a reductive group G over the ring of adeles $\mathbf{A}_K$ of a number field K and by $G(K)$ its discrete subgroup over K. The Banach algebra $L^1(G(K)\backslash G(\mathbf{A}_K))$ consists of the integrable complex-valued functions endowed with the operator norm. The addition of functions $f_1, f_2 \in L^1(G(K)\backslash G(\mathbf{A}_K))$ is defined pointwise and multiplication is given by the convolution product

$$(f_1 * f_2)(g) = \int_{G(K)\backslash G(\mathbf{A}_K)} f_1(gh^{-1}) f_2(h) dh.$$

Consider an enveloping C^*-algebra, $\mathcal{A}_G$, of the algebra $L^1(G(K)\backslash G(\mathbf{A}_K))$; we refer the reader to [69, Section 13.9] for details of this construction. The algebra $\mathcal{A}_G$ encodes all unitary irreducible representations of the locally compact group $G(\mathbf{A}_K)$ induced by $G(K)$. Such representations are related to the automorphic cusp forms and nonabelian class field theory [91]. The algebra $\mathcal{A}_G$ has an amazingly simple structure. Namely, if $G \cong \mathrm{GL}_n(\mathbf{A}_K)$ then $\mathcal{A}_G$ is a stationary AF-algebra defined by a positive integer matrix $B \in \mathrm{SL}_n(\mathbf{Z})$ (Section 3.5.2).

Let V be the complex projective variety over a number field K. Denote by $\mathcal{A}_V$ the corresponding Serre C^*-algebra (Section 5.3.1). It is known that the Langlands philosophy does not distinguish between the arithmetic and automorphic objects [146].

https://doi.org/10.1515/9783110788709-008

Therefore one can expect a regular map between the C^*-algebras $\mathcal{A}_V$ and $\mathcal{A}_G$. To formulate our result, we shall need the following definitions. Recall [195] that the ith trace cohomology $\{H^i_{tr}(V) \mid 0 \le i \le 2\dim_{\mathbf{C}} V\}$ of an arithmetic variety V is an additive abelian subgroup of $\mathbf{R}$ obtained from a canonical trace on the Serre C^*-algebra of V (Section 8.1.1). Likewise, the group $K_0(\mathcal{A}_G)$ of the stationary AF-algebra $\mathcal{A}_G$ is an additive abelian subgroup of $\mathbf{R}$ (Section 3.5.2).

Definition 8.1.1. The arithmetic variety V is called G-coherent if

$$H^i_{tr}(V) \subseteq K_0(\mathcal{A}_G) \quad \text{for all } 0 \le i \le 2\dim_{\mathbf{C}} V.$$

Remark 8.1.1. If V is the Shimura variety of the group $G(\mathbf{A}_K)$ [67], then V is a G-coherent variety. To put it simple, the arithmetic variety V is G-coherent if for all $0 \le i \le 2\dim_{\mathbf{C}} V$ the number fields $\mathbf{k}_i := H^i_{tr}(V) \otimes \mathbf{Q}$ are subfields of (or coincide with) a number field $\mathbf{K} := K_0(\mathcal{A}_G) \otimes \mathbf{Q}$. For a simple example, we refer the reader to Proposition 8.1.2.

Theorem 8.1.1. *There exists a canonical embedding $\mathcal{A}_V \hookrightarrow \mathcal{A}_G$, where V is a G-coherent variety.*

Remark 8.1.2. Theorem 8.1.1 can be viewed as an analog of the Langlands reciprocity for C^*-algebras. In other words, the coordinate ring $\mathcal{A}_V$ of a G-coherent variety V is a subalgebra of the algebra $\mathcal{A}_G$.

An application of Theorem 8.1.1 is as follows. Recall that to each arithmetic variety V one can attach the Hasse–Weil (motivic) L-function. Likewise, to each irreducible representation of the group $G(\mathbf{A}_K)$ one can attach an automorphic (standard) L-function [91] and [146]. Theorem 8.1.1 implies one of the Langlands Conjectures [146].

Corollary 8.1.1. *The Hasse–Weil L-function of a G-coherent variety V is a product of the automorphic L-functions.*

8.1.1 Trace cohomology

For the sake of clarity, let us review the trace cohomology [195], see also Section 6.7.1. Let V be an n-dimensional complex projective variety endowed with an automorphism $\sigma : V \to V$ and denote by $B(V, \mathcal{L}, \sigma)$ its twisted homogeneous coordinate ring (Section 5.3.1). Let R be a commutative graded ring such that $V = \mathrm{Spec}(R)$. Denote by $R[t, t^{-1}; \sigma]$ the ring of skew Laurent polynomials defined by the commutation relation $b^\sigma t = tb$ for all $b \in R$, where b^σ is the image of b under automorphism σ. It is known that $R[t, t^{-1}; \sigma] \cong B(V, \mathcal{L}, \sigma)$.

Let $\mathcal{H}$ be a Hilbert space and $\mathcal{B}(\mathcal{H})$ the algebra of all bounded linear operators on $\mathcal{H}$. For a ring of skew Laurent polynomials $R[t, t^{-1}; \sigma]$, consider a homomorphism

$$\rho : R[t, t^{-1}; \sigma] \longrightarrow \mathcal{B}(\mathcal{H}).$$

Recall that $\mathcal{B}(\mathcal{H})$ is endowed with a $*$-involution; the involution comes from the scalar product on the Hilbert space $\mathcal{H}$. We shall call the representation ρ $*$-coherent, if (i) $\rho(t)$ and $\rho(t^{-1})$ are unitary operators such that $\rho^*(t) = \rho(t^{-1})$ and (ii) for all $b \in R$ it holds $(\rho^*(b))^{\sigma(\rho)} = \rho^*(b^\sigma)$, where $\sigma(\rho)$ is an automorphism of $\rho(R)$ induced by σ. Whenever $B = R[t,t^{-1};\sigma]$ admits a $*$-coherent representation, $\rho(B)$ is a $*$-algebra; the norm closure of $\rho(B)$ is a C^*-algebra. We shall denote it by $\mathcal{A}_V$ and refer to $\mathcal{A}_V$ as the Serre C^*-algebra of variety V.

Let $\mathcal{K}$ be the C^*-algebra of all compact operators on $\mathcal{H}$. We shall write $\tau : \mathcal{A}_V \otimes \mathcal{K} \to \mathbf{R}$ to denote the canonical normalized trace on $\mathcal{A}_V \otimes \mathcal{K}$, i. e., a positive linear functional of norm 1 such that $\tau(yx) = \tau(xy)$ for all $x, y \in \mathcal{A}_V \otimes \mathcal{K}$, see [32, p. 31]. Because $\mathcal{A}_V$ is a crossed product C^*-algebra of the form $\mathcal{A}_V \cong C(V) \rtimes \mathbf{Z}$, one can use the Pimsner–Voiculescu six-term exact sequence for the crossed products, see, e. g., [32, p. 83] for the details. Thus one gets the short exact sequence of the algebraic K-groups, $0 \to K_0(C(V)) \xrightarrow{i_*} K_0(\mathcal{A}_V) \to K_1(C(V)) \to 0$, where map i_* is induced by the natural embedding of $C(V)$ into $\mathcal{A}_V$. We have $K_0(C(V)) \cong K^0(V)$ and $K_1(C(V)) \cong K^{-1}(V)$, where K^0 and K^{-1} are the topological K-groups of V, see [32, p. 80]. By the Chern character formula, one gets $K^0(V) \otimes \mathbf{Q} \cong H^{\mathrm{even}}(V;\mathbf{Q})$ and $K^{-1}(V) \otimes \mathbf{Q} \cong H^{\mathrm{odd}}(V;\mathbf{Q})$, where H^{even} (H^{odd}) is the direct sum of even (resp. odd) cohomology groups of V. Notice that $K_0(\mathcal{A}_V \otimes \mathcal{K}) \cong K_0(\mathcal{A}_V)$ because of a stability of the K_0-group with respect to tensor products by the algebra $\mathcal{K}$, see, e. g., [32, p. 32]. One gets the commutative diagram in Fig. 6.7, where τ_* denotes a homomorphism induced on K_0 by the canonical trace τ on the C^*-algebra $\mathcal{A}_V \otimes \mathcal{K}$. Since $H^{\mathrm{even}}(V) := \bigoplus_{i=0}^n H^{2i}(V)$ and $H^{\mathrm{odd}}(V) := \bigoplus_{i=1}^n H^{2i-1}(V)$, one gets for each $0 \le i \le 2n$ an injective homomorphism $\tau_* : H^i(V) \longrightarrow \mathbf{R}$.

Definition 8.1.2. By an ith trace cohomology group $H^i_{tr}(V)$ of variety V one understands the abelian subgroup of $\mathbf{R}$ defined by the map τ_*.

8.1.2 Langlands reciprocity

Let V be an n-dimensional complex projective variety over a number field K; consider its reduction $V(\mathbf{F}_p)$ modulo the prime ideal $\mathfrak{P} \subset K$ over a good prime p. Recall that the Weil zeta function is defined as

$$Z_p(t) = \exp\left(\sum_{r=1}^{\infty} |V(\mathbf{F}_{p^r})| \frac{t^r}{r}\right), \quad r \in \mathbf{C},$$

where $|V(\mathbf{F}_{p^r})|$ is the number of points of variety $V(\mathbf{F}_{p^r})$ defined over the field with p^r elements. It is known that

$$Z_p(t) = \frac{P_1(t) \cdots P_{2n-1}(t)}{P_0(t) \cdots P_{2n}(t)},$$

where $P_0(t) = 1 - t, P_{2n} = 1 - p^n t$, and each $P_i(t)$ for $1 \le i \le 2n-1$ is a polynomial with integer coefficients such that $P_i(t) = \prod(1 - \alpha_{ij}t)$ for some algebraic integers α_{ij} of the absolute value $p^{\frac{i}{2}}$. Consider an infinite product

$$L(s,V) := \prod_p Z_p(p^{-s}) = \frac{L^1(s,V)\cdots L^{2n-1}(s,V)}{L^0(s,V)\cdots L^{2n}(s,V)},$$

where $L^i(s,V) = \prod_p P_i(p^{-s})$; the product $L(s,V)$ is called the Hasse–Weil (or motivic) L-function of V.

On the other hand, if K is a number field then the adele ring $\mathbf{A}_K$ of K is a locally compact subring of the direct product $\prod K_v$ taken over all places v of K; $\mathbf{A}_K$ is endowed with a canonical topology. Consider a reductive group $G(\mathbf{A}_K)$ over $\mathbf{A}_K$; the latter is a topological group with a canonical discrete subgroup $G(K)$. Denote by $L^2(G(K)\backslash G(\mathbf{A}_K))$ the Hilbert space of all square-integrable complex-valued functions on the homogeneous space $G(K)\backslash G(\mathbf{A}_K)$ and consider the right regular representation $\mathcal{R}$ of the locally compact group $G(\mathbf{A}_K)$ by linear operators on the space $L^2(G(K)\backslash G(\mathbf{A}_K))$ given by formula $(f_1 * f_2)(g) = \int_{G(K)\backslash G(\mathbf{A}_K)} f_1(gh^{-1})f_2(h)dh$. It is well known that each irreducible component π of the unitary representation $\mathcal{R}$ can be written in the form $\pi = \otimes\pi_v$, where v are all unramified places of K. Using the spherical functions, one gets an injection $\pi_v \mapsto [A_v]$, where $[A_v]$ is a conjugacy class of matrices in the group $\mathrm{GL}_n(\mathbf{C})$. The automorphic L-function is given by the formula

$$L(s,\pi) = \prod_v \left(\det[I_n - [A_v](Nv)^{-s}]\right)^{-1}, \quad s \in \mathbf{C},$$

where Nv is the norm of place v; we refer the reader to [146, p. 170] and [91, p. 201] for details of this construction. The following conjecture relates the Hasse–Weil and automorphic L-functions.

Conjecture 8.1.1 ([146]). *For each $0 \le i \le 2n$, there exists an irreducible representation π_i of the group $G(\mathbf{A}_K)$ such that $L^i(s,V) \equiv L(s,\pi_i)$.*

8.1.3 Proofs

8.1.3.1 Proof of Theorem 8.1.1

We shall split the proof in two lemmas.

Lemma 8.1.1. *The algebra $\mathcal{A}_G$ is isomorphic to a stationary AF-algebra.*

Proof. Let $\mathbf{A}_K^\times$ be the idele group, i. e., a group of invertible elements of the adele ring $\mathbf{A}_K^\times$. Denote by $\mathrm{Gal}(K^{ab}|K)$ the Galois group of the maximal abelian extension K^{ab} of the number field K. The Artin reciprocity says that there exists a continuous isomorphism

$$\mathbf{A}_K^\times \longrightarrow \mathrm{Gal}(K^{ab}|K).$$

Recall that $\mathrm{Gal}(K^{ab}|K)$ is a profinite abelian group, i. e., a topological group isomorphic to the inverse limit of finite abelian groups. By the Artin reciprocity, the idele group $\mathbf{A}_K^{\times}$ is also a profinite abelian group. Since every finite abelian group is a product of the cyclic groups $\mathbf{Z}/p_i^{k_i}\mathbf{Z}$, we can write the $\mathbf{A}_K^{\times}$ in the form

$$\mathbf{A}_K^{\times} \cong \varprojlim \prod_{i=1}^{l_m} (\mathbf{Z}/p_i^{k_i}\mathbf{Z}),$$

where $m \to \infty$. Notice that the cyclic group $\mathbf{Z}/p_i^{k_i}\mathbf{Z}$ can be embedded into the finite field $\mathbf{F}_{q_i}$, where $q_i = p_i^{k_i}$. Thus the group $GL_n(\mathbf{Z}/p_i^{k_i}\mathbf{Z})$ is correctly defined and one gets an isomorphism

$$GL_n(\mathbf{A}_K) \cong \varprojlim \prod_{i=1}^{l_m} GL_n(\mathbf{F}_{q_i}),$$

where $GL_n(\mathbf{F}_{q_i})$ is a finite group of order $\prod_{j=0}^{n-1}(q_i^n - q_i^j)$ and such a group is no longer abelian. In particular, it follows from the above that the $GL_n(\mathbf{A}_K)$ is a profinite group.

(i) Let us show that the group $GL_n(\mathbf{A}_K)$ being profinite implies that the $\mathcal{A}_G$ is an AF-algebra. Indeed, if G is a finite group then the group algebra $\mathbf{C}[G]$ has the form

$$\mathbf{C}[G] \cong M_{n_1}(\mathbf{C}) \oplus \cdots \oplus M_{n_h}(\mathbf{C}),$$

where n_i are degrees of the irreducible representations of G and h is the total number of such representations [251, Proposition 10]. In view of the above, we have

$$GL_n(\mathbf{A}_K) \cong \varprojlim G_i,$$

where G_i is a finite group. Consider a group algebra

$$\mathbf{C}[G_i] \cong M_{n_1}^{(i)}(\mathbf{C}) \oplus \cdots \oplus M_{n_h}^{(i)}(\mathbf{C})$$

corresponding to G_i. Notice that the $\mathbf{C}[G_i]$ is a finite-dimensional C^*-algebra. The above inverse limit defines an ascending sequence of the finite-dimensional C^*-algebras of the form

$$\varprojlim M_{n_1}^{(i)}(\mathbf{C}) \oplus \cdots \oplus M_{n_h}^{(i)}(\mathbf{C}).$$

Since $\mathcal{A}_G$ is the norm closure of the group algebra $\mathbf{C}[GL_n(\mathbf{A}_K)]$ [69, Section 13.9], we conclude that the limit converges to the algebra $\mathcal{A}_G$. In other words, the $\mathcal{A}_G$ is an AF-algebra. Item (i) is proved.

Example 8.1.1. If $G = \mathrm{GL}_1(\mathbf{A_Q}) \cong \mathbf{A}_{\mathbf{Q}}^{\times}$, then $\mathcal{A}_G$ is an UHF-algebra of the form $\{M_{p_1^{\infty}p_2^{\infty}\dots} \mid p_i \in \mathcal{P}\}$, where $\mathcal{P}$ is the set of all prime numbers. We refer the reader to Section 3.6 for a definition of the UHF-algebra.

(ii) It remains to prove that $\mathcal{A}_G$ is a stationary AF-algebra. Indeed, denote by Fr_q the Frobenius map, i. e., an endomorphism of the finite field $\mathbf{F}_q$ acting by the formula $x \mapsto x^q$. The map Fr_{q_i} induces an automorphism of the group $\mathrm{GL}_n(\mathbf{F}_{q_i})$. Thus one gets an automorphism of the group $\mathrm{GL}_n(\mathbf{A}_K)$ and the corresponding group algebra $\mathbf{C}[\mathrm{GL}_n(\mathbf{A}_K)]$. Taking the norm closure of the algebra $\mathbf{C}[\mathrm{GL}_n(\mathbf{A}_K)]$, we conclude that there exists a nontrivial automorphism ϕ of the AF-algebra $\mathcal{A}_G$. But the AF-algebra admits an automorphism $\phi \neq \pm\mathrm{Id}$ if and only if it is a stationary AF-algebra [74, p. 37]. Thus the algebra $\mathcal{A}_G$ is a stationary AF-algebra. Lemma 8.1.1 is proved. □

Remark 8.1.3. Recall that the AF-algebra $\mathcal{A}_G$ is determined by a partial multiplicity matrix B of rank n, and one can always assume that $B \in \mathrm{SL}_n(\mathbf{Z})$. Consider an isomorphism

$$\mathcal{A}_G \rtimes \mathbf{Z} \cong \mathcal{O}_B \otimes \mathcal{K},$$

where the crossed product is taken by the shift automorphism of $\mathcal{A}_G$, $\mathcal{O}_B$ is the Cuntz–Krieger algebra (Section 3.7) defined by matrix B and $\mathcal{K}$ is the C^*-algebra of compact operators [32, Exercise 10.11.9]. Consider a continuous group of modular automorphisms $\{\sigma^t : \mathcal{O}_B \to \mathcal{O}_B \mid t \in \mathbf{R}\}$ acting on the generators $s_1, \dots, s_n$ of the algebra $\mathcal{O}_B$ by the formula $s_k \mapsto e^{it}s_k$. Then a pull back of σ^t corresponds to the action of continuous symmetry group $\mathrm{GL}_n(\mathbf{A}_K)$ on the homogeneous space $\mathrm{GL}_n(K)\backslash\mathrm{GL}_n(\mathbf{A}_K)$. This observation can be applied to prove Weil's conjecture on the Tamagawa numbers.

Lemma 8.1.2. *The algebra $\mathcal{A}_V$ embeds into the AF-algebra $\mathcal{A}_G$, where V is a G-coherent variety.*

Proof. We shall use Pimsner's Theorem [229, Theorem 7] about an embedding of the crossed product algebra $\mathcal{A}_V$ into an AF-algebra. It will develop that the G-coherence of V implies that the AF-algebra is Morita equivalent to the algebra $\mathcal{A}_G$. We pass to a detailed argument.

Let V be a complex projective variety. Following [229], we shall think of V as a compact metrizable topological space X. Recall that for a homeomorphism $\varphi : X \to X$ the point $x \in X$ is called nonwandering if for each neighborhood U of x and every $N > 0$ there exists $n > N$ such that

$$\varphi^n(U) \cap U \neq \emptyset.$$

(In other words, the point x does not "wander" too far from its initial position in the space X.) If each point $x \in X$ is a nonwandering point, then the homeomorphism φ is called nonwandering.

Let $\sigma : V \to V$ be an automorphism of finite order of the G-coherent variety V such that the representation ρ is $*$-coherent. Then the crossed product

$$\mathcal{A}_V = C(V) \rtimes_\sigma \mathbf{Z}$$

is the Serre C^*-algebra of V. Since σ is of a finite order, it is a nonwandering homeomorphism of X. In particular, the σ is a pseudo-nonwandering homeomorphism [229, Definition 2]. Then there exists a unital (dense) embedding

$$\mathcal{A}_V \hookrightarrow \mathcal{A},$$

where $\mathcal{A}$ is an AF-algebra defined by the homeomorphism φ [229, Theorem 7].

Let us show that the algebra $\mathcal{A}$ is Morita equivalent to the AF-algebra $\mathcal{A}_G$. Indeed, the above embedding induces an injective homomorphism of the K_0-groups

$$K_0(\mathcal{A}_V) \hookrightarrow K_0(\mathcal{A}).$$

As explained in Section 8.1.1, the above map defines an inclusion

$$H^i_{\mathrm{tr}}(V) \subseteq K_0(\mathcal{A}).$$

On the other hand, the trace cohomology of a G-coherent variety V must satisfy an inclusion

$$H^i_{\mathrm{tr}}(V) \subseteq K_0(\mathcal{A}_G).$$

Let $b^* = \max_{0\le i\le 2n} b_i$ be the maximal Betti number of variety V. Then the inclusion is an isomorphism, i. e., $H^*_{\mathrm{tr}}(V) \cong K_0(\mathcal{A})$ and $H^*_{\mathrm{tr}}(V) \cong K_0(\mathcal{A}_G)$. One concludes that

$$K_0(\mathcal{A}) \cong K_0(\mathcal{A}_G).$$

In other words, the AF-algebras $\mathcal{A}$ and $\mathcal{A}_G$ are Morita equivalent. The embedding $\mathcal{A}_V \hookrightarrow \mathcal{A}_G$ follows from the above formulas. Lemma 8.1.2 is proved. □

Theorem 8.1.1 follows from Lemma 8.1.2.

8.1.3.2 Proof of Corollary 8.1.1

Corollary 8.1.1 follows from an observation that the Frobenius action $\sigma(Fr^i_p) : H^i_{\mathrm{tr}}(V) \to H^i_{\mathrm{tr}}(V)$ extends to a Hecke operator $T_p : K_0(\mathcal{A}_G) \to K_0(\mathcal{A}_G)$, whenever $H^i_{\mathrm{tr}}(\mathcal{A}_V) \subseteq K_0(\mathcal{A}_G)$. Let us pass to a detailed argument.

Recall that the Frobenius map on the ith trace cohomology of variety V is given by an integer matrix $\sigma(\mathrm{Fr}^i_p) \in \mathrm{GL}_{b_i}(\mathbf{Z})$, where b_i is the ith Betti number of V. Moreover, it holds that

$$|V(\mathbf{F}_p)| = \sum_{i=0}^{2n} (-1)^i \operatorname{tr} \sigma(\mathrm{Fr}_p^i),$$

where $V(\mathbf{F}_p)$ is the reduction of V modulo a good prime p (Section 6.7). Notice that the above formula is sufficient to calculate the Hasse–Weil L-function $L(s, V)$ of variety V; hence the map $\sigma(\mathrm{Fr}_p^i) : H_{\mathrm{tr}}^i(V) \to H_{\mathrm{tr}}^i(V)$ is motivic.

Definition 8.1.3. Denote by T_p^i an endomorphism of $K_0(\mathcal{A}_G)$ such that the diagram in Fig. 8.1 is commutative, where ι is the Pimsner's embedding. By $\mathfrak{H}_i$ we understand an algebra over $\mathbf{Z}$ generated by the $T_p^i \in \mathrm{End}\,(K_0(\mathcal{A}_G))$, where p runs through all but a finite set of primes.

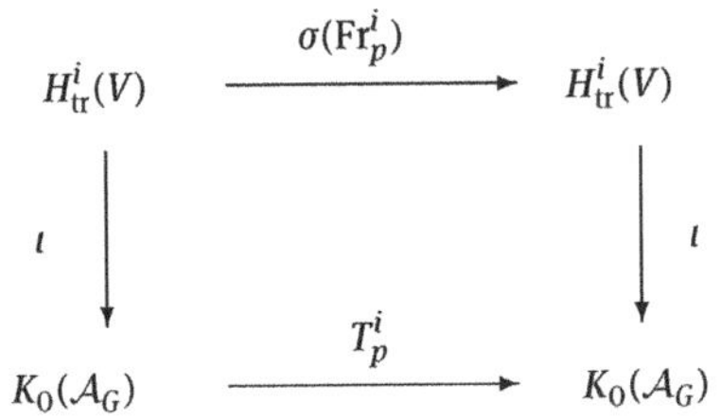

Figure 8.1: The Hecke operator T_p^i.

Remark 8.1.4. The algebra $\mathfrak{H}_i$ is commutative. Indeed, the endomorphisms T_p^i correspond to multiplication of the group $K_0(\mathcal{A}_G)$ by the real numbers; the latter commute with each other. We shall call the $\{\mathfrak{H}_i \mid 0 \le i \le 2n\}$ an ith Hecke algebra.

Lemma 8.1.3. *The algebra $\mathfrak{H}_i$ defines an irreducible representations π_i of the group $G(\mathbf{A}_K)$.*

Proof. Let $f \in L^2(G(K)\backslash G(\mathbf{A}_K))$ be an eigenfunction of the Hecke operators T_p^i; in other words, the Fourier coefficients c_p of function f coincide with the eigenvalues of the Hecke operators T_p up to a scalar multiple. Such an eigenfunction is defined uniquely by the algebra $\mathfrak{H}_i$.

Let $\mathcal{L}_f \subset L^2(G(K)\backslash G(\mathbf{A}_K))$ be a subspace generated by the right translates of f by the elements of the locally compact group $G(\mathbf{A}_K)$. It is immediate that the $\mathcal{L}_f$ is an irreducible subspace of the space $L^2(G(K)\backslash G(\mathbf{A}_K))$; therefore it gives rise to an irreducible representation π_i of the locally compact group $G(\mathbf{A}_K)$. Lemma 8.1.3 follows. □

Lemma 8.1.4. $L(s, \pi_i) \equiv L^i(s, V)$.

Proof. Recall that the function $L^i(s, V)$ can be written as

$$L^i(s, V) = \prod_p (\det[I_n - \sigma(\mathrm{Fr}_p^i)p^{-s}])^{-1},$$

where $\sigma(\mathrm{Fr}_p^i) \in GL_{b_i}(\mathbf{Z})$ is a matrix form of the action of Fr_p^i on the trace cohomology $H_{tr}^i(V)$.

On the other hand, we have

$$L(s, \pi_i) = \prod_p (\det[I_n - [A_p^i]p^{-s}])^{-1},$$

where $[A_p^i] \subset GL_n(\mathbf{C})$ is a conjugacy class of matrices corresponding to the irreducible representation π_i of the group $G(\mathbf{A}_K)$. As explained, for such a representation one gets an inclusion $T_p^i \in [A_p^i]$. But the action of the Hecke operator T_p^i is an extension of the action of $\sigma(\mathrm{Fr}_p^i)$ on $H_{tr}^i(V)$, see Fig. 8.1. Therefore

$$\sigma(\mathrm{Fr}_p^i) = [A_p^i]$$

for all but a finite set of primes p. Comparing the above formulas, we conclude that $L(s, \pi_i) \equiv L^i(s, V)$. Lemma 8.1.4 follows. □

Corollary 8.1.1 follows from Lemma 8.1.4.

8.1.4 Pimsner–Voiculescu embedding

We shall illustrate Theorem 8.1.1 and Corollary 8.1.1 for the group $G \cong GL_2(\mathbf{A}_K)$, where $K = \mathbf{Q}(\sqrt{D})$ is a real quadratic field.

Proposition 8.1.1. *$K_0(\mathcal{A}_G) \cong \mathbf{Z} + \mathbf{Z}\omega$, where*

$$\omega = \begin{cases} \frac{1+\sqrt{D}}{2}, & \text{if } D \equiv 1 \mod 4, \\ \sqrt{D}, & \text{if } D \equiv 2, 3 \mod 4. \end{cases}$$

Proof. By Lemma 8.1.1, $\mathcal{A}_G$ is a stationary AF-algebra given by partial multiplicity matrix $B \in SL_2(\mathbf{Z})$. In particular, $K_0(\mathcal{A}_G) \cong \mathbf{Z} + \mathbf{Z}\omega$, where $\omega \in \mathbf{Q}(\lambda_B)$, where λ_B is the Perron–Frobenius eigenvalue of matrix B. Moreover, by the construction $\mathrm{End}(K) \cong \mathrm{End}(K_0(\mathcal{A}_G))$, where End is the endomorphism ring of the corresponding $\mathbf{Z}$-module. But $\mathrm{End}(K) \cong O_K$, where O_K is the ring of integers of K. Thus, $\lambda_B \in K$ and ω is as given above. Proposition 8.1.1 follows. □

Proposition 8.1.2. *Let $\mathcal{E}_{CM} \cong \mathbf{C}/O_k$ be an elliptic curve with complex multiplication by the ring of integers of the imaginary quadratic field $k = \mathbf{Q}(\sqrt{-D})$. Then $\mathcal{E}_{CM}$ is a G-coherent variety of the group $G \cong GL_2(\mathbf{A}_K)$.*

Proof. Recall that the Serre C^*-algebra of an elliptic curve $\mathcal{E}_\tau \cong \mathbf{C}/(\mathbf{Z}+\mathbf{Z}\tau)$ is isomorphic to the noncommutative torus $\mathcal{A}_\theta$ for any $\{\tau \mid \mathrm{Im}\,\tau > 0\}$ (Section 1.3). In particular, if

$\tau \in O_k$ then

$$\begin{cases} H^0_{\mathrm{tr}}(\mathcal{E}_{CM}) = H^2_{\mathrm{tr}}(\mathcal{E}_{CM}) \cong \mathbf{Z}, \\ H^1_{\mathrm{tr}}(\mathcal{E}_{CM}) \cong \mathbf{Z} + \mathbf{Z}\omega. \end{cases}$$

Comparing the above formulas, one concludes that $H^i_{\mathrm{tr}}(\mathcal{E}_{CM}) \subseteq K_0(\mathcal{A}_G)$, i. e., the $\mathcal{E}_{CM}$ is a G-coherent variety of the group $G \cong \mathrm{GL}_2(\mathbf{A}_K)$. Proposition 8.1.2 is proved. □

Remark 8.1.5 (Pimsner–Voiculescu). The embedding of $\mathcal{A}_\theta$ into an AF-algebra was first constructed in [230].

Proposition 8.1.3. $L(s, \mathcal{E}_{CM}) \equiv \frac{L(s,\pi_1)}{L(s,\pi_0)L(s,\pi_2)}$, *where* π_i *are irreducible representations of the locally compact group* $\mathrm{GL}_2(\mathbf{A}_K)$.

Proof. The Hasse–Weil L-function of $\mathcal{E}_{CM}$ has the form

$$L(s, \mathcal{E}_{CM}) = \frac{\prod_p [\det(I_2 - \sigma(\mathrm{Fr}^1_p)p^{-s}]^{-1}}{\zeta(s)\zeta(s-1)}, \qquad s \in \mathbf{C},$$

where $\zeta(s)$ is the Riemann zeta function and the product is taken over the set of good primes. It is immediate that

$$\begin{cases} L(s, \pi_0) = \zeta(s), \\ L(s, \pi_2) = \zeta(s-1), \end{cases}$$

where $L(s, \pi_0)$ and $L(s, \pi_2)$ are the automorphic L-functions corresponding to the irreducible representations π_0 and π_2 of the group $\mathrm{GL}_2(\mathbf{A}_K)$. An irreducible representation π_1 gives rise to an automorphic L-function

$$L(s, \pi_1) = \prod_p \left(\det[I_2 - [A^1_p]p^{-s}]\right)^{-1}.$$

But it was proved earlier that $[A^1_p] = \sigma(\mathrm{Fr}^1_p)$ and therefore the numerator in the above equation coincides with the $L(s, \pi_1)$. Proposition 8.1.3 is proved. □

Remark 8.1.6. Proposition 8.1.3 can be proved in terms of the Grössencharacters [259, Chapter II, § 10].

Guide to the literature

The Langlands Program is a series of influental conjectures linking arithmetic geometry and representation theory [145, 146]. We refer the reader to the excellent survey by Gelbart [91]. Our exposition follows [202].

8.2 K-theory of rational quadratic forms

The binary quadratic forms $q(u,v) = \{au^2 + buv + cv^2 \mid a,b,c \in \mathbf{Z}\}$ were studied by C. F. Gauss. Recall that two forms q and q' are said to be equivalent if a substitution $\{u = \alpha u' + \beta v',\ v = \gamma u' + \delta v' \mid \alpha,\beta,\gamma,\delta \in \mathbf{Z},\ \alpha\delta - \beta\gamma = 1\}$ transforms q into q'. It is easy to see that the discriminant $\Delta = b^2 - 4ac$ of the form $q(u,v)$ is an invariant of the substitution and, therefore, equivalent forms have the same discriminant. But the converse is false, in general. Gauss showed that there exists a finite number of the pairwise nonequivalent binary quadratic forms having the same discriminant. Moreover, the equivalence classes make an abelian group $\mathcal{G}$ under a composition defined on these forms [50, Chapter 14]. The cardinality $g = |\mathcal{G}|$ of such a group is called the genus of $q(u,v)$. An extension of the Gauss composition to the general quadratic forms $q(x) := q(x_1,\dots,x_n) = \{\sum_{i=1}^n \sum_{j=1}^n a_{ij}x_ix_j \mid a_{ij} \in \mathbf{Z},\ n \geq 1\}$ is a difficult and important problem [27–30], see also the survey [31].

Let $G(K)$ be a noncompact reductive algebraic group defined over a number field K. Denote by $\mathbb{A}$ the ring of adeles of K and by $\mathbb{A}_\infty$ a subring of the integer adeles. It is well known that $G(K)$ is a discrete subgroup of $G(\mathbb{A})$ and the double cosets $G(\mathbb{A}_\infty)\backslash G(\mathbb{A})/G(K)$ form a finite set. The set $G(\mathbb{A}_\infty)\backslash G(\mathbb{A})/G(K)$ is an important arithmetic invariant of the group $G(K)$. For instance, if $G \cong O(q)$ is the orthogonal group of a quadratic form, then $|G(\mathbb{A}_\infty)\backslash G(\mathbb{A})/G(K)| = g$ [232, Chapter 8]. Consider a Banach algebra $L^1(G(\mathbb{A})/G(K))$ of the integrable complex-valued functions on the homogeneous space $G(\mathbb{A})/G(K)$ endowed with the operator norm. Recall that the addition of functions $f_1, f_2 \in L^1(G(\mathbb{A})/G(K))$ is defined pointwise and multiplication is given by the convolution product

$$(f_1 * f_2)(g) = \int\limits_{G(\mathbb{A})/G(K)} f_1(gh^{-1})f_2(h)dh.$$

By $\mathcal{A}$ we understand an enveloping C^*-algebra of the algebra $L^1(G(\mathbb{A})/G(K))$; we refer the reader to [69, Section 13.9] for details of this construction.

In this section we construct a map from the set $G(\mathbb{A}_\infty)\backslash G(\mathbb{A})/G(K)$ to the K-theory of the C^*-algebra $\mathcal{A}$. Since the K-groups are abelian, one gets the structure of an abelian group on the set $G(\mathbb{A}_\infty)\backslash G(\mathbb{A})/G(K)$. As an application, we consider the case $G \cong O(q)$, where $O(q)$ is the orthogonal group of the rational quadratic form $q(x)$. In this case, one gets a higher composition law for the quadratic forms and a formula for their genera. To formulate our results, we shall need the following definitions.

It is known that $\mathcal{A}$ is a stationary AF-algebra of rank $n = \mathrm{rk}\, G(K)$, see Lemma 8.1.1. Recall that such an algebra can be given by an $n \times n$ integer matrix A with $\det A = 1$ (Section 3.5.2). Denote by $\mathcal{A} \rtimes_\sigma \mathbf{Z}$ the crossed product of the C^*-algebra $\mathcal{A}$ by the shift automorphism σ of $\mathcal{A}$ [32, Exercise 10.11.9]. It is known that $K_0(\mathcal{A} \rtimes_\sigma \mathbf{Z}) \cong \mathbf{Z}^n/(I - A)\mathbf{Z}^n$ and $|K_0(\mathcal{A} \rtimes_\sigma \mathbf{Z})| = |\det(I - A)|$, where I is the identity matrix and $K_0(\mathcal{A} \rtimes_\sigma \mathbf{Z})$ is the

K_0-group of the algebra $\mathcal{A} \rtimes_\sigma \mathbf{Z}$, see Theorems 3.7.1 and 3.7.2. Our main results can be formulated as follows.

Theorem 8.2.1. *There exists a one-to-one map*

$$\phi : G(\mathbb{A}_\infty)\backslash G(\mathbb{A})/G(K) \to K_0(\mathcal{A} \rtimes_\sigma \mathbf{Z}).$$

In particular, the map ϕ defines the structure of an abelian group on the set $G(\mathbb{A}_\infty)\backslash G(\mathbb{A})/G(K)$, so that ϕ becomes an isomorphism of the abelian groups.

Corollary 8.2.1. *If $G \cong O(q)$ is the orthogonal group of a rational quadratic form q, then ϕ defines a composition law for the equivalence classes of the quadratic forms. In particular, the genus g of the quadratic form q is given by the formula*

$$g = |\det(I - A)|.$$

8.2.1 Algebraic groups over adeles

We briefly review basic facts on the algebraic groups following [36] and [232].

8.2.1.1 Algebraic groups

An algebraic group is an algebraic variety G together with (i) an element $e \in G$, (ii) a morphism $\mu : G \times G \to G$ given by the formula $(x, y) \mapsto xy$, and (iii) a morphism $i : G \to G$ given by the formula $x \mapsto x^{-1}$ with respect to which the set G is a group. If G is a noncompact variety, the algebraic group G is said to be noncompact. Variety G is called a K-group if G is a variety defined over the field K and if μ and i are defined over K. In what follows, we deal with the linear algebraic groups, i. e., the subgroups of the general linear group GL_n. An algebraic group G is called reductive if the unipotent radical of G is trivial. An informal equivalent definition says that G is reductive if and only if a representation of G is a direct sum of the irreducible representations.

8.2.1.2 Orthogonal group of a quadratic form

Let $q(x) = \{\sum_{i=1}^n \sum_{j=1}^n a_{ij}x_ix_j \mid a_{ij} \in \mathbf{Z},\ n \geq 1\}$ be a quadratic form. Roughly speaking, the orthogonal group is a subgroup of the GL_n which preserves the form $q(x)$. Namely, let $A = (a_{ij})$ be a symmetric matrix attached to the quadratic form $q(x)$. The $O(q) = \{g \in \mathrm{GL}_n \mid gAg = A\}$ is called an orthogonal group and $\mathrm{SO}(q) = \{g \in O(q) \mid \det g = 1\}$ is called a special orthogonal group of the form $q(x)$.

8.2.1.3 Ring of adeles

The adeles is a powerful tool describing the Artin reciprocity for the abelian extensions of a number field K. In intrinsic terms, the ring of adeles $\mathbb{A}$ of a field K is a subset of the direct product $\prod K_v$ taken over almost all places K_v of K endowed with the natural topology. The embeddings $K \hookrightarrow K_v$ induce a discrete diagonal embedding $K \hookrightarrow \mathbb{A}$; the image of such is a ring of the principal adeles of $\mathbb{A}$. The ring of integral adeles $\mathbb{A}_\infty := \prod \mathcal{O}_v$, where $\mathcal{O}_v$ is a localization of the ring $\mathcal{O}_K$ of the integers of the field K. The Artin reciprocity says that there exists a continuous homomorphism $\mathbb{A}^\times \to \mathrm{Gal}(K^{ab}|K)$, where $\mathbb{A}^\times$ is a group of the invertible adeles (the idele group) and $\mathrm{Gal}(K^{ab}|K)$ is the absolute Galois group of the abelian extensions of K endowed with a profinite topology. On the other hand, there exists a canonical isomorphism $\mathbb{A}_\infty^\times \backslash \mathbb{A}^\times / K^\times \to \mathrm{Cl}(K)$, where $\mathbb{A}_\infty^\times$ ($\mathbb{A}^\times$ and $K^\times$, resp.) is a group of units of the ring $\mathbb{A}_\infty$ ($\mathbb{A}$ and $\mathcal{O}_K$, resp.) and $\mathrm{Cl}(K)$ is the ideal class group of the field K.

8.2.1.4 Algebraic groups over adeles

The algebraic groups over the ring of adeles can be viewed as an analog of the Artin recipricity for the nonabelian extensions of the field K and, therefore, we deal with a noncommutative arithmetic [232, p. 243]. Such groups are a starting point of the Langlands Program [146].

Let G be an algebraic group. It is known that the double cosets $G(\mathbb{A}_\infty)\backslash G(\mathbb{A})/G(K)$ form a finite set [36]. In particular, if $G \cong GL_n$ then $|G(\mathbb{A}_\infty)\backslash G(\mathbb{A})/G(K)| = h_K$ and if $G \cong O(q)$ then $|G(\mathbb{A}_\infty)\backslash G(\mathbb{A})/G(K)| = g$, where $h_K = |\mathrm{Cl}(K)|$ is the class number of the field K and g is the genus of quadratic form $q(x)$, respectively. The cardinality of the set $G(\mathbb{A}_\infty)\backslash G(\mathbb{A})/G(K)$ is a difficult open problem.

8.2.2 Proofs

8.2.2.1 Proof of Theorem 8.2.1

We shall split the proof in a series of lemmas.

Lemma 8.2.1. *Consider the group* $\mathrm{Ext}(\mathcal{A} \rtimes_\sigma \mathbf{Z})$ *consisting of the equivalence classes of extensions*

$$1 \to \mathbf{C} \to E \to \mathcal{A} \rtimes_\sigma \mathbf{Z} \to 1$$

of the $\mathcal{A} \rtimes_\sigma \mathbf{Z}$ *by the complex numbers* $\mathbf{C}$. *Then* $\mathrm{Ext}(\mathcal{A} \rtimes_\sigma \mathbf{Z}) \cong K_0(\mathcal{A} \rtimes_\sigma \mathbf{Z})$.

Proof. The result follows from the equivalence of the **Ext** and the $\mathbf{K}_0$-functors on the category of C^*-algebras. The special case of the C^*-algebra $\mathcal{A} \rtimes_\sigma \mathbf{Z} := O_A$ can be found in [32, Section 16.4.5]. □

Lemma 8.2.2. *Denote by* $E_0 \cong (\mathcal{A} \rtimes_\sigma \mathbf{Z}) \oplus \mathbf{C}$ *the trivial extension of the crossed product* $\mathcal{A} \rtimes_\sigma \mathbf{Z}$ *corresponding to the null element of the group* $\mathrm{Ext}(\mathcal{A} \rtimes_\sigma \mathbf{Z})$. *Then* E_0 *is a noncommutative coordinate ring of the group* $G(\mathbb{A})$, *i. e., there exists a covariant functor between* *the respective categories, such that the morphisms of $G(\mathbb{A})$ correspond to the morphisms of E_0.*

Proof. (i) It follows from the proof of Lemma 8.1.1 that the AF-algebra $\mathcal{A}$ is a noncommutative coordinate ring of the group $G(\mathbb{A})$. Consider the crossed product $\mathcal{A} \rtimes_\sigma \mathbf{Z}$ by the shift automorphism σ of $\mathcal{A}$. It is well known that there exists a canonical embedding of the C^*-algebras

$$\mathcal{A} \hookrightarrow \mathcal{A} \rtimes_\sigma \mathbf{Z}.$$

Thus one can extend each morphism of $\mathcal{A}$ to such of the crossed product $\mathcal{A} \rtimes_\sigma \mathbf{Z}$. Such an extension is well defined and unique. Thus one gets a functor between the category of groups $G(\mathbb{A})$ and the category of crossed products $\mathcal{A} \rtimes_\sigma \mathbf{Z}$. In other words, the $\mathcal{A} \rtimes_\sigma \mathbf{Z}$ is a noncommutative coordinate ring of the group $G(\mathbb{A})$.

(ii) On the other hand, each morphism h of $\mathcal{A} \rtimes_\sigma \mathbf{Z}$ can be extended to $E_0 \cong (\mathcal{A} \rtimes_\sigma \mathbf{Z}) \oplus \mathbf{C}$ by the formula

$$h(E_0) = h(\mathcal{A} \rtimes_\sigma \mathbf{Z}) \oplus \mathbf{C}.$$

Therefore we get a functor from the category of groups $G(\mathbb{A})$ and such of the C^*-algebras $(\mathcal{A} \rtimes_\sigma \mathbf{Z}) \oplus \mathbf{C}$. In other words, the E_0 is a noncommutative coordinate ring of the group $G(\mathbb{A})$. □

Lemma 8.2.3. *There exists a one-to-one map*

$$\phi : G(\mathbb{A}_\infty)\backslash G(\mathbb{A})/G(K) \to K_0^{\mathbf{set}}(\mathcal{A} \rtimes_\sigma \mathbf{Z}),$$

where $K_0^{\mathbf{set}}(\mathcal{A} \rtimes_\sigma \mathbf{Z})$ is the set of elements of the group $K_0(\mathcal{A} \rtimes_\sigma \mathbf{Z})$.

Proof. Consider an exact sequence of the C^*-algebras

$$1 \to \mathbf{C} \to E_0 \to \mathcal{A} \rtimes_\sigma \mathbf{Z} \to 1,$$

where $E_0 \cong (\mathcal{A} \rtimes_\sigma \mathbf{Z}) \oplus \mathbf{C}$. This exact sequence gives rise to an exact sequence of the K_0-groups

$$1 \to \mathbf{Z} \to K_0(E_0) \to K_0(\mathcal{A} \rtimes_\sigma \mathbf{Z}) \to 1,$$

where $K_0(\mathbf{C}) \cong \mathbf{Z}$.

In view of Lemma 8.2.2, we have a functor F acting by the formula $G(\mathbb{A}) \mapsto E_0$. On the other hand, the K_0 is a covariant functor from the category of C^*-algebras to the category of abelian groups. Thus, the composition $K_0 \circ F$ defines a functor acting by the formula $G(\mathbb{A}) \mapsto K_0(E_0)$.

$$\begin{array}{ccc} G(\mathbb{A}) & \longrightarrow & G(\mathbb{A}_\infty)\backslash G(\mathbb{A})/G(K) \\ \Big\downarrow K_0 \circ F & & \Phi \Big\downarrow \\ K_0(E_0) & \longrightarrow & K_0^{\mathbf{set}}(\mathcal{A} \rtimes_\sigma \mathbf{Z}) \end{array}$$

Figure 8.2: Double cosets $G(\mathbb{A}_\infty)\backslash G(\mathbb{A})/G(K)$.

It follows from the above argument that one can define a map $K_0(E_0) \to K_0^{\mathbf{set}}(\mathcal{A} \rtimes_\sigma \mathbf{Z})$ from the abelian group to a finite set. On the other hand, we have a double coset map $G(\mathbb{A}) \to G(\mathbb{A}_\infty)\backslash G(\mathbb{A})/G(K)$ from an algebraic group to a finite set, see Section 8.2.1.4. Bringing these facts together, one gets a commutative diagram in Fig. 8.2.

It remains to notice that Φ is a functor from the category of finite sets to itself. This observation implies that Φ is a trivial functor, i. e., $|G(\mathbb{A}_\infty)\backslash G(\mathbb{A})/G(K)| = |K_0^{\mathbf{set}}(\mathcal{A} \rtimes_\sigma \mathbf{Z})|$. In particular, there exists a one-to-one map $\phi : G(\mathbb{A}_\infty)\backslash G(\mathbb{A})/G(K) \to K_0^{\mathbf{set}}(\mathcal{A} \rtimes_\sigma \mathbf{Z})$ between the corresponding finite sets. Lemma 8.2.3 follows. □

Corollary 8.2.2. *The map ϕ^{-1} defines the structure of an abelian group on the set $G(\mathbb{A}_\infty)\backslash G(\mathbb{A})/G(K)$, thus extending ϕ to an isomorphism of the abelian groups.*

Proof. Recall that

$$K_0(\mathcal{A} \rtimes_\sigma \mathbf{Z}) \cong \frac{\mathbf{Z}^n}{(I - A)\mathbf{Z}^n} \cong \frac{\mathbf{Z}}{p_1^{n_1}\mathbf{Z}} \oplus \cdots \oplus \frac{\mathbf{Z}}{p_k^{n_k}\mathbf{Z}},$$

where A is an $n \times n$ integer matrix defining the AF-algebra $\mathcal{A}$, p_i are prime and n_i are positive integer numbers. One can take a generator x_i in each cyclic group $\mathbf{Z}/p_k^{n_k}\mathbf{Z}$ and let $\phi^{-1}(x_i)$ be a generator of the abelian group structure on the set $G(\mathbb{A}_\infty)\backslash G(\mathbb{A})/G(K)$. It is clear that the map ϕ defines an isomorphism between the two abelian groups. Corollary 8.2.2 is proved. □

Theorem 8.2.1 follows form Lemma 8.2.3 and Corollary 8.2.2.

8.2.2.2 Proof of Corollary 8.2.1

Recall that if $G \cong O(q)$ is the orthogonal group of a quadratic form $q(x)$, then the equivalence classes of $q(x)$ correspond one-to-one to the double cosets $G(\mathbb{A}_\infty)\backslash G(\mathbb{A})/G(K)$. In view of Corollary 8.2.2, the set $G(\mathbb{A}_\infty)\backslash G(\mathbb{A})/G(K)$ has the natural structure of an abelian group. In particular, the group operation defines a composition law for the equivalence classes of quadratic form $q(x)$. The first part of Corollary 8.2.1 is proved.

To express the genus g of the form $q(x)$ in terms of an invariant of the algebra $\mathcal{A}$, recall that $g = |G(\mathbb{A}_\infty)\backslash G(\mathbb{A})/G(K)|$. But $|G(\mathbb{A}_\infty)\backslash G(\mathbb{A})/G(K)| = |K_0(\mathcal{A} \rtimes_\sigma \mathbf{Z})|$ and one gets

$$g = |K_0(\mathcal{A} \rtimes_\sigma \mathbf{Z})| = \left|\frac{\mathbf{Z}^n}{(I-A)\mathbf{Z}^n}\right| = \left|\frac{\mathbf{Z}}{p_1^{n_1}\mathbf{Z}} \oplus \cdots \oplus \frac{\mathbf{Z}}{p_k^{n_k}\mathbf{Z}}\right|$$
$$= p_1^{n_1} \cdots p_k^{n_k} = |\det(I-A)|.$$

The genus formula of Corollary 8.2.1 follows from the above equations.

8.2.3 Binary quadratic forms

Let $n = 2$ and $\Delta > 0$ be a square-free integer, i. e., $q_\Delta(x)$ is an indefinite binary quadratic form. In view of Corollary 8.2.1, we are looking for a matrix A with $\det A = 1$ such that

$$g(q_{f^2\Delta}(x)) = |\det(I - A^k)|,$$

where $f \geq 1$ is a conductor of the quadratic form and $k \geq 1$ is an integer.

Remark 8.2.1. The left- and right-hand sides of the genus equation depend only on the integers f and k, respectively. Indeed, the left-hand side depends only on the conductor f if the discriminant Δ is a fixed square-free integer. For the right-hand side, we have the easily verified equalities $|\det(I - A^k)| = \operatorname{tr}(A^k) - 2 = 2T_k(\frac{1}{2}\operatorname{tr}(A)) - 2 = 2T_k(\frac{1}{2}\sqrt{\Delta+4}) - 2$, where $T_k(x)$ is the Chebyshev polynomial of degree k.

Let f and k be the least integers satisfying the genus equation. Recall that the equivalence classes of the quadratic form $q_{f^2\Delta}(x)$ correspond one-to-one to the (narrow) ideal classes of the order $R_f := \mathbf{Z} + fO_K$ in the real quadratic field $K := \mathbf{Q}(\sqrt{\Delta})$. The number h_{R_f} of such classes is calculated by the formula

$$h_{R_f} = h_K \frac{f}{e_f} \prod_{p|f} \left(1 - \left(\frac{\Delta}{p}\right)\frac{1}{p}\right),$$

where h_K is the class number of the field K, e_f is the index of the group of units of the order R_f in the group of units of the ring O_K, p is a prime number, and $(\frac{\Delta}{p})$ is the Legendre symbol [111, pp. 297 and 351]. Thus the genus of the binary quadratic form of discriminant $\Delta > 0$ is given by the formula

$$g(q_\Delta(x)) = \frac{|\det(I - A^k)|}{\frac{f}{e_f} \prod_{p|f}(1 - (\frac{\Delta}{p})\frac{1}{p})}.$$

Example 8.2.1. Let $G \cong O(q)$ be the orthogonal group of the quadratic form

$$q(u, v) = u^2 + 3uv + v^2.$$

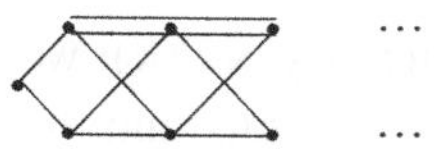

Figure 8.3: Bratteli diagram of the AF-algebra $\mathcal{A}$.

The discriminant is $\Delta = 5$ and therefore $q(u, v)$ is an indefinite quadratic form. The AF-algebra $\mathcal{A}$ corresponding to the group $G(\mathbb{A})$ is represented by the matrix

$$A = \begin{pmatrix} 2 & 1 \\ 1 & 1 \end{pmatrix}.$$

The Bratteli diagram of algebra $\mathcal{A}$ is shown in Fig. 8.3. To calculate the genus of $q(u, v)$, we shall use the genus formula with $f = k = 1$. Namely, one gets

$$g = |\det(I - A)| = \left|\det \begin{pmatrix} -1 & -1 \\ -1 & 0 \end{pmatrix}\right| = 1.$$

Guide to the literature

Our exposition is based on [208].

8.3 Quantum dynamics of elliptic curves

Let K be a number field and let $\mathcal{E}(K)$ be an elliptic curve over K. Here we consider a functor F between elliptic curves $\mathcal{E}(K)$ and the C^*-algebras $\mathcal{A}_{RM}$, i. e., noncommutative tori with real multiplication (Section 1.4.1). Such a functor maps K-isomorphic elliptic curves $\mathcal{E}(K)$ and $\mathcal{E}'(K)$ to isomorphic C^*-algebras $\mathcal{A}_{RM}$ and $\mathcal{A}'_{RM}$, respectively. It is useful to think of the $\mathcal{A}_{RM}$ as a noncommutative analog of the coordinate ring of $\mathcal{E}(K)$.

Recall that $\mathcal{E}(K)$ is an algebraic group over K; such a group is compact and thus abelian. Mordell–Weil Theorem says that $\mathcal{E}(K) \cong \mathbf{Z}^r \oplus \mathcal{E}_{tors}(K)$, where $r = \operatorname{rk} \mathcal{E}(K) \geq 0$ is the rank of $\mathcal{E}(K)$ and $\mathcal{E}_{tors}(K)$ is a finite abelian group. The group operation $\mathcal{E}(K) \times \mathcal{E}(K) \to \mathcal{E}(K)$ defines an action of the group $\mathcal{E}(K)$ by the K-automorphisms of $\mathcal{E}(K)$. Recall that each K-automorphism of $\mathcal{E}(K)$ gives rise to an automorphism of $\mathcal{A}_{RM}$. Thus one gets an action of $\mathcal{E}(K)$ on $\mathcal{A}_{RM}$ by automorphisms of the algebra $\mathcal{A}_{RM}$. The object of our study is a crossed product C^*-algebra coming from such an action, i. e., the C^*-algebra

$$\mathcal{A}_{RM} \rtimes \mathcal{E}(K).$$

The above crossed product is often identified with a dynamical system on the noncommutative topological space $\mathcal{A}_{RM}$. In other words, we deal with a "quantum" dynamical system; hence the title of this section. We refer the reader to [32, Chapter V] and [222] for an excellent introduction to the quantum dynamics.

The aim of this section is the K-theory of the crossed product $\mathcal{A}_{RM} \rtimes \mathcal{E}(K)$; we refer the reader to Theorem 8.3.1. We apply this result to evaluate the rank and the Shafarevich–Tate group of elliptic curve $\mathcal{E}(K)$; see Corollaries 8.3.1 and 8.3.2, respectively. To formalize our results, let us recall the following facts.

Denote by τ the canonical tracial state on the crossed product $\mathcal{A}_{RM} \rtimes \mathcal{E}(K)$; existence of τ follows from [228, Theorem 3.4]. It is well known that $K_0(\mathcal{A}_\theta) \cong \mathbf{Z}^2$. Moreover, τ defines an embedding $K_0(\mathcal{A}_\theta) \hookrightarrow \mathbf{R}$ given by the formula $\tau(K_0(\mathcal{A}_\theta)) = \mathbf{Z} + \theta\mathbf{Z} \subset \mathbf{R}$ [32, Exercise 10.11.6]. Following Yu. I. Manin, we shall call $\mathbf{Z} + \theta\mathbf{Z}$ a "pseudo-lattice". By Λ we denote the ring of endomorphisms of the pseudo-lattice $\tau(K_0(\mathcal{A}_{RM}))$. Since θ is an irrational quadratic number, the ring Λ is an order in the real quadratic field $k = \mathbf{Q}(\theta)$. Thus $\Lambda \cong \mathbf{Z} + fO_k$, where O_k is the ring of integers of k and $f \geq 1$ is a conductor of the order. We shall write $\mathrm{Cl}(\Lambda)$ to denote the class group of the ring Λ. By $h_\Lambda = |\mathrm{Cl}(\Lambda)|$ we denote the class number of Λ. Denote by $\mathcal{K}_{ab}$ the maximal abelian extension of the field k modulo conductor f. If $f = 1$, then the extension $\mathcal{K}_{ab}$ is unramified, i. e., the Hilbert class field of k. It is known that $\mathrm{Gal}(\mathcal{K}_{ab}|k) \cong \mathrm{Cl}(\Lambda)$, where $\mathrm{Gal}(\mathcal{K}_{ab}|k)$ is the Galois group of the extension $k \subseteq \mathcal{K}_{ab}$. Let $\{\alpha_i \mid 1 \leq i \leq h_\Lambda\}$ be generators of the field $\mathcal{K}_{ab}$ such that α_i are conjugate algebraic numbers. Consider a normalization of α_i given by the formula $\{\lambda_i = \alpha_i \alpha_{h_\Lambda}^{-1} \mid 1 \leq i \leq h_\Lambda - 1\}$. Our main result can be formulated as follows.

Theorem 8.3.1. *The K-theory of the crossed product C^*-algebra $\mathcal{A}_{RM} \rtimes \mathcal{E}(K)$ is described by the following formulas:*

$$\begin{cases} K_0(\mathcal{A}_{RM} \rtimes \mathcal{E}(K)) \cong \mathbf{Z}^{h_\Lambda + 1}, \\ \tau(K_0(\mathcal{A}_{RM} \rtimes \mathcal{E}(K))) = \mathbf{Z} + \theta\mathbf{Z} + \lambda_1\mathbf{Z} + \cdots + \lambda_{h_\Lambda - 1}\mathbf{Z}. \end{cases}$$

Denote by $\mathrm{rk}\,\mathcal{E}(K)$ the rank of elliptic curve $\mathcal{E}(K)$. Let $Ш(\mathcal{E}(K))$ be the Shafarevich–Tate group of $\mathcal{E}(K)$. Theorem 8.3.1 implies the following formulas.

Corollary 8.3.1. $\mathrm{rk}\,\mathcal{E}(K) = h_\Lambda - 1$.

Corollary 8.3.2. $Ш(\mathcal{E}(K)) \cong \mathrm{Cl}(\Lambda) \oplus \mathrm{Cl}(\Lambda)$.

Remark 8.3.1. For the sake of simplicity, we treat the case of elliptic curves only. However, Theorem 8.3.1 and Corollaries 8.3.1, 8.3.2 can be extended to any abelian variety over a number field K, see Section 8.4 for the details.

Remark 8.3.2. It follows from Corollaries 8.3.1 and 8.3.2 that

$$|Ш(\mathcal{E}(K))| = (1 + \mathrm{rk}\,\mathcal{E}(K))^2.$$

It is hard to verify this formula directly, since the group $Ш(\mathcal{E}(K))$ is unknown for a single $\mathcal{E}(K)$ [274, p. 193]. Indirectly, one can predict the "analytic" values of $\mathrm{rk}\,\mathcal{E}(K)$ and $|Ш(\mathcal{E}(K))|$ assuming the Birch and Swinnerton-Dyer Conjecture [270]. While many of such values satisfy the above formula, the other do not [61]. We do not know an exact relation between the analytic values and those described by our formula.

8.3.1 C^*-dynamical systems

We briefly review the crossed product C^*-algebras, see also Section 3.2. Let $\mathcal{A}$ be a C^*-algebra and G a locally compact group. We shall consider a continuous homomorphism α from G to the group $\mathrm{Aut}\,\mathcal{A}$ of $*$-automorphisms of A endowed with the topology of pointwise norm-convergence. Roughly speaking, the idea of the crossed product construction is to embed $\mathcal{A}$ into a larger C^*-algebra in which the automorphism becomes the inner automorphism. A covariant representation of the triple $(\mathcal{A}, G, \alpha)$ is a pair of representations (π, ρ) of $\mathcal{A}$ and G on the same Hilbert space $\mathcal{H}$, such that $\rho(g)\pi(a)\rho(g)^* = \pi(\alpha_g(a))$ for all $a \in \mathcal{A}$ and $g \in G$. Each covariant representation of $(\mathcal{A}, G, \alpha)$ gives rise to a convolution algebra $C(G, \mathcal{A})$ of continuous functions from G to $\mathcal{A}$; the completion of $C(G, \mathcal{A})$ in the norm topology is a C^*-algebra $\mathcal{A} \rtimes_\alpha G$ called a crossed product of $\mathcal{A}$ by G. If α is a single automorphism of $\mathcal{A}$, one gets an action of $\mathbf{Z}$ on $\mathcal{A}$; the crossed product in this case is called simply the crossed product of $\mathcal{A}$ by α.

8.3.2 Abelian extensions of quadratic fields

Let D be a square-free integer and let $k = \mathbf{Q}(\sqrt{D})$ be a quadratic number field, i. e., an extension of degree two of the field of rationals. Denote by O_k the ring of integers of k and by Λ an order in O_k, i. e., a subring of the ring O_k containing 1. The order Λ can be written in the form $\Lambda = \mathbf{Z} + fO_k$, where the integer $f \geq 1$ is a conductor of Λ. If $f = 1$, then $\Lambda \cong O_k$ is the maximal order.

Denote by $\mathrm{Cl}(\Lambda)$ the ideal class group and by $h_\Lambda = |\mathrm{Cl}(\Lambda)|$ the class number of the ring Λ. If $\Lambda \cong O_k$, then h_Λ coincides with the class number h of the field k. The integer $h \leq h_\Lambda$ is always a divisor of h_Λ given by the formula

$$h_\Lambda = h\frac{f}{e_f}\prod_{p|f}\left(1 - \left(\frac{D}{p}\right)\frac{1}{p}\right),$$

where e_f is the index of the group of units of Λ in the group of units of O_k, p is a prime number, and $(\frac{D}{p})$ is the Legendre symbol.

Let $\mathcal{K}_{ab}$ be the maximal abelian extension of the field k modulo conductor $f \geq 1$. The class field theory says that

$$\mathrm{Gal}(\mathcal{K}_{ab}|k) \cong \mathrm{Cl}(\Lambda),$$

where $\mathrm{Gal}(\mathcal{K}_{ab}|k)$ is the Galois group of the extension $(\mathcal{K}_{ab}|k)$. The $\mathcal{K}_{ab}$ is the Hilbert class field (i. e., a maximal unramified abelian extension) of k if and only if $f = 1$.

For $D < 0$, an explicit construction of generators of the field $\mathcal{K}_{ab}$ is realized by elliptic curves with complex multiplication, see, e. g., [173, Theorem 6.10]. For $D > 0$, an explicit construction of generators of the field $\mathcal{K}_{ab}$ is realized by noncommutative tori with real multiplication (Theorem 6.4.1).

8.3.3 Shafarevich–Tate group of elliptic curves

The Shafarevich–Tate group $Ш(\mathcal{E}(K))$ is a measure of failure of the Hasse principle for the elliptic curve $\mathcal{E}(K)$. Recall that if $\mathcal{E}(K)$ has a K-rational point, then it has also a K_v-point for every completion K_v of the number field K. The converse of this statement is called the Hasse principle. In general, the Hasse principle fails for the elliptic curve $\mathcal{E}(K)$.

Denote by $H^1(K,\mathcal{E})$ the first Galois cohomology group of $\mathcal{E}(K)$ [258, Appendix B]. There exists a natural homomorphism

$$\omega : H^1(K,\mathcal{E}) \to \prod_v H^1(K_v,\mathcal{E}),$$

where $H^1(K_v,\mathcal{E})$ is the first Galois cohomology over the field K_v. The Shafarevich–Tate group of an elliptic curve $\mathcal{E}(K)$ is

$$Ш(\mathcal{E}(K)) := \operatorname{Ker}\omega.$$

The group $Ш(\mathcal{E}(K))$ is trivial if and only if elliptic curve $\mathcal{E}(K)$ satisfies the Hasse principle.

Remark 8.3.3. The Shafarevich–Tate group $Ш(A(K))$ of an abelian variety $A(K)$ over the number field K is defined similarly and has the same properties as $Ш(\mathcal{E}(K))$.

8.3.4 Proofs

8.3.4.1 Proof of Theorem 8.3.1

For the sake of clarity, let us outline the main ideas. Our proof is based on a "rigidity principle" for extensions of the pseudo-lattice $\mathbf{Z}+\theta\mathbf{Z}$ corresponding to the algebra $\mathcal{A}_{RM}$. Such a rigidity follows from the class field theory for the real quadratic field $k = \mathbf{Q}(\theta)$. Namely, the canonical embedding $\mathcal{A}_{RM} \hookrightarrow \mathcal{A}_{RM} \rtimes \mathcal{E}(K)$ implies an inclusion $K_0(\mathcal{A}_{RM}) \subseteq K_0(\mathcal{A}_{RM} \rtimes \mathcal{E}(K))$. Using the canonical tracial state τ on $\mathcal{A}_{RM} \rtimes \mathcal{E}(K)$ [228, Theorem 3.4], one gets an inclusion

$$\mathbf{Z} + \theta\mathbf{Z} \subseteq \lambda_1\mathbf{Z} + \cdots + \lambda_m\mathbf{Z},$$

where λ_i are generators of the pseudo-lattice $\tau(K_0(\mathcal{A}_{RM} \rtimes \mathcal{E}(K)))$. It is easy to see that each $\lambda_i \in \mathbf{R}$ is an integer algebraic number. But the crossed product $\mathcal{A}_{RM} \rtimes \mathcal{E}(K)$ depends solely on the algebra $\mathcal{A}_{RM}$. Therefore the above field extension must satisfy a "rigidity principle." In other words, the arithmetic of the number field $k(\lambda_i)$ is controlled by such of the field k. It is well known that this happens if and only if $k(\lambda_i) \cong \mathcal{K}_{ab}$, where $\mathcal{K}_{ab}$ is the maximal abelian extension of the field k modulo conductor $f \geq 1$. Thus

$m = 1 + h_\Lambda$, where h_Λ is the class number of the order $\Lambda \subseteq O_k$. We pass to a detailed argument by splitting the proof in a series of lemmas and corollaries.

Lemma 8.3.1. *The real numbers λ_i are algebraic integers.*

Proof. Recall that the endomorphism ring Λ of the pseudo-lattice $\mathbf{Z} + \theta\mathbf{Z}$ is an order $\mathbf{Z}+fO_k$ in the number field k. In particular, since $f \neq 0$ we conclude that Λ is a nontrivial ring, i. e., $\Lambda \not\equiv \mathbf{Z}$. On the other hand, the inclusion $\mathbf{Z} + \theta\mathbf{Z} \subseteq \lambda_1\mathbf{Z} + \cdots + \lambda_m\mathbf{Z}$ implies that

$$\Lambda \subseteq \mathrm{End}(\lambda_1\mathbf{Z} + \cdots + \lambda_m\mathbf{Z}),$$

where $\mathrm{End}(\lambda_1\mathbf{Z} + \cdots + \lambda_m\mathbf{Z})$ is the endomorphism ring of the pseudo-lattice $\lambda_1\mathbf{Z} + \cdots + \lambda_m\mathbf{Z} \subset \mathbf{R}$. We conclude that the ring $\mathrm{End}(\lambda_1\mathbf{Z} + \cdots + \lambda_m\mathbf{Z})$ is nontrivial, i. e., bigger than the ring $\mathbf{Z}$.

Recall that the endomorphisms of pseudo-lattice $\lambda_1\mathbf{Z} + \cdots + \lambda_m\mathbf{Z} \subset \mathbf{R}$ coincide with multiplication by the real numbers. In other words, the ring $\mathrm{End}(\lambda_1\mathbf{Z} + \cdots + \lambda_m\mathbf{Z})$ is the coefficient ring of a $\mathbf{Z}$-module $\lambda_1\mathbf{Z} + \cdots + \lambda_m\mathbf{Z} \subset \mathbf{R}$ [38, p. 87]. Up to a multiple, any such ring must be an order in a real number field $\mathcal{K}$. Thus we have a field extension $\mathcal{K}|k$ and the following inclusions:

$$\Lambda \subseteq O_k \subseteq O_{\mathcal{K}},$$

where $O_{\mathcal{K}}$ is the ring of integers of the field $\mathcal{K}$.

On the other hand, it is known that the full $\mathbf{Z}$-module $\lambda_1\mathbf{Z} + \cdots + \lambda_m\mathbf{Z}$ is contained in its coefficient ring $O_{\mathcal{K}}$ [38, Lemma 1, p. 88]. In particular, each λ_i is an algebraic integer. Lemma 8.3.1 is proved. □

Remark 8.3.4. It is useful to scale the RHS of the above inclusion dividing it by the real number $\lambda_m \neq 1$. Such a normalization is always possible, since the embedding $\tau : K_0(\mathcal{A}_{RM} \rtimes \mathcal{E}(K)) \to \mathbf{R}$ is defined up to a scalar multiple. Thus we can rewrite our inclusion in the form

$$\tau(K_0(\mathcal{A}_{RM} \rtimes \mathcal{E}(K))) = \mathbf{Z} + \theta\mathbf{Z} + \lambda_1\mathbf{Z} + \cdots + \lambda_{m-1}\mathbf{Z}.$$

Lemma 8.3.2. *The number field $k(\lambda_1, \ldots, \lambda_m)$ is the maximal abelian extension of the field k modulo conductor $f \geq 1$.*

Proof. Let $\mathcal{E}(K)$ be an elliptic curve over the number field K and let $\mathcal{A}_{RM} = F(\mathcal{E}(K))$ be the corresponding noncommutative torus with real multiplication (Section 1.3). The functor F is faithful on the category of K-rational elliptic curves and therefore F has a correctly defined inverse F^{-1}. Thus $\mathcal{E}(K) = F^{-1}(\mathcal{A}_{RM})$ and one can write the crossed product in the form

$$\mathcal{A}_{RM} \rtimes F^{-1}(\mathcal{A}_{RM}).$$

Consider an endomorphism ring M of the pseudo-lattice $\lambda_1\mathbf{Z}+\cdots+\lambda_m\mathbf{Z} \subset \mathbf{R}$. Denote by $\mathcal{K} \cong M \otimes \mathbf{Q}$ a number field, such that M is an order in the ring of integers $O_{\mathcal{K}}$ of the field $\mathcal{K}$. In view of the above, one gets an inclusion $\Lambda \subseteq M$, where Λ is an order in the ring O_k. Since $\Lambda \subsetneq O_k$ and $M \subsetneq O_{\mathcal{K}}$, we get the inclusions

$$O_k \subseteq O_{\mathcal{K}} \quad \text{and} \quad k \subseteq \mathcal{K}.$$

On the other hand, it follows from the formulas above that the crossed product $\mathcal{A}_{RM} \rtimes \mathcal{E}(K)$ depends only on the inner structure of algebra $\mathcal{A}_{RM}$. The same is true for the inclusions of groups $K_0(\mathcal{A}_{RM}) \subsetneq K_0(\mathcal{A}_{RM} \rtimes \mathcal{E}(K))$, the inclusion of pseudo-lattices $\tau(K_0(\mathcal{A}_{RM})) \subsetneq \tau(K_0(\mathcal{A}_{RM} \rtimes \mathcal{E}(K))) \subset \mathbf{R}$ and the inclusion of rings $\operatorname{End}(\tau(K_0(\mathcal{A}_{RM}))) \subsetneq \operatorname{End}(\tau(K_0(\mathcal{A}_{RM} \rtimes \mathcal{E}(K))))$. In particular, the last inclusion says that arithmetic of the number field $\mathcal{K}$ is predetermined by the arithmetic of the field k. In other words, there exists an isomorphism

$$\operatorname{Gal}(\mathcal{K}|k) \cong \operatorname{Cl}(\Lambda),$$

where $\operatorname{Gal}(\mathcal{K}|k)$ is the Galois group of the extension $k \subseteq \mathcal{K}$. Therefore $\mathcal{K}$ is the maximal abelian extension of the field k modulo conductor $f \geq 1$ (Section 8.3.2).

Let us show that $\mathcal{K}$ is an extension of the real quadratic field k by the values $\{\lambda_i\}_{i=1}^m$. Indeed, since the coefficient ring of the full $\mathbf{Z}$-module $\lambda_1\mathbf{Z}+\cdots+\lambda_m\mathbf{Z}$ is isomorphic to the order $M \subsetneq O_{\mathcal{K}}$, we conclude that $\lambda_i \in O_{\mathcal{K}}$ [38, Lemma 1, p. 88]. Moreover, by a change of basis in the $\mathbf{Z}$-module, we can always arrange the generators $\{\lambda_i\}_{i=1}^m$ to be algebraically conjugate numbers of the field extension $\mathcal{K}|k$. In particular, one gets $\mathcal{K} = k(\lambda_1, \ldots, \lambda_m)$. Lemma 8.3.2 is proved. □

Corollary 8.3.3. *The cardinality of the set of generators λ_i is $m = h_\Lambda$.*

Proof. Indeed, since $\{\lambda_i\}_{i=1}^m$ are algebraically conjugate numbers of the field extension $\mathcal{K}|k$, we conclude that $m = |\operatorname{Gal}(\mathcal{K}|k)|$. But $\operatorname{Gal}(\mathcal{K}|k) \cong \operatorname{Cl}(\Lambda)$ and therefore $m = |\operatorname{Cl}(\Lambda)| = h_\Lambda$. Corollary 8.3.3 follows. □

Remark 8.3.5. To prove our results, we do not need an explicit formula for the values of generators λ_i in terms of $\theta \in k$; however, we refer an interested reader to Theorem 6.4.1 for such a formula.

Corollary 8.3.4. $\tau(K_0(\mathcal{A}_{RM} \rtimes \mathcal{E}(K))) = \mathbf{Z} + \theta\mathbf{Z} + \lambda_1\mathbf{Z} + \cdots + \lambda_{h_\Lambda - 1}\mathbf{Z}$.

Proof. The formula follows from Remark 8.3.4 and Corollary 8.3.3. □

Corollary 8.3.5. $K_0(\mathcal{A}_{RM} \rtimes \mathcal{E}(K)) \cong \mathbf{Z}^{h_\Lambda+1}$.

Proof. Indeed, the rank of the abelian group $K_0(\mathcal{A}_{RM} \rtimes \mathcal{E}(K))$ is equal to the number of generators of the pseudo-lattice $\tau(K_0(\mathcal{A}_{RM} \rtimes \mathcal{E}(K))) \subset \mathbf{R}$. It follows from Corollary 8.3.4 that such a number is equal to $h_\Lambda + 1$. Corollary 8.3.5 follows. □

Theorem 8.3.1 follows from Corollaries 8.3.4 and 8.3.5.

8.3.4.2 Proof of Corollary 8.3.1

Let $\mathcal{E}(K)$ be an elliptic curve over the number field K. Mordell–Weil Theorem says that $\mathcal{E}(K) \cong \mathbf{Z}^r \oplus \mathcal{E}_{tors}(K)$, where $r = \operatorname{rk} \mathcal{E}(K)$ and $\mathcal{E}_{tors}(K)$ is a finite abelian group.

Consider again the pseudo-lattice $\tau(K_0(\mathcal{A}_{RM} \rtimes \mathcal{E}(K))) \subset \mathbf{R}$ and substitute $\mathcal{E}(K) \cong \mathbf{Z}^r \oplus \mathcal{E}_{tors}(K)$ to obtain

$$\begin{aligned}\tau[K_0(\mathcal{A}_{RM} \rtimes \mathcal{E}(K))] &= \tau[K_0(\mathcal{A}_{RM} \rtimes (\mathbf{Z}^r \oplus \mathcal{E}_{tors}(K)))] \\ &= \tau[K_0(\mathcal{A}_{RM} \rtimes \mathbf{Z}^r) \oplus K_0(\mathcal{A}_{RM} \rtimes \mathcal{E}_{tors}(K))] \\ &= \tau[K_0(\mathcal{A}_{RM} \rtimes \mathbf{Z}^r)] + \tau[K_0(\mathcal{A}_{RM} \rtimes \mathcal{E}_{tors}(K))].\end{aligned}$$

In the last line, we have the following two terms:

(i) $\tau[K_0(\mathcal{A}_{RM} \rtimes \mathcal{E}_{tors}(K))] = \frac{1}{k}(\mathbf{Z} + \theta\mathbf{Z})$, where $k \geq 2$ is an integer depending on the order of finite group $\mathcal{E}_{tors}$; we refer the reader to [72, Theorem 0.1] for the proof of this fact.

(ii) $\tau[K_0(\mathcal{A}_{RM} \rtimes \mathbf{Z}^r)] = \mathbf{Z} + \theta\mathbf{Z} + \lambda_1\mathbf{Z} + \cdots + \lambda_r\mathbf{Z}$. For $r = 0$, this formula follows from [72, Theorem 0.1] after one rescales pseudo-lattice $\frac{1}{k}(\mathbf{Z} + \theta\mathbf{Z}) \subset \mathbf{R}$ to the pseudo-lattice $\mathbf{Z} + \theta\mathbf{Z} \subset \mathbf{R}$. For $r = 1$, the formula follows from [83, Proposition 19]. For $r \geq 2$, the formula is proved by an induction. Namely, it is verified directly that the case $i + 1$ adds an extra generator λ_{i+1} of the pseudo-lattice $\mathbf{Z} + \theta\mathbf{Z} + \lambda_1\mathbf{Z} + \cdots + \lambda_i\mathbf{Z}$ corresponding to the case i.

It follows from (i) and (ii) that after a scaling, one gets the following inclusion of the pseudo-lattices:

$$\tau[K_0(\mathcal{A}_{RM} \rtimes \mathcal{E}_{tors}(K))] \subseteq \tau[K_0(\mathcal{A}_{RM} \rtimes \mathbf{Z}^r)].$$

From formulas above, we get the following equality:

$$\tau[K_0(\mathcal{A}_{RM} \rtimes \mathcal{E}(K))] = \tau[K_0(\mathcal{A}_{RM} \rtimes \mathbf{Z}^r)].$$

Using calculations of item (ii), one obtains the following equality:

$$\mathbf{Z} + \theta\mathbf{Z} + \lambda_1\mathbf{Z} + \cdots + \lambda_{h_\Lambda - 1}\mathbf{Z} = \mathbf{Z} + \theta\mathbf{Z} + \lambda_1\mathbf{Z} + \cdots + \lambda_r\mathbf{Z}.$$

It is easy to see that above equation is solvable if and only if $r = h_\Lambda - 1$. In other words, $r := \operatorname{rk} \mathcal{E}(K) = h_\Lambda - 1$. Corollary 8.3.1 follows.

8.3.4.3 Proof of Corollary 8.3.2

Let us make general remarks and outline the main ideas of the proof. A relation between the quadratic number fields and the ranks of elliptic curves has been known for a while [98]. In fact, the famous Birch and Swinnerton-Dyer Conjecture uses the relation to compare (special values of) the Dirichlet L-functions of a number field with the Hasse–Weil L-function of an elliptic curve [270]. Let us mention a recent generalization of this idea by Bloch and Kato [33].

Our goal is to show that there exists a natural correspondence between the arithmetic of ideals of the real quadratic fields and the Hasse principle for elliptic curves. Namely, denote by $\mathcal{A}_{RM}^{(1)}, \dots, \mathcal{A}_{RM}^{(h_\Lambda)}$ the companion noncommutative tori corresponding to the $\mathcal{A}_{RM}$ (Section 6.4.3); simply speaking, these are pairwise nonisomorphic algebras $\mathcal{A}_{RM}^{(i)}$ such that $\operatorname{End} K_0(\mathcal{A}_{RM}^{(i)}) \cong \Lambda$ for all $1 \le i \le h_\Lambda$. Since the companion algebras $\mathcal{A}_{RM}^{(i)}$ have the same endomorphisms, so will be their "quantum dynamics", i. e., the crossed product $\mathcal{A}_{RM}^{(i)} \rtimes \mathcal{E}(K)$.

On the other hand, we establish a natural isomorphism between the abelian groups $K_0(\mathcal{A}_{RM}) \cong H^1(K, \mathcal{E})$ and $K_0(\mathcal{A}_{RM} \rtimes \mathcal{E}(K)) \cong \prod_v H^1(K_v, \mathcal{E})$. This means that the preimage of each cocycle in $\prod_v H^1(K_v, \mathcal{E})$ under the homomorphism ω consists of the $h_\Lambda \ge 1$ distinct cocycles of the $H^1(K, \mathcal{E})$. In other words, we get an inclusion $\mathrm{Cl}(\Lambda) \subset Ш(\mathcal{E}(K))$.

A precise formula is derived from Atiyah's pairing between the K-theory and the K-homology of C^*-algebras (Section 10.2). Namely, it is known that the $K^0(\mathcal{A}_\theta) \cong K_0(\mathcal{A}_\theta)$, where $K^0(\mathcal{A}_\theta)$ is the zero K-homology group of the noncommutative torus $\mathcal{A}_\theta$ [103, Proposition 4]. Repeating the argument for the group $K^0(\mathcal{A}_{RM})$, we get another subgroup $\mathrm{Cl}(\Lambda) \subset Ш(\mathcal{E}(K))$. In view of the Atiyah pairing, one gets $Ш(\mathcal{E}(K)) \cong \mathrm{Cl}(\Lambda) \oplus \mathrm{Cl}(\Lambda)$. We pass to a detailed argument by splitting the proof in a series of lemmas.

Lemma 8.3.3. *The $\mathcal{A}_{RM}^{(i)}$ and $\mathcal{A}_{RM}^{(j)}$ are companion noncommutative tori if and only if the crossed products $\mathcal{A}_{RM}^{(i)} \rtimes \mathcal{E}(K) \cong \mathcal{A}_{RM}^{(j)} \rtimes \mathcal{E}(K)$ are Morita equivalent C^*-algebras.*

Proof. (i) Let $\mathcal{A}_{RM}^{(i)}$ and $\mathcal{A}_{RM}^{(j)}$ be companion noncommutative tori. In this case we have

$$\operatorname{End} K_0(\mathcal{A}_{RM}^{(i)}) \cong \operatorname{End} K_0(\mathcal{A}_{RM}^{(j)}) \cong \Lambda.$$

Denote by $\widetilde{\operatorname{End}}(\mathcal{A}_{RM}^{(i)}) \cong \widetilde{\operatorname{End}}(\mathcal{A}_{RM}^{(j)})$ the pullback of the above isomorphism to the category of noncommutative tori.

Recall that the crossed product $\mathcal{A}_{RM}^{(i)} \rtimes \mathcal{E}(K)$ is an extension of the algebra $\mathcal{A}_{RM}^{(i)}$ by the elements $v \in \mathcal{A}_{RM}^{(i)} \rtimes \mathcal{E}(K)$ such that each $\phi \in \widetilde{\operatorname{End}}(\mathcal{A}_{RM}^{(i)})$ becomes an inner endomorphism, i. e., $\phi(u) = v^{-1}uv$ for every $u \in \mathcal{A}_{RM}^{(i)}$. Therefore the algebras $\mathcal{A}_{RM}^{(i)}$ and $\mathcal{A}_{RM}^{(j)}$ must have $*$-isomorphic extensions. In other words, the corresponding crossed products $\mathcal{A}_{RM}^{(i)} \rtimes \mathcal{E}(K)$ and $\mathcal{A}_{RM}^{(j)} \rtimes \mathcal{E}(K)$ are isomorphic up to an adjustment of generators of the extension, i. e., the Morita equivalence. The "only if" part of lemma 8.3.3 is proved.

(ii) Let $\mathcal{A}_{RM}^{(i)} \rtimes \mathcal{E}(K) \cong \mathcal{A}_{RM}^{(j)} \rtimes \mathcal{E}(K)$ be Morita equivalent crossed products. Let us show that $\mathcal{A}_{RM}^{(i)}$ and $\mathcal{A}_{RM}^{(j)}$ are companion noncommutative tori. Indeed, our assumption implies immediately an isomorphism $\widetilde{\operatorname{End}}(\mathcal{A}_{RM}^{(i)}) \cong \widetilde{\operatorname{End}}(\mathcal{A}_{RM}^{(j)})$. We apply the K_0-functor and we get an isomorphism

$$\mathrm{End}\, K_0(\mathcal{A}_{RM}^{(i)}) \cong \mathrm{End}\, K_0(\mathcal{A}_{RM}^{(j)}) \cong \Lambda.$$

In other words, the $\mathcal{A}_{RM}^{(i)}$ and $\mathcal{A}_{RM}^{(j)}$ are companion algebras. The "if" part of Lemma 8.3.3 is proved. □

Lemma 8.3.4. *Let $H^1(K,\mathcal{E})$ and $H^1(K_v,\mathcal{E})$ be the first Galois cohomology over the field K and over the completion K_v of K, respectively. There exists a natural isomorphism between the following groups:*

$$\begin{cases} H^1(K,\mathcal{E}) \cong K_0(\mathcal{A}_{RM}), \\ \prod_v H^1(K_v,\mathcal{E}) \cong K_0(\mathcal{A}_{RM} \rtimes \mathcal{E}(K)). \end{cases}$$

Proof. (i) Let us show that $H^1(K,\mathcal{E}) \cong K_0(\mathcal{A}_{RM})$. Indeed, such an isomorphism is a special case of [176, Theorem 1.1] saying that $H^1(\mathrm{Gal}(\mathbf{C}|K), \mathrm{Aut}_{\mathbf{C}}^{ab}(V)) \cong (K_0(\mathcal{A}_V), K_0^+(\mathcal{A}_V))$. For that, one has to restrict to the case $V = \mathcal{E}(K)$ and notice that $\mathcal{A}_V = \mathcal{A}_{RM}$. On the other hand, since $\mathcal{E}(K)$ is an algebraic group, one gets $\mathrm{Aut}_{\mathbf{C}}^{ab}(\mathcal{E}(K)) \cong \mathcal{E}(K)$. The rest of the formula follows from the definition of the group $H^1(K,\mathcal{E})$.

(ii) Let us prove that $\prod_v H^1(K_v,\mathcal{E}) \cong K_0(\mathcal{A}_{RM} \rtimes \mathcal{E}(K))$. An idea of the proof is to construct an AF-algebra, $\mathbb{A}$, connected to the profinite group $\prod_v H^1(K_v,\mathcal{E})$. Next we show that the crossed product $\mathcal{A}_{RM} \rtimes \mathcal{E}(K)$ embeds into $\mathbb{A}$, so that $K_0(\mathcal{A}_{RM} \rtimes \mathcal{E}(K)) \cong K_0(\mathbb{A})$. The rest of the proof will follow from the properties of the AF-algebra $\mathbb{A}$. We pass to a detailed argument.

Recall that $\prod_v H^1(K_v,\mathcal{E})$ is a profinite group, i. e.,

$$\prod_v H^1(K_v,\mathcal{E}) \cong \varprojlim G_k,$$

where $G_k = \prod_{i=1}^k H^1(K_{v_i},\mathcal{E})$ is a finite group. Consider a group algebra

$$\mathbf{C}[G_k] \cong M_{n_1}^{(k)}(\mathbf{C}) \oplus \cdots \oplus M_{n_h}^{(k)}(\mathbf{C})$$

corresponding to G_k. Notice that the $\mathbf{C}[G_k]$ is a finite-dimensional C^*-algebra. The above inverse limit defines an ascending sequence of the finite-dimensional C^*-algebras:

$$\mathbb{A} := \varprojlim M_{n_1}^{(k)}(\mathbf{C}) \oplus \cdots \oplus M_{n_h}^{(k)}(\mathbf{C}).$$

In other words, the limit $\mathbb{A}$ is an AF-algebra, such that $K_0(\mathbb{A}) \cong \prod_v H^1(K_v,\mathcal{E})$.

To prove that $K_0(\mathcal{A}_{RM} \rtimes \mathcal{E}(K)) \cong K_0(\mathbb{A})$, we shall use the "rigidity principle" mentioned above. Namely, the extension $H^1(K,\mathcal{E}) \subset \prod_v H^1(K_v,\mathcal{E})$ is defined solely by the group $H^1(K,\mathcal{E})$ [258, Appendix B]. Since $H^1(K,\mathcal{E}) \cong K_0(\mathcal{A}_{RM})$ and $\prod_v H^1(K_v,\mathcal{E}) \cong K_0(\mathbb{A})$, we conclude that the extension $K_0(\mathcal{A}_{RM}) \subset K_0(\mathbb{A})$ is defined by the group $K_0(\mathcal{A}_{RM})$ alone. But the extension $K_0(\mathcal{A}_{RM}) \subset K_0(\mathcal{A}_{RM} \rtimes \mathcal{E}(K))$ is the only extension with such a property. Thus $K_0(\mathbb{A}) \cong K_0(\mathcal{A}_{RM} \rtimes \mathcal{E}(K))$ and the crossed product $\mathcal{A}_{RM} \rtimes \mathcal{E}(K)$ embeds into the AF-algebra $\mathbb{A}$.

To finish the proof of Lemma 8.3.4, we recall that $K_0(\mathbb{A}) \cong \prod_v H^1(K_v, \mathcal{E})$ and therefore $\prod_v H^1(K_v, \mathcal{E}) \cong K_0(\mathcal{A}_{RM} \rtimes \mathcal{E}(K))$. □

Lemma 8.3.5. $\text{Ш}(\mathcal{E}(K)) \cong \text{Cl}(\Lambda) \oplus \text{Cl}(\Lambda)$.

Proof. Let $\{\mathcal{A}_{RM}^{(i)}\}_{i=1}^{h_\Lambda}$ be the companion noncommutative tori of the $\mathcal{A}_{RM}$. Consider a group homomorphism

$$h : K_0(\mathcal{A}_{RM}) \to K_0(\mathcal{A}_{RM} \rtimes \mathcal{E}(K)),$$

induced by the natural embedding $\mathcal{A}_{RM} \hookrightarrow \mathcal{A}_{RM} \rtimes \mathcal{E}(K)$. It follows from Lemma 8.3.3 that

$$h(K_0(\mathcal{A}_{RM}^{(i)})) = \mathbf{Z} + \theta\mathbf{Z} \quad \text{for all } 1 \le i \le h_\Lambda.$$

In other words, one gets $\text{Ker}\, h \cong \text{Cl}(\Lambda)$, where $\text{Cl}(\Lambda)$ is the class group of the order Λ in the real quadratic field $\mathbf{Q}(\theta)$.

But $K_0(\mathcal{A}_{RM}) \cong H^1(K, \mathcal{E})$ and $K_0(\mathcal{A}_{RM} \rtimes \mathcal{E}(K)) \cong \prod_v H^1(K_v, \mathcal{E})$. Therefore the abelian group $\text{Cl}(\Lambda)$ is an obstacle to the Hasse principle for the elliptic curve $\mathcal{E}(K)$. In other words, $\text{Cl}(\Lambda) \subset \text{Ш}(\mathcal{E}(K))$.

To calculate an exact relation between the groups $\text{Cl}(\Lambda)$ and $\text{Ш}(\mathcal{E}(K))$, recall that the K-homology is the dual theory to the K-theory [32, Section 16.3]. Roughly speaking, cocycles in K-theory are represented by vector bundles. Atiyah proposed using elliptic operators to represent the K-homology cycles. An elliptic operator can be twisted by a vector bundle, and the Fredholm index of the twisted operator defines a pairing between the K-homology and the K-theory with values in $\mathbf{Z}$.

In particular, it is known that for the algebra $\mathcal{A}_\theta$, one has $K^0(\mathcal{A}_\theta) \cong K_0(\mathcal{A}_\theta)$, where $K^0(\mathcal{A}_\theta)$ is the zero K-homology group of $\mathcal{A}_\theta$ [103, Proposition 4]. Repeating the argument for the group $K^0(\mathcal{A}_{RM})$, one can prove an analog of Theorem 8.3.1 for such a group. In other words, we get another subgroup $\text{Cl}(\Lambda) \subset \text{Ш}(\mathcal{E}(K))$. Since there are no other duals to the K-theory of C^*-algebras, we conclude from the Atiyah pairing that $\text{Ш}(\mathcal{E}(K)) \cong \text{Cl}(\Lambda) \oplus \text{Cl}(\Lambda)$. Lemma 8.3.5 is proved. □

Corollary 8.3.2 follows from Lemma 8.3.5.

Remark 8.3.6. The reader can observe that the construction of a generator of $\mathcal{E}(K)$ is similar to the construction of an "ideal number" (i. e., a principal ideal) of the number field k. Namely, it is well known that not every ideal of the ring $\Lambda \subset O_k$ is principal; an obstruction is a nontrivial group $\text{Cl}(\Lambda)$. However, this can be repaired in a bigger field $\mathcal{K} = \mathcal{K}_{ab}$; there exists a finite extension $k \subseteq \mathcal{K}$ such that every ideal of Λ is principal in the ring $O_{\mathcal{K}}$. Likewise, one cannot construct a generator of $\mathcal{E}(K)$ by a *finite* descent in general; an obstruction is a nontrivial group $Ш(\mathcal{E}(K))$. However, in an extension $\mathcal{A}_{RM} \rtimes \mathcal{E}(K)$ of the coordinate ring $\mathcal{A}_{RM}$ of $\mathcal{E}(K)$, the descent will be always finite and give a generator of the $\mathcal{E}(K)$. Such an analogy explains the formula $Ш(\mathcal{E}(K)) \cong \mathrm{Cl}(\Lambda) \oplus \mathrm{Cl}(\Lambda)$ on an intuitive level. Notice also that the $\mathcal{A}_{RM} \rtimes \mathcal{E}(K)$ is the coordinate ring of an abelian variety $A(K)$, which is related to the Euler variety V_E coming from the continued fraction of θ (Section 6.2.1).

Guide to the literature

A relation between the quadratic number fields and the ranks of elliptic curves is well known [98]. The famous Birch and Swinnerton-Dyer Conjecture uses this relation to compare the special values of the Dirichlet L-functions of a number field with the Hasse–Weil L-function of an elliptic curve [270]. A generalization of this idea can be found in [33]. Our exposition follows [209].

8.4 Shafarevich–Tate groups of abelian varieties

The study of diophantine equations is the oldest part of mathematics. If such an equation has an integer solution, then using the reduction modulo any prime p, one gets a solution of the equation lying in the finite field $\mathbf{F}_p$ and a solution in the field of real numbers $\mathbf{R}$. The equation is said to satisfy the Hasse principle, if the converse is true. For instance, the quadratic equations satisfy the Hasse principle, while the equation $x^4 - 17 = 2y^2$ has a solution over $\mathbf{R}$ and each $\mathbf{F}_p$, but no rational solutions. Measuring the failure of the Hasse principle is a difficult open problem of number theory.

Let $\mathcal{A}_K$ be an abelian variety over the number field K which we assume to be simple, i. e., the $\mathcal{A}_K$ has no proper subabelian varieties over K. Denote by K_v the completion of K at the (finite or infinite) place v. Consider the Weil–Châtelet group $\mathrm{WC}(\mathcal{A}_K)$ of the abelian variety $\mathcal{A}_K$ and the group homomorphism

$$\mathrm{WC}(\mathcal{A}_K) \to \prod_v \mathrm{WC}(\mathcal{A}_{K_v}).$$

The Shafarevich–Tate group $Ш(\mathcal{A}_K)$ of $\mathcal{A}_K$ is defined as the kernel of the above homomorphism. The variety $\mathcal{A}_K$ satisfies the Hasse principle if and only if the group $Ш(\mathcal{A}_K)$ is trivial. Little is known about the $Ш(\mathcal{A}_K)$, in general. The existing methods include an evaluation of the analytic order of $Ш(\mathcal{A}_K)$ based on the second part of the Birch and Swinnerton-Dyer Conjecture and an exact calculation of the p-part of $Ш(\mathcal{A}_K)$ [247].

In this section we calculate the group $Ш(\mathcal{A}_K)$ based on a functor F between the n-dimensional abelian varieties $\mathcal{A}_K$ and the $2n$-dimensional noncommutative tori $\mathcal{A}_\Theta$, i. e., the C^*-algebras generated by the unitary operators $U_1, \dots, U_{2n}$ satisfying the commutation relations $\{U_jU_i = e^{2\pi i\theta_{ij}}U_iU_j \mid 1 \le i,j \le 2n\}$ described by a skew-symmetric matrix

$$\Theta = \begin{pmatrix} 0 & \theta_{12} & \dots & \theta_{1,2n} \\ -\theta_{12} & 0 & \dots & \theta_{2,2n} \\ \vdots & \vdots & \ddots & \vdots \\ -\theta_{1,2n} & -\theta_{2,2n} & \dots & 0 \end{pmatrix} \in GL_{2n}(\mathbf{R}).$$

The functor maps isomorphic abelian varieties $\mathcal{A}$ and $\mathcal{A}'$ to the Morita equivalent algebras $\mathcal{A}_\Theta = F(\mathcal{A})$ and $\mathcal{A}'_\Theta = F(\mathcal{A}')$ (Section 1.3). Restricting F to the simple abelian varieties $\mathcal{A}_K$, one gets the noncommutative tori $\mathcal{A}_{\Theta(k)} = F(\mathcal{A}_K)$, where $\Theta(k)$ is the matrix defined over a number field $k \subset \mathbf{R}$.

Roughly speaking, the idea is this. To calculate the $Ш(\mathcal{A}_K)$, we recast the homomorphism $WC(\mathcal{A}_K) \to \prod_v WC(\mathcal{A}_{K_v})$ in terms of the K-theory of algebra $\mathcal{A}_{\Theta(k)} = F(\mathcal{A}_K)$. Recall that an isomorphism class of $\mathcal{A}_{\Theta(k)}$ is defined by the triple $(K_0(\mathcal{A}_{\Theta(k)}), K_0^+(\mathcal{A}_{\Theta(k)}), \Sigma(\mathcal{A}_{\Theta(k)}))$ consisting of the K_0-group, the positive cone K_0^+ and the scale Σ of the algebra $\mathcal{A}_{\Theta(k)}$ [32, Section 6], see also Theorem 3.5.2. We prove that $\Sigma(\mathcal{A}_{\Theta(k)})$ is a torsion group such that $WC(\mathcal{A}_K) \cong \Sigma(\mathcal{A}_{\Theta(k)})$. The RHS $\prod_v WC(\mathcal{A}_{K_v})$ corresponds to the crossed product C^*-algebra $\mathcal{A}_{\Theta(k)} \rtimes_{L_v} \mathbf{Z}$. It is proved that $\prod_v WC(\mathcal{A}_{K_v}) \cong \prod_v K_0(\mathcal{A}_{\Theta(k)} \rtimes_{L_v} \mathbf{Z})$. Thus $WC(\mathcal{A}_K) \to \prod_v WC(\mathcal{A}_{K_v})$ can be written in the form

$$\Sigma(\mathcal{A}_{\Theta(k)}) \to \prod_v K_0(\mathcal{A}_{\Theta(k)} \rtimes_{L_v} \mathbf{Z}).$$

Both sides of the above equation are functions of a single positive matrix $B \in GL(2n, \mathbf{Z})$. However, the LHS depends on the similarity class of B, while the RHS depends on the characteristic polynomial of B. This observation is critical, since it puts the elements of $Ш(\mathcal{A}_K)$ into a one-to-one correspondence with the similarity classes of matrices having the same characteristic polynomial. The latter is an old problem of linear algebra and it is known that the number of such classes is finite. They correspond to the ideal classes $Cl(\Lambda)$ of an order Λ in the field $k = \mathbf{Q}(\lambda_B)$, where λ_B is the Perron–Frobenius eigenvalue of matrix B [147]. Let $Cl(\Lambda) \cong (\mathbf{Z}/2^k\mathbf{Z}) \oplus Cl_{\mathbf{odd}}(\Lambda)$ for $k \geq 0$. Using the Atiyah pairing between the K-theory and K-homology, one gets the following result.

Theorem 8.4.1.

$$Ш(\mathcal{A}_K) \cong \begin{cases} Cl(\Lambda) \oplus Cl(\Lambda), & \textit{if } k \textit{ is even}, \\ (\mathbf{Z}/2^k\mathbf{Z}) \oplus Cl_{\mathbf{odd}}(\Lambda) \oplus Cl_{\mathbf{odd}}(\Lambda), & \textit{if } k \textit{ is odd}. \end{cases}$$

8.4.1 Abelian varieties

Let $\mathbf{C}^n$ be the n-dimensional complex space and Λ be a lattice in the underlying $2n$-dimensional real space $\mathbf{R}^{2n}$. The complex torus $\mathbf{C}^n/\Lambda$ is said to be an n-dimensional (principally polarized) abelian variety $\mathcal{A}$ if it admits an embedding into a projective space. In other words, the set $\mathcal{A}$ is both an additive abelian group and a complex projective variety. Recall that the Siegel upper half-space $\mathbb{H}_n$ consists of the symmetric $n \times n$ matrices with complex entries τ_i, such that $\mathbb{H}_n := \{\tau = (\tau_i) \in \mathbf{C}^{\frac{n(n+1)}{2}} \mid \Im(\tau_i) > 0\}$. The points of $\mathbb{H}_n$ parametrize the n-dimensional abelian varieties, i. e., $\mathcal{A} \cong \mathcal{A}_\tau$ for some $\tau \in \mathbb{H}_n$. Denote by $Sp(2n, \mathbf{R})$ the $2n$-dimensional symplectic group, i. e., a subgroup of the linear group defined by the equation

$$\begin{pmatrix} A & B \\ C & D \end{pmatrix}^T \begin{pmatrix} 0 & I \\ -I & 0 \end{pmatrix} \begin{pmatrix} A & B \\ C & D \end{pmatrix} = \begin{pmatrix} 0 & I \\ -I & 0 \end{pmatrix},$$

where A, B, C, D are $n \times n$ matrices with real entries and I is the identity matrix. It is well known that the abelian varieties $\mathcal{A}_\tau$ and $\mathcal{A}_{\tau'}$ are isomorphic whenever

$$\tau' = \frac{A\tau + B}{C\tau + D}, \quad \text{where} \begin{pmatrix} A & B \\ C & D \end{pmatrix} \in Sp(2n, \mathbf{Z}).$$

8.4.2 Weil–Châtelet group

Let $K \subset \mathbf{C}$ be a number field and let $\mathcal{A}_K$ be an abelian variety over K. The principal homogeneous space of $\mathcal{A}_K$ is a variety $\mathcal{C}$ over K with a map $\mu : \mathcal{C} \times \mathcal{A}_K \to \mathcal{C}$ satisfying (i) $\mu(x, 0) = x$, (ii) $\mu(x, a + b) = \mu(\mu(x, a), b)$ for all $x \in \mathcal{C}$ and $a, b \in \mathcal{A}_K$, and (iii) for all $x \in \mathcal{C}$ the map $a \mapsto \mu(x, a)$ is a K-isomorphism between $\mathcal{A}_K$ and $\mathcal{C}$. The homogeneous spaces $(\mathcal{C}, \mu)$ and $(\mathcal{C}', \mu')$ are equivalent if there is a K-isomorphism $\mathcal{C} \to \mathcal{C}'$ compatible with the maps μ and μ'. If $\mathcal{C}$ has K-points, then the equivalence class $(\mathcal{C}, \mu)$ is said to be trivial. The nontrivial classes $(\mathcal{C}, \mu)$ correspond to the varieties $\mathcal{C}$ without K-points, but with the L-points, where L is a Galois extension of K. The Weil–Châtelet group $WC(\mathcal{A}_K)$ is an additive abelian group of the equivalence classes $(\mathcal{C}, \mu)$, where the trivial class corresponds to the zero element of the group [49]. The $WC(\mathcal{A}_K)$ is a torsion group, i. e., each element of the group has finite order. It is not hard to see that if $K \subset K_v$ is an extension of K by completion at the place v, then there exists a natural group homomorphism

$$WC(\mathcal{A}_K) \to WC(\mathcal{A}_{K_v}).$$

The kernel of the homomorphism consists of those homogeneous spaces having the K_v-points, but no K-points. The Shafarevich–Tate group $Ш(\mathcal{A}_K)$ of $\mathcal{A}_K$ is defined as the kernel of homomorphism $WC(\mathcal{A}_K) \to \prod_v WC(\mathcal{A}_{K_v})$. Thus the abelian variety $\mathcal{A}_K$ satisfies the Hasse principle if and only if the group $Ш(\mathcal{A}_K)$ is trivial.

8.4.3 Localization formulas

Recall that the even-dimensional noncommutative torus $\mathcal{A}_\Theta$ is the universal C^*-algebra generated by the unitary operators $U_1, \dots, U_{2n}$ and satisfying the relations $\{U_j U_i = e^{2\pi i \theta_{ij}} U_i U_j \mid 1 \le i,j \le 2n\}$, where $\Theta = (\theta_{ij})$ is a skew-symmetric matrix. Let $SO(n,n;\mathbf{Z})$ be a subgroup of the group $GL(2n,\mathbf{Z})$ preserving the quadratic form $x_1x_{n+1} + x_2x_{n+2} + \dots + x_nx_{2n}$. The matrix $\left(\begin{smallmatrix} A & B \\ C & D \end{smallmatrix}\right) \in SO(n,n;\mathbf{Z})$ if and only if

$$A^TD + C^TB = I, \quad A^TC + C^TA = 0 = B^TD + D^TB,$$

where $A, B, C, D \in GL(n,\mathbf{Z})$ and I is the identity matrix. The noncommutative tori $\mathcal{A}_\Theta$ and $\mathcal{A}_{\Theta'}$ are Morita equivalent whenever

$$\Theta' = \frac{A\Theta + B}{C\Theta + D}, \quad \text{where } \begin{pmatrix} A & B \\ C & D \end{pmatrix} \in SO(n,n;\mathbf{Z}).$$

The reader can verify an inclusion $Sp(2n,\mathbf{Z}) \subseteq SO(n,n;\mathbf{Z})$. Such an inclusion is an isomorphism if and only if $n = 1$. This observation prompts the construction of a functor F from the n-dimensional abelian varieties $\mathcal{A}$ to the $2n$-dimensional noncommutative tori $\mathcal{A}_\Theta$ such that if $\mathcal{A}$ and $\mathcal{A}'$ are isomorphic abelian varieties, $\mathcal{A}_\Theta = F(\mathcal{A})$ and $\mathcal{A}'_\Theta = F(\mathcal{A}')$ will be the Morita equivalent noncommutative tori. The restriction of F to the simple abelian varieties $\mathcal{A}_K$ corresponds to the noncommutative tori $\mathcal{A}_{\Theta(k)} = F(\mathcal{A}_K)$, where $\Theta(k)$ is the matrix over a number field k of $\deg(k|\mathbf{Q}) = 2n$.

8.4.3.1 Crossed product $\mathcal{A}_{\Theta(k)} \rtimes_{L_v} \mathbf{Z}$

Let $\mathcal{A}_\Theta$ be a $2n$-dimensional noncommutative torus endowed with the canonical trace $\tau : \mathcal{A}_\Theta \to \mathbf{C}$. Since $K_0(\mathcal{A}_\Theta) \cong \mathbf{Z}^{2^{2n-1}}$ and $\tau_* : K_0(\mathcal{A}_\Theta) \to \mathbf{R}$ is a homomorphism induced by τ, one gets a $\mathbf{Z}$-module $\Lambda := \tau_*(K_0(\mathcal{A}_\Theta)) \subset \mathbf{R}$ of rank 2^{2n-1}. The generators of Λ belong to the ring $\mathbf{Z}[\theta_{ij}]$. In what follows, we assume that $\theta_{ij} \in k$. In this case one gets the algebraic constraints and the rank of Λ is equal to $2n$ (Remark 6.6.1). In other words,

$$\Lambda \cong \mathbf{Z} + \mathbf{Z}\theta_1 + \dots + \mathbf{Z}\theta_{2n-1}, \quad \text{where } \theta_i \in k.$$

Definition 8.4.1. We denote by $B \in GL(2n,\mathbf{Z})$ a positive matrix such that $(1, \theta_1, \dots, \theta_{2n-1})$ is the normalized Perron–Frobenius eigenvector of B.

Let $\pi(n)$ be an integer-valued function defined in Section 6.5.3. Consider the characteristic polynomial $\mathrm{Char}(B^{\pi(p)}) = x^{2n} - a_1x^{2n-1} - \dots - a_{2n-1}x - 1$ and a matrix

$$L_v = \begin{pmatrix} a_1 & 1 & 0 & \dots & 0 & 0 \\ a_2 & 0 & 1 & \dots & 0 & 0 \\ \vdots & \vdots & \vdots & \ddots & \vdots & \vdots \\ a_{2n-1} & 0 & 0 & \dots & 0 & 1 \\ p & 0 & 0 & \dots & 0 & 0 \end{pmatrix},$$

where p is the prime underlying v. Matrix L_v defines an endomorphism of the algebra $\mathcal{A}_{\Theta(k)}$ by its action on the generators $U_1, \dots, U_{2n}$. We consider the crossed product C^*-algebra $\mathcal{A}_{\Theta(k)} \rtimes_{L_v} \mathbf{Z}$ associated to such an action.

Remark 8.4.1. There exits an isomorphism

$$K_0(\mathcal{A}_{\Theta(k)} \rtimes_{L_v} \mathbf{Z}) \cong \mathcal{A}_{\mathbf{F}_p},$$

where $\mathcal{A}_{\mathbf{F}_p}$ is a localization of the $\mathcal{A}_K$ at the prime ideal $\mathcal{P} \subset K$ over p (Section 6.6). The above formula implies that the crossed product $\mathcal{A}_{\Theta(k)} \rtimes_{L_v} \mathbf{Z}$ is an analog of the variety $\mathcal{A}_{K_v}$.

8.4.4 Proof of Theorem 8.4.1

An outline of the proof was given at the top of Section 8.4. The detailed argument is given below. We shall split the proof in a series of lemmas.

Lemma 8.4.1. *The scale* $\Sigma(\mathcal{A}_{\Theta(k)})$ *has the structure of a torsion group, so that* $\mathrm{WC}(\mathcal{A}_K) \cong \Sigma(\mathcal{A}_{\Theta(k)})$, *where* $\mathcal{A}_{\Theta(k)} = F(\mathcal{A}_K)$.

Proof. (i) For the sake of clarity, we restrict to the case $n = 1$, i. e., when the variety $\mathcal{A}$ is an elliptic curve $\mathcal{E}$. In this case matrix Θ can be written as

$$\Theta = \begin{pmatrix} 0 & \theta \\ -\theta & 0 \end{pmatrix}.$$

We shall denote the corresponding noncommutative torus by $\mathcal{A}_\theta$. One gets $\deg(k|\mathbf{Q}) = 2$, i. e., k is a real quadratic field. Therefore, $F(\mathcal{E}_K) = \mathcal{A}_\theta$, where $\theta \in k$ is a quadratic irrationality.

To prove that the scale $\Sigma(\mathcal{A}_\theta)$ has the structure of a torsion group, we shall use the Minkowski question-mark function $?(x)$ [161, p. 172]. Such a function is known to map quadratic irrational numbers of the unit interval to the rational numbers of the same interval preserving their natural order. This observation implies $\Sigma(\mathcal{A}_\theta) \subset \mathbf{Q}/\mathbf{Z}$, i. e., the scale $\Sigma(\mathcal{A}_\theta)$ is a subgroup of the torsion group $\mathbf{Q}/\mathbf{Z}$. Let us pass to a detailed argument.

Let τ be the canonical trace on the algebra $\mathcal{A}_\theta$. Since $K_0(\mathcal{A}_\theta) \cong \mathbf{Z}^2$, one gets $\tau_*(K_0(\mathcal{A}_\theta)) = \mathbf{Z} + \mathbf{Z}\theta$ and $\tau_*(K_0^+(\mathcal{A}_\theta)) = \{m + n\theta \geq 0 \mid m, n \in \mathbf{Z}\}$. It is known that

$\tau_*(u) = 1$, where $u \in K_0^+(\mathcal{A}_\theta)$ is an order unit. Therefore the traces on the scale $\Sigma(\mathcal{A}_\theta)$ are given by the formula

$$\tau_*(\Sigma(\mathcal{A}_\theta)) = [0,1] \cap \mathbf{Z} + \mathbf{Z}\theta.$$

Recall that the Minkowski question-mark function is defined by the convergent series

$$?(x) := a_0 + 2\sum_{k=1}^{\infty} \frac{(-1)^{k+1}}{2^{a_1+\cdots+a_k}},$$

where $x = [a_0, a_1, a_2, \ldots]$ is the continued fraction of the irrational number x. The function $?(x) : [0,1] \to [0,1]$ is a monotone continuous function with the following properties: (i) $?(0) = 0$ and $?(1) = 1$; (ii) $?(\mathbf{Q}) = \mathbf{Z}[\frac{1}{2}]$ are dyadic rationals; and (iii) $?(\mathcal{Q}) = \mathbf{Q} - \mathbf{Z}[\frac{1}{2}]$, where $\mathcal{Q}$ are quadratic irrational numbers [161, p. 172].

Consider the subset $\tau_*(\Sigma(\mathcal{A}_\theta))$ of the interval $[0,1]$. Since θ is a quadratic irrationality, we conclude that $\tau_*(\Sigma(\mathcal{A}_\theta)) \subset \mathcal{Q}$. By property (iii) of the Minkowski question-mark function, one gets an inclusion $?(\tau_*(\Sigma(\mathcal{A}_\theta))) \subset \mathbf{Q}/\mathbf{Z}$. Thus we constructed a one-to-one map

$$\Sigma(\mathcal{A}_\theta) \to \mathcal{Y} \subset \mathbf{Q}/\mathbf{Z},$$

where $\mathcal{Y} :=?(\tau_*(\Sigma(\mathcal{A}_\theta)))$. It follows from the above formula that $\Sigma(\mathcal{A}_\theta)$ is a torsion group.

Remark 8.4.2. Formula $\Sigma(\mathcal{A}_\theta) \to \mathcal{Y} \subset \mathbf{Q}/\mathbf{Z}$ is part of a duality between the K-theory of noncommutative tori and the Galois cohomology of abelian varieties.

(ii) Let us show that $\mathrm{WC}(\mathcal{E}_K) \cong \Sigma(\mathcal{A}_\theta)$, where $\cong$ is an isomorphism of the torsion groups. Indeed, let $\mathcal{C}$ be the principal homogeneous space of the elliptic curve $\mathcal{E}_K$. It is known that $\mathcal{C} \cong \mathcal{E}_K'$ is the twist of $\mathcal{E}_K$, i. e., an isomorphic but not K-isomorphic elliptic curve $\mathcal{E}_K'$. It follows that the $\mathcal{A}_\theta$ and $\mathcal{A}_\theta' = F(\mathcal{E}_K')$ are Morita equivalent, but not isomorphic C^*-algebras. This would imply $(K_0(\mathcal{A}_\theta), K_0^+(\mathcal{A}_\theta)) \cong (K_0(\mathcal{A}_\theta'), K_0^+(\mathcal{A}_\theta'))$, but $(K_0(\mathcal{A}_\theta), K_0^+(\mathcal{A}_\theta), \Sigma(\mathcal{A}_\theta)) \ncong (K_0(\mathcal{A}_\theta'), K_0^+(\mathcal{A}_\theta'), \Sigma(\mathcal{A}_\theta'))$. In other words, the principal homogeneous spaces of elliptic curve $\mathcal{E}_K$ are classified by the scales of the algebra $\mathcal{A}_\theta$.

On the other hand, each element of $K_0^+(\mathcal{A}_\theta)$ can be taken for an order unit u of the dimension group $(K_0(\mathcal{A}_\theta), K_0^+(\mathcal{A}_\theta))$. Since $\Sigma(\mathcal{A}_\theta) = \{a \in K_0^+(\mathcal{A}_\theta) \mid 0 \le a \le u\}$, we conclude that the elements of $K_0^+(\mathcal{A}_\theta)$ classify all scales of the algebra $\mathcal{A}_\theta$. We can always restrict to the generating subset $\Sigma(\mathcal{A}_\theta) \subset K_0^+(\mathcal{A}_\theta)$, since all other elements of the positive cone $K_0^+(\mathcal{A}_\theta)$ correspond to the finite unions $\mathcal{C}_1 \cup \cdots \cup \mathcal{C}_k$ of the generating homogeneous spaces $\mathcal{C}_i$. It remains to notice that equivalence classes of the principal homogeneous spaces $(\mathcal{C}, \mu)$ coincide with the isomorphism classes of the algebra $\mathcal{A}_\theta$. In other words, one gets an isomorphism of the torsion groups

$$\mathrm{WC}(\mathcal{E}_K) \cong \Sigma(\mathcal{A}_\theta).$$

(iii) The general case $n > 1$ is proved by an adaption of the argument for the case $n = 1$. Notice that one must use the Perron–Frobenius n-dimensional continued fractions and a higher-dimensional analog of the Minkowski question-mark function [221]. The details are left to the reader.

Lemma 8.4.1 is proved. □

Lemma 8.4.2. $\mathrm{WC}(\mathcal{A}_{K_v}) \cong K_0(\mathcal{A}_{\Theta(k)} \rtimes_{L_v} \mathbf{Z})$, *where* $\mathcal{A}_{\Theta(k)} = F(\mathcal{A}_K)$.

Proof. The abelian group $\mathrm{WC}(\mathcal{A}_{K_v})$ is known to be finite [37]. Specifically,

$$\mathrm{WC}(\mathcal{A}_{K_v}) \cong (\widehat{\mathcal{A}}_{K_v})^*,$$

where $\widehat{\mathcal{A}}_{K_v}$ are the rational points of K_v on the dual abelian variety of $\mathcal{A}_{K_v}$ and $(\widehat{\mathcal{A}}_{K_v})^*$ is the character group of $\widehat{\mathcal{A}}_{K_v}$ [272].

Recall that $K_0(\mathcal{A}_{\Theta(k)} \rtimes_{L_v} \mathbf{Z}) \cong \mathcal{A}(\mathbf{F}_p)$, where $\mathcal{A}_{\Theta(k)} = F(\mathcal{A}_K)$ and $\mathcal{A}(\mathbf{F}_p)$ is a localization of the abelian variety $\mathcal{A}_K$ at the prime ideal $\mathcal{P}$ over prime p (Remark 8.4.1). In view of Hensel Lemma, the points of $\mathcal{A}(\mathbf{F}_p)$ are the rational points of $\mathcal{A}_{K_v}$, where v is the place over prime p. Since the torsion points of $\mathcal{A}_{K_v}$ and the torsion points of $\widehat{\mathcal{A}}_{K_v}$ coincide, we conclude that $\widehat{\mathcal{A}}_{K_v} \cong \mathcal{A}(\mathbf{F}_p)$. On the other hand, since finite abelian groups are Pontryagin self-dual, one gets $(\widehat{\mathcal{A}}_{K_v})^* \cong \mathcal{A}(\mathbf{F}_p)$. Therefore isomorphism $\mathrm{WC}(\mathcal{A}_{K_v}) \cong (\widehat{\mathcal{A}}_{K_v})^*$ can be written as

$$\mathrm{WC}(\mathcal{A}_{K_v}) \cong K_0(\mathcal{A}_{\Theta(k)} \rtimes_{L_v} \mathbf{Z}).$$

Lemma 8.4.2 is proved. □

Corollary 8.4.1. $\prod_v \mathrm{WC}(\mathcal{A}_{K_v}) \cong \prod_v K_0(\mathcal{A}_{\Theta(k)} \rtimes_{L_v} \mathbf{Z})$.

Proof. One takes the direct sum of finite groups $\mathrm{WC}(\mathcal{A}_{K_v})$ over all nonarchimedean places v of the field K. The corollary follows from isomorphism $\mathrm{WC}(\mathcal{A}_{K_v}) \cong K_0(\mathcal{A}_{\Theta(k)} \rtimes_{L_v} \mathbf{Z})$. □

Lemma 8.4.3.

$$Ш(\mathcal{A}_K) \cong \begin{cases} \mathrm{Cl}(\Lambda) \oplus \mathrm{Cl}(\Lambda), & \textit{if } k \textit{ is even,} \\ (\mathbf{Z}/2^k\mathbf{Z}) \oplus \mathrm{Cl}_{\mathbf{odd}}(\Lambda) \oplus \mathrm{Cl}_{\mathbf{odd}}(\Lambda), & \textit{if } k \textit{ is odd.} \end{cases}$$

Proof. For the sake of clarity, let us outline the main idea. In view of Lemma 8.4.2 and Corollary 8.4.1, one can substitute the group homomorphism $\mathrm{WC}(\mathcal{A}_K) \to \prod_v \mathrm{WC}(\mathcal{A}_{K_v})$ by that of the form

$$\Sigma(\mathcal{A}_{\Theta(k)}) \to \prod_v K_0(\mathcal{A}_{\Theta(k)} \rtimes_{L_v} \mathbf{Z}). \tag{8.1}$$

To calculate the kernel $Ш(\mathcal{A}_K)$, recall that the algebra $\mathcal{A}_{\Theta(k)}$ at the LHS depends on the similarity class of a single matrix $B \in GL(2n, \mathbf{Z})$. On the other hand, the RHS of (8.1) depends solely on the characteristic polynomial of B. Roughly speaking, the problem of kernel $\text{Ш}(\mathcal{A}_K)$ reduces to an old question of the linear algebra: how many non-similar matrices with the same characteristic polynomial are there in $GL(2n, \mathbf{Z})$? This was solved by Latimer and MacDuffee [147]. Their theorem says that the similarity classes of matrices $B \in GL(2n, \mathbf{Z})$ with the fixed polynomial $\mathrm{Char}(B)$ are in one-to-one correspondence with the ideal classes $Cl(\Lambda)$ of an order Λ of a number field k generated by the eigenvalues of B. Coupled with the fact that the scale $\Sigma(\mathcal{A}_{\Theta(k)})$ contains the scales corresponding to all similarity classes, Latimer–MacDuffee Theorem implies that (half of) the kernel of (8.1) is isomorphic to the group $Cl(\Lambda)$. The rest of the proof follows from the Atiyah pairing between the K-theory and K-homology. We pass to a step by step argument.

(i) Recall that the positive cone $K_0^+(\mathcal{A}_{\Theta(k)})$ and the scale $\Sigma(\mathcal{A}_{\Theta(k)})$ can be recovered from $\Lambda \cong \mathbf{Z}+\mathbf{Z}\theta_1+\dots+\mathbf{Z}\theta_{2n-1}$ by solving the inequality $\Lambda \geq 0$ and $0 \leq \Lambda \leq 1$, respectively. Since $\theta_i \in k$ are components of the normalized Perron–Frobenius eigenvector of a matrix $B \in GL(2n, \mathbf{Z})$, we conclude that the similarity class of B defines the algebra $\mathcal{A}_{\Theta(k)}$ up to an isomorphism, and vice versa. The set of all pairwise nonsimilar matrices with the characteristic polynomial $\mathrm{Char}(B)$ will be denoted by $B_1, \dots, B_h$.

(ii) Let us show, that $\Lambda_1 \subset \dots \subset \Lambda_h$, where Λ_i are $\mathbf{Z}$-modules corresponding to the matrices B_i. Indeed, since B_i has the same characteristic polynomial as B, we conclude that the components of the normalized Perron–Frobenius eigenvector of B_i must lie in the number field k. Therefore Λ_i is a full $\mathbf{Z}$-module in the field k. It is well known, that set $\{\Lambda\}_{i=1}^h$ of such modules can be ordered by inclusion $\Lambda_1 \subset \dots \subset \Lambda_h$, where Λ_h is the maximal $\mathbf{Z}$-module.

Remark 8.4.3. The inclusion of scales $\Sigma_1 \subset \dots \subset \Sigma_h$ follows from the inclusion $\Lambda_1 \subset \dots \subset \Lambda_h$ and the inequality $0 \leq \Lambda_i \leq 1$.

(iii) Let us prove that the matrix L_v does not depend on $\{B_i \mid 1 \leq i \leq h\}$. Indeed, a_k are coefficients of the characteristic polynomial of the matrix $B^{\pi(p)}$, where $\pi(p)$ is a positive integer. Notice that the spectrum $\mathrm{Spec}(B_i) = \{\lambda_1^i, \dots, \lambda_{2n}^i\}$ does not depend on the matrix B_i, since $\mathrm{Char}(B_i)$ does not vary. Because $\mathrm{Spec}(B_i^{\pi(p)}) = \{(\lambda_1^i)^{\pi(p)}, \dots, (\lambda_{2n}^i)^{\pi(p)}\}$, we conclude that the spectrum of the matrix $B_i^{\pi(p)}$ is independent of B_i. Therefore the polynomial $\mathrm{Char}(B_i^{\pi(p)})$ and the coefficients a_k are the same for all matrices B_i. Therefore the L_v is independent of the choice of matrix $\{B_i \mid 1 \leq i \leq h\}$.

(iv) Let us calculate the kernel of homomorphism $\Sigma(\mathcal{A}_{\Theta(k)}) \to \prod_v K_0(\mathcal{A}_{\Theta(k)} \rtimes_{L_v} \mathbf{Z})$. Without loss of generality, we assume $B \cong B_h$. By Remark 8.4.3, we have an inclusion of the torsion groups

$$\Sigma_1 \subset \dots \subset \Sigma_{h-1} \subset \Sigma(\mathcal{A}_{\Theta(k)}).$$

Let $\mathbf{e} \in \prod_v K_0(\mathcal{A}_{\Theta(k)} \rtimes_{L_v} \mathbf{Z})$ be the trivial element of torsion group. From item (iii), it is known that the preimage of $\mathbf{e}$ under the homomorphism consists of h distinct elements of the group $\Sigma(\mathcal{A}_{\Theta(k)})$. Indeed, each Σ_i has a unique such element lying in $\Sigma_i \backslash \Sigma_{i-1}$, since otherwise the corresponding abelian variety would have two different reductions modulo p. We conclude, therefore, that the kernel is an abelian group of order h.

(v) Let us calculate the Shafarevich–Tate group $Ш(\mathcal{A}_K)$. It follows from item (iv) that $h = |\mathrm{Cl}(\Lambda)|$, where $\mathrm{Cl}(\Lambda)$ is the ideal class group of Λ [147]. It is easy to see that the kernel is isomorphic to the group $\mathrm{Cl}(\Lambda)$. To express $Ш(\mathcal{A}_K)$ in terms of the group $\mathrm{Cl}(\Lambda)$, recall that the K-homology is dual to the K-theory [32, Section 16.3]. Roughly speaking, the cocycles in K-theory are represented by the vector bundles. Atiyah proposed using elliptic operators to represent the K-homology cycles. An elliptic operator can be twisted by a vector bundle, and the Fredholm index of the twisted operator defines a pairing between the K-homology and the K-theory. In particular, $K^0(\mathcal{A}_{\Theta(k)}) \cong K_0(\mathcal{A}_{\Theta(k)})$, where $K^0(\mathcal{A}_{\Theta(k)})$ is the zero K-homology group of the algebra $\mathcal{A}_{\Theta(k)}$. Thus one gets $\Sigma_0(\mathcal{A}_{\Theta(k)}), \Sigma^0(\mathcal{A}_{\Theta(k)}) \hookrightarrow \mathbf{Q}/\mathbf{Z}$ and, therefore, a pair of embeddings

$$\mathrm{Cl}(\Lambda_0),\ \mathrm{Cl}(\Lambda^0) \hookrightarrow \mathbf{Q}/\mathbf{Z}.$$

Since $\mathrm{Cl}(\Lambda_0) \cong \mathrm{Cl}(\Lambda^0)$ are finite abelian groups, the Atiyah pairing gives rise to a bilinear form $Q_{\mathbf{F}_p}(x,y)$ over the finite field $\mathbf{F}_p$ for each prime p dividing $|\mathrm{Cl}(\Lambda)|$. The map $\mathrm{Cl}(\Lambda_0),\ \mathrm{Cl}(\Lambda^0) \hookrightarrow \mathbf{Q}/\mathbf{Z}$ is a group homomorphism if and only if the $Q_{\mathbf{F}_p}(x,y)$ is an alternating form, i. e., $\{Q_{\mathbf{F}_p}(x,x) = 0 \mid \forall x \in \mathrm{Cl}(\Lambda_0) \oplus \mathrm{Cl}(\Lambda^0)\}$. Recall that the alternating bilinear forms $Q_{\mathbf{F}_p}(x,y)$ exist if and only if $p \neq 2$. Namely, the form $Q_{\mathbf{F}_2}(x,y)$ is always symmetric, i. e., $Q_{\mathbf{F}_2}(x,x) \neq 0$ unless $x = 0$. Thus there are no Atiyah pairing in the case $p = 2$. The rest of the proof follows the Basis Theorem for finite abelian groups. Lemma 8.4.3 is proved. □

Remark 8.4.4. Both cases of Lemma 8.4.3 are realized by the concrete abelian varieties, see [247, Examples (B) and (C)] and [235, Proposition 27], respectively.

Theorem 8.4.1 follows from Lemma 8.4.3.

8.4.5 Abelian varieties with complex multiplication

The $Ш(\mathcal{A}_K)$ can be calculated using Theorem 8.4.1 if the functor F is given explicitly. To illustrate the idea, we consider the abelian varieties with complex multiplication. Denote by $\mathbf{K}$ a CM-field, i. e., the totally imaginary quadratic extension of a totally real number field $\mathbf{k}$. The abelian variety $\mathcal{A}_{CM}$ is said to have complex multiplication if the endomorphism ring of $\mathcal{A}_{CM}$ contains the field $\mathbf{K}$, i. e., $\mathbf{K} \subset \mathrm{End}\, \mathcal{A}_{CM} \otimes \mathbf{Q}$. The $\deg(\mathbf{K}|\mathbf{Q}) = 2n$, where n is the complex dimension of the $\mathcal{A}_{CM}$. For a canonical basis in $\mathbf{K}$ the lifting of the Frobenius endomorphism has the matrix form

$$\mathrm{Fr}_v = \begin{pmatrix} a_1 & 1 & 0 & \dots & 0 & 0 \\ -a_2 & 0 & 1 & \dots & 0 & 0 \\ \vdots & \vdots & \vdots & \ddots & \vdots & \vdots \\ a_{2n-1} & 0 & 0 & \dots & 0 & 1 \\ -p & 0 & 0 & \dots & 0 & 0 \end{pmatrix}.$$

The functor F is acting by the formula $\mathrm{Fr}_v \mapsto L_v$, see Section 6.6.1 for the details. Since $L_v \in \mathrm{End}\,\Lambda$, one can recover the ring Λ from the eigenvalues of matrix L_v. Let us consider the simplest case $n = 1$, i. e., the elliptic curves with complex multiplication.

8.4.5.1 Elliptic curves

Denote by $\mathcal{E}_{\mathrm{CM}}$ an elliptic curve with complex multiplication by the ring $R = \mathbf{Z} + fO_{\mathbf{K}}$, where $\mathbf{K} = \mathbf{Q}(\sqrt{-D})$ is an imaginary quadratic field with the square free discriminant $D > 1$ and $f \geq 1$ is the conductor of R. The ring $\Lambda = F(R)$ is given by the formula $\Lambda = \mathbf{Z} + f'O_k$, where $k = \mathbf{Q}(\sqrt{D})$ is the real quadratic field and the conductor $f' \geq 1$ is the least integer solution of the equation $|\mathrm{Cl}(\mathbf{Z} + f'O_k)| = |\mathrm{Cl}(R)|$ (Theorem 1.4.1). Moreover, there exists a group isomorphism $\mathrm{Cl}(\Lambda) \cong \mathrm{Cl}(R)$. Using Theorem 8.4.1, one gets the following corollary.

Corollary 8.4.2. $Ш(\mathcal{E}_{\mathrm{CM}}) \cong \mathrm{Cl}(R) \oplus \mathrm{Cl}(R)$.

Example 8.4.1 ([247, Example (B)]). Let $\mathcal{E}_{\mathrm{CM}}$ be the Fermat cubic $x^3 + y^3 = z^3$. The $\mathcal{E}_{\mathrm{CM}}$ has complex multiplication by the ring $R \cong \mathbf{Z} + O_{\mathbf{K}}$, where $\mathbf{K} \cong \mathbf{Q}(\sqrt{-3})$. It is well known that in this case $\mathrm{Cl}(R)$ is trivial. We conclude from Corollary 8.4.2 that $Ш(\mathcal{E}_{\mathrm{CM}})$ is a trivial group. An alternative proof of this fact is based on the exact calculation of the p-part of $Ш(\mathcal{E}_{\mathrm{CM}})$ and can be found in [247, Theorem 1 and Example (B)].

Example 8.4.2 ([247, Example (C)]). Let $\mathcal{E}_{\mathrm{CM}}$ be the modular curve $X_0(49)$ given by the equation $y^2 + xy = x^3 - x^2 - 2x - 1$. The curve $\mathcal{E}_{\mathrm{CM}}$ has complex multiplication by the ring $R \cong \mathbf{Z} + O_{\mathbf{K}}$, where $\mathbf{K} \cong \mathbf{Q}(\sqrt{-7})$. It is well known that the group $\mathrm{Cl}(R)$ is trivial. Using Corollary 8.4.2, one concludes that $Ш(\mathcal{E}_{\mathrm{CM}})$ is a trivial group. A different proof of this result can be found in [247, Theorem 1 and Example (C)].

Remark 8.4.5. It is tempting to compare Corollary 8.4.2 with a vast database of the analytic values for $|Ш(\mathcal{E}_{\mathrm{CM}})|$, see, e. g., https://www.lmfdb.org. An exact relation between the database and Corollary 8.4.2 is unknown to the author.

Remark 8.4.6. Theorem 8.4.1 is true for the simple abelian varieties $\mathcal{A}_K$. If $\mathcal{A}_K$ is not simple, then the functor F splits and one gets different formulas for the group $Ш(\mathcal{A}_K)$, see the corresponding calculations in the excellent paper [267, Theorem 3.1].

Guide to the literature

The Shafarevich–Tate group is notoriously hard to compute even for elliptic curves $\mathcal{E}$ [274, § 7]. It was conjectured that Ш($\mathcal{E}$) is finite [274, Conjecture 1]. The existing methods include an evaluation of the analytic order of Ш($\mathcal{E}$) based on the second part of the Birch and Swinnerton-Dyer Conjecture and an exact calculation of the p-part of Ш($\mathcal{E}$) [247]. Our exposition follows [212].

8.5 Noncommutative geometry of elliptic surfaces

Elliptic surfaces play an outstanding rôle in algebraic geometry, number theory, and topology. Roughly speaking, they are bundles over the base curve of genus $g \geq 0$ with the fibers of genus one. We assume further that $g = 0$ and denote by $\mathbf{Q}(t)$ the field of rational functions on the base $\mathbf{C}P^1$. The elliptic curve over $\mathbf{Q}(t)$ can be identified with an elliptic surface [256, Chapter 5]; we shall write such a surface as $\mathcal{E}(\mathbf{Q}(t))$. The generic fiber of $\mathcal{E}(\mathbf{Q}(t))$ will be denoted by $\mathcal{E}_t$. Silverman's specialization theorem says that the homomorphism $\mathcal{E}(\mathbf{Q}(t)) \to \mathcal{E}_t$ is injective for all but finitely many values of t. In particular, the rank of $\mathcal{E}_t$ is related to the rank of $\mathcal{E}(\mathbf{Q}(t))$.

Recall that the noncommutative torus $\mathcal{A}_\theta$ is defined as a C^*-algebra generated by the unitary operators u and v satisfying the relation $vu = e^{2\pi i\theta}uv$. The $\mathcal{A}_\theta$ is said to have real multiplication (RM) if θ is a quadratic irrationality represented by the k-periodic continued fraction $[b_1, \dots, b_N; \overline{a_1, \dots, a_k}]$. We denote by F a functor mapping the fibers $\mathcal{E}_t$ into the noncommutative tori $\mathcal{A}_\theta$ (Section 1.3). If $t \in \mathbf{Q}$, then the $F(\mathcal{E}_t)$ is a noncommutative torus with RM.

Example 8.5.1 ([196]). Let $\mathcal{E}(\mathbf{Q}(t))$ be defined by the affine equation

$$y^2 = x(x-1)\left(x - \frac{t-2}{t+2}\right),$$

where $t \in \mathbf{C}P^1$. Then F is given the formula

$$F(\mathcal{E}(\mathbf{Q}(t))) = \mathcal{A}_{[t-1;\, \overline{1,\, t-2}]}.$$

In this section we extend the above formula describing a "noncommutative" elliptic surface $\mathcal{A}_{[t-1;\, \overline{1,\, t-2}]}$ to arbitrary $\mathcal{E}(\mathbf{Q}(t))$ (Theorem 8.5.1). Such a result allows calculating the Picard number, rank, and minimal model for the $\mathcal{E}(\mathbf{Q}(t))$ (Theorem 8.5.2). The following notation will be used.

Definition 8.5.1 ([43]). We denote by $V_{N,k}(\mathbf{C})$ an affine variety defined by the polynomials in variables $a_{1\leq i\leq k}$ and $b_{1\leq j\leq N}$ satisfying the following obvious equality:

$$[b_1, \dots, b_N, a_1, \dots, a_k; \overline{a_1, \dots, a_k}] = [b_1, \dots, b_N; \overline{a_1, \dots, a_k}].$$

Remark 8.5.1. The equations of the $V_{N,k}(\mathbf{C})$ date back to Euler [80]. The corresponding projective variety was studied in Section 6.2.1.

The integer points of $V_{N,k}(\mathbf{C})$ parametrize the k-periodic continued fractions $[b_1,\dots,b_N;\overline{a_1,\dots,a_k}]$ and $\dim V_{N,k}(\mathbf{C}) = N+k-2$. Variety $V_{N,k}(\mathbf{C})$ is a fiber bundle over the Fermat–Pell conic $\mathcal{P} : Cx^2 - Bxy + Ay^2 = (-1)^k A$ with the fiber map $\pi : V_{N,k}(\mathbf{C}) \to \mathcal{P}$ [43]. By $t \in \mathbf{CP}^1$ we denote a rational parametrization of $\mathcal{P}$. Our main result can be formulated as follows.

Theorem 8.5.1. *For each elliptic surface $\mathcal{E}(\mathbf{Q}(t))$, there exists section σ of a subbundle $(U_{b_1,\dots,b_N;a_1,\dots,a_k}, \mathcal{P}, \pi')$ of $(V_{N,k}(\mathbf{C}), \mathcal{P}, \pi)$ such that*

$$F(\mathcal{E}(\mathbf{Q}(t))) = \{\mathcal{A}_{[b_1(t),\dots,b_N(t);\ \overline{a_1(t),\dots,a_k(t)}]} \mid a_i(t), b_j(t) \in \mathbf{Z}[t],\ t \in \mathcal{P}\}.$$

Denote by $\mathcal{F}$ a fiber of the bundle $(U_{a_1,\dots,a_N;b_1,\dots,b_k}, \mathcal{P}, \pi')$. For $p(t) \in \mathbf{Z}[t]$, let $\mathcal{E}_{CM}(p(t))$ be an elliptic surface whose fibers over $t \in \mathbf{Q}$ have complex multiplication (CM) by the ring of integers of the imaginary quadratic field $\mathbf{Q}(\sqrt{-1-p^2(t)})$. Denote by $\mathcal{E}_{\min}(\mathbf{Q}(t))$ the minimal model of the surface $\mathcal{E}(\mathbf{Q}(t))$. An application of Theorem 8.5.1 is as follows.

Theorem 8.5.2. *Let $\mathcal{E}(\mathbf{Q}(t))$ be a surface as in Theorem* 8.5.1 *and let $\rho(\mathcal{E}(\mathbf{Q}(t)))$ be its Picard number over $\mathbf{Q}$. The following is true:*
(i) *the Picard number $\rho(\mathcal{E}(\mathbf{Q}(t))) = N + k$;*
(ii) *the rank* $\mathrm{rk}(\mathcal{E}(\mathbf{Q}(t))) = \dim \mathcal{F}$;
(iii) *$\mathcal{E}_{\min}(\mathbf{Q}(t))$ is $\mathbf{C}$-isomorphic to $\mathcal{E}_{CM}(p(t))$ for some $p(t) > 0$.*

8.5.1 Brock–Elkies–Jordan variety

By an infinite continued fraction one understands an expression of the form

$$[c_1, c_2, c_3, \dots] := c_1 + \cfrac{1}{c_2 + \cfrac{1}{c_3 + \cdots}},$$

where c_1 is an integer and $c_2, c_3, \dots$ are positive integers. The continued fraction converges to an irrational number and each irrational number has a unique representation by its continued fraction. The continued fraction is called k-periodic, if $c_{i+k} = c_i$ for all $i \geq N$ and a minimal index $k \geq 1$. We shall denote the k-periodic continued fraction by

$$[b_1, \dots, b_N, \overline{a_1, \dots, a_k}],$$

where $(a_1, \dots, a_k)$ is the minimal period. Such a continued fraction converges to one of the irrational roots of a quadratic polynomial

$$Ax^2 + Bx + C \in \mathbf{Z}[x].$$

Conversely, an irrational root of any quadratic polynomial has a representation by the periodic continued fraction. Notice that the following two continued fractions define the same irrational number:

$$[b_1, \dots, b_N, \overline{a_1, \dots, a_k}] = [b_1, \dots, b_N, a_1, \dots, a_k, \overline{a_1, \dots, a_k}].$$

But it is well known that two infinite continued fractions with at most finite number of distinct entries must be related by a linear fractional transformation given by a matrix $\mathcal{E} \in GL_2(\mathbf{Z})$. Therefore the above equation can be written in the form

$$x = \frac{E_{11}x + E_{12}}{E_{21}x + E_{22}},$$

where $\mathcal{E} = (E_{ij}) \in GL_2(\mathbf{Z})$ and $x = [b_1, \dots, b_N, \overline{a_1, \dots, a_k}]$.

Remark 8.5.2. It is easy to see that x is a root of the quadratic polynomial with $A = E_{21}, B = E_{22} - E_{11}$, and $C = -E_{12}$.

Definition 8.5.2. The Brock–Elkies–Jordan variety $V_{N,k}(\mathbf{C}) \subset \mathbb{A}^{N+k}$ is an affine variety over $\mathbf{Z}$ defined by the three equations:

$$\begin{cases} A[E_{22} - E_{11}](y_1, \dots, y_N, x_1, \dots, x_k) = BE_{21}(y_1, \dots, y_N, x_1, \dots, x_k), \\ -AE_{12}(y_1, \dots, y_N, x_1, \dots, x_k) = CE_{21}(y_1, \dots, y_N, x_1, \dots, x_k), \\ -BE_{12}(y_1, \dots, y_N, x_1, \dots, x_k) = C[E_{22} - E_{11}](y_1, \dots, y_N, x_1, \dots, x_k). \end{cases}$$

It is verified directly from Remark 8.5.2 and the equality $E_{11}E_{22} - E_{12}E_{21} = (-1)^k$ that

$$CE_{21}^2 - BE_{21}E_{22} + AE_{22}^2 = (-1)^k A.$$

Definition 8.5.3. By the Fermat–Pell conic $\mathcal{P}$ one understands the plane curve

$$Cu^2 - Buv + Av^2 = (-1)^k A.$$

Theorem 8.5.3 ([43]). *The affine variety $V_{N,k}(\mathbf{C})$ fibers over the Fermat–Pell conic $\mathcal{P}$, i. e., there exists a map $\pi : V_{N,k}(\mathbf{C}) \to \mathcal{P}$ such that*

$$\pi(y_1, \dots, y_N, x_1, \dots, x_k) = (E_{21}, E_{22}).$$

8.5.2 Elliptic surfaces

An algebraic surface S is a variety of dimension two. An elliptic surface S over a curve C is a smooth projective surface with an elliptic fibration over C, i. e., a surjective morphism $f : S \to C$ such that almost all fibers are smooth elliptic curves.

8.5.2.1 Blow-ups

A map $\phi : S \dashrightarrow S'$ is called rational if it is given by a rational function defined everywhere except for the poles of ϕ. The map ϕ is birational if the inverse ϕ^{-1} is a rational map.

A birational map $\epsilon : S \dashrightarrow S'$ is called a blow-up if it is defined everywhere except for a point $p \in S$ and a rational curve $C \subset S'$ such that $\epsilon^{-1}(C) = p$. Every birational map $\phi : S \dashrightarrow S'$ is a composition of a finite number of the blow-ups, i. e., $\phi = \epsilon_1 \circ \cdots \circ \epsilon_k$.

8.5.2.2 Minimal models

The surface S is called a minimal model if any birational map $S \dashrightarrow S'$ is an isomorphism. Minimal models exist and are unique unless S is a ruled surface. By Castelnuovo Theorem, the surface S is a minimal model if and only if S does not contain rational curves C with the self-intersection index -1.

8.5.3 Proofs

8.5.3.1 Proof of Theorem 8.5.1

For the sake of clarity, let us outline the main ideas. Let $(V_{N,k}, \mathcal{P}, \pi)$ be a fiber bundle defined by the map in Theorem 8.5.3. Using an exclusion process described below, we construct a subbundle

$$(U_{b_1,\dots,b_N;\, a_1,\dots,a_k}, \mathcal{P}, \pi') \subset (V_{N,k}, \mathcal{P}, \pi)$$

depending on the point $(b_1, \dots, b_N;\ a_1, \dots, a_k) \in V_{N,k}$ and whose fibers are r-dimensional. Each section σ of the bundle defines a family of the noncommutative tori $\mathcal{A}_{[b_1(t),\dots,b_N(t);\ \overline{a_1(t),\dots,a_k(t)}]}$, where $t \in \mathcal{P}$ so that $a_i(t_0) = a_i$ and $b_j(t_0) = b_j$. By the construction,

$$F(\mathcal{E}(\mathbf{Q}(t))) = \mathcal{A}_{[b_1(t),\dots,b_N(t);\ \overline{a_1(t),\dots,a_k(t)}]}.$$

Let us pass to a detailed argument.

Proof. (i) Let $(b_1, \dots, b_N;\ a_1, \dots, a_k)$ be an integer point of the variety $V_{N,k}$. To construct a subvariety $U_{b_1,\dots,b_N;\, a_1,\dots,a_k}$, we denote by $(u_1, \dots, u_m)$ the variables and by $(c_1, \dots c_m)$

the constants. Unless stated otherwise, it is assumed that $c_1 = b_1, \dots, c_m = a_k$, where $m = N + k$.

The polynomial equations defining the variety $V_{N,k}$ allow excluding two variables, say, u_{m-1} and u_m, i. e., they become algebraically dependent on the variables $\{u_i \mid 1 \le i \le m-2\}$. Namely, one can write

$$\begin{cases} u_{m-1} = \dfrac{P_{m-1}(u_1, \dots, u_{m-2})}{Q_{m-1}(u_1, \dots, u_{m-2})}, \\ u_m = \dfrac{P_m(u_1, \dots, u_{m-2})}{Q_m(u_1, \dots, u_{m-2})}, \end{cases}$$

for some polynomials $P_{m-1}, Q_{m-1}, P_m, Q_m \in \mathbf{Z}[u_1, \dots, u_{m-2}]$.

Since $(c_1, \dots, c_m) \in V_{N,k}$, one obtains a subvariety of $V_{N.k}$ consisting of the points $(u_1, \dots, u_{m-2}, c_{m-1}, c_m)$. Such a subvariety is given by the system of equations

$$\begin{cases} P_m(u_1, \dots, u_{m-2}) = c_m Q_m(u_1, \dots, u_{m-2}), \\ P_{m-1}(u_1, \dots, u_{m-2}) = c_{m-1} Q_{m-1}(u_1, \dots, u_{m-2}). \end{cases}$$

It follows from the above equations that again two variables, say, u_{m-3} and u_{m-2}, become algebraically dependent on the variables $\{u_i \mid 1 \le i \le m-4\}$. We repeat the argument, obtaining a subvariety made of the points $(u_1, \dots, u_{m-4}, c_{m-3}, c_{m-2}, c_{m-1}, c_m)$.

It is clear that the algorithm will stop when the following system of the polynomial equations in the variables u_i is satisfied:

$$\begin{cases} P_m(u_1, \dots, u_{m-2}) = c_m Q_m(u_1, \dots, u_{m-2}), \\ P_{m-1}(u_1, \dots, u_{m-2}) = c_{m-1} Q_{m-1}(u_1, \dots, u_{m-2}), \\ \vdots \\ P_2(u_1, u_2) = c_2 Q_2(u_1, u_2), \\ P_1(u_1, u_2) = c_1 Q_1(u_1, u_2). \end{cases}$$

Remark 8.5.3. Notice that above system always has a solution, e. g., the trivial solution $(c_1, \dots, c_m)$. Example 8.5.1 shows that, in fact, such solutions can be a variety of dimension 1. Below we consider the general case in terms of the Krull dimension of a polynomial ring.

(ii) Let $\mathcal{I}_{c_1,\dots,c_m}$ be an ideal generated by the above equations in the polynomial ring $\mathbf{C}[u_1, \dots, u_m]$. Consider the ring $\mathbf{C}[u_1, \dots, u_m]/\mathcal{I}_{c_1,\dots,c_m}$. The Krull dimension of such a ring will be denoted by $r + 1$.

(iii) Consider an $(r+1)$-dimensional affine subvariety $U_{c_1,\dots,c_m}$ of $V_{N,k}$ given by the above equations. As it was shown in item (ii), the points of the variety $U_{c_1,\dots,c_m}$ can be written in the form

$$(R_1, \dots, R_{r+1}, c_{r+2}, \dots, c_m),$$

where $R_i \in \mathbf{Z}[u_1, \dots, u_{r+1}]$.

(iv) Theorem 8.5.3 says that variety $V_{N,k}$ fibers over the Fermat–Pell conic $\mathcal{P}$. We denote by π' a restriction the map $\pi : V_{N,k} \to \mathcal{P}$ to the subvariety $U_{c_1,\dots,c_m}$. Assuming $c_1 = b_1, \dots, c_m = a_k$, one gets a fiber bundle

$$(U_{b_1,\dots,b_N;\, a_1,\dots,a_k}, \mathcal{P}, \pi')$$

consisting of the r-dimensional fibers, the 1-dimensional base, and the $(r+1)$-dimensional total space.

(v) Consider a global section $\sigma : \mathcal{P} \to U_{b_1,\dots,b_N;\, a_1,\dots,a_k}$. Section σ is given by the polynomials

$$b_1(t), b_2(t), \dots, b_N(t) \in \mathbf{Z}[t]; \quad a_1(t), a_2(t), \dots, a_k(t) \in \mathbf{Z}[t],$$

where all but $r+1$ polynomials are constants.

Remark 8.5.4. The number r is equal to the rank of the $\mathcal{E}(\mathbf{Q}(t))$, see Theorem 8.5.2(ii).

(vi) It remains to show that $F(\mathcal{E}(\mathbf{Q}(t))) = \mathcal{A}_{[b_1(t),\dots,b_N(t);\, \overline{a_1(t),\dots,a_k(t)}]}$. We shall prove this fact adapting the argument of [196] to the case of the $\mathcal{E}(\mathbf{Q}(t))$. Namely, suppose that $\mathcal{E}(\mathbf{Q}(t))$ is given in the Legendre form

$$y^2 = x(x-1)(x-\alpha(t)), \quad \alpha(t) \in \mathbf{Q}(t).$$

(vii) Recall that if $F(\mathcal{E}(\mathbf{Q}(t))) = \mathcal{A}_\theta$, then

$$\begin{pmatrix} b-1 & 1 \\ b-2 & 1 \end{pmatrix} \begin{pmatrix} \theta \\ 1 \end{pmatrix} = \begin{pmatrix} \theta \\ 1 \end{pmatrix}, \quad \text{where } \frac{b-2}{b+2} = \alpha(t),$$

see [196, Theorem 1 and Corollary 1.2]. Since $b = \frac{2(1+\alpha(t))}{1-\alpha(t)}$, one can write this equation in the form

$$\begin{pmatrix} \frac{3\alpha(t)+1}{1-\alpha(t)} & 1 \\ \frac{4\alpha(t)}{1-\alpha(t)} & 1 \end{pmatrix} \begin{pmatrix} \theta \\ 1 \end{pmatrix} = \begin{pmatrix} \theta \\ 1 \end{pmatrix}.$$

(viii) On the other hand, it is known that

$$\begin{pmatrix} E_{11} & E_{12} \\ E_{21} & E_{22} \end{pmatrix} \begin{pmatrix} \theta \\ 1 \end{pmatrix} = \begin{pmatrix} \theta \\ 1 \end{pmatrix},$$

where $\theta = [b_1, \dots, b_N, \overline{a_1, \dots, a_k}]$.

(ix) One can factorize the above matrix as follows:

$$\begin{pmatrix} E_{11} & E_{12} \\ E_{21} & E_{22} \end{pmatrix} = \begin{pmatrix} b_1 & 1 \\ 1 & 0 \end{pmatrix} \cdots \begin{pmatrix} b_N & 1 \\ 1 & 0 \end{pmatrix} \begin{pmatrix} a_1 & 1 \\ 1 & 0 \end{pmatrix} \cdots \begin{pmatrix} a_k & 1 \\ 1 & 0 \end{pmatrix} \begin{pmatrix} b_N & 1 \\ 1 & 0 \end{pmatrix}^{-1} \cdots \begin{pmatrix} b_1 & 1 \\ 1 & 0 \end{pmatrix}^{-1},$$

see, e. g., [43, Definition 2.4].

(x) It remains to compare the above equations, i. e.,

$$E_{11} = \frac{3\alpha(t)+1}{1-\alpha(t)}, \quad E_{21} = \frac{4\alpha(t)}{1-\alpha(t)}, \quad E_{12} = E_{22} = 1.$$

(xi) Since $\alpha(t) \in \mathbf{Q}(t)$, one gets $E_{12}(t), E_{21}(t) \in \mathbf{Q}(t)$. Moreover, clearing the denominators, one can always assume $E_{ij}(t) \in \mathbf{Z}[t]$. Therefore one obtains $a_i(t), b_j(t) \in \mathbf{Z}[t]$.

Theorem 8.5.1 is proved. □

8.5.3.2 Proof of Theorem 8.5.2

Proof. Let us prove item (i) of Theorem 8.5.2. Recall that the Neron–Severi group $\mathrm{NS}(\mathcal{E}(\mathbf{Q}(t)))$ is the abelian group of divisors on $\mathcal{E}(\mathbf{Q}(t))$ modulo algebraic equivalence. The Picard number $\rho(\mathcal{E}(\mathbf{Q}(t)))$ is defined as the rank of the $\mathrm{NS}(\mathcal{E}(\mathbf{Q}(t)))$. Such a number is always finite.

The idea of the proof is based on an identification of the $F(\mathrm{NS}(\mathcal{E}(\mathbf{Q}(t))))$ with the convergents of the continued fraction $[b_1(t),\dots,b_N(t);\ \overline{a_1(t),\dots,a_k(t)}]$. By an elementary property of the continued fractions, all such convergents are rational functions of the first $N+k$ convergents. Let us pass to a detailed argument.

(i) Consider a global section $\sigma_i : \mathcal{P} \to \mathcal{E}(\mathbf{Q}(t))$ of the elliptic surface $\mathcal{E}(\mathbf{Q}(t))$ with the base curve $\mathcal{P}$. Section $\sigma_i(\mathcal{P}) := \mathcal{P}_i$ is a genus zero curve on the surface $\mathcal{E}(\mathbf{Q}(t))$. Thus one can identify $\mathcal{P}_i$ with a divisor of the $\mathcal{E}(\mathbf{Q}(t))$.

(ii) Denote by $\{\mathcal{A}_{\frac{p}{q}} \mid p,q \in \mathbf{Z}\}$ the noncommutative tori with rational values of the parameter $\theta = \frac{p}{q}$; $\mathcal{A}_{\frac{p}{q}}$ correspond to the degenerate elliptic curves $\mathcal{P}$, i. e., $F(\mathcal{P}) = \mathcal{A}_{\frac{p}{q}}$ (Section 1.3).

(iii) On the other hand, any noncommutative torus $\mathcal{A}_\theta$ is the inductive limit of an ascending sequence of $\mathcal{A}_{\frac{p_i}{q_i}}$, where $\frac{p_i}{q_i}$ are convergents of the continued fraction of θ (Section 3.5). Consider a commutative diagram in Fig. 8.4, where $\mathcal{P}_i$ is the union of all divisors of the $\mathcal{E}(\mathbf{Q}(t))$ obtained as a pullback of F as t runs through all admissible values.

(iv) The ascending sequence of the rational noncommutative tori

$$\mathcal{A}_{\frac{p_1(t)}{q_1(t)}} \subset \mathcal{A}_{\frac{p_2(t)}{q_2(t)}} \subset \cdots$$

$$\begin{array}{ccc} \mathcal{P}_i & \xrightarrow{\text{embedding}} & \mathcal{E}(\mathbf{Q}(t)) \\ F\downarrow & & \downarrow F \\ \mathcal{A}_{\frac{p_i(t)}{q_i(t)}} & \xrightarrow{\text{embedding}} & \mathcal{A}_{[b_1(t),\dots,b_N(t);\ \overline{a_1(t),\dots,a_k(t)}]} \end{array}$$

Figure 8.4: Rational approximation.

gives rise to an infinite inclusion sequence of the divisors $\mathcal{P}_1 \subset \mathcal{P}_2 \subset \cdots$. It is easy to see that

$$\mathrm{NS}(\mathcal{E}(\mathbf{Q}(t))) = \lim_{i\to\infty} \mathcal{P}_i.$$

(v) Let us evaluate the number of generators of the group $\mathrm{NS}(\mathcal{E}(\mathbf{Q}(t)))$. In view of the above argument, this question can be reduced to the number of generators of the sequence $\mathcal{A}_{\frac{p_1(t)}{q_1(t)}} \subset \mathcal{A}_{\frac{p_2(t)}{q_2(t)}} \subset \cdots$. Namely, given the periodic continued fraction $[b_1(t),\dots,b_N(t);\ \overline{a_1(t),\dots,a_k(t)}]$, how many algebraically independent convergents $\frac{p_i(t)}{q_i(t)}$ are there?

(vi) It is easy to see that the total number of the independent convergents is equal to $N+k$. The idea is simple: we recover $a_i(t)$ and $b_j(t)$ from the first $N+k$ convergents, and then express the remaining convergents $\{\frac{p_i(t)}{q_i(t)} \mid i > N+k\}$ as the rational functions of $a_i(t)$ and $b_j(t)$.

Namely, the first convergent $\frac{p_1(t)}{q_1(t)}$ coincides with $b_1(t)$. The second convergent is $\frac{p_2(t)}{q_2(t)} = \frac{p_1(t)}{q_1(t)} + \frac{1}{b_2(t)}$ and, therefore, $b_2(t) = (\frac{p_2(t)}{q_2(t)} - \frac{p_1(t)}{q_1(t)})^{-1}$. Similarly, one gets $b_3(t)$ as a rational function of $\frac{p_1(t)}{q_1(t)}, \frac{p_2(t)}{q_2(t)}$, and $\frac{p_3(t)}{q_3(t)}$. Finally, $a_k(t)$ can be written as a rational function of the convergents $\{\frac{p_i(t)}{q_i(t)} \mid 1 \le i \le N+k\}$.

Clearly, the remaining convergents $\{\frac{p_i(t)}{q_i(t)} \mid i > N+k\}$ depend algebraically on $a_i(t)$ and $b_j(t)$ and, therefore, on the convergents $\{\frac{p_i(t)}{q_i(t)} \mid 1 \le i \le N+k\}$.

(vii) We conclude from (i)–(vi) that the free abelian group $\mathrm{NS}(\mathcal{E}(\mathbf{Q}(t)))$ has $N+k$ generators. In particular, the Picard number of the surface $\mathcal{E}(\mathbf{Q}(t))$ is given by the formula

$$\rho(\mathcal{E}(\mathbf{Q}(t))) = N + k.$$

Item (i) of Theorem 8.5.2 is proved. □

Proof. Let us prove item (ii) of Theorem 8.5.2. Roughly speaking, to calculate the rank $\mathrm{rk}(\mathcal{E}(\mathbf{Q}(t)))$, we need to know the number of the convergents $\{\frac{p_i(t)}{q_i(t)} \mid 1 \le i \le N+k\}$ independent of the parameter $t \in \mathcal{P}$. Those correspond to the "horizontal" and "vertical" divisors $\{\mathcal{P}_i \mid F(\mathcal{P}_i) = \mathcal{A}_{\frac{p_i(t)}{q_i(t)}}\}$, see [256, Section 6.1] for the terminology. We pass to a detailed argument.

(i) Let $(U_{b_1,\dots,b_N;\, a_1,\dots,a_k}, \mathcal{P}, \pi')$ be the fiber bundle constructed in Theorem 8.5.1 and let $\mathcal{F}$ be a fiber of the $(U_{b_1,\dots,b_N;\, a_1,\dots,a_k}, \mathcal{P}, \pi')$ having dimension $\dim \mathcal{F} = r$. As shown above, one can express $a_i(t)$ and $b_j(t)$ in terms of $\frac{p_i(t)}{q_i(t)}$ and write points of the variety $U_{b_1,\dots,b_N;a_1,\dots,a_k}$ in the form

$$(R_1, \dots, R_{r+1}, c_{r+2}, \dots, c_{N+k}),$$

where $R_i \in \mathbf{Z}[\frac{p_1(t)}{q_1(t)}, \dots, \frac{p_{r+1}(t)}{q_{r+1}(t)}]$.

(ii) Recall the Tate–Shioda formula

$$\rho(\mathcal{E}(\mathbf{Q}(t))) = \mathrm{rk}(\mathcal{E}(\mathbf{Q}(t))) + 2 + \sum_{v \in R} (m_v - 1),$$

where R is the finite set of singular fibers of the $\mathcal{E}(\mathbf{Q}(t))$ and m_v is the number of components of the fiber $v \in R$ [255, Corollary 1.5]. In the above formula, the number of horizontal divisors of $\mathcal{E}(\mathbf{Q}(t))$ is equal 2 and such of the vertical divisors is equal to $\sum_{v \in R}(m_v - 1)$ [256, Section 6.1].

(iii) Since the horizontal and vertical divisors are not generic, they cannot depend on $t \in \mathcal{P}$. Thus the constants c_i represent the horizontal and vertical divisors. Comparing with the Tate–Shioda formula, one gets

$$\mathrm{rk}(\mathcal{E}(\mathbf{Q}(t))) = r = \dim \mathcal{F}.$$

Item (ii) of Theorem 8.5.2 is proved. □

Proof. Let us prove item (iii) of Theorem 8.5.2. Roughly speaking, the minimal model corresponds to the surface with the least Picard number among the surfaces in the birational equivalence class of $\mathcal{E}(\mathbf{Q}(t))$ [256, Section 4.5]. But from item (i) of Theorem 8.5.2, such a surface must minimize the sum $N + k$. We show that for the minimal model $N = k = 1$, so that $F(\mathcal{E}_{CM}(p(t))) = \mathcal{A}_{[p(t);\, \overline{2p(t)}]}$ for a positive definite polynomial $p(t) \in \mathbf{Z}[t]$. The latter formula follows from a symmetry between the complex and real multiplication (Section 1.4.1). We pass to a detailed argument.

(i) Let B be the birational equivalence class of $\mathcal{E}(\mathbf{Q}(t))$. Recall that a birational map $\{\mathcal{E} \dashrightarrow \mathcal{E}' \mid \mathcal{E}, \mathcal{E}' \in B\}$ is called dominant if $\rho(\mathcal{E}) > \rho(\mathcal{E}')$. The surface $\mathcal{E}_{\min} \in B$ is a minimal model if $\mathcal{E}_{\min}$ is dominated by any other $\mathcal{E} \in B$.

(ii) One gets from item (i) of Theorem 8.5.2 that

$$\rho(\mathcal{E}_{\min}) = \min_{\mathcal{E} \in B}(N + k),$$

where $N \geq 0$ and $k \geq 1$. Let us find the minimal values of N and k separately.

(iii) The minimal value of k is equal to 1, since the period of continued fraction cannot vanish. Thus we have $k_{min} = 1$.

(iv) The minimal value of N is equal to 0, since a continued fraction can be purely periodic. Thus we have $N_{min} = 0$.

(v) Using (iii) and (iv), we can write

$$F(\mathcal{E}_{min}) = \{\mathcal{A}_{[\overline{a(t)}]} \mid a(t) \in \mathbf{Z}[t]\}.$$

(vi) For an explicit construction of $\mathcal{E}_{min}$, recall that one can add a finite tail $\frac{1}{2}a(t)$ to the purely periodic fraction $[\overline{a(t)}]$. The obtained surface $\mathcal{E}'_{min}$ will be isomorphic over $\mathbf{C}$ to the original surface $\mathcal{E}_{min}$ (Theorem 1.3.1). In other words, one gets

$$F(\mathcal{E}'_{min}) = \{\mathcal{A}_{[p(t);\, \overline{2p(t)}]} \mid p(t) \in \mathbf{Z}[t]\},$$

where $p(t) := \frac{1}{2}a(t)$.

(vii) Since $[p(t); \overline{2p(t)}] = (1+p^2(t))^{\frac{1}{2}}$, one can use an explicit formula for the functor F, saying that fibers of the surface $\mathcal{E}'_{min}$ must have complex multiplication by the ring of integers of the imaginary quadratic field $\mathbf{Q}(\sqrt{-1-p^2(t)})$ (Theorem 1.4.1). Thus $\mathcal{E}'_{min} \cong \mathcal{E}_{CM}(p(t))$.

This argument finishes the proof of item (iii) of Theorem 8.5.2. □

8.5.4 Picard numbers

In this section we apply item (i) of Theorem 8.5.2 to the elliptic surfaces with a generic fiber having complex multiplication.

Let $D > 1$ be a square-free integer. Denote by $\mathcal{E}_{CM}^{(-D)}$ an elliptic curve with CM by the ring of integers of the imaginary quadratic field $\mathbf{Q}(\sqrt{-D})$. It is known that

$$F(\mathcal{E}_{CM}^{(-D)}) = \mathcal{A}_{RM}^{(D,f)},$$

where $\mathcal{A}_{RM}^{(D,f)}$ is a noncommutative torus with RM by the order of conductor $f \geq 1$ in the ring of integers of the real quadratic field $\mathbf{Q}(\sqrt{D})$ (Theorem 1.4.1). In other words, one gets $\mathcal{A}_\theta$, where

$$\theta = \begin{cases} \sqrt{f^2D} = [b_1; \overline{a_1, a_2, \dots, a_2, a_1, 2b_1}], & \text{if } D \equiv 2,3 \mod 4, \\ \frac{1+\sqrt{f^2D}}{2} = [b_1; \overline{a_1, a_2, \dots, a_2, a_1, 2b_1 - 1}], & \text{if } D \equiv 1 \mod 4. \end{cases}$$

As usual, denote by $(V_{1,k}(\mathbf{C}), \mathcal{P}, \pi)$ the fiber bundle corresponding to the periodic continued fractions.

Definition 8.5.4. By $\mathcal{E}(D(t))$ we denote an elliptic surface such that for a section $\sigma : \mathcal{P} \to U_{b_1;a_1,\dots,a_k}$ of the bundle $(U_{b_1;a_1,\dots,a_k}, \mathcal{P}, \pi') \subset (V_{1,k}(\mathbf{C}), \mathcal{P}, \pi)$,

$$F(\mathcal{E}(D(t))) = \begin{cases} \mathcal{A}_{[b_1(t);\ \overline{a_1(t),\dots,a_1(t),2b_1(t)}]}, & \text{if } D \equiv 2,3 \mod 4, \\ \mathcal{A}_{[b_1(t);\ \overline{a_1(t),\dots,a_1(t),2b_1(t)-1}]}, & \text{if } D \equiv 1 \mod 4. \end{cases}$$

Corollary 8.5.1. *The Picard number of the surface $\mathcal{E}(D(t))$ is given by the formula*

$$\rho(\mathcal{E}(D(t))) = 1 + k,$$

where k is the length of period of the continued fraction.

Example 8.5.2. Let $D = 2, 3, 7, 11, 19, 43, 67$, or 163. The imaginary quadratic field $\mathbf{Q}(\sqrt{-D})$ has class number one. Since such a number for the real quadratic field $\mathbf{Q}(\sqrt{D})$ is also one, one gets $f = 1$ from a symmetry equation (Theorem 1.4.1). In view of $D \equiv 2, 3 \mod 4$, we use the first line in the formulas for the functor F. The Picard numbers of the surface $\mathcal{E}(D(t))$ are shown in Fig. 8.5.

D	θ	Picard number of surface $\mathcal{E}(D(t))$
2	$[1, \overline{2}]$	2
3	$[1, \overline{1, 2}]$	3
7	$[2, \overline{1, 1, 1, 4}]$	5
11	$[3, \overline{3, 6}]$	3
19	$[4, \overline{2, 1, 3, 1, 2, 8}]$	7
43	$[6, \overline{1, 1, 3, 1, 5, 1, 3, 1, 1, 12}]$	11
67	$[8, \overline{5, 2, 1, 1, 7, 1, 1, 2, 5, 16}]$	11
163	$[12, \overline{1, 3, 3, 2, 1, 1, 7, 1, 11, 1, 7, 1, 1, 2, 3, 3, 1, 24}]$	19

Figure 8.5: Picard numbers.

Example 8.5.3. Let the surface $\mathcal{E}(\mathbf{Q}(t))$ be given by equation in Example 8.5.1. There is no complex multiplication on the fibers in this case. However, one gets from the above $N = 1$ and $k = 2$. Therefore the Picard number $\rho(\mathcal{E}(\mathbf{Q}(t))) = 3$.

Guide to the literature

Our exposition follows [214].

8.6 Class field towers and minimal models

Let k be a number field and let $\mathcal{H}(k)$ be the Hilbert class field of k, i. e., the maximal abelian unramified extension of k. The class field tower is a sequence of the field extensions

$$k \subseteq \mathcal{H}(k) \subseteq \mathcal{H}^2(k) \subseteq \mathcal{H}^3(k) \subseteq \cdots,$$

where $\mathcal{H}^2(k) = \mathcal{H}(\mathcal{H}(k))$, $\mathcal{H}^3(k) = \mathcal{H}(\mathcal{H}^2(k))$, etc. Whether there exists an integer $m \geq 1$ such that $\mathcal{H}^m(k) \cong \mathcal{H}^{m+1}(k) \cong \cdots$ is known as the class field tower problem [89]. It says that the class field tower is finite if $m < \infty$ and infinite otherwise. It is easy to see that the tower is finite if and only if the ring of integers of the field $\mathcal{H}^m(k)$ is the principal ideal domain, i. e., has the class number 1.

Golod–Shafarevich Theorem says that the tower can be infinite for some fields k [95]. This result solves in the negative the class field tower problem. On the other hand, many fields k have finite class field towers. The sorting of k by the finite and infinite towers is a difficult open problem.

In this section we study the real class field towers, i. e., when all fields $\mathcal{H}^i(k)$ are real. It is shown that such towers are always finite (Corollary 8.6.1). To outline the idea, we prove that the blow-up map of an algebraic surface induces a Hilbert class field extension of a field coming from the K_0-group of the corresponding Etesi C^*-algebra (Section 7.5). Castelnuovo Theorem says that it takes a finite number of the blow-ups of an algebraic surface to get the minimal model, hence the tower is finite. To formalize our results, let us recall some definitions.

Recall that an algebraic surface S is a variety of complex dimension 2. The rational map $\phi : S \dashrightarrow S'$ is called birational if its inverse ϕ^{-1} is a rational map. A birational map $\phi : S \dashrightarrow S'$ is a blow-up if it is defined everywhere except for a point $p \in S$ and a rational curve $C \subset S'$ such that $\phi^{-1}(C) = p$. Each birational map is composition of a finite number of the blow-ups. Surface S is called a minimal model if any birational map $S \dashrightarrow S'$ is an isomorphism. Castelnuovo Theorem says that S is a minimal model if and only if S does not contain rational curves C with the self-intersection index -1. In particular, the minimal model is obtained from S by a finite number of the blow-ups along C.

The surface S can be identified with a smooth 4-dimensional manifold. We denote by $\mathrm{Diff}(S)$ a group of the orientation-preserving diffeomorphisms of S and by $\mathrm{Diff}_0(S)$ the connected component of $\mathrm{Diff}(S)$ containing the identity. The Etesi C^*-algebra $\mathbb{E}_S$ is a group C^*-algebra of the locally compact group $\mathrm{Diff}(S)/\mathrm{Diff}_0(S)$ (Section 7.5). The $\mathbb{E}_S$ is a stationary AF-algebra depending on a constant integer matrix $A \in GL(n, \mathbf{Z})$ (Theorem 7.5.1). Let $\lambda_A > 1$ be the Perron–Frobenius eigenvalue of A and let $k = \mathbf{Q}(\lambda_A)$ be a real number field generated by the algebraic number λ_A. Our main result can be formulated as follows.

Theorem 8.6.1. *The birational map $S \dashrightarrow S'$ is a blow-up if and only if $k' \cong \mathcal{H}(k)$.*

As explained above, Theorem 8.6.1 can be used to study the class field towers. Namely, in view of Castelnuovo's theory of the minimal models, one gets the following application of Theorem 8.6.1.

Corollary 8.6.1. *The real class field towers are always finite.*

Remark 8.6.1. The real class field towers and the class field towers over the real fields are not the same, in general. The latter admit imaginary Hilbert class fields and can be infinite.

8.6.1 Algebraic surfaces

An algebraic surface is a variety S of the complex dimension 2. One can identify S with a complex surface and therefore with a smooth 4-dimensional manifold $\mathcal{M}$. In what follows, we denote by $\mathbb{E}_S$ the Etesi C^*-algebra of $\mathcal{M}$ corresponding to the surface S.

The map $\phi : S \dashrightarrow S'$ is called rational if it is given by a rational function. The rational maps cannot be composed unless they are dominant, i. e., the image of ϕ is Zariski dense in S'. The map ϕ is birational if the inverse ϕ^{-1} is a rational map. A birational map $\epsilon : S \dashrightarrow S'$ is called a blow-up if it is defined everywhere except for a point $p \in S$ and a rational curve $C \subset S'$, such that $\epsilon^{-1}(C) = p$. Every birational map $\phi : S \dashrightarrow S'$ is composition of a finite number of the blow-ups, i. e., $\phi = \epsilon_1 \circ \cdots \circ \epsilon_k$.

Surface S is called a minimal model if any birational map $S \dashrightarrow S'$ is an isomorphism. The minimal models exist and are unique unless S is a ruled surface. By Castelnuovo Theorem, the surface S is a minimal model if and only if S does not contain rational curves C with the self-intersection index -1.

8.6.2 Proofs

8.6.2.1 Proof of Theorem 8.6.1

For the sake of clarity, let us outline the main ideas. Let $S \to S'$ be a regular (polynomial) map between the surfaces S and S'. It is known that such a map induces a homomorphism $\mathbb{E}_S \to \mathbb{E}_{S'}$ of the Etesi C^*-algebras (Section 7.2) and an extension of the fields $k \subseteq k'$ in the corresponding Handelman triples $(\Lambda, [\mathfrak{m}], k) \subseteq (\Lambda', [\mathfrak{m}'], k')$ (Section 3.5.2). Recall that the rational map $S \dashrightarrow S'$ is regular only on an open subset $U \subset S$ such that the Zariski closure of U coincides with S. Roughly speaking, we prove that an operation corresponding to the Zariski closure of U consists in passing from the field k to its Hilbert class field $\mathcal{H}(k)$ (Lemma 8.6.1). The rest of the proof follows from the inclusion of fields $\mathcal{H}(k) \subseteq \mathcal{H}(k')$ induced by the rational map $S \dashrightarrow S'$.

We shall split the proof in a series of lemmas.

Lemma 8.6.1. *If $S \dashrightarrow S'$ is a rational map, then $\mathcal{H}(k) \subseteq \mathcal{H}(k')$.*

Proof. In short, an open set $U \subset S$ is a smooth 4-dimensional manifold with boundary. Taking a connected sum with the copies of S^4, one gets a compact smooth manifold S_0 and a regular map $S_0 \to S$. Such a map defines a field extension $k_0 \subseteq k$. Since U is

Zariski dense in S, we conclude that the surface S_0 determines S up to an isomorphism. Therefore the field extension $k_0 \subseteq k$ must depend solely on the arithmetic of the field k_0. In other words, the intrinsic invariants of k_0 control the Galois group $\mathrm{Gal}(k|k_0)$. This can happen if and only if $k \cong \mathcal{H}(k_0)$, so that $\mathrm{Gal}(k|k_0) \cong \mathrm{Cl}(k_0)$, where $\mathrm{Cl}(k_0)$ is the ideal class group of k_0. We pass to a detailed argument.

(i) Let $\phi : S \dashrightarrow S'$ be a rational map. Then there exist the open sets $U \subset S$ and $U' \subset S'$ such that

$$\phi : U \longrightarrow U'$$

is a regular map.

(ii) Since $U \subset S$ is an open set, it is Zariski dense in S. The set U is a 4-dimensional manifold with a boundary corresponding to the poles of the rational map $S \dashrightarrow S'$. Let n be the total number of the boundary components of U. Consider a compact smooth 4-dimensional manifold

$$S_0 := U \#_n S^4$$

coming from the connected sum of U with the n copies of the 4-dimensional sphere S^4 equipped with the standard smooth structure.

(iii) Notice that S_0 can be endowed with a complex structure and, by Chow's Theorem, S_0 is an algebraic surface. It is not hard to see that there exists a regular map $S_0 \to S$ and the corresponding field extension $k_0 \subseteq k$ given by the commutative diagram in Fig. 8.6. Since k_0 and k are totally real number fields, $k_0 \subseteq k$ is a Galois extension. In particular, the Galois group $\mathrm{Gal}(k|k_0)$ is correctly defined.

(iv) Recall that the Zariski closure of U coincides with the surface S. Since S_0 contains U, the Zariski closure of S_0 will coincide with S as well. Using the diagram in Fig. 8.6, we conclude that the Galois extension $k_0 \subseteq k$ depends only on the arithmetic of the ground field k_0. This means that the Galois group $\mathrm{Gal}(k|k_0)$ must be an invariant of the field k_0. The only extension with such a property is the Hilbert class field $\mathcal{H}(k_0)$, i. e., $\mathrm{Gal}(k|k_0) \cong \mathrm{Cl}(k_0)$, where $\mathrm{Cl}(k_0)$ is the ideal class group of k_0. Thus $k \cong \mathcal{H}(k_0)$.

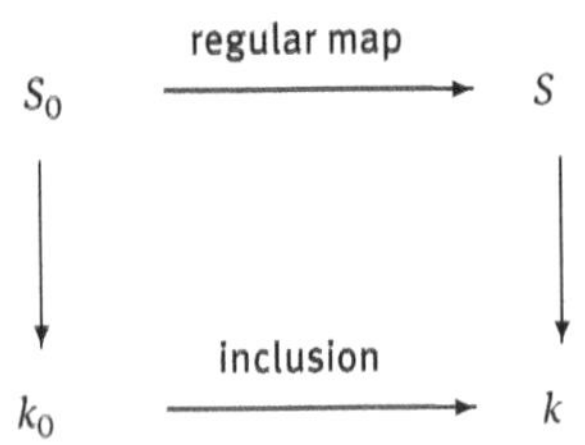

Figure 8.6: Field extension.

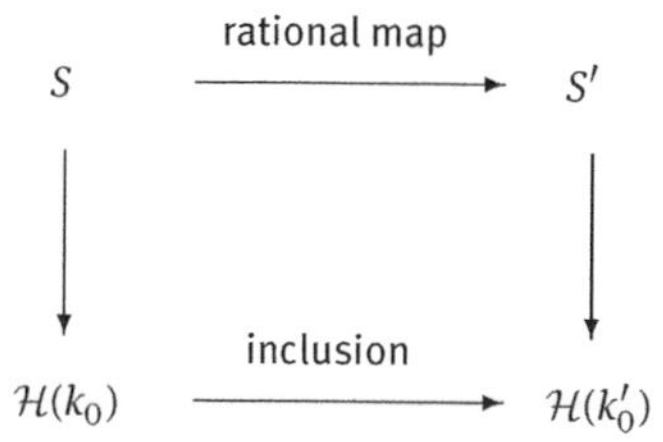

Figure 8.7: Hilbert class field extension.

(v) Using a regular map $\phi : U \to U'$, one gets an inclusion of the number fields $k_0 \subseteq k_0'$ and therefore an inclusion $\mathcal{H}(k_0) \subseteq \mathcal{H}(k_0')$. In other words, the diagram in Fig. 8.6 implies a commutative diagram in Fig. 8.7.

(vi) Lemma 8.6.1 follows from Fig. 8.7 after an adjustment of the notation, i. e., dropping the subscript zero for the number field k_0. □

Lemma 8.6.2. *If $S \dashrightarrow S'$ is a birational map, then $\mathcal{H}(k) \cong \mathcal{H}(k')$.*

Proof. In view of Lemma 8.6.1, the rational map $S \dashrightarrow S'$ implies an inclusion of the number fields $\mathcal{H}(k) \subseteq \mathcal{H}(k')$. Since $S \dashrightarrow S'$ is birational, the inverse rational map $S' \dashrightarrow S$ gives an inclusion of the number fields $\mathcal{H}(k') \subseteq \mathcal{H}(k)$. Clearly, the above inclusions are compatible if and only if $\mathcal{H}(k') \cong \mathcal{H}(k)$. Lemma 8.6.2 is proved. □

Lemma 8.6.3. *If $S \dashrightarrow S'$ is a blow-up, then $k' \cong \mathcal{H}(k)$.*

Proof. We shall use the same argument as in Lemma 8.6.1. Since the blow-up is a dominant rational map, the image of surface S must be Zariski dense in S'. As it was shown earlier, one gets an isomorphism between the field k' and the Hilbert class field of k. We pass to a detailed argument.

(i) Recall that any birational map $\phi : S \dashrightarrow S'$ is a composition

$$\phi = \epsilon_1 \circ \cdots \circ \epsilon_m,$$

where ϵ_i is a blow-up and $m < \infty$. In particular, ϵ_i must be dominant rational maps, see Fig. 8.8. The latter means that the image $\epsilon_i(S)$ is Zariski dense in S'.

(ii) Consider a dominant rational map $\phi : S \dashrightarrow S'$ and the corresponding Galois extension of the number fields $k \subseteq k'$ shown in Fig. 8.9. Since S' is the closure of a Zariski dense subset $\phi(S)$, we conclude that the extension $k \subseteq k'$ depends only on the arithmetic of the ground field k. In particular, the Galois group $\mathrm{Gal}(k'|k)$ is an invariant of the field k. The only extension with such a property is the Hilbert class field $\mathcal{H}(k)$ for which $\mathrm{Gal}(k|k_0) \cong \mathrm{Cl}(k_0)$, where $\mathrm{Cl}(k_0)$ is the ideal class group of k_0. Thus $k' \cong \mathcal{H}(k)$. Lemma 8.6.3 is proved. □

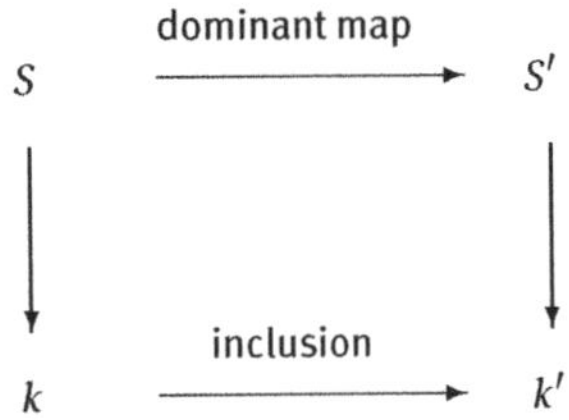

Figure 8.8: Dominant rational map.

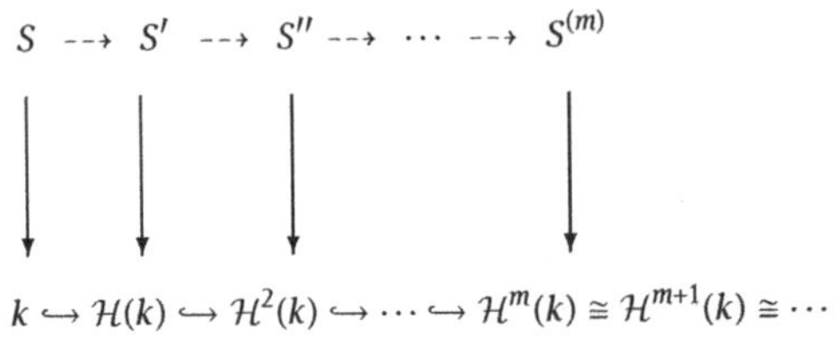

Figure 8.9: Class field tower.

The "if" part of Theorem 8.6.1 follows from Lemma 8.6.3. Let us show that if $k' \cong \mathcal{H}(k)$, then the corresponding birational map $\phi : S \dashrightarrow S'$ is a blow-up. Indeed, since ϕ is a birational map, one can apply Lemma 8.6.2 to obtain an isomorphism of the number fields

$$\mathcal{H}(k) \cong \mathcal{H}(k').$$

The substitution $k' \cong \mathcal{H}(k)$ into the above formula will imply that $\mathcal{H}(k) \cong \mathcal{H}^2(k)$. In other words, the class field tower of k is stable after the first step. In particular, such a tower cannot be decomposed into the subtowers, i. e., the birational map ϕ cannot be decomposed into a composition of the blow-ups ϵ_i. The latter means that ϕ is a blow-up itself. The "only if" part of Theorem 8.6.1 follows.

This argument finishes the proof of Theorem 8.6.1.

8.6.2.2 Proof of Corollary 8.6.1

Our proof is based on the Castelnuovo theory of the minimal models for algebraic surfaces. Namely, such a theory says that it takes a finite number of the blow-ups to get the minimal model of S. The rest of the proof follows from Theorem 8.6.1 applied to the corresponding class field tower. We pass to a detailed argument.

(i) Let us prove that if S is an algebraic surface, then its minimal model gives rise to a finite real class field tower. We denote by $S^{(m)}$ the minimal model obtained from S by composition of the blow-ups

$$\epsilon_i : S^{(i-1)} \dashrightarrow S^{(i)},$$

where $1 \leq i \leq m$. In view of Theorem 8.6.1, each blow-up ϵ_i defines a Hilbert class field extension of the k_{i-1}, i. e.,

$$k_i = \mathcal{H}(k_{i-1}),$$

where all $\{k_i \mid 1 \leq i \leq m\}$ are real number fields.

(ii) After a finite number of the blow-ups, one gets a commutative diagram in Fig. 8.9. Thus the finite real class field tower corresponding to the minimal model $S^{(m)}$ has the form

$$k \subset \mathcal{H}(k) \subset \mathcal{H}^2(k) \subset \cdots \subset \mathcal{H}^m(k) \cong \mathcal{H}^{m+1}(k) \cong \cdots.$$

(iii) Let us show that if

$$k := \mathcal{H}^0(k) \subset \mathcal{H}^1(k) \subset \mathcal{H}^2(k) \subset \cdots$$

is a real class field tower, then it is finite, i. e., there exists an integer $m < \infty$ such that $\mathcal{H}^m(k) \cong \mathcal{H}^{m+1}(k) \cong \cdots$. Indeed, since $k_i := \{\mathcal{H}^i(k) \mid i \geq 0\}$ is a real number field, one can construct a Handelman triple $(\Lambda_i, [\mathfrak{m}_i], k_i)$ and a smooth 4-dimensional manifold $\mathcal{M}_i$. Since the triple $(\Lambda_i, [\mathfrak{m}_i], k_i)$ is a topological invariant of $\mathcal{M}_i$, we can always assume that $\mathcal{M}_i$ is a complex surface by choosing a proper smoothing of $\mathcal{M}_i$ if necessary. By Chow's Theorem, one can identify $\mathcal{M}_i$ with an algebraic surface $S^{(i)}$.

(iv) Since $k_i = \mathcal{H}(k_{i-1})$, one can apply Theorem 8.6.1 saying that $S^{(i)}$ is a blow-up of the surface $S^{(i-1)}$. Using Castelnuovo Theorem, one concludes that the class field tower must stabilize for an integer $m < \infty$, i. e., such a tower is always finite.

This argument finishes the proof of Corollary 8.1.1.

Guide to the literature

The class field tower problem was raised in 1916 by Furtwängler [89]. Golod and Shafarevich proved that such towers can be infinite [95]. Our exposition is based on [215].

Exercises

1. Find an exact relation between the analytic values of $|\text{Ш}(\mathcal{E}(K))|$ and those given by formula in Remark 8.3.2.
2. Find an analytic interpretation to the formula $\text{Ш}(\mathcal{E}_{CM}) \cong \mathrm{Cl}(R) \oplus \mathrm{Cl}(R)$ given in Corollary 8.4.2.
3. Prove that the surface $\mathcal{E}(D(t))$ in Example 8.5.2 is a minimal model.

Part III: **Brief survey of NCG**

9 Finite geometries

Denote by D a skew field, i. e., an associative but in general noncommutative division ring. Desargues Theorem is true in the projective spaces $\mathbb{P}^n(D)$ for any $n \geq 3$. On the other hand, Pappus Theorem in $\mathbb{P}^n(D)$ is true only when D is commutative, i. e., a field. Thus the geometry of the space $\mathbb{P}^n(D)$ depends on the skew field D. There exists a bijection between the rational identities in D and configurations of the points, lines, etc., in the projective space $\mathbb{P}^n(D)$. In particular, when $D \cong \mathbb{F}_p$ is a finite field, there exists only a finite set of such configurations. We call the space $\mathbb{P}^n(D)$ a *finite geometry*. We refer the reader to the monograph [116] for a detailed account. For an algebraic approach, we recommend [7] and [19]. Our exposition follows [252, pp. 84–87].

9.1 Axioms of projective geometry

Recall that all statements of plane geometry can be deduced from a finite set of axioms. Namely,

(a) Through any two distinct points there is one and only one line;
(b) Given any line and a point not on it, there exists one and only one line through the point and not intersecting the line, i. e., parallel to it;
(c) There exist three points not on any line.

Are there geometries different from the plane geometry satisfying the set of axioms (a)–(c)? Below is an example of such a geometry.

Example 9.1.1. Consider a model with four points A, B, C, and D, and six lines AB, BC, CD, DA, AC, and BD, see Fig. 9.1. The reader can verify that the axioms (a)–(c) are satisfied. The parallel lines are $AB \parallel CD$, $BC \parallel AD$, and, counterintuitively, $AC \parallel BD$.

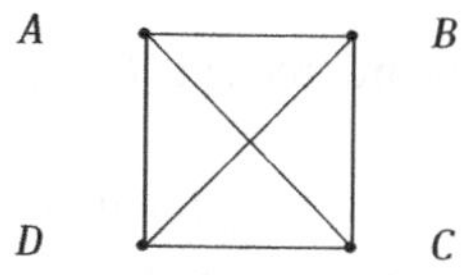

Figure 9.1: Finite geometry on four points and six lines.

Following the seminal idea of Descartes, one can introduce a coordinate system (X, Y) in the plane and write the points of a line as solutions to the linear equation

$$aX + bY = c,$$

for some constants a, b, c in a skew field D; the axioms (a)–(c) will reduce to a system of algebraic equations over D.

https://doi.org/10.1515/9783110788709-009

Example 9.1.2. Let $D \cong \mathbb{F}_2$ be the field of characteristic 2. The reader can verify that a finite geometry on four points and six lines can be written as

$$\begin{cases} A = (0,0), \\ B = (0,1), \\ C = (1,0), \\ D = (1,1), \end{cases}$$

and

$$\begin{cases} AB: & X = 0, \\ CD: & X = 1, \\ AD: & X + Y = 0, \\ BC: & X + Y = 1, \\ AC: & Y = 0, \\ BD: & Y = 1. \end{cases}$$

9.2 Projective spaces over skew fields

Let D be a skew field; consider its matrix ring $M_n(D)$. The left ideals of $M_n(D)$ are in a one-to-one correspondence with subspaces V_i of the n-dimensional space $M_n(D^{op})$, where D^{op} is the opposite skew field of D. The set of all V_i has a natural (partial) order structure coming from the inclusions of V_i into each other. Is it possible to recover the matrix ring $M_n(D)$, its dimension n, and the skew field D itself from the partially ordered set $\{V_i\}$? The set $\{V_i\}$ corresponds to a set of all linear subspaces of the projective space $\mathbb{P}^{n-1}(D)$ and, therefore, our question reduces to that about the axiomatic structure of the projective space $\mathbb{P}^{n-1}(D)$. Thus one arrives at the following

Definition 9.2.1. By a projective space one understands a partially ordered set $(\mathcal{P}, \geq)$, which satisfies the following conditions:

(i) For any set of elements $x_\alpha \in \mathcal{P}$, there exists an element y such that $y \geq x_\alpha$ for all α; if $z \geq x_\alpha$, then $z \geq y$. The element y is called the sum of the x_α and is denoted by $\cup x_\alpha$. The sum of all $x \in \mathcal{P}$ exists and is called the whole projective space $I(\mathcal{P})$.

(ii) For any set of elements $x_\alpha \in \mathcal{P}$, there exists an element y' such that $y' \leq x_\alpha$ for all α; if $z' \leq x_\alpha$, then $z' \leq y'$. The element y' is called the intersection of the x_α and is denoted by $\cap x_\alpha$. The intersection of all $x \in \mathcal{P}$ exists and is called the empty set $\emptyset(\mathcal{P})$.

(iii) For any $x, y \in \mathcal{P}$ and $a \in x/y$, there exists an element $b \in x/y$ such that $a \cup b = I(x/y)$ and $a \cap b = \emptyset(x/y)$, where x/y is the partially ordered set of all $z \in \mathcal{P}$ such that $y \leq z \leq x$. If $b' \in x/y$ is another element with the same properties and if $b \leq b'$, then $b = b'$.

(iv) The length of all chains $a_1 \leq a_2 \leq \cdots \leq a_r$ with $a_1 \neq a_2 \neq a_3 \neq \cdots \neq a_r$ is bounded.
(v) The element $a \in \mathcal{P}$ is called a point if $b \leq a$ and $b \neq a$ imply that $b = \emptyset(\mathcal{P})$. For any two points a and b, there exists a point c such that $c \neq a$, $c \neq b$ and $c \leq a \cup b$.

Definition 9.2.2. By a dimension function on the projective space $(\mathcal{P}, \geq)$ one understands the function $d(a)$ equal to the maximum length of a chain $\emptyset \leq \cdots \leq a$; such a function satisfies the equality

$$d(a \cap b) + d(a \cup b) = d(a) + d(b)$$

for all $a, b \in \mathcal{P}$. The number $d(I(\mathcal{P}))$ is called the dimension of projective space $(\mathcal{P}, \geq)$.

Example 9.2.1. If $(\mathcal{P}, \geq)$ is the projective space of all linear subspaces of the space $\mathbb{P}^n(D)$, then $n = d(I(\mathcal{P}))$.

The following important result says that the skew field D can be recovered from the corresponding partially ordered set $\mathbb{P}^n(D)$. In other words, a given projective geometry determines the skew field D.

Theorem 9.2.1 (Fundamental Theorem of Projective Geometry).
(i) *If $n \geq 2$ and $(\mathcal{P}, \geq)$ is a projective space coming from the linear subspaces of the space $\mathbb{P}^n(D)$ for a skew field D, then $(\mathcal{P}, \geq)$ determines the number n and the skew field D;*
(ii) *If $(\mathcal{P}, \geq)$ is an arbitrary projective space of dimension $n \geq 3$, then it is isomorphic to the space $\mathbb{P}^n(D)$ over some skew field D.*

Remark 9.2.1. Not every projective space $(\mathcal{P}, \geq)$ of dimension 2 is isomorphic to $\mathbb{P}^2(D)$, unless one adds the Desargues axiom to the list defining the projective space.

9.3 Desargues and Pappus axioms

Theorem 9.3.1 (Desargues). *If the three lines AA', BB', and CC' joining corresponding vertices of two triangles ABC and $A'B'C'$ intersect in a point O, then the points of intersection of the corresponding sides are collinear.*

Theorem 9.3.2 (Pappus). *If the vertices of a hexagon $P_1, \ldots, P_6$ lie three by three on two lines, then the points of intersection of the opposite sides P_1P_2 and P_4P_5, P_2P_3 and P_5P_6, as well as P_3P_4 and P_6P_1 are collinear, see Fig. 9.2.*

Definition 9.3.1. By a Desargues projective plane one understands a two-dimensional projective plane $(\mathcal{P}, \geq)$ which satisfies Desargues Theorem.

Theorem 9.3.3. *The Desargues projective plane $(\mathcal{P}, \geq)$ is isomorphic to $\mathbb{P}^2(D)$ for some skew field D.*

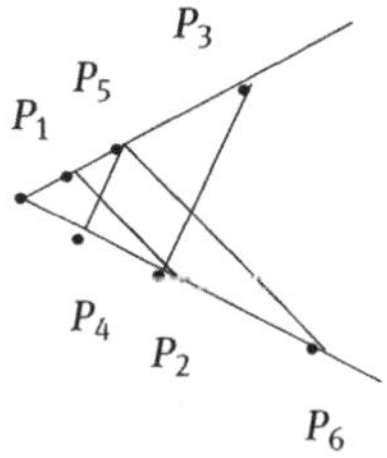

Figure 9.2: Pappus Theorem.

Definition 9.3.2. By a Pappus projective space one understands an n-dimensional Desargues projective space $(\mathcal{P}, \geq)$ which satisfies Pappus Theorem.

The following elegant result links the geometry of projective spaces to an intrinsic property of the skew field D.

Theorem 9.3.4. *The n-dimensional Desargues projective space* $(\mathcal{P}, \geq) \cong \mathbb{P}^n(D)$ *is isomorphic to a Pappus projective space if and only if D is commutative, i. e., a field.*

Guide to the literature

The subject of finite geometries is covered in the classical book [116]. We refer the reader to the monographs [7] and [19] for a more algebraic treatment. Our exposition follows the guidelines of [252, pp. 84–87].

10 Continuous geometries

Continuous geometry is an extension of the finite geometry $\mathbb{P}^n(D)$ to the case $n = \infty$ and $D \cong \mathbf{C}$. Geometry $\mathbb{P}^\infty(\mathbf{C})$ is identified with the ring of bounded linear operators on a Hilbert space $\mathcal{H}$ called factor of a von Neumann (or W^*-) algebra $\mathcal{A}$, which is a $*$-subalgebra of the algebra $B(\mathcal{H})$ of bounded linear operators on $\mathcal{H}$ such that its weak topological closure can be expressed in purely algebraic terms as a double commutant of the $*$-subalgebra. The W^*- algebras and factors are reviewed in Section 10.1, and corresponding continuous geometries are reviewed in Section 10.2.

10.1 W^*-algebras

Definition 10.1.1. The weak topology on the space $B(\mathcal{H})$ is defined by the seminorm $\|\bullet\|$ on each $S \in B(\mathcal{H})$ given by the formula

$$\|S\| = |(Sx, y)|$$

as x and y run through all of $\mathcal{H}$.

Definition 10.1.2. By a commutant of a subset $\mathcal{S} \subset B(\mathcal{H})$ one understands a unital subalgebra $\mathcal{S}'$ of $B(\mathcal{H})$ given by the formula

$$\mathcal{S}' = \{T \in B(\mathcal{H}) \mid TS = ST \text{ for all } S \in \mathcal{S}\}.$$

The double commutant of $\mathcal{S}$ is denoted by $\mathcal{S}'' := (\mathcal{S}')'$.

Remark 10.1.1. The reader can verify that $\mathcal{S}'$ is closed in the weak operator topology.

Definition 10.1.3. Let $\mathcal{S}$ be a subset of $B(\mathcal{H})$. Then:
(i) **Alg** $\mathcal{S}$ denotes the subalgebra of $B(\mathcal{H})$ generated by finite linear combinations of the finite products of elements of $\mathcal{S}$;
(ii) $\mathcal{S}$ is called self-adjoint if for each $S \in \mathcal{S}$ we have $S^* \in \mathcal{S}$;
(iii) $\mathcal{S}$ is called nondegenerate if the only $x \in \mathcal{H}$ for which $Sx = 0$ for all $S \in \mathcal{S}$ is $x = 0$.

Theorem 10.1.1 (Murray–von Neumann). *If $\mathcal{S} \subset B(\mathcal{H})$ is self-adjoint and nondegenerate, then* **Alg** $\mathcal{S}$ *is dense in $\mathcal{S}''$.*

Corollary 10.1.1 (Double Commutant Theorem). *For any unital $*$-algebra $\mathcal{A} \subset B(\mathcal{H})$, the double commutant $\mathcal{A}''$ coincides with the weak closure* **Clos** $\mathcal{A}$ *of the algebra $\mathcal{A}$.*

Definition 10.1.4. By a (concrete) W^*-algebra one understands a unital $*$-subalgebra $\mathcal{A}$ of $B(\mathcal{H})$ such that

$$\mathcal{A} \cong \mathcal{A}'' \cong \mathbf{Clos}\, \mathcal{A}.$$

https://doi.org/10.1515/9783110788709-010

Example 10.1.1. The algebra $B(\mathcal{H})$ is closed in the weak topology. Therefore $B(\mathcal{H})$ is a W^*-algebra.

Example 10.1.2. Ring $M_n(\mathbf{C})$ is a W^*-algebra and also a C^*-algebra, because the norm topology and the weak topology coincide on the finite-dimensional Hilbert space $\mathbf{C}^n$.

Example 10.1.3. If G is a group and $g \mapsto u_g$ is a unitary representation of G, then the commutant $\{u_g\}'$ is a W^*-algebra.

Example 10.1.4. Let S^1 be the unit circle and $L^\infty(S^1)$ the algebra of measurable functions on S^1; consider the crossed product $L^\infty(S^1) \rtimes \mathbf{Z}$ by the automorphisms of $L^\infty(S^1)$ induced by rotation of S^1 through an angle α. The $L^\infty(S^1) \rtimes \mathbf{Z}$ is a W^*-algebra.

An analogue of the AF-algebras for the W^*-algebras is given by the following definition.

Definition 10.1.5. A W^*-algebra $\mathcal{A}$ is called hyper-finite if there exists an increasing sequence $\mathcal{A}_n$ of finite-dimensional W^*-subalgebras of $\mathcal{A}$ whose union is weakly dense in $\mathcal{A}$.

Example 10.1.5. The W^*-algebra $L^\infty(S^1) \rtimes \mathbf{Z}$ is hyper-finite.

Definition 10.1.6. A W^*-algebra $\mathcal{A}$ is called a factor if the center of $\mathcal{A}$ (i. e., abelian algebra $Z(\mathcal{A}) \subset \mathcal{A}$ which commutes with everything in $\mathcal{A}$) is isomorphic to the scalar multiple of the identity operator.

Example 10.1.6. The W^*-algebra $B(\mathcal{H})$ is a factor.

Example 10.1.7. The W^*-algebra $L^\infty(S^1) \rtimes \mathbf{Z}$ is a factor.

Each W^*-algebra decomposes into a direct integral of factors; to classify the latter, one needs the following definition.

Definition 10.1.7. A pair of projections $p, q \in \mathcal{A}$ is written $p \leq q$ if there exists a partial isometry $u \in \mathcal{A}$ such that $p = uu^*$ and $u^*uq = u^*u$; we say that $p \sim q$ are equivalent if $p = uu^*$ and $q = u^*u$.

Theorem 10.1.2 (Murray–von Neumann). *If $\mathcal{A}$ is a factor and p, q are projections in $\mathcal{A}$, then either $p \leq q$ or $q \leq p$.*

Definition 10.1.8. If $p \in \mathcal{A}$ is a projection, then
(i) p is called finite if $p \sim q \leq p$ implies $q = p$ for all projections $q \in \mathcal{A}$;
(ii) p is called infinite if there is projection $q \sim p$ such that $p < q$;
(iii) $p \neq 0$ is called minimal if p dominates no other projection in $\mathcal{A}$.

Definition 10.1.9. A factor $\mathcal{A}$ is said belong to the type:
(i) **I** if $\mathcal{A}$ has a minimal projection;
(ii) $\mathbf{II}_1$ if $\mathcal{A}$ has no minimal projections and every projection is finite;

(iii) $\mathbf{II}_\infty$ if $\mathcal{A}$ has no minimal projections, but $\mathcal{A}$ has both finite and infinite projections;
(iv) $\mathbf{III}$ if $\mathcal{A}$ has no finite projections except 0.

Remark 10.1.2. The factors of types $\mathbf{I}$, $\mathbf{II}_1$, $\mathbf{II}_\infty$, and $\mathbf{III}$ can be realized as concrete W^*-algebras.

Example 10.1.8. The W^*-algebra $L^\infty(S^1) \rtimes \mathbf{Z}$ is a type $\mathbf{II}_1$ factor whenever the angle α is irrational.

Theorem 10.1.3 (Murray–von Neumann). *All hyper-finite type $\mathbf{II}_1$ factors are isomorphic.*

Remark 10.1.3. Unlike the irrational rotation C^*-algebras, the corresponding W^*-algebras $L^\infty(S^1) \rtimes \mathbf{Z}$ are isomorphic to each other; it follows from Murray–von Neumann Theorem.

10.2 Von Neumann geometry

Murray and von Neumann observed that projections in the type $\mathbf{I}_n$, $\mathbf{I}_\infty$, $\mathbf{II}_1$, $\mathbf{II}_\infty$, and $\mathbf{III}$-factors in W^*-algebras are partially ordered sets $(\mathcal{P}, \geq)$ by Theorem 10.1.2. The set $(\mathcal{P}, \geq)$ satisfies axioms (i)–(v) of projective space $\mathbb{P}^n(D)$; we refer the reader to Definition 9.2.1. However, the range of dimension function d introduced in Definition 9.2.2 is no longer a discrete but a continuous subset of $\mathbf{R}$ unless $\mathbb{P}^n(D)$ is a type $\mathbf{I}_n$ factor. According to Theorem 9.2.1, type $\mathbf{I}_\infty$, $\mathbf{II}_1$, $\mathbf{II}_\infty$, and $\mathbf{III}$-factors correspond to the infinite-dimensional projective spaces $\mathbb{P}^\infty(\mathbf{C})$. Geometry $\mathbb{P}^\infty(\mathbf{C})$ is called a *von Neumann geometry*.

Theorem 10.2.1 (von Neumann). *The von Neumann geometry belongs to one of the following types:*
(i) $\mathbf{I}_n$ *if the dimension function takes values* $0, 1, 2, \dots, n$, *and the corresponding factor is isomorphic to the ring* $M_n(\mathbf{C})$;
(ii) $\mathbf{I}_\infty$ *if the dimension function takes values* $0, 1, 2, \dots, \infty$, *and the corresponding factor is isomorphic to the ring* $\mathcal{B}(\mathcal{H})$ *of all bounded operators on a Hilbert space* $\mathcal{H}$;
(iii) $\mathbf{II}_1$ *if the dimension function takes values in the interval* $[0, 1]$;
(iv) $\mathbf{II}_\infty$ *if the dimension function takes values in the interval* $[0, \infty]$;
(v) $\mathbf{III}$ *if the dimension function takes values* 0 *and* ∞.

Guide to the literature

The W^*-algebras were introduced in the seminal paper [171]. The textbooks [248] and [86, Chapter 4] cover the W^*-algebras. Continuous geometries were introduced and studied in the monograph [174].

11 Connes geometries

The Connes' method uses geometry as a tool to classify noncommutative algebras; a similar approach prevails in the noncommutative algebraic geometry of M. Artin, J. Tate, and M. van den Bergh, see Chapter 13. The early works of A. Connes indicated that a difficult problem of classification of the W^*-algebras admits an elegant solution in terms of the dynamics of flows and geometry of foliations [53]. This approach became a vast program, where one seeks to recast geometry of the space X in terms of the commutative algebra $C(X)$ and to use the acquired formal language to classify the noncommutative algebras. The program is known as noncommutative geometry in the sense of A. Connes, or *Connes geometry* for brevity.

11.1 Classification of type III factors

The first successful application of Connes' method was classification of the type III factors of W^*-algebras in terms of the flow of weights. A precursor of the classification was the Tomita–Takesaki theory of the modular automorphism group connected to a W^*-algebra $\mathcal{M}$.

11.1.1 Tomita–Takesaki theory

Definition 11.1.1. Let $\mathcal{M}$ be a W^*-algebra. By a weight φ on $\mathcal{M}$ one understands a linear map

$$\varphi : \mathcal{M}_+ \longrightarrow [0, \infty],$$

where $\mathcal{M}_+$ is the positive cone of $\mathcal{M}$. The trace on $\mathcal{M}$ is a weight φ which is invariant under the action of the unitary group $U(\mathcal{M})$ of $\mathcal{M}$, i. e.,

$$\varphi(uxu^{-1}) = \varphi(x), \quad \forall u \in U(\mathcal{M}).$$

Remark 11.1.1. Every W^*-algebra on a separable Hilbert space $\mathcal{H}$ has a weight but not a trace; for instance, type III factors have a weight but no traces.

Remark 11.1.2. Recall that the GNS construction is well defined for the traces. In the case of a weight φ on $\mathcal{M}$, the representation of $\mathcal{M}$ using the GNS construction does not yield a unique scalar product on the Hilbert space $\mathcal{H}_\varphi$ because in general $\varphi(x^*x) \neq \varphi(xx^*)$ for $x \in \mathcal{M}$. Thus one gets an unbounded operator

$$S_\varphi : \mathcal{H}_\varphi \to \mathcal{H}_\varphi$$

https://doi.org/10.1515/9783110788709-011

which measures a discrepancy between the scalar products $\varphi(x^*x)$ and $\varphi(xx^*)$. While dealing with the unbounded operator S_φ it was suggested by Tomita and Takesaki to consider the corresponding polar decomposition

$$S_\varphi = J\Delta_\varphi^{\frac{1}{2}},$$

where $J^2 = I$ is an isometric involution on $\mathcal{H}_\varphi$ such that $J\mathcal{M}J = \mathcal{M}'$ replaces the $*$-involution and Δ_φ a positive operator.

Theorem 11.1.1 (Tomita–Takesaki). *For every $t \in \mathbf{R}$, the positive operator Δ_φ satisfies an isomorphism*

$$\Delta_\varphi^{it}\, \mathcal{M}\, \Delta_\varphi^{-it} \cong \mathcal{M}.$$

Remark 11.1.3. The Tomita–Takesaki formula defines a one-parameter group of automorphisms σ_φ^t of the W^*-algebra $\mathcal{M}$; σ_φ^t is called a modular automorphism group of $\mathcal{M}$.

Remark 11.1.4. The lack of trace property $\varphi(xy) = \varphi(yx)$ for general weights φ on $\mathcal{M}$ has an elegant phrasing as the Kubo–Martin–Schwinger (KMS) condition

$$\varphi(x\sigma_\varphi^{-i\beta}(y)) = \varphi(yx),$$

where $t = -i\beta$ and β is known in the quantum statistical physics as an inverse temperature.

11.1.2 Connes invariants

Theorem 11.1.2 (Connes). *The modular automorphism group σ_φ^t of the W^*-algebra $\mathcal{M}$ is independent of the weight φ modulo an inner automorphism of $\mathcal{M}$.*

Remark 11.1.5. The canonical modular automorphism group $\sigma^t : \mathcal{M} \to \mathcal{M}$ is called a flow of weights on $\mathcal{M}$.

Definition 11.1.2. By the Connes invariants $T(\mathcal{M})$ and $S(\mathcal{M})$ of the W^*-algebra $\mathcal{M}$ one understands the following subsets of the real line $\mathbf{R}$:

$$\begin{cases} T(\mathcal{M}) := \{t \in \mathbf{R} \mid \sigma_t \in \mathrm{Aut}(\mathcal{M}) \text{ is inner}\}, \\ S(\mathcal{M}) := \{\cap_\varphi \mathbf{Spec}(\Delta_\varphi) \subset \mathbf{R} \mid \varphi \text{ a weight on } \mathcal{M}\}, \end{cases}$$

where $\mathbf{Spec}(\Delta_\varphi)$ is the spectrum of the self-adjoint linear operator Δ_φ.

Theorem 11.1.3 (Connes). *If $\mathcal{M}$ is a type III factor, then $T(\mathcal{M}) \cong \mathbf{R}$ and $S(\mathcal{M})$ is a closed multiplicative semigroup of the interval $[0, \infty]$.*

Corollary 11.1.1. *The type III factors $\mathcal{M}$ are isomorphic to one of the following:*

$$S(\mathcal{M}) = \begin{cases} [0,\infty) & \textit{type } \mathrm{III}_0, \\ \{\lambda^{\mathbf{Z}} \cup \{0\} \mid 0 < \lambda < 1\} & \textit{type } \mathrm{III}_\lambda, \\ \{0,1\} & \textit{type } \mathrm{III}_1. \end{cases}$$

Remark 11.1.6 (Connes–Takesaki). The type III factors have the form

$$\begin{cases} \mathbf{III}_0 \cong \mathbf{II}_\infty \rtimes \mathbf{Z}, \\ \mathbf{III}_\lambda \cong \mathbf{II}_\infty \rtimes_\lambda \mathbf{Z}, \\ \mathbf{III}_1 \cong \mathbf{II}_\infty \rtimes \mathbf{R}, \end{cases}$$

where the automorphism of type $\mathbf{II}_\infty$ factor is given by scaling of the trace by constant λ.

11.2 Noncommutative differential geometry

Let $\mathcal{A} \cong C(X)$ be a commutative C^*-algebra. Gelfand–Naimark Theorem says that such algebras are in a one-to-one correspondence with the Hausdorff topological spaces X. If X is an algebraic variety, then its coordinate ring $A(X)$ consists of smooth (even meromorphic!) complex-valued functions on X; the Nullstellensatz says that the rings $A(X)$ are in bijection with the algebraic varieties X. Thus to study the Hausdorff spaces with the differentiable structure, we must take a dense subalgebra of $\mathcal{A}$ consisting of the smooth complex-valued functions on X. An analog of the de Rham theory for such subalgebras is known as a cyclic homology. The cyclic homology is a modification of the Hochschild homology known in algebraic geometry. Unlike the Hochschild homology, the cyclic homology is linked to the K-theory by a homomorphism called the Chern–Connes character.

11.2.1 Hochschild homology

Definition 11.2.1. For a commutative algebra A, by the Hochschild homology $HH_*(A)$ we mean the homology of the complex $C_*(A)$,

$$\cdots \xrightarrow{b_{n+1}} A^{\otimes n+1} \xrightarrow{b_n} A^{\otimes n} \xrightarrow{b_{n-1}} \cdots \xrightarrow{b_1} A,$$

where

$$A^{\otimes n} := \underbrace{A \otimes \cdots \otimes A}_{n \text{ times}},$$

and the boundary map b is defined by the formula

$$b_n(a_0 \otimes a_1 \otimes \cdots \otimes a_n) = \sum_{i=0}^{n-1} (-1)^i a_0 \otimes \cdots \otimes a_i a_{i+1} \otimes \cdots \otimes a_n$$
$$+ (-1)^n a_n a_0 \otimes a_1 \otimes \cdots \otimes a_{n-1}.$$

Example 11.2.1 (Hochschild–Kostant–Rosenberg). Let A be the algebra of regular functions on a smooth affine variety X; denote by $\Omega^n(A)$ the module of differential n-forms over an algebraic de Rham complex of A. For each $n \geq 0$, there exists an isomorphism

$$\Omega^n(A) \cong HH_n(A).$$

Remark 11.2.1. If A is a commutative algebra, the Hochschild homology is isomorphic to the space of differential forms over the algebra. Note that the groups $HH_n(A)$ are well defined for a noncommutative algebra A.

Remark 11.2.2. Let $K_*(A)$ be the K-theory of algebra A. Although there exist natural maps

$$K_i(A) \to HH_i(A),$$

they are rarely homomorphisms. One needs to modify the Hochschild homology to get the desired homomorphisms, i. e., an analog of the Chern character. Thus one arrives at the notion of a cyclic homology.

11.2.2 Cyclic homology

The cyclic homology comes from a subcomplex of the Hochschild complex $C_*(A)$ of algebra A; the subcomplex is known as a cyclic complex.

Definition 11.2.2. An n-chain of the Hochschild complex $C_*(A)$ of algebra A is called cyclic if

$$a_0 \otimes a_1 \otimes \cdots \otimes a_n = (-1)^n a_n \otimes a_0 \otimes a_1 \otimes \cdots \otimes a_{n-1}$$

for all $a_0, \ldots, a_n$ in A; in other words, the cyclic n-chains are invariant of a map $\lambda : C_*(A) \to C_*(A)$ such that $\lambda^n = \mathrm{Id}$. We shall denote the space of all cyclic n-chains by $C_n^\lambda(A)$. The subcomplex of the Hochschild complex $C_*(A)$ of the form

$$\cdots \xrightarrow{b_{n+1}} C_{n+1}^\lambda(A) \xrightarrow{b_n} C_n^\lambda(A) \xrightarrow{b_{n-1}} \cdots \xrightarrow{b_1} C_1^\lambda(A)$$

is called the cyclic complex of algebra A. The homology of the cyclic complex is denoted $HC_*(A)$ and is called a cyclic homology.

Remark 11.2.3 (Connes). The inclusion map of the cyclic complex C_*^λ into the Hochschild complex $C_*(A)$,

$$\iota : C_*^\lambda(A) \longrightarrow C_*(A),$$

induces a map

$$\iota_* : HH_n(A) \longrightarrow HC_n(A).$$

The map is an isomorphism for $n = 0$ and a surjection for $n = 1$.

Example 11.2.2. Let $\mathcal{A}_\theta^0 \subset \mathcal{A}_\theta$ be the dense subalgebra of smooth functions of noncommutative torus $\mathcal{A}_\theta$, where $\theta \in \mathbf{R}\backslash\mathbf{Q}$; let $HC^*(\mathcal{A}_\theta^0)$ denote the cyclic cohomology of $\mathcal{A}_\theta^0$, i. e., a dual of the cyclic homology $HC_*(\mathcal{A}_\theta^0)$. Then

$$HC^0(\mathcal{A}_\theta^0) \cong \mathbf{C}$$

and the map

$$\iota^* : HC^1(\mathcal{A}_\theta^0) \longrightarrow HH^1(\mathcal{A}_\theta^0)$$

is an isomorphism.

Theorem 11.2.1 (Chern–Connes character). *For an associative algebra A and each integer $n \geq 0$, there exist natural maps*

$$\begin{cases} \chi^{2n} : K_0(A) \longrightarrow HC_{2n}(A), \\ \chi^{2n+1} : K_1(A) \longrightarrow HC_{2n+1}(A). \end{cases}$$

Moreover, in case $n = 0$, the map χ^0 is a homomorphism.

11.2.3 Novikov Conjecture for hyperbolic groups

Similar to Kasparov's KK-theory for C^*-algebras, the cyclic homology can be used to prove certain cases of the Novikov Conjecture on the higher signatures of smooth manifolds. We refer the reader to Section 12.4.1 for a review of the Novikov Conjecture.

Definition 11.2.3. Let $G = \langle g_1, \dots, g_n \mid r_1, \dots, r_s\rangle$ be a group on n generators and s relations. By the length $l(w)$ of a word $w \in G$ one understands the minimal number of g_i (counted with powers) which is necessary to write w; the function

$$l : G \to \mathbf{R}$$

turns G into a metric space (G, d), where d is the distance function. The group G is called hyperbolic whenever (G, d) is a hyperbolic metric space, i. e., for some $w_0 \in G$ there exists $\delta_0 > 0$ such that

$$\delta(w_1, w_3) \geq \inf(\delta(w_1, w_2), \delta(w_2, w_3)) - \delta_0, \quad \forall w_1, w_2, w_3 \in G,$$

where

$$\delta(w, w') := \frac{d(w, w_0) + d(w', w_0) - d(w, w')}{2}.$$

Theorem 11.2.2 (Connes–Moscovici). *If G is a hyperbolic group, then G satisfies the Strong Novikov Conjecture.*

Proof. The proof uses a homotopy invariance of the cyclic homology. □

11.3 Connes' Index Theorem

The Index Theory says that the index of a Fredholm operator on a closed smooth manifold M is an integer number; we refer the reader to Chapter 10 for a review. A noncommutative analog of Atiyah–Singer Index Theorem due to A. Connes says that the index can be any real number. The idea is to consider Fredholm operators not on M but on a foliated M, i. e., on the leaves of a foliation $\mathcal{F}$ of M. The values of the index are equal to the trace of elements of the group $K_0(C^*_{red}(\mathcal{F}))$, where $C^*_{red}(\mathcal{F})$ is a reduced C^*-algebra of foliation $\mathcal{F}$. A precursor of Connes' Index Theorem was Atiyah–Singer Index Theorem for the families of elliptic operators on closed manifolds.

11.3.1 Atiyah–Singer Theorem for families of elliptic operators

We refer the reader to Chapter 10 for a brief introduction to the Index Theory. Unlike the classical Atiyah–Singer Theorem, the analytic index of a family of elliptic operators on a compact manifold M no longer belongs to $\mathbf{Z}$ but to the K-homology group of the corresponding parameter space. It was the way M. Atiyah constructed his realization of the K-homology, which is a theory dual to the topological K-theory. Below we briefly review the construction.

Let M be an oriented compact smooth manifold and let

$$D : \Gamma^\infty(E) \to \Gamma^\infty(F)$$

be an elliptic operator defined on the cross-sections of the vector bundles E, F on M. Let B be a locally compact space called a parameter space and consider a continuous family of elliptic operators on M,

$$\mathcal{D} := \{D_b \mid b \in B\},$$

parametrized by B. It is not hard to see that the maps

$$\begin{cases} b \longrightarrow \operatorname{Ker} D_b, \\ b \longrightarrow \operatorname{Ker} D_b^* \end{cases}$$

define two vector bundles E, F over B with a compact support.

Definition 11.3.1. By an analytic index $\mathbf{Ind}(\mathcal{D})$ of the family $\mathcal{D}$ one understands the difference of equivalence classes $[E]$, $[F]$ of the vector bundles E, F, i. e.,

$$\mathbf{Ind}(\mathcal{D}) := [E] - [F] \in K^0(B),$$

where $K^0(B)$ is the K-homology group of the topological space B.

Theorem 11.3.1 (Atiyah–Singer). *For any continuous family $\mathcal{D}$ of elliptic operators on a compact oriented differentiable manifold M, the analytic index $\mathbf{Ind}(\mathcal{D})$ is given by the formula*

$$\mathbf{Ind}(\mathcal{D}) = \langle \mathbf{ch}(\mathcal{D})\, \mathbf{Td}(M) \rangle [M] \in K_0(B),$$

where $\mathbf{ch}(\mathcal{D})$ is the Chern class of family $\mathcal{D}$, $\mathbf{Td}(M)$ is the Todd genus of M, $[M]$ the fundamental homology class of M, and $K_0(B)$ the K_0-group of the parameter space B.

11.3.2 Foliated spaces

A foliated space (or a foliation) is a decomposition of the space X into a disjoint union of subspaces of X called leaves such that a neighborhood of each point $x \in X$ is a trivial fibration. The global behavior of leaves can be rather complicated. Each global fiber bundle (E, p, B) is a foliation of the total space E, yet the converse is false. Thus foliations is a generalization of the notion of the locally trivial fiber bundles.

Definition 11.3.2. By a p-dimensional class C^r foliation $\mathcal{F}$ of an m-dimensional manifold M one understands a decomposition of M into a union of disjoint connected subsets $\{\mathcal{L}_\alpha\}_{\alpha \in A}$ called leaves of the foliation, so that every point of M has a neighborhood U and a system of local class C^r coordinates $x = (x_1, \dots, x_m) : U \to \mathbf{R}^m$ such that for each leaf $\mathcal{L}_\alpha$ the components $U \cap \mathcal{L}_\alpha$ are described by the equations

$$\begin{cases} x^{p+1} = \text{const}, \\ x^{p+2} = \text{const}, \\ \quad\vdots \\ x^m = \text{const}. \end{cases}$$

The integer number $m - p$ is called the codimension of foliation $\mathcal{F}$. If x, y are two points on the leaf $\mathcal{L}_\alpha$ of a foliation $\mathcal{F}$, then one can consider $(m-p)$-dimensional planes T_x and T_y through x and y transversal to $\mathcal{F}$; a map $H : T_x \to T_y$ defined by the shift of points along the nearby leaves of $\mathcal{F}$ is called a holonomy of the foliation $\mathcal{F}$. Whenever the holonomy map $H : T_x \to T_y$ preserves a measure on T_x and T_y for all $x, y \in M$, a foliation will be called a foliation with the measure-preserving holonomy, or a measured foliation for short.

Example 11.3.1. Let M and M' be manifolds of dimension m and $m' \leq m$ respectively; let $f : M \to M'$ be a submersion of $\text{rank}(df) = m'$. It follows from the Implicit Function Theorem that f induces a codimension m' foliation on M whose leaves are defined to be the components of $f^{-1}(x)$, where $x \in M'$.

Example 11.3.2. The dimension-one foliation $\mathcal{F}$ is given by the orbits of a nonsingular flow $\varphi^t : M \to M$ on the manifold M; the holonomy H of $\mathcal{F}$ coincides with the Poincaré map of the first return along the flow. Whenever φ^t admits an invariant measure, foliation $\mathcal{F}$ will be a measured foliation. For instance, for the flow $\frac{dx}{dt} = \theta = \text{const}$ on the torus T^2, the corresponding foliation $\mathcal{F}_\theta$ is a measured foliation for any $\theta \in \mathbf{R}$.

11.3.3 Index Theorem for foliated spaces

Notice that Atiyah–Singer Theorem for the families of elliptic operators can be rephrased as an index theorem for a foliation $\mathcal{F}$ which is defined by the fiber bundle (E, p, B) whose base space coincides with the parameter space B. In this case each elliptic operator $D_b \in \mathcal{D}$ acts on the leaf M, where M is the manifold associated to the family $\mathcal{D}$. What happens if $\mathcal{F}$ is no longer a fiber bundle but a general foliation? An analog of the base space B in this case will be the leaf space of foliation $\mathcal{F}$, i. e., $E/\mathcal{F}$. Such a space is far from being a Hausdorff topological space. The innovative idea of A. Connes is to replace B by a C^*-algebra $C^*_{\text{red}}(\mathcal{F})$ attached naturally to a foliation $\mathcal{F}_\mu$ endowed with measure μ. The analytic index of the family $\mathcal{D}$ is given by the canonical trace τ (defined by measure μ) of an element of group $K_0(C^*_{\text{red}}(\mathcal{F}))$ described by family $\mathcal{D}$. The topological index is obtained from the Ruelle–Sullivan current C_μ attached to the measure μ. The so-defined index is no longer an integer number or an element of the group $K^0(B)$, but an arbitrary real number.

Definition 11.3.3. By a C^*-algebra $C^*_{\mathrm{red}}(\mathcal{F})$ of a measured foliation $\mathcal{F}$ on a manifold M one understands the reduced groupoid C^*-algebra $C^*_{\mathrm{red}}(H_\mu)$, where H_μ is the holonomy groupoid of foliation $\mathcal{F}$.

Example 11.3.3. Let $\mathcal{F}_\theta$ be a foliation given by the flow $\frac{dx}{dt} = \theta = \mathrm{const}$ on the torus T^2. Then $C^*_{\mathrm{red}}(\mathcal{F}) \cong \mathcal{A}_\theta$, where $\mathcal{A}_\theta$ is the noncommutative torus.

Remark 11.3.1 (Connes). If $\mathcal{D}$ is a collection of elliptic operators defined on each leaf $\mathcal{F}_\alpha$ of foliation $\mathcal{F}$, then $\mathcal{D}$ gives rise to an element $p_{\mathcal{D}} \in K_0(C^*_{\mathrm{red}}(\mathcal{F}))$; moreover, if τ is the canonical trace on $C^*_{\mathrm{red}}(\mathcal{F})$ coming from the invariant measure μ on $\mathcal{F}$, then the analytic index of $\mathcal{D}$ is defined by the formula

$$\mathbf{Ind}(\mathcal{D}) := \tau(p_{\mathcal{D}}) \in \mathbf{R}.$$

Theorem 11.3.2 (Connes). *Let M be a compact smooth manifold and let $\mathcal{F}$ be an oriented foliation of M endowed with invariant transverse measure μ; suppose that $C_\mu \in H_*(M;\mathbf{R})$ is the Ruelle–Sullivan current associated to $\mathcal{F}$. Then*

$$\mathbf{Ind}(\mathcal{D}) = \langle \mathbf{ch}(\mathcal{D})\, \mathbf{Td}(M), [C_\mu] \rangle \in \mathbf{R},$$

where $\mathbf{ch}(\mathcal{D})$ is the Chern class of family $\mathcal{D}$ and $\mathbf{Td}(M)$ is the Todd genus of M.

11.4 Bost–Connes dynamical system

Weil's Conjectures and Delignes' proof of the Riemann Hypothesis for the zeta function of projective varieties over finite fields rely heavily on the spectral data of a linear operator known as the Frobenius endomorphism. J. P. Bost and A. Connes considered the following related problem: Characterize an operator $T : \mathcal{H} \to \mathcal{H}$ on a Hilbert space $\mathcal{H}$ such that

$$\mathbf{Spec}(T) = \{2, 3, 5, 7, \ldots\} := \mathcal{P},$$

where $\mathcal{P}$ is the set of prime numbers. The C^*-algebra $C^*(\mathbf{N}^*)$ generated by T has a spate of remarkable properties, e. g., the flow of weights $(C^*(\mathbf{N}^*), \sigma_t)$ on $C^*(\mathbf{N}^*)$ gives a partition function $\zeta(s) = \mathrm{Tr}(e^{-sH})$, where H is a self-adjoint element of $C^*(\mathbf{N}^*)$. The function $\zeta(s)$ coincides with the Riemann zeta function. Below we give a brief review of a more general C^*-algebra with the same properties.

11.4.1 Hecke C^*-algebra

Definition 11.4.1. Let Γ be a discrete group and $\Gamma_0 \subset \Gamma$ a subgroup such that Γ/Γ_0 is a finite set. By the Hecke C^*-algebra $C^*_{\mathrm{red}}(\Gamma, \Gamma_0)$ one understands the norm closure of a convolution algebra on the Hilbert space $\ell^2(\Gamma_0\backslash\Gamma)$, where the convolution of $f, f' \in C^*_{\mathrm{red}}(\Gamma, \Gamma_0)$ is given by the formula

$$(f * f')(\gamma) = \sum_{\gamma_1 \in \Gamma_0\backslash\Gamma} f(\gamma\gamma_1^{-1}) f'(\gamma_1), \quad \forall \gamma \in \Gamma.$$

Remark 11.4.1. It is not hard to see that $C^*_{\mathrm{red}}(\Gamma, \Gamma_0)$ is a reduced group C^*-algebra; hence the notation.

Example 11.4.1. Let $\Gamma \cong GL(2, \mathbf{Q})$ and $\Gamma_0 \simeq GL(2, \mathbf{Z})$. Then $C^*_{\mathrm{red}}(\Gamma, \Gamma_0)$ contains a (commutative) subalgebra of the classical Hecke operators acting on the space of automorphic cusp forms.

Remark 11.4.2 (Bost–Connes). There exists a unique one-parameter group of automorphisms $\sigma_t \in \mathrm{Aut}(C^*_{\mathrm{red}}(\Gamma, \Gamma_0))$ such that

$$(\sigma_t(f))(\gamma) = \left(\frac{L(\gamma)}{R(\gamma)}\right)^{-it} f(\gamma), \quad \forall \gamma \in \Gamma_0\backslash\Gamma/\Gamma_0,$$

where

$$\begin{cases} L(\gamma) = |\Gamma_0\gamma\Gamma_0| & \text{in } \Gamma/\Gamma_0, \\ R(\gamma) = |\Gamma_0\gamma\Gamma_0| & \text{in } \Gamma_0\backslash\Gamma. \end{cases}$$

11.4.2 Bost–Connes Theorem

Definition 11.4.2. We define a matrix group over a ring R by

$$P_R := \left\{ \begin{pmatrix} 1 & b \\ 0 & a \end{pmatrix} \,\middle|\, aa^{-1} = a^{-1}a = 1,\ a \in R \right\}$$

and let P_R^+ denote a restriction of the group to the case $a > 0$.

Theorem 11.4.1 (Bost–Connes). *Let (A, σ_t) be the C^*-dynamical system associated to the Hecke C^*-algebra*

$$A := C^*_{\mathrm{red}}(P^+_{\mathbf{Q}}, P^+_{\mathbf{Z}}).$$

Then:

(i) *for $0 < \beta \le 1$ (with $\beta = it$), there exists a unique KMS state φ_β on (A, σ_t) and each φ_β is a factor state so that the associated factor is the hyper-finite factor of type $\mathbf{III}_1$;*

(ii) *for $\beta > 1$, the KMS states on (A, σ_t) form a simplex whose extreme points φ^χ_β are parameterized by the complex embeddings*

$$\chi : \mathbf{Q}_{ab} \to \mathbf{C},$$

where $\mathbf{Q}_{ab}$ *the abelian extension of the field* $\mathbf{Q}$ *by the roots of unity, and such states correspond to the type* $\mathbf{I}_\infty$ *factors;*

(iii) *the partition function* $\mathrm{Tr}(e^{-\beta H})$ *of the* C^**-dynamical system* (A, σ_t) *is given by the formula*

$$\mathrm{Tr}(e^{-sH}) = \zeta(s),$$

where

$$\zeta(s) = \sum_{n=1}^{\infty} \frac{1}{n^s}$$

is the Riemann zeta function.

Guide to the literature

Tomita–Takesaki Modular Theory was developed in [271]. Connes Invariants $T(\mathcal{M})$ and $S(\mathcal{M})$ were introduced and studied in [53]; see also an excellent survey [54]. The Hochschild homology was introduced in [117]; Hochschild–Kostant–Rosenberg Theorem (Example 11.2.1) was proved in [118]. The cyclic homology was introduced in [56] and, independently, in [278] in the context of matrix Lie algebras. The Novikov Conjecture for hyperbolic groups was proved in [60]. The Index Theorem for families of elliptic operators was proved in [17]. The Index Theorem for foliation is due to [55]; see also the monograph [162] for a detailed account. The Bost–Connes dynamical system was introduced and studied in [39].

12 Index Theory

Such a theory grew from Atiyah–Singer Theorem saying that an index of a Fredholm operator on a manifold M can be expressed in purely topological terms. The Index Theory is a covariant version of the topological K-theory. M. Atiayh himself and later L. Brown, R. Douglas, and P. Fillmore elaborated a C^*-algebra realization of Atiyah–Singer Theorem known as the K-homology. The KK-theory of G. Kasparov fuses the K-theory and K-homology into a spectacular bifunctor $KK(\mathcal{A}, \mathcal{B})$ on the category of pairs of the C^*-algebras $\mathcal{A}$ and $\mathcal{B}$. The so far topological applications of the Index Theory include the proof of certain cases of the Novikov Conjecture which would remain out of reach otherwise.

12.1 Atiyah–Singer Theorem

12.1.1 Fredholm operators

Definition 12.1.1. An operator $F \in B(\mathcal{H})$ is called Fredholm if $F(\mathcal{H})$ is a closed subspace of the Hilbert space $\mathcal{H}$ and the subspaces $\operatorname{Ker} F$ and $\operatorname{Ker} F^*$ are finite-dimensional. The integer number

$$\mathbf{Ind}(F) := \dim(\operatorname{Ker} F) - \dim(\operatorname{Ker} F^*)$$

is called an index of the Fredholm operator F.

Example 12.1.1. If $\{e_i\}$ is a basis of the Hilbert space $\mathcal{H}$, then the unilateral shift $S(e_n) = e_{n+1}$ is a Fredholm operator; the reader can verify that $\mathbf{Ind}(S) = -1$.

Theorem 12.1.1 (Atkinson). *The following conditions on an operator $F \in B(\mathcal{H})$ are equivalent:*
(i) *F is Fredholm;*
(ii) *there exists an operator $G \in B(\mathcal{H})$ such that $GF - I$ and $FG - I$ are compact operators;*
(iii) *the image of F in the Calkin algebra $B(\mathcal{H})/\mathcal{K}(\mathcal{H})$ is invertible.*

Corollary 12.1.1. *If F is a Fredholm operator, then F^* is a Fredholm operator; moreover,*

$$\mathbf{Ind}(F^*) = -\mathbf{Ind}(F).$$

Corollary 12.1.2. *If F_1 and F_2 are Fredholm operators, then their product F_1F_2 is a Fredholm operator; moreover,*

$$\mathbf{Ind}(F_1F_2) = \mathbf{Ind}(F_1) + \mathbf{Ind}(F_2).$$

https://doi.org/10.1515/9783110788709-012

Remark 12.1.1. If $\mathcal{F}_n$ is the set of all Fredholm operators on a Hilbert space $\mathcal{H}$ having index n, then the map

$$\mathbf{Ind} : \bigcup_{n=-\infty}^{n=\infty} \mathcal{F}_n := \mathcal{F} \longrightarrow \mathbf{Z}$$

is locally constant, i. e., $\mathcal{F}_n$ is an open subset of $B(\mathcal{H})$. Two Fredholm operators are connected by a norm-continuous path of Fredholm operators in the space $B(\mathcal{H})$ if and only if they have the same index; in particular, the connected components of the space $\mathcal{F}$ coincide with $\mathcal{F}_n$ for $n \in \mathbf{Z}$.

Definition 12.1.2. By an essential spectrum $\mathbf{SpEss}(T)$ of an operator $T \in B(\mathcal{H})$ one understands a subset of complex plane given by the formula

$$\mathbf{SpEss}(T) := \{\lambda \in \mathbf{C} \mid T - \lambda I \notin \mathcal{F}\}.$$

Remark 12.1.2. It is not hard to see that

$$\mathbf{SpEss}(T) \subseteq \mathbf{Sp}(T),$$

where $\mathbf{Sp}(T)$ is the usual spectrum of T.

12.1.2 Elliptic operators on manifolds

Let M be a compact oriented smooth n-dimensional manifold. Let E be a smooth complex vector bundle over M. Denote by $\Gamma^\infty(E)$ the space of all smooth sections of E. We shall consider linear differential operators

$$D : \Gamma^\infty(E) \to \Gamma^\infty(F)$$

for a pair E, F of vector bundles over M. Denote by T^*M the cotangent vector bundle of M, by $S(M)$ the unit vector subbundle of T^*M and by $\pi : S(M) \to M$ the natural projection.

Definition 12.1.3. By a symbol $\sigma(D)$ of a linear differential operator D one understands a vector bundle homomorphism

$$\sigma(D) : \pi^*E \to \pi^*F.$$

An operator D is called elliptic if $\sigma(D)$ is an isomorphism.

Remark 12.1.3. If the linear differential operator $D : \Gamma^\infty(E) \to \Gamma^\infty(F)$ is elliptic, then D is a Fredholm operator. We shall denote by $\mathbf{Ind}(D)$ the corresponding index.

Denote by $B(M)$ the unit ball subbundle of T^*M; clearly, $S(M) \subset B(M)$. Recall that two vector bundles E, F on a topological space X with an isomorphism σ on a subspace

$X_0 \subset X$ define an element

$$d(E, F, \sigma) \in K(X \backslash X_0),$$

where $K(X\backslash X_0)$ is the commutative ring (under the tensor product of vector bundles) coming from the topological K-theory of the space X minus X_0. Thus if $p : B(M) \to M$ is the natural projection, then the elliptic operator D defines an element

$$d(p^*E, p^*F, \sigma(D)) \in K(B(M)\backslash S(M)).$$

But the Chern character provides us with a ring homomorphism

$$\mathbf{ch} : K(B(M)\backslash S(M)) \to H^*(B(M)\backslash S(M); \mathbf{Q}),$$

where $H^*(B(M)\backslash S(M); \mathbf{Q})$ is the rational cohomology ring of the topological space $B(M)\backslash S(M)$; thus one gets an element

$$\mathbf{ch}\, d(p^*E, p^*F, \sigma(D)) \in H^*(B(M)\backslash S(M); \mathbf{Q}).$$

Definition 12.1.4. By a Chern character $\mathbf{ch}(D)$ of an elliptic operator D on a manifold M one understands the element

$$\mathbf{ch}(D) := \phi_*^{-1}\, \mathbf{ch}\, d(p^*E, p^*F, \sigma(D)) \in H^*(M; \mathbf{Q}),$$

where $\phi_* : H^k(M, \mathbf{Q}) \to H^{n+k}(B(M)\backslash S(M); \mathbf{Q})$ is the Thom isomorphism of the cohomology rings.

12.1.3 Index Theorem

Theorem 12.1.2 (Atiyah–Singer). *For any elliptic differential operator D on a compact oriented differentiable manifold M, the index* $\mathbf{Ind}(D)$ *is given by the formula*

$$\mathbf{Ind}(D) = \langle \mathbf{ch}(D)\, \mathbf{Td}(M)\rangle [M],$$

where $\mathbf{Td}(M) \in H^*(M; \mathbf{Q})$ *is the Todd genus of the complexification of the tangent bundle over M and $\langle\alpha\rangle[M]$ is the value of the top-dimensional component of an element $\alpha \in H^*(M; \mathbf{Q})$ on the fundamental homology class $[M]$ of M.*

Example 12.1.2. Let $M \cong S^1$ be the circle and let E, F be one-dimensional trivial vector bundles over S^1; in this case $\Gamma^\infty(E) \cong \Gamma^\infty(F) \cong C^\infty(S^1)$. It is not hard to see that elliptic operators $D : C^\infty(S^1) \to C^\infty(S^1)$ correspond to the multiplication operators M_f acting by the pointwise multiplication of functions of $C^\infty(S^1)$ by a function $f \in C^\infty(S^1)$;

therefore

$$\mathbf{Ind}(D) = -w(f),$$

where $w(f)$ is the winding number of f, i. e., the degree of map f taken with the plus or minus sign.

Example 12.1.3 (Hirzebruch–Riemann–Roch Formula). Let M be a complex manifold of dimension l endowed with a holomorphic vector bundle W of dimension n. Then the spaces $\Gamma^\infty(E)$ and $\Gamma^\infty(F)$ can be interpreted as the differential harmonic forms of type $(0,p)$ on M, where $0 \le p \le l$. Using the Dolbeault isomorphism, one can show that $\mathbf{Ind}(D) = \sum_{p=0}^{l}(-1)^p \dim H^p(M,W)$, where $H^p(M,W)$ denotes the p-dimensional cohomology group of M with coefficients in the sheaf of germs of holomorphic sections of W. Thus for any compact complex manifold M and any holomorphic bundle W, one gets

$$\sum_{p=0}^{l}(-1)^p \dim H^p(M,W) = \langle \mathbf{ch}(W)\,\mathbf{Td}(M)\rangle[M].$$

12.2 K-homology

The K-theory is a covariant functor on algebras and a contravariant functor on spaces. The category theory tells us that there exists an abstract dual functor, which is contravariant on algebras and covariant on spaces; such a functor will be called a K-homology. Surprisingly, all realizations of the K-homology involve the index of Fredholm operators. In particular, Atiyah–Singer Index Theorem becomes a pairing statement between K-homology and K-theory.

12.2.1 Topological K-theory

Definition 12.2.1. For a compact Hausdorff topological space X, a vector bundle over X is a topological space E, a continuous map $p : E \to X$, and a finite-dimensional vector space $E_x = p^{-1}(X)$ such that E is locally trivial; usually, the vector space is taken over the field of complex numbers **C**. An isomorphism of vector bundles E and F is a homeomorphism from E to F which takes fiber E_x to fiber F_x for each $x \in X$ and which is linear on the fibers. A trivial bundle over X is a bundle of the form $X \times V$, where V is a fixed finite-dimensional vector space and p is projection onto the first coordinate. The Whitney sum $E \oplus F$ of the vector bundles E and F is a vector bundle of the form

$$E \oplus F = \{(e,f) \in E \times F \mid p(e) = q(f)\},$$

where $p : E \to X$ and $q : F \to X$ are the corresponding projections.

Remark 12.2.1. The Whitney sum makes the set of isomorphism classes of (complex) vector bundles over X into an abelian semigroup $V_{\mathbf{C}}(X)$ with an identity; the identity is the isomorphism class of the 0-dimensional trivial bundle. A continuous map $\phi : Y \to X$ induces a homomorphism $\phi^* : V_{\mathbf{C}}(X) \to V_{\mathbf{C}}(Y)$; thus we have a contravariant functor from topological spaces to abelian semigroups.

Definition 12.2.2. By a K-group $K(X)$ of the compact Hausdorff topological space X one understands the Grothendieck group of the abelian semigroup $V_{\mathbf{C}}(X)$.

Remark 12.2.2. A continuous map $\phi : Y \to X$ induces a homomorphism $\phi^* : K(X) \to K(Y)$; thus we have a contravariant functor from topological spaces to the abelian groups, see Fig. 12.1.

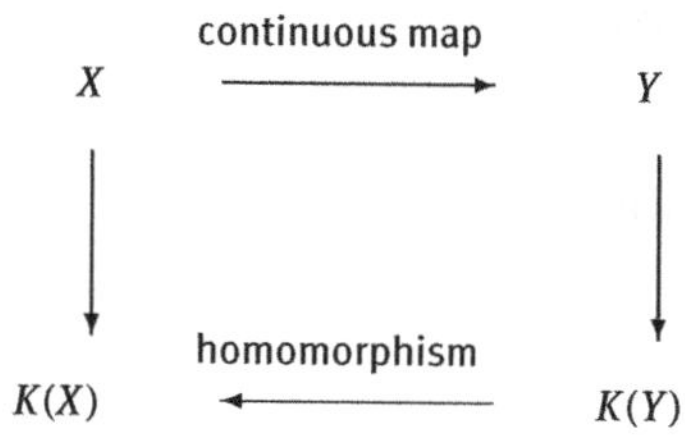

Figure 12.1: Contravariant functor $K(\bullet)$.

Example 12.2.1. If $X \cong S^2$ is a two-dimensional sphere, then $K(S^2) \cong \mathbf{Z}^2$.

Example 12.2.2. If X is a compact Hausdorff topological space, then

$$K(X) \cong K_0(C(X)),$$

where $C(X)$ is the abelian C^*-algebra of continuous complex-valued function on X and $K_0(C(X))$ is the respective K_0-group; we refer the reader to Chapter 3.

Definition 12.2.3. If X is a compact Hausdorff space, then by a suspension of X one understands a topological space SX given by the formula $SX := X \times \mathbf{R}$. The higher-order K-theory of X can be defined according to the formulas:

$$\begin{cases} K^0(X) = K(X), \\ K^{-1}(X) = K(SX) \cong K(X \times \mathbf{R}), \\ \vdots \\ K^{-n}(X) = K(S^n X) \cong K(X \times \mathbf{R}^n). \end{cases}$$

Theorem 12.2.1 (Bott Periodicity). *There exists an isomorphism (a natural transformation, resp.) between the groups (functors, resp.) $K(X)$ and $K^{-2}(X)$ so that one gets a gen-*

eral isomorphism

$$K^{-n} \cong K^{-n-2}(X).$$

Remark 12.2.3. The topological K-theory is an extraordinary cohomology theory, i. e., a sequence of the homotopy invariant contravariant functors from compact Hausdorff spaces to abelian groups satisfying all the Eilenberg–Steenrod axioms but the dimension axiom.

Theorem 12.2.2 (Chern Character). *If X is a compact Hausdorff topological space, then*

$$\begin{cases} K^0(X) \rtimes \mathbf{Q} \cong \bigoplus_k H^{2k}(X;\mathbf{Q}), \\ K^{-1}(X) \rtimes \mathbf{Q} \cong \bigoplus_k H^{2k-1}(X;\mathbf{Q}), \end{cases}$$

where $H^k(X;\mathbf{Q})$ denotes the kth cohomology group of X with coefficients in $\mathbf{Q}$.

12.2.2 Atiayh's realization of K-homology

An abstract covariant functor on a compact Hausdorff topological space X will be denoted by $K_0(X)$. While looking for a realization of $K_0(X)$, it was observed by M. Atiyah that the index map

$$\mathbf{Ind} : \mathcal{F} \to \mathbf{Z}$$

behaves much like a homology on the space $\mathcal{F}$ of all Fredholm operators on a Hilbert space $\mathcal{H}$. To substantiate such a homology, the following definition was introduced.

Definition 12.2.4. Let X is a compact Hausdorff topological space and let $C(X)$ be the C^*-algebra of continuous complex-valued functions on X. Let

$$\begin{cases} \sigma_1 : C(X) \to B(\mathcal{H}_1), \\ \sigma_2 : C(X) \to B(\mathcal{H}_2) \end{cases}$$

be a pair of representations of $C(X)$ by bounded linear operators on the Hilbert spaces $\mathcal{H}_1$ and $\mathcal{H}_2$, respectively. Denote by $F : \mathcal{H}_1 \to \mathcal{H}_2$ a Fredholm operator such that the operator

$$F \circ \sigma_1(f) \quad \sigma_2(f) \circ F$$

is a compact operator for all $f \in C(X)$. By $\mathbf{Ell}(X)$ one understands the set of all triples (σ_1, σ_2, F). A binary operation of addition on the set $\mathbf{Ell}(X)$ is defined as the orthogonal direct sum of the Hilbert spaces $\mathcal{H}_i$; the operation turns $\mathbf{Ell}(X)$ into an abelian semigroup and the Grothendieck completion of the semigroup turns $\mathbf{Ell}(X)$ into an abelian group.

Example 12.2.3. The basic example is F being an elliptic differential operator between two smooth vector bundles E and F on a smooth manifold M; it can be proved that F is a Fredholm operator which intertwines the action of $C(M)$ modulo a compact operator.

Theorem 12.2.3 (Atiyah). *There exists a surjective map*

$$\mathbf{Ell}(X) \longrightarrow K_0(X)$$

whenever X is a finite CW-complex.

Remark 12.2.4. To finalize Atiyah's realization of the K-homology by the group $\mathbf{Ell}(X)$, one needs a proper notion of the equivalence relation on $\mathbf{Ell}(X)$ to make the quotient equal to $K_0(X)$; it was an open problem solved by L. Brown, R. Douglas, and P. Fillmore.

12.2.3 Brown–Douglas–Fillmore Theory

Larry Brown, Ron Douglas, and Peter Fillmore introduced a realization of the K-homology inadvertently (or not) while working on the problem of classification of the essentially normal operators on a Hilbert space $\mathcal{H}$ raised by J. von Neumann. Problem solution depends on the classification of the extensions of the C^*-algebra $\mathcal{K}$ of all compact operators by a C^*-algebra $C(X)$ for some compact metrizable space X.

Definition 12.2.5. By an extension of the C^*-algebra $\mathcal{A}$ by a C^*-algebra $\mathcal{B}$ one understands a C^*-algebra $\mathcal{E}$ which fits into the short exact sequence

$$0 \to \mathcal{A} \to \mathcal{E} \to \mathcal{B} \to 0,$$

i. e., such that $\mathcal{A} \cong \mathcal{E}/\mathcal{B}$. Two extensions τ and τ' are said to be stably equivalent if they differ by a trivial extension. If τ_1 and τ_2 are two extensions of $\mathcal{A}$ by $\mathcal{B}$, then there is a naturally defined extension $\tau_1 \oplus \tau_2$ called the sum of τ_1 and τ_2. With the addition operation the set of all extensions of $\mathcal{A}$ by $\mathcal{B}$ is an abelian semigroup; the Grothendieck completion of the semigroup yields an abelian group denoted by $\mathbf{Ext}(\mathcal{A}, \mathcal{B})$.

Theorem 12.2.4 (Brown–Douglas–Fillmore). *If X is a compact metrizable topological space, then the map*

$$X \mapsto \mathbf{Ext}(\mathcal{K}, C(X))$$

is a homotopy invariant covariant functor $K_1(X)$ from the category of compact metrizable spaces to the category of abelian groups.

Corollary 12.2.1. *The suspension SX of X gives a realization of the K-homology according to the formula*

$$K_0(X) \cong K_1(SX).$$

Remark 12.2.5. The existence of the functor $K_1(X)$ yields a solution to the following problem of J. von Neumann: Given two operators $T_1, T_2 \in B(\mathcal{H})$, when is it true that

$$T_1 = UT_2U^* + K,$$

where U is a unitary and K is a compact operator? Below we give a brief account of the solution, which was the first successful application of the K-theory to an open problem in analysis.

Definition 12.2.6. An operator $T \in B(\mathcal{H})$ is called essentially normal whenever

$$TT^* - T^*T \in \mathcal{K}.$$

Let $T \in B(\mathcal{H})$ be an operator and denote by $\mathcal{E}_T := C^*(I, T, \mathcal{K})$ a C^*-algebra generated by the identity operator I, operator T, and the C^*-algebra $\mathcal{K}$. It follows from the definition that $\mathcal{E}_T/\mathcal{K}$ is a commutative C^*-algebra if and only if T is an essentially normal operator. Thus one gets a short exact sequence

$$0 \to \mathcal{K} \to \mathcal{E}_T \to C(\mathbf{SpEss}(T)) \to 0,$$

where T is an essentially normal operator and $\mathbf{SpEss}(T)$ its essential spectrum. Therefore the von Neumann problem for the essentially normal operators reduces to a classification up to stable equivalence of the extensions of $\mathcal{K}$ by $C(X)$, where X is a subset of the complex plane. A precise statement is this.

Theorem 12.2.5 (Brown–Douglas–Fillmore). *Two essentially normal operators T_1 and T_2 are related by the formula $T_1 = UT_2U^* + K$ if and only if*

$$\begin{cases} \mathbf{SpEss}(T_1) = \mathbf{SpEss}(T_1), \\ \mathbf{Ind}(T_1 - \lambda I) = \mathbf{Ind}(T_2 - \lambda I), \end{cases}$$

for all $\lambda \in \mathbf{C} \backslash \mathbf{SpEss}(T_i)$.

12.3 Kasparov's KK-theory

Kasparov's bifunctor $KK(\mathcal{A}, \mathcal{B})$ on pairs of the C^*-algebras $\mathcal{A}$ and $\mathcal{B}$ blends K-homology with the K-theory. Such a functor was designed to solve a concrete open problem of topology – the higher signatures hypothesis of S. P. Novikov. The $KK(\mathcal{A}, \mathcal{B})$ achieves the goal, albeit in certain special cases. The original definition of the KK-groups uses the notion of a Hilbert module.

12.3.1 Hilbert modules

A Hilbert module is a Hilbert space whose inner product takes values not in $\mathbf{C}$ but in an arbitrary C^*-algebra $\mathcal{A}$. One can define bounded and compact operators acting on such modules and the construction appears to be very useful.

Definition 12.3.1. For a C^*-algebra $\mathcal{A}$, by a Hilbert module over $\mathcal{A}$ one understands a right $\mathcal{A}$-module E endowed with a $\mathcal{A}$-valued inner product

$$\langle \bullet, \bullet \rangle : E \times E \to \mathcal{A},$$

which satisfies the following properties:
(i) $\langle \bullet, \bullet \rangle$ is sesquilinear;
(ii) $\langle x, ya \rangle = \langle x, y \rangle a$ for all $x, y \in E$ and $a \in \mathcal{A}$;
(iii) $\langle y, x \rangle = \langle x, y \rangle^*$ for all $x, y \in E$;
(iv) $\langle x, x \rangle \geq 0$ and if $\langle x, x \rangle = 0$ then $x = 0$

and the norm $\|x\| = \sqrt{\|\langle x, x \rangle\|}$ under which E is complete.

Example 12.3.1. The C^*-algebra $\mathcal{A}$ is itself a Hilbert $\mathcal{A}$-module with the inner product $\langle a, b \rangle := a^* b$.

Example 12.3.2. Any closed right ideal of the C^*-algebra $\mathcal{A}$ is a Hilbert $\mathcal{A}$-module.

Definition 12.3.2. For a Hilbert $\mathcal{A}$-module E, by $B(E)$ one understands the set of all module homomorphisms

$$T : E \longrightarrow E$$

for which there is an adjoint module homomorphism $T^* : E \to E$ such that $\langle Tx, y \rangle = \langle x, T^* y \rangle$ for all $x, y \in E$.

Remark 12.3.1. Each homomorphism in $B(E)$ is bounded and $B(E)$ itself is a C^*-algebra with respect to the operator norm.

Definition 12.3.3. By the set $\mathcal{K}(E) \subset B(E)$ one understands the closure of the linear span of all bounded operators $T(x, y)$ acting by the formula

$$z \longmapsto x\langle y, z \rangle, \quad z \in E,$$

where $x, y \in E$ and $T^*(x, y) = T(y, x)$.

Remark 12.3.2. The set $\mathcal{K}(E)$ is a closed ideal in $B(E)$.

12.3.2 KK-groups

Definition 12.3.4. For a pair of the C^*-algebras $\mathcal{A}$ and $\mathcal{B}$, by a Kasparov module $\mathbf{E}(\mathcal{A},\mathcal{B})$ one understands the set of all triples (E,ϕ,F), where E is a countably generated graded Hilbert module over $\mathcal{B}$,

$$\phi : \mathcal{A} \to B(E)$$

is an $*$-homomorphism, and $F \in B(E)$ is a bounded operator such that

$$\begin{cases} F \circ \phi(a) - \phi(a) \circ F \in K(E), \\ (F^2 - I) \circ \phi(a) \in K(E), \\ (F - F^*) \circ \phi(a) \in K(E), \end{cases}$$

for all $a \in \mathcal{A}$.

Example 12.3.3 (Basic example). Let M be a compact oriented smooth manifold endowed with a Riemannian metric. Consider a pair of vector bundles V_1 and V_2 over M and let

$$P : \Gamma^\infty(V_1) \to \Gamma^\infty(V_2)$$

be an elliptic operator on M which extends to a Fredholm operator from $L^2(V_1)$ to $L^2(V_2)$. Denote by Q the parametrix for operator P. Let $E = L^2(V_1)\oplus L^2(V_2)$ be the Hilbert module (over $\mathbf{C}$) and consider a $*$-homomorphism

$$\phi : C(M) \to B(E)$$

realized by the multiplication operators on $B(E)$. The triple

$$\left(E, \phi, \begin{pmatrix} 0 & Q \\ P & 0 \end{pmatrix}\right)$$

is an element of the Kasparov module $\mathbf{E}(C(M), \mathbf{C})$.

Definition 12.3.5. Let $\mathcal{A}$ and $\mathcal{B}$ be a pair of the C^*-algebras and consider the set of equivalence classes of the Kasparov modules $\mathbf{E}(\mathcal{A},\mathcal{B})$ under a homotopy equivalence relation. The direct sum of the Kasparov modules turns the equivalence classes of $\mathbf{E}(\mathcal{A},\mathcal{B})$ into an abelian semigroup; the Grothendieck completion of the semigroup is an abelian group denoted by $KK(\mathcal{A},\mathcal{B})$.

Example 12.3.4. If $\mathcal{A} \cong \mathcal{B} \cong \mathbf{C}$, then $KK(\mathbf{C},\mathbf{C}) \cong \mathbf{Z}$.

Remark 12.3.3. The KK-groups is a bifunctor defined on the pairs of C^*-algebras and with values in the abelian groups; the functor is homotopy-invariant in each variable.

A powerful new feature of the KK-functor is an intersection product acting by the formula

$$KK(\mathcal{A},\mathcal{B}) \times KK(\mathcal{B},\mathcal{C}) \to KK(\mathcal{A},\mathcal{C}).$$

The intersection product generalizes the cap and cup products of algebraic topology and turns $KK(\mathcal{A},\mathcal{B})$ into a ring with multiplication given by the intersection product.

Example 12.3.5. If $\mathcal{A}$ is an AF-algebra, then

$$KK(\mathcal{A},\mathcal{A}) \cong_{\text{ring}} \operatorname{End}(K_0(\mathcal{A})),$$

where $\cong_{\text{ring}}$ is a ring isomorphism and $\operatorname{End}(K_0(\mathcal{A}))$ is the endomorphism ring of K_0-group of the AF-algebra $\mathcal{A}$.

12.4 Applications of Index Theory

Index Theory is known to have successful applications in topology (higher signatures) and geometry (positive scalar curvature). The Baum–Connes Conjecture can be viewed as a generalization of Index Theory in topology.

12.4.1 Novikov Conjecture

It is well known that the homology and cohomology groups are invariants of homeomorphisms and also more general continuous maps called homotopies. For instance, a homotopy affects the dimension of topological space while a homeomorphism always preserves it. The numerical invariants obtained by pairing homology and cohomology groups (i. e., integration of cycles against the cocycles) are invariants of the homeomorphisms but not of the homotopies. It is a difficult problem to find all numerical invariants of the topological space preserved by the homotopies.

Example 12.4.1 (Hirzebruch Signature). Suppose M^{4k} is a smooth $4k$-dimensional manifold. Denote by $H^*(M^{4k};\mathbf{Q})$ its cohomology ring with the coefficients in $\mathbf{Q}$ and let $L(M^{4k}) \in H^*(M^{4k};\mathbf{Q})$ be the Hirzebruch characteristic class (an L-class) of the manifold M^{4k}. If $[M^{4k}] \in H_{4k}(M^{4k};\mathbf{Z})$ is the fundamental class of M^{4k}, then one can consider a rational number

$$\mathbf{Sign}(M^{4k}) = \langle L(M^{4k}), [M^{4k}]\rangle \in \mathbf{Q}$$

obtained by integration of the Hirzebruch characteristic class against the fundamental class of manifold M^{4k}. The number coincides with the signature of a nondegenerate symmetric bilinear form on the space $H^{2k}(M^{4k};\mathbf{Q})$ and for this reason is called a signature $\mathbf{Sign}(M^{4k})$. The signature is a homotopy invariant of the manifold M^{4k}.

Example 12.4.2. Let M^n be an n-dimensional manifold and let $p_k(M^n) \in H^{4k}(M^n;\mathbf{Q})$ be its Pontryagin characteristic classes, i. e., such classes for the tangent bundle over M^n. The integrals

$$\langle p_k(M^n), [c_{n-4k}]\rangle \in \mathbf{Q}$$

of the Pontryagin classes against the cycles $[c_{n-4k}] \in H_{n-4k}(M^n;\mathbf{Z})$ are homeomorphism invariants but not homotopy invariants.

Remark 12.4.1. Let K denote a CW-complex. It is known that K has a distinguished set of cohomology classes determined by its fundamental group $\pi_1(K)$. Namely, such classes come from the canonical continuous map

$$f : K \to K(\pi, 1),$$

where $K(\pi,1)$ is the classifying space of the group $\pi_1(K)$; the distinguished set is given by the induced homomorphism

$$f^* : H^*(K(\pi,1)) \to H^*(K)$$

with any coefficients. In the case of the rational coefficients, the corresponding class of distinguished cycles will be denoted by

$$Df^*H^*(\pi;\mathbf{Q}) \subset H_*(K;\mathbf{Q}).$$

Conjecture 12.4.1 (Novikov Conjecture on the Higher Signatures). *For each cycle $z \in Df^*H^*(\pi;\mathbf{Q})$, the integral*

$$\langle L_k(p_1,\dots,p_k), z\rangle \in \mathbf{Q}$$

of the Pontryagin–Hirzebruch characteristic class $L_k(p_1,\dots,p_k) \in H^*(M^n;\mathbf{Q})$ against z is a homotopy invariant.

Theorem 12.4.1 (Kasparov). *Whenever $\pi_1(M^n)$ is isomorphic to a discrete subgroup of a connected Lie group, the Novikov Conjecture for the manifold M^n is true.*

Proof. The proof is based on the KK-theory. The idea is to introduce a pair of the C^*-algebras $\mathcal{A} \cong C^*(\pi_1(M^n))$ and $\mathcal{B} \cong C(M^n)$, where $C^*(\pi_1(M^n))$ is a group C^*-algebra attached to the group $\pi_1(M^n)$. (Such an approach can recapitulated in terms of representation of $\pi_1(M^n)$ by linear operators on a Hilbert space $\mathcal{H}$; thus one gets a functor from topological spaces to the C^*-algebras.) The bilinear form $\langle L_k(p_1,\dots,p_k), z\rangle$ can be written in terms of the intersection product on the Kasparov's KK-groups. Since the *KK*-functor is a homotopy invariant, so will be the bilinear form and its signature whenever $\pi_1(M^n)$ satisfies assumptions of the theorem. □

Remark 12.4.2. Novikov's Conjecture can be rephrased as a question about general discrete groups and their representations by the C^*-algebras. A positive answer to such a question would imply Novikov's Conjecture. Namely, one has the following

Conjecture 12.4.2 (Strong Novikov Conjecture). *Let $B\pi$ be the classifying space of a discrete group π; then the map*

$$\beta : K_*(B\pi) \to K_*(C^*(\pi))$$

is rationally injective.

Remark 12.4.3. Kasparov's proof of Novikov's Conjecture is valid for the Strong Novikov Conjecture.

12.4.2 Baum–Connes Conjecture

The Baum–Connes Conjecture can be viewed as a strengthening to an isomorphism and generalization to the Lie groups of the Strong Novikov Conjecture.

Conjecture 12.4.3 (Baum–Connes Conjecture). *Let G be a Lie group or a countable discrete group; let BG be the corresponding classifying space. Let $C^*_{\mathrm{red}}(G)$ be the reduced group C^*-algebra of G. Then there exists an isomorphism*

$$\mu : K^0(BG) \to K_0(C^*_{\mathrm{red}}(G)),$$

where $K^0(BG)$ is the K-homology of the topological space BG.

Remark 12.4.4. The Baum–Connes Conjecture is proved in many cases, e. g., for the hyperbolic, amenable, etc., groups. We refer the reader to the respective literature. It seems that the cyclic cohomology is an appropriate replacement for the *KK*-theory in this case.

12.4.3 Positive scalar curvature

The topology of a manifold imposes constraints on the type of metric one can realize on the manifold. For instance, there are no zero scalar curvature Riemannian metrics on any surface of genus $g > 1$. It is remarkable that Index Theory detects the topological obstructions for having a metric of positive scalar curvature on a given manifold. Below we give a brief account of this theory.

Let M be a smooth manifold having an oriented spin structure; let D be the canonical Dirac operator corresponding to the spin structure. Suppose that M has even dimension and let $D = D^+ \oplus D^-$ be the corresponding canonical decomposition of the Dirac operator on positive and negative components. For a unital C^*-algebra B, one can define a flat B-vector bundle V over M and consider the Dirac operators D_V^+ and D_V^-. The Dirac operator D_V^+ is a Fredholm operator and

$$\mathbf{Ind}_B(D_V^+) \in K_0(B) \otimes \mathbf{Q}.$$

Theorem 12.4.2 (Rosenberg). *If M admits a metric of positive scalar curvature, then*

$$\mathbf{Ind}_B(D_V^+) = 0$$

for any flat B-vector bundle V.

Remark 12.4.5 (Mischenko–Fomenko). One can express the left-hand side of the last equality in purely topological terms; namely,

$$\mathbf{Ind}_B(D_V^+) = \langle \mathbf{A}(M) \cup \mathbf{ch}(V), [M] \rangle,$$

where $\mathbf{A}(M) \in H^*(M, \mathbf{Q})$, $\mathbf{ch}(V)$ is the Chern class of V and $[M]$ is the fundamental class of M.

Corollary 12.4.1. *If M admits a metric of positive scalar curvature, then*

$$\langle \mathbf{A}(M) \cup \mathbf{ch}(V), [M] \rangle = 0$$

for any flat B-vector bundle V.

Conjecture 12.4.4 (Gromov–Lawson Conjecture). *If M has positive scalar curvature, then for all $x \in H^*(B\pi; \mathbf{Q})$ it holds that*

$$\langle \mathbf{A}(M) \cup f^*(x), [M] \rangle = 0,$$

where $f : M \to B\pi$ is the classifying map.

Remark 12.4.6. The Strong Novikov Conjecture implies the Gromov–Lawson Conjecture.

12.5 Coarse geometry

Index Theory gives only rough topological invariants of manifolds. The idea of coarse geometry is to replace usual homeomorphisms and homotopies between topological spaces by more general coarse maps so that Index Theory will classify manifolds modulo the equivalence relation defined by such maps. The coarse geometry is present in

Mostow's proof of the Rigidity Theorem [164]. It has been studied in the geometric group theory. We refer the reader to the books [242] and [216] for a detailed account.

Definition 12.5.1. If X and Y are metric spaces, then a (generally discontinuous) map $f : X \to Y$ is called coarse if:
(i) for each $R > 0$ there exists $S > 0$ such that $d(x, x') \le R$ implies $d(f(x), f(x')) \le S$;
(ii) for each bounded subset $B \subseteq Y$ the preimage $f^{-1}(B)$ is bounded in X.

Definition 12.5.2. Two coarse maps $f_0 : X \to Y$ and $f_1 : X \to Y$ are coarsely equivalent if there is a constant K such that

$$d(f_0(x), f_1(x)) \le K$$

for all $x \in X$.

Definition 12.5.3. Two metric spaces X and Y are said to be coarsely equivalent if there are maps from X to Y and from Y to X whose composition is coarsely equivalent to the identity map; the coarse equivalence class of metrics on X is called a coarse structure (or coarse geometry) on X.

Remark 12.5.1. It is useful to think of coarse geometry as a "blurry version" of usual geometry; all the local data is washed out and only the large scale features are preserved. Because the index of a Fredholm operator on a Hilbert space $\mathcal{H}$ cannot see the local geometry, one can think of the index as an abstract topological invariant of coarse equivalence.

Remark 12.5.2. Any complete Riemannian manifold is a metric space; thus it can be endowed with a coarse structure so that Index Theory becomes a functor on such a structure.

Example 12.5.1. The natural inclusion $\mathbf{Z}^n \to \mathbf{R}^n$ is a coarse equivalence.

Example 12.5.2. The natural inclusion $\mathbf{R}^n \to \mathbf{R}^n \times [0, \infty)$ is a coarse equivalence.

Remark 12.5.3. Since the tools of algebraic topology can be applied to the coarse manifolds, one can reformulate all content of Sections 12.1–12.4 in terms of the coarse geometry. Thus one arrives at the notion of a coarse index, coarse assembly map, coarse Baum–Connes Conjecture, etc. We refer the interested reader to the monographs [242] and [216].

Guide to the literature

Atiyah–Singer Theorem was announced in [16] and proved in the series of papers [17]. For the foundation of topological K-theory, we refer the reader to monograph [14]. Atiyah's realization of K-homology can be found in [15]. The Brown–Douglas–Fillmore realization of K-homology appeared in [44]; see also the monograph [70]. Kasparov's

KK-theory can be found in [130]; see the book [32, Chapter VIII] for a detailed account. For the proof of special cases of Novikov's Conjecture using the KK-theory, see, e. g., [131]. The published version of Baum–Connes Conjecture can be found in [22]. Rosenberg's Theorem on positive scalar curvature appears in [246]. The ideas and methods of coarse geometry are covered in the books [242] and [216]. An interesting link between topology and operator algebras was studied by [121].

13 Jones polynomials

In the 1980s V. F. R. Jones discovered a representation $\rho : B_n \to A_n$, where B_n is the braid group on n strings and A_n is an n-dimensional W^*-algebra. Algebra A_n admits a trace which is coherent with respect to the second Markov move of the braid,

$$\mathrm{tr}(e_n x) = \frac{1}{[\mathcal{M} : \mathcal{N}]} \mathrm{tr}(x), \quad \forall x \in A_n.$$

(We refer the reader to Section 13.3 for the notation and details.) The lack of such a trace for other known representations of B_n did not allow defining topological invariants of the closure of a braid $b \in B_n$, i. e., a knot or a link. The trace invariant $V_L(t)$ of a link L is a Laurent polynomial obtained by a normalization of the above trace equation.

13.1 Subfactors

Subfactors are an analog of the Galois theory for the W^*-algebras. We refer the reader to Chapter 8 for a short review of such algebras.

Definition 13.1.1. By a subfactor of a factor $\mathcal{M}$ one understands the unital W^*-subalgebra of $\mathcal{M}$ which is a factor itself.

Theorem 13.1.1 (Galois Theory for type II_1 factors). *Let G be a finite group of outer automorphisms of a type $\mathbf{II}_1$ factor $\mathcal{M}$ and let $H \subseteq G$ be its subgroup. Then the formula*

$$H \leftrightarrow \mathcal{M}^H := \{x \in \mathcal{M} \mid \alpha(x) = x,\ \forall \alpha \in H\}$$

gives a Galois correspondence between subgroups of G and subfactors of $\mathcal{M}$ containing the fixed point subalgebra $\mathcal{M}^G \subseteq \mathcal{M}$.

To each subgroup $H \subseteq G$ one can assign an integer $[G : H]$, called the index of the subgroup. The following definition extends the notion of an index to the subfactors.

Definition 13.1.2. Let $\mathcal{N} \subseteq \mathcal{M}$ be two factors of type $\mathbf{II}_1$; then the index of $\mathcal{N}$ in $\mathcal{M}$ is defined by the formula

$$[\mathcal{M} : \mathcal{N}] = \frac{\dim L^2(\mathcal{N})}{\dim L^2(\mathcal{M})},$$

where $L^2(\mathcal{N})$ and $L^2(\mathcal{M})$ are the Hilbert spaces associated to the GNS construction for W^*-algebras $\mathcal{N}$ and $\mathcal{M}$, respectively.

Unlike the case of groups, the index of subfactors is not always an integer. The following theorem lists all possible values of the index for subfactors of type $\mathbf{II}_1$ factors.

https://doi.org/10.1515/9783110788709-013

Theorem 13.1.2 (V. F. R. Jones). *If $\mathcal{M}$ is a type $\mathbf{II}_1$ factor, then:*

(i) *if $\mathcal{N}$ is a subfactor with the same identity and $[\mathcal{M}:\mathcal{N}] < 4$, then*

$$[\mathcal{M}:\mathcal{N}] = 4\cos^2\left(\frac{\pi}{n}\right)$$

for some $n = 3, 4, \ldots$;

(ii) *if $\mathcal{M}$ is the hyper-finite type $\mathbf{II}_1$ factor, then the index of a subfactor $\mathcal{N}$ takes the following values:*

$$[\mathcal{M}:\mathcal{N}] = \left\{4\cos^2\left(\frac{\pi}{n}\right) : n \geq 3\right\} \cup [4, \infty].$$

Remark 13.1.1. The proof of Jones Theorem exploits the following basic construction. Take $\mathcal{M}$ and its GNS representation on the Hilbert space $L^2(\mathcal{M})$; suppose that $\mathcal{N}$ is a subfactor. By uniqueness, $\mathrm{tr}(\mathcal{N})$ is the restriction of $\mathrm{tr}(\mathcal{M})$ and thus $L^2(\mathcal{N})$ is a subspace of $L^2(\mathcal{M})$. The projection

$$e : L^2(\mathcal{M}) \to L^2(\mathcal{N})$$

has the following properties: (i) $\mathcal{M} \cap \{e\}' = \mathcal{N}$; (ii) W^*-algebra $\langle \mathcal{M}, e\rangle := (\mathcal{M} \cup \{e\})''$ is a factor; (iii) $[\langle \mathcal{M}, e\rangle : \mathcal{M}] = [\mathcal{M}:\mathcal{N}]$; and (iv) $\mathrm{tr}(ae) = \frac{1}{[\mathcal{M}:\mathcal{N}]}\mathrm{tr}(a)$.

Definition 13.1.3. By the Jones projections one understands a sequence $\{e_i\}_{i=1}^{\infty}$ obtained as the result of iteration of the basic construction, i. e.,

$$\mathcal{N} \subset \mathcal{M} \subset \langle \mathcal{M}, e_1\rangle \subset \langle\langle \mathcal{M}, e_1\rangle, e_2\rangle \subset \cdots.$$

Corollary 13.1.1 (V. F. R. Jones). *The Jones projections satisfy the following relations and a trace formula:*

$$\begin{cases} e_i e_{i\pm1} e_i = \dfrac{1}{[\mathcal{M}:\mathcal{N}]}\, e_i, & \\ e_i e_j = e_j e_i, & \text{if } |i-j| \geq 2, \\ \mathrm{tr}(e_n x) = \dfrac{1}{[\mathcal{M}:\mathcal{N}]}\, \mathrm{tr}(x), & \forall x \in \langle \mathcal{M}, e_1, \ldots, e_{n-1}\rangle. \end{cases}$$

13.2 Braids

Motivated by a topological classification of knots and links in the three-dimensional space, E. Artin introduced the notion of a braid. The braid is a simpler concept than the knot or link, and braids can be composed with each other so that one obtains a finitely generated group.

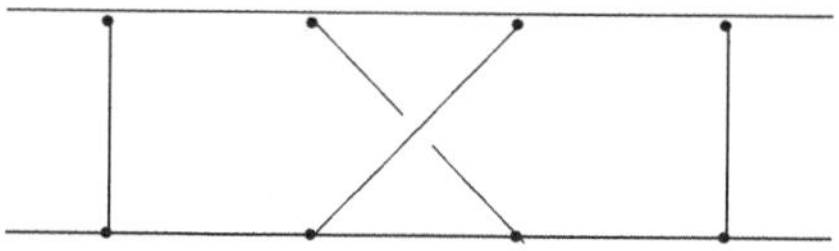

Figure 13.1: The diagram of a braid b_4.

Definition 13.2.1. By an n-string braid b_n one understands two parallel copies of the plane $\mathbf{R}^2$ in $\mathbf{R}^3$ with n distinguished points taken together with n disjoint smooth paths ("strings") joining pairwise the distinguished points of the planes; the tangent vector to each string is never parallel to the planes. The braid b_n can be given by a diagram by projecting its strings into a generic plane in $\mathbf{R}^3$ and indicating the over- and underpasses of the strings, see Fig. 13.1.

Remark 13.2.1. The braids b_n are endowed with a natural equivalence relation: two braids b_n and b_n' are equivalent if b_n can be deformed into b_n' without intersection of the strings and so that at each moment of the deformation b_n remains a braid.

Definition 13.2.2. By a closure $\hat{b}_n$ of the braid b_n one understands a link or knot in $\mathbf{R}^3$ obtained by gluing the endpoints of strings at the top of the braid with such at the bottom of the braid.

Definition 13.2.3. By an n-string braid group B_n one understands the set of all n-string braids b_n endowed with a multiplication operation of the concatenation of $b_n \in B_n$ and $b_n' \in B_n$, i. e., the identification of the bottom of b_n with the top of b_n'. The group is noncommutative and the identity is given by the trivial braid, see Fig. 13.2.

Theorem 13.2.1 (E. Artin). *The braid group B_n is isomorphic to a group on the standard generators $\sigma_1, \sigma_2, \ldots, \sigma_{n-1}$ satisfying the following relations:*

$$\begin{cases} \sigma_i\sigma_{i+1}\sigma_i = \sigma_{i+1}\sigma_i\sigma_{i+1} & \\ \sigma_i\sigma_j = \sigma_j\sigma_i, & \text{if } |i-j| \geq 2. \end{cases}$$

Definition 13.2.4. By a Markov move one understand the following two types of transformations of the braid b_n:

(i) **Type I:** $b_n \mapsto a_n b_n a_n^{-1}$ for a braid $a_n \in B_n$;
(ii) **Type II:** $b_n \mapsto b_n \sigma_n^{\pm 1} \in B_{n+1}$, where $\sigma_n \in B_{n+1}$.

Theorem 13.2.2 (A. Markov). *The closure of any two braids $b_n \in B_n$ and $b_m \in B_m$ gives the same link $\hat{b}_n \cong \hat{b}_m$ in $\mathbf{R}^3$ if and only if b_n and b_m can be connected by a sequence of the Markov moves of types I and II.*

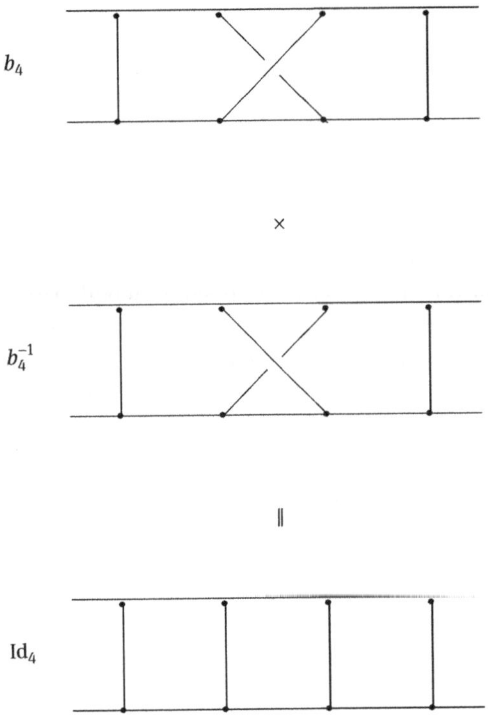

Figure 13.2: The concatenation of a braids b_4 and b_4^{-1}.

13.3 Trace invariant

Definition 13.3.1. By the Jones algebra A_n we shall understand an n-dimensional W^*-algebra generated by the identity and the Jones projections $e_1, e_2, \ldots, e_n$ satisfying the relations:

$$\begin{cases} e_i e_{i\pm 1} e_i = \dfrac{1}{[\mathcal{M} : \mathcal{N}]} e_i, \\ e_i e_j = e_j e_i, & \text{if } |i - j| \geq 2, \end{cases}$$

and the trace formula

$$\mathrm{tr}(e_n x) = \frac{1}{[\mathcal{M} : \mathcal{N}]} \mathrm{tr}(x), \quad \forall x \in A_n.$$

Remark 13.3.1. The reader can verify that the relations for the Jones projections e_i coincide with such for the generators σ_i of the braid group after a minor adjustment of the notation:

$$\begin{cases} \sigma_i \mapsto \sqrt{t}[(t+1)e_i - 1], \\ [\mathcal{M} : \mathcal{N}] = 2 + t + \frac{1}{t}. \end{cases}$$

Thus one gets a family ρ_t of representations of the braid group B_n into the Jones algebra A_n. To get a topological invariant of the closed braid $\hat{b}$ of $b \in B_n$ coming from the trace of the representation, one needs to choose a representation whose trace is invariant under the first and the second Markov moves of the braid b. There is no problem with the first move because two similar matrices have the same trace for any representation from the family ρ_t. For the second Markov move, we have the trace formula which after a substitution takes the form

$$\begin{cases} \operatorname{tr}(b\sigma_n) = -\dfrac{1}{t+1}\operatorname{tr}(b), \\ \operatorname{tr}(b\sigma_n^{-1}) = -\dfrac{t}{t+1}\operatorname{tr}(b). \end{cases}$$

In general, $\operatorname{tr}(b\sigma_n^{\pm 1}) \neq \operatorname{tr}(b)$, but one can always rescale the trace to get the equality. Indeed, the second Markov move takes the braid from B_i and replaces it by a braid from B_{i-1}. There is a finite number of such replacements since the algorithm stops for B_1. Therefore rescaling by the constants $-\frac{1}{t+1}$ and $-\frac{t}{t+1}$ will give a quantity invariant under the second Markov move. Such a quantity is called the *Jones polynomial* of $\hat{b}$.

Theorem 13.3.1 (V. F. R. Jones). *Let $b \in B_n$ be a braid and* $\exp(b)$ *be the sum of all powers of generators σ_i and σ_i^{-1} in the word presentation of b; let $L := \hat{b}$ be the closure of b. Then the number*

$$V_L(t) := \left(-\frac{t+1}{\sqrt{t}}\right)^{n-1} (\sqrt{t})^{\exp(b)} \operatorname{tr}(b)$$

is an isotopy invariant of the link L.

Example 13.3.1. Let $b \in B_2$ be a braid whose closure is isotopic to the trefoil knot. Then

$$V_L(t) := \left(-\frac{t+1}{\sqrt{t}}\right)(\sqrt{t})^3\left(\frac{t^3 - t^2 - 1}{t+1}\right) = -t^4 + t^3 + t.$$

Example 13.3.2. Let $b \in B_2$ be a braid whose closure is isotopic to a pair of linked circles $S^1 \cup S^1$. Then

$$V_L(t) := \left(-\frac{t+1}{\sqrt{t}}\right)(\sqrt{t})^2\left(\frac{t^2+1}{t+1}\right) = -\sqrt{t}(t^2+1).$$

Guide to the literature

E. Artin introduced the braid groups in the seminal paper [5], see also more accessible [6]. The equivalence classes of braids were studied in [157]. For the index of subfactors the reader is referred to [124]. The trace invariant was introduced in [125]. For an extended exposition of subfactors and knots, we refer the reader to the monograph [126].

14 Quantum groups

Quantum groups is a family of the noncommutative rings depending on a parameter q so that $q = 1$ corresponds to a commutative ring. Such groups are closely related to the quantum mechanics and a noncommutative version of the Pontryagin duality for abelian groups. We consider the simplest example of such a group.

Example 14.0.1. Let $A(k)$ be an associative algebra over a field k and let

$$M_q(2) = \begin{pmatrix} a & b \\ c & d \end{pmatrix}$$

be the set of two-by-two matrices with entries $a, b, c, d \in A(k)$ satisfying the commutation relations

$$\begin{cases} ab = q\, ba, \\ ac = q\, ca, \\ bd = q\, db, \\ cd = q\, dc, \\ bc = cb, \\ ad - da = \left(q - \dfrac{1}{q}\right) bc, \end{cases}$$

where $q \in k$ is a parameter. Algebra $A(k)$ itself can be viewed as an algebra of functions $\mathcal{F}(M_q(2))$ on $M_q(2)$. If $q = 1$ then $\mathcal{F}(M_q(2))$ is commutative and the product of two matrices gives a matrix with the entries in $\mathcal{F}(M_q(2))$. If $q \neq 1$ then such a product gives a matrix whose entries do not satisfy the commutation relations. However, if $\mathcal{F}(M_q(2))$ is a bialgebra endowed with multiplication

$$\mu : \mathcal{F}(M_q(2)) \otimes \mathcal{F}(M_q(2)) \to \mathcal{F}(M_q(2)),$$

and a comultiplication

$$\Delta : \mathcal{F}(M_q(2)) \to \mathcal{F}(M_q(2)) \otimes \mathcal{F}(M_q(2)),$$

then $M_q(2)$ becomes a semigroup; the bialgebra $\mathcal{F}(M_q(2))$ is an example of a *Hopf algebra*. To get the structure of a group, one needs to invert the quantum determinant. The resulting group $M_q(2)$ is called a quantum group.

https://doi.org/10.1515/9783110788709-014

14.1 Manin's quantum plane

Definition 14.1.1. By Manin's quantum plane P_q we understand the Euclidean plane whose coordinates (x, y) satisfy the commutation relation

$$xy = q\,yx$$

Such a plane is shown in Fig. 14.1.

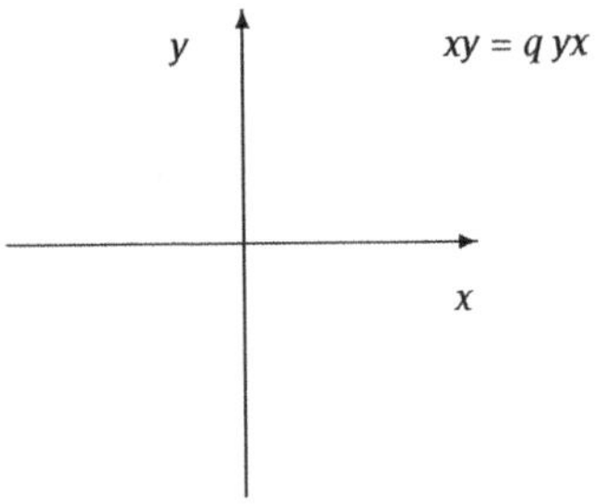

Figure 14.1: Quantum plane P_q.

Theorem 14.1.1 (Manin). *The group of linear transformations of the quantum plane P_q is isomorphic to the quantum group $M_q(2)$.*

Proof. A linear transformation of the quantum plane P_q can be written in the form

$$\begin{pmatrix} x' \\ y' \end{pmatrix} = \begin{pmatrix} a & b \\ c & d \end{pmatrix} \begin{pmatrix} x \\ y \end{pmatrix}.$$

Using this transformation, one gets the following system of equations:

$$\begin{cases} x'y' = ac\,x^2 + bc\,yx + ad\,xy + bd\,y^2, \\ y'x' = ca\,x^2 + da\,yx + cb\,xy + db\,y^2. \end{cases}$$

Because $x'y' = q\,y'x'$, we have the following identities

$$\begin{cases} ac = q\,ca, \\ bc = q\,da, \\ ad = q\,cb, \\ bd = q\,db. \end{cases}$$

The first and last equations are exactly the commutation relations for the quantum group $M_q(2)$; it is easy to see that, whenever $bc = cb$, it follows from $bc = q\,da$ and

$ad = q\,cb$ that

$$ad - da = \left(q - \frac{1}{q}\right)bc.$$

We leave it as exercise to the reader to prove the remaining relations $bc = cb$, $ab = q\,ba$, and $cd = q\,dc$. □

Remark 14.1.1. The quantum group $M_q(2)$ can be viewed as a group of symmetries of Manin's quantum plane P_q.

14.2 Hopf algebras

Definition 14.2.1. By a Hopf algebra one understands a bialgebra A endowed with a product

$$\mu : A \otimes A \to A$$

and coproduct

$$\Delta : A \to A \otimes A$$

such that μ and Δ are homomorphisms of (graded) algebras.

Remark 14.2.1. If $A \cong \mathcal{F}(M_q(2))$ then the product structure μ on the Hopf algebra A ensures that $M_q(2)$ is a semigroup, while the coproduct structure Δ (together with an inverse of the determinant) turns $M_q(2)$ into a group.

Example 14.2.1. Let A be an algebra of regular functions on an affine algebraic group G; let us define a product μ on A coming from multiplication

$$G \times G \to G,$$

and a coproduct Δ coming from the embedding

$$\{e\} \to G,$$

where e is the unit of G. Then the algebra A is a Hopf algebra.

Example 14.2.2. Let X be a connected topological group with multiplication

$$m : X \times X \to X$$

and comultiplication

$$\delta : X \to X \times X$$

defined by the embedding $\iota : \{e\} \to X$. Denote by $H^*(X)$ the cohomology ring of X; let

$$\mu := m^* : H^*(X) \otimes H^*(X) \to H^*(X)$$

be a product induced by m on the cohomology ring $H^*(X)$ and let

$$\Delta := \delta^* : H^*(X) \to H^*(X) \otimes H^*(X)$$

be the corresponding coproduct. The algebra $A \cong H^*(X)$ is a Hopf algebra.

Remark 14.2.2. The algebra $H^*(X)$ has been introduced in algebraic topology by Heinz Hopf; hence the name.

14.3 Operator algebras and quantum groups

An analytic approach to quantum groups is based on an observation that locally compact abelian groups G correspond to commutative C^*-algebras $C_0(G)$; therefore a noncommutative version of the Pontryagin duality must involve more general C^*-algebras. It was suggested in [289] that such C^*-algebras come from an embedding of the Hopf algebra into a C^*-algebra coherent with the coproduct structure of the Hopf algebra. This C^*-algebra is known as a compact quantum group. To give some details, recall that if G is a locally compact group then multiplication is a continuous map from $G \times G$ to G and therefore can be translated to a morphism

$$\Delta : C_0(G) \to M(C_0(G \times G)),$$

where $M(C_0(G\times G))$ is the multiplier algebra of $C_0(G\times G)$, i. e., the maximal C^*-algebra containing $C_0(G\times G)$ as an essential ideal [32, Section 12]. The translation itself is given by the formula $\Delta(f)(p,q) = f(pq)$. Define a morphism ι by a coassociativity formula $(\Delta \otimes \iota)\Delta = (\iota \otimes \Delta)\Delta$.

Definition 14.3.1 (Woronowicz). If A is a unital C^*-algebra together with a unital $*$-homomorphism $\Delta : A \to A \otimes A$ such that $(\Delta \otimes \iota)\Delta = (\iota \otimes \Delta)\Delta$ and such that the spaces $\Delta(A)(A \otimes 1)$ and $\Delta(A)(1 \otimes A)$ are dense in $A \otimes A$, then the pair (A, Δ) is called a compact quantum group.

Let φ be a weight on a C^*-algebra A; we refer the reader to Section 11.1.1. We denote by $\mathcal{M}\varphi^+$ the space of all integrable elements of A, i. e., $\mathcal{M}_\varphi^+ := \{x \in A^+ \mid \varphi(x) < \infty\}$. The following definition generalizes the compact quantum groups to the locally compact case.

Definition 14.3.2 (Kustermans–Vaes). Suppose that A is a unital C^*-algebra together with a unital $*$-homomorphism $\Delta : A \to A \otimes A$ such that $(\Delta \otimes \iota)\Delta = (\iota \otimes \Delta)\Delta$ and such that the spaces $\Delta(A)(A \otimes 1)$ and $\Delta(A)(1 \otimes A)$ are dense in $A \otimes A$. Moreover, assume that

(i) there exists a faithful KMS weight φ (see Section 11.1.1) on the compact quantum group (A, Δ) such that $\varphi((\omega \otimes \iota)\Delta(x)) = \varphi(x)\omega(1)$ for $\omega \in A_+$ and $x \in \mathcal{M}_\varphi^+$;
(ii) there exists a KMS weight ψ on the compact quantum group (A, Δ) such that $\psi((\iota \otimes \omega)\Delta(x)) = \psi(x)\omega(1)$ for $\omega \in A_+$ and $x \in \mathcal{M}_\psi^+$.

Then the pair (A, Δ) is called a locally compact quantum group.

Guide to the literature

The main references to quantum groups are the books [132] and [277]. Hopf algebras were introduced in [119]. Compact quantum groups were defined in [289] and locally compact quantum groups in [141]. Locally compact quantum groups in the context of W^*-algebras were considered in [142].

15 Noncommutative algebraic geometry

There exist noncommutative rings behaving much like the polynomial rings used in algebraic geometry. To explain the idea, let **CRng** $\cong k[x_1, \dots, x_m]$ be a polynomial ring over an algebraically closed field k and let **Grp** be the multiplicative group consisting of the $n \times n$ matrices with entries in $k[x_1, \dots, x_m]$; consider the functor $\mathbf{GL}_n : \mathbf{CRng} \to \mathbf{Grp}$, see Example 2.2.4. Let $\mathbf{U}_R : R \to R^*$ be a functor taking the commutative ring R to its group of units R^*. Suppose that our matrix group can be given by a finite number g of generators satisfying the relations $w_1 = w_2 = \cdots = w_r = I$. Taking the determinants of the matrices corresponding to such relations, one gets the following system of polynomial equations:

$$\begin{cases} \det(w_1) = 1, \\ \det(w_2) = 1, \\ \quad\vdots \\ \det(w_r) = 1. \end{cases}$$

The zero set of these polynomials is a variety V and let R be the coordinate ring of V. The category **CRng** of all such rings turns $\mathbf{det}_R$ into a natural transformation (an isomorphism) between the functors $\mathbf{GL}_n$ and $\mathbf{U}_R$, see Fig. 15.1 and Example 2.3.1.

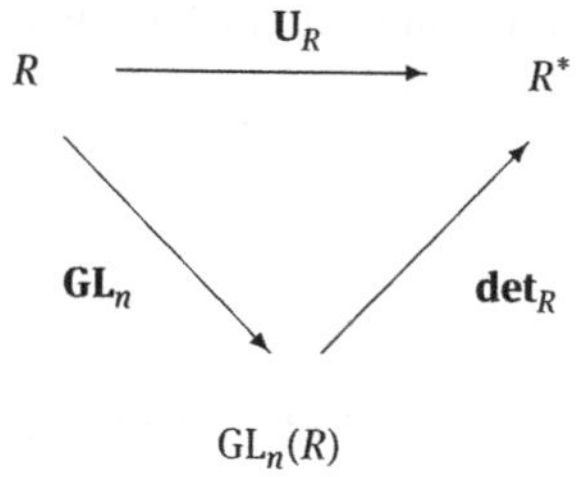

Figure 15.1: The natural transformation $\mathbf{det}_R$.

Remark 15.0.1. An example of functor $\mathbf{GL}_2$ with $g = 4$ and $r = 6$ was first constructed in [261]; the variety V in this case is isomorphic to the product $\mathcal{E}(\mathbf{C}) \times \mathcal{E}(\mathbf{C})$, where $\mathcal{E}(\mathbf{C})$ is an elliptic curve over the field of complex numbers.

15.1 Serre isomorphism

Let X be a projective variety, $\mathbf{Coh}(X)$ be a category of the quasi-coherent sheaves on X and $\mathbf{Mod}(B)$ be a category of the finitely generated graded modules over B factored by a torsion **Tors**. The following Serre isomorphism is true

$$\mathbf{Coh}(X) \cong \mathbf{Mod}(B)/\mathbf{Tors},$$

https://doi.org/10.1515/9783110788709-015

see [249]. Category **Mod**(B) is correctly defined for noncommutative rings. A noncommutative ring B satisfying the Serre isomorphism is called the twisted homogeneous coordinate ring of X. Consider the simplest example. If X is a compact Hausdorff space and $C(X)$ the commutative algebra of continuous functions from X to $\mathbf{C}$, then the topology of X is determined by algebra $C(X)$; in terms of the K-theory this can be written as $K_0^{top}(X) \cong K_0^{alg}(C(X))$. Taking the two-by-two matrices with entries in $C(X)$, one gets an algebra $C(X) \otimes M_2(\mathbf{C})$; in view of stability of K-theory under tensor products, it holds that $K_0^{top}(X) \cong K_0^{alg}(C(X)) \cong K_0^{alg}(C(X) \otimes M_2(\mathbf{C}))$. In other words, the topology of X is defined by algebra $C(X) \otimes M_2(\mathbf{C})$, which is no longer a commutative algebra. In algebraic geometry, one replaces X by a projective variety, $C(X)$ by its coordinate ring, $C(X) \otimes M_2(\mathbf{C})$ by a twisted homogeneous coordinate ring of X, and $K^{top}(X)$ by a category of the quasi-coherent sheaves on X. The simplest examples of rings B are given below.

Example 15.1.1. Let k be a field and $U_\infty(k)$ the algebra of polynomials over k in two noncommuting variables x_1 and x_2 satisfying the quadratic relation

$$x_1x_2 - x_2x_1 - x_1^2 = 0.$$

If $\mathbb{P}^1(k)$ is the projective line over k, then $B = U_\infty(k)$ and $X = \mathbb{P}^1(k)$ satisfy the fundamental duality **Coh**$(X) \cong$ **Mod**(B)/**Tors**.

Example 15.1.2. Let k be a field and $U_q(k)$ the algebra of polynomials over k in two noncommuting variables x_1 and x_2 satisfying the quadratic relation

$$x_1x_2 = qx_2x_1,$$

where $q \in k^*$ is a nonzero element of k. If $\mathbb{P}^1(k)$ is the projective line over k, then $B = U_q(k)$ and $X = \mathbb{P}^1(k)$ satisfy the fundamental duality **Coh**$(X) \cong$ **Mod**(B)/**Tors** for all $q \in k^*$.

There exists a canonical noncommutative ring B attached to the projective variety X and an automorphism $\alpha : X \to X$; we refer the reader to [265, pp. 180–182]. To give an idea, let $X = \mathrm{Spec}(R)$ for a commutative graded ring R. One considers the ring $B := R[t, t^{-1}; \alpha]$ of skew Laurent polynomials defined by the commutation relation

$$b^\alpha t = tb,$$

for all $b \in R$, where $b^\alpha \in R$ is the image of b under automorphism α; then B satisfies the isomorphism **Coh**$(X) \cong$ **Mod**(B)/**Tors**. The ring B is noncommutative, unless α is the trivial automorphism of X.

Example 15.1.3. The ring $B = U_\infty(k)$ corresponds to the automorphism $\alpha(u) = u + 1$ of the projective line $\mathbb{P}^1(k)$. Indeed, $u = x_2x_1^{-1} = x_1^{-1}x_2$ and, therefore, α maps x_2 to $x_1 + x_2$;

if one substitutes $t = x_1$, $b = x_2$, and $b^\alpha = x_1 + x_2$ in equation $b^\alpha t = tb$, then one gets the defining relation $x_1x_2 - x_2x_1 - x_1^2 = 0$ for the algebra $U_\infty(k)$.

Example 15.1.4. The ring $B = U_q(k)$ corresponds to the automorphism $\alpha(u) = qu$ of the projective line $\mathbb{P}^1(k)$. Indeed, $u = x_2x_1^{-1} = x_1^{-1}x_2$ and, therefore, α maps x_2 to qx_2; if one substitutes $t = x_1$, $b = x_2$, and $b^\alpha = qx_2$ in equation $b^\alpha t = tb$, then one gets the defining relation $x_1x_2 = qx_2x_1$ for the algebra $U_q(k)$.

15.2 Twisted homogeneous coordinate rings

Let X be a projective scheme over a field k, and let $\mathcal{L}$ be the invertible sheaf $\mathcal{O}_X(1)$ of linear forms on X. Recall that the homogeneous coordinate ring of X is a graded k-algebra, which is isomorphic to the algebra

$$B(X, \mathcal{L}) = \bigoplus_{n\geq 0} H^0(X, \mathcal{L}^{\otimes n}).$$

Denote by **Coh** the category of quasi-coherent sheaves on a scheme X and by **Mod** the category of graded left modules over a graded ring B. If $M = \oplus M_n$ and $M_n = 0$ for $n \gg 0$, then the graded module M is called right bounded. The direct limit $M = \lim M_\alpha$ is called a torsion, if each M_α is a right bounded graded module. Denote by **Tors** the full subcategory of **Mod** of the torsion modules. The following result is basic about the graded ring $B = B(X, \mathcal{L})$.

Theorem 15.2.1 (Serre). $\mathbf{Mod}(B)/\mathbf{Tors} \cong \mathbf{Coh}(X)$.

Let α be an automorphism of X. The pullback of sheaf $\mathcal{L}$ along α will be denoted by $\mathcal{L}^\alpha$, i. e., $\mathcal{L}^\alpha(U) := \mathcal{L}(\alpha U)$ for every $U \subset X$. We shall set

$$B(X, \mathcal{L}, \alpha) = \bigoplus_{n\geq 0} H^0(X, \mathcal{L} \otimes \mathcal{L}^\alpha \otimes \cdots \otimes \mathcal{L}^{\alpha^n}).$$

The multiplication of sections is defined by the rule

$$ab = a \otimes b^{\alpha^m},$$

whenever $a \in B_m$ and $b \in B_n$.

Definition 15.2.1. Given a pair (X, α) consisting of a Noetherian scheme X and an automorphism α of X, an invertible sheaf $\mathcal{L}$ on X is called *α-ample* if, for every coherent sheaf $\mathcal{F}$ on X, the cohomology group

$$H^q(X, \mathcal{L} \otimes \mathcal{L}^\alpha \otimes \cdots \otimes \mathcal{L}^{\alpha^{n-1}} \otimes \mathcal{F})$$

vanishes for $q > 0$ and $n \gg 0$.

Remark 15.2.1. If α is trivial, then one gets a definition of the ample invertible sheaf [249].

A noncommutative generalization of the Serre theorem is as follows.

Theorem 15.2.2 (Artin–van den Bergh). *Let $\alpha : X \to X$ be an automorphism of a projective scheme X over k and let $\mathcal{L}$ be a α-ample invertible sheaf on X. If $B(X, \mathcal{L}, \alpha) \cong \bigoplus_{n\geq 0} H^0(X, \mathcal{L} \otimes \mathcal{L}^{\alpha} \otimes \cdots \otimes \mathcal{L}^{\alpha^n})$, then*

$$\mathbf{Mod}(B(X, \mathcal{L}, \alpha))/\mathbf{Tors} \cong \mathbf{Coh}(X).$$

Remark 15.2.2. The question of which invertible sheaves are α-ample is fairly subtle, and there is no characterization of the automorphisms α for which such an invertible sheaf exists. However, in many important special cases this problem is solvable, see [9, Corollary 1.6].

Definition 15.2.2. For an automorphism $\alpha : X \to X$ of a projective scheme X and α-ample invertible sheaf $\mathcal{L}$ on X, the ring

$$B(X, \mathcal{L}, \alpha) \cong \bigoplus_{n\geq 0} H^0(X, \mathcal{L} \otimes \mathcal{L}^{\alpha} \otimes \cdots \otimes \mathcal{L}^{\alpha^n})$$

is called a twisted homogeneous coordinate ring of X.

15.3 Sklyanin algebras

Definition 15.3.1. If k is a field of char $k \neq 2$, then by a Sklyanin algebra $\mathfrak{S}_{\alpha,\beta,\gamma}(k)$ one understands a free k-algebra on four generators x_i and six quadratic relations

$$\begin{cases} x_1x_2 - x_2x_1 = \alpha(x_3x_4 + x_4x_3), \\ x_1x_2 + x_2x_1 = x_3x_4 - x_4x_3, \\ x_1x_3 - x_3x_1 = \beta(x_4x_2 + x_2x_4), \\ x_1x_3 + x_3x_1 = x_4x_2 - x_2x_4, \\ x_1x_4 - x_4x_1 = \gamma(x_2x_3 + x_3x_2), \\ x_1x_4 + x_4x_1 = x_2x_3 - x_3x_2, \end{cases}$$

where $\alpha, \beta, \gamma \in k$ are such that

$$\alpha + \beta + \gamma + \alpha\beta\gamma = 0.$$

Let $\alpha \notin \{0; \pm 1\}$ and consider a nonsingular elliptic curve $\mathcal{E}(k)$ in the Jacobi form

$$\begin{cases} u^2 + v^2 + w^2 + z^2 = 0, \\ \frac{1-\alpha}{1+\beta} v^2 + \frac{1+\alpha}{1-\gamma} w^2 + z^2 = 0, \end{cases}$$

and an automorphism $\sigma : \mathcal{E}(k) \to \mathcal{E}(k)$ acting on the points of $\mathcal{E}(k)$ according to the formula

$$\begin{cases} \sigma(u) = -2\alpha\beta\gamma vwz - u(-u^2 + \beta\gamma v^2 + \alpha\gamma w^2 + \alpha\beta z^2), \\ \sigma(v) = 2\alpha uwz + v(u^2 - \beta\gamma v^2 + \alpha\gamma w^2 + \alpha\beta z^2), \\ \sigma(w) = 2\beta vwz + w(u^2 + \beta\gamma v^2 - \alpha\gamma w^2 + \alpha\beta z^2), \\ \sigma(z) = 2\gamma uvw + z(u^2 + \beta\gamma v^2 + \alpha\gamma w^2 - \alpha\beta z^2). \end{cases}$$

Theorem 15.3.1 (Sklyanin). *The algebra $\mathfrak{S}_{\alpha,\beta,\gamma}(k)$ is a twisted homogeneous coordinate ring of the elliptic curve $\mathcal{E}(k)$ and automorphism $\sigma : \mathcal{E}(k) \to \mathcal{E}(k)$, i. e.,*

$$\mathbf{Mod}(\mathfrak{S}_{\alpha,\beta,\gamma}(k)/\Omega) \cong \mathbf{Coh}(\mathcal{E}(k)),$$

where **Mod** *is the category of the quotient graded modules over the algebra $\mathfrak{S}_{\alpha,\beta,\gamma}(k)$ modulo torsion,* **Coh** *the category of the quasi-coherent sheaves on $\mathcal{E}(k)$, and $\Omega \subset \mathfrak{S}_{\alpha,\beta,\gamma}(k)$ the two-sided ideal generated by the central elements*

$$\begin{cases} \Omega_1 = x_1^2 + x_2^2 + x_3^2 + x_4^2, \\ \Omega_2 = x_2^2 + \frac{1+\beta}{1-\gamma} x_3^2 + \frac{1-\beta}{1+\alpha} x_4^2. \end{cases}$$

Guide to the literature

Noncommutative algebraic geometry grew from the seminal works and [261]. Namely, the Serre isomorphism has been established in [249]. The noncommutative ring of an elliptic curve was constructed in [261]; for a subsequent development, see [262]. For a generalization of the Sklyanin algebras, we refer the reader to the papers [84] and [85]. A general approach to noncommutative algebraic geometry can be found in [8] and [9]. The topic inspired geometers [12] and grew further [264] and [11]. We refer the reader to the survey and [217]. The goal of noncommutative algebraic geometry is a classification of the related algebras in terms of curves, surfaces, etc. Such an approach is similar to the Connes classification program.

16 Trends in noncommutative geometry

The noncommutative geometry has been treated independently in [34, 128, 139, 219, 220]; we review their work below. A new trend appears in the remarkable paper [148]. The powerful *reconstruction theory* of the late A. L. Rosenberg has been developed in [244]; a collection of Rosenberg's papers on the noncommutative geometry can be found in [245].

16.1 Derived categories

Recall that in algebraic geometry the category of finitely generated projective modules over the coordinate ring of variety X is isomorphic to the category of coherent sheaves over X (Serre's Theorem); in other words, variety X can be recovered from a derived category of modules over commutative rings. This observation has been formalized by A. Bondal and D. Orlov and can be adapted to the case of noncommutative rings.

Definition 16.1.1. Let $\mathbf{A}$ be an abelian category and let $\mathbf{Kom}(\mathbf{A})$ be the category of complexes over $\mathbf{A}$, i. e., a category whose objects are cochain complexes of abelian groups and arrows are morphisms of the complexes. By a derived category $\mathcal{D}(\mathbf{A})$ of abelian category $\mathbf{A}$ one understands a category such that a functor

$$Q : \mathbf{Kom}(\mathbf{A}) \longrightarrow \mathcal{D}(\mathbf{A})$$

has the following properties:
(i) $Q(f)$ is an isomorphism for any quasi-isomorphism f in $\mathbf{Kom}(\mathbf{A})$, i. e., an arrow of $\mathbf{A}$ which gives an isomorphism of $\mathbf{Kom}(\mathbf{A})$;
(ii) any functor $F : \mathbf{Kom}(\mathbf{A}) \to \mathcal{D}$ transforming a quasi-isomorphism into isomorphism can be uniquely factored through $\mathcal{D}(\mathbf{A})$.

Example 16.1.1. If $\mathbf{CRng}$ is the category of commutative rings and $\mathbf{Mod}$ the category of modules over the rings, then

$$\mathbf{Mod} \cong \mathcal{D}(\mathbf{CRng}).$$

Definition 16.1.2. If X is the category of algebraic varieties, then by $\mathcal{D}_{coh}(X)$ one understands the derived category of coherent sheaves on X.

The following result is known as the Reconstruction Theorem for smooth projective varieties; this remarkable theorem has been proved in 1990s by Alexei Bondal and Dmitri Orlov.

Theorem 16.1.1 (Bondal–Orlov). *Let X be a smooth irreducible projective variety with ample canonical and anticanonical sheaf. If for two projective varieties X and X' the*

https://doi.org/10.1515/9783110788709-016

categories

$$\mathcal{D}_{\mathrm{coh}}(X) \cong \mathcal{D}_{\mathrm{coh}}(X')$$

are equivalent, then $X \cong X'$ are isomorphic projective varieties.

Remark 16.1.1. It is known that certain projective abelian varieties and $K3$ surfaces without restriction on the sheaves can be nonisomorphic but have equivalent derived categories of coherent sheaves; thus the requirements on the sheaves cannot be dropped.

Remark 16.1.2. The Reconstruction Theorem extends to the category of modules over the noncommutative rings; therefore such a theorem can be regarded as a categorical foundation of the noncommutative algebraic geometry of M. Artin, J. Tate, and M. van den Bergh.

16.2 Noncommutative thickening

Let R be an associative algebra over $\mathbf{C}$ and $R_{ab} = R/[R,R]$ its abelianization by the commutator $[R,R] = \{xyx^{-1}y^{-1} \mid x,y \in R\}$. Denote by

$$X_{ab} = \mathbf{Spec}(R_{ab})$$

the space of all prime ideals of R_{ab} endowed with the Zariski topology.

Remark 16.2.1. The naive aim of noncommutative algebraic geometry consists in construction of an embedding

$$X_{ab} \hookrightarrow X \cong \mathbf{Spec}(R),$$

where X is some noncommutative space associated to the homomorphism of rings $R \to R_{ab}$.

Definition 16.2.1 (Kapranov). By a noncommutative thickening of the space X_{ab} one understands a sheaf of noncommutative rings $\mathcal{O}^{NC}$ (corresponding to a formal small neighborhood of X_{ab} in X) which is the completion of the algebra R by the iterated commutators of the form

$$\lim_{n\to\infty} [a_1, [a_2, \dots [a_{n-1}, a_n]]], \quad a_i \in R,$$

such that the commutators are small in a completion of the Zariski topology.

Definition 16.2.2 (Kapranov). By a noncommutative scheme X one understands the glued ringed spaces of the form $(X_{ab}, \mathcal{O}^{NC})$.

Theorem 16.2.1 (Uniqueness of noncommutative thickening). *If R is a finitely generated commutative algebra, then its noncommutative thickening $(X_{ab}, \mathcal{O}^{NC})$ is unique up to an isomorphism of R.*

Remark 16.2.2. If $X_{ab} = M$ is a smooth manifold, then its noncommutative thickening $(X_{ab}, \mathcal{O}^{NC})$ induces to a new structure on M corresponding to additional characteristic classes of M.

16.3 Deformation quantization of Poisson manifolds

The deformation quantization extends Rieffel's construction of the n-dimensional noncommutative tori to arbitrary smooth manifolds. It was shown by M. Kontsevich that the Poisson bracket is all it takes to obtain a family of noncommutative associative algebras from the algebra of functions on a manifold; manifolds with such a bracket are called Poisson manifolds. Below we give details of the construction.

Let M be a manifold endowed with a smooth (or analytic, or algebraic) structure; let $\mathcal{F}(M)$ be the (commutative) algebra of function on M. The idea is to keep the pointwise addition of functions on M and replace the pointwise multiplication by a noncommutative (but associative) binary operation $*_\hbar$ depending on a deformation parameter $\hbar$ so that $\hbar = 0$ corresponds to the usual pointwise multiplication of functions on M. One can unfold the noncommutative multiplication in an infinite series in parameter $\hbar$,

$$f *_\hbar g = fg + \{f,g\}\hbar + O(\hbar),$$

where the bracket

$$\{\bullet,\bullet\} : \mathcal{F}(M) \otimes \mathcal{F}(M) \to \mathcal{F}(M)$$

is bilinear and, without loss of generality, one can assume that $\{f,g\} = -\{g,f\}$. It follows from the associativity of operation $*_\hbar$ that

$$\begin{cases} \{f,gh\} = \{f,g\}h + \{f,h\}g, \\ \{f,\{g,h\}\} + \{h,\{f,g\}\} + \{g,\{h,f\}\} = 0. \end{cases}$$

The first equation implies that $\{f,f\} = 0$ and the second is the *Jacobi identity*; in other words, the noncommutative multiplication $*_\hbar$ gives rise to a *Poisson bracket* $\{\bullet,\bullet\}$ on the manifold M. It is remarkable that the converse is true.

Theorem 16.3.1 (Kontsevich). *If M is a manifold endowed with the Poisson bracket $\{\bullet,\bullet\}$, then there exists an associative noncommutative multiplication on $\mathcal{F}(M)$ defined by the* *formula*

$$f *_\hbar g := fg + \{f, g\} + \sum_{n\geq 2} B_n(f, g)\hbar^n,$$

where $B_n(f, g)$ are certain explicit operators defined by the Poisson bracket.

Example 16.3.1 (Rieffel). Let $M \cong T^n$ and $\mathcal{F}(M) \cong C^\infty(T^n)$; let us define the Poisson bracket by the formula

$$\{f, g\} = \sum \theta_{ij} \frac{\partial f}{\partial x_i} \frac{\partial g}{\partial x_j},$$

where $\Theta = (\theta_{ij})$ is a real skew-symmetric matrix. Then the noncommutative multiplication $*_{\hbar=1}$ defines a (smooth) n-dimensional noncommutative torus $\mathcal{A}_\Theta$.

16.4 Algebraic geometry of noncommutative rings

In the 1980s Fred van Oystaeyen and Alain Verschoren initiated a vast program aimed to develop a noncommutative version of algebraic geometry based on the notion of a spectrum of a noncommutative ring R; such spectra consist of the maximal (left and right) ideals of R. A difficulty here is that noncommutative rings occurring in practice are simple rings, i. e., have no ideals whatsoever; the difficulty can be overcome, see the original works of the above authors. Below we briefly review some of the constructions.

Definition 16.4.1. For an (associative) algebra A, the maximal ideal spectrum $\mathbf{Max}(A)$ of A is the set of all maximal two-sided ideals $M \subset A$ equipped with the noncommutative Zariski topology, i. e., a topology with the typical open set of $\mathbf{Max}(A)$ given by the formula

$$\mathbb{X}(I) = \{M \in \mathbf{Max}(A) \mid I \not\subset M\}.$$

Noncommutative sheaves $\mathcal{O}_A^{nc}$ associated to A can be defined as follows.

Definition 16.4.2. Sheave $\mathcal{O}_A^{nc}$ is defined by taking the sections over a typical open set $\mathbb{X}(I)$ of a two-sided ideal $I \subset \mathbf{Max}(A)$ according to the formula

$$\Gamma(\mathbb{X}(I), \mathcal{O}_A^{nc}) := \{\delta \in \Sigma \mid \exists l \in \mathbf{N} : I^l\delta \subset A\},$$

where Σ is the central simple algebra of A.

Remark 16.4.1. One can develop all essential features which the noncommutative scheme $(\mathbf{Max}(A), \mathcal{O}_A^{nc})$ must have; we refer the interested reader to the corresponding literature.

Guide to the literature

The derived categories were studied in [34], see also [35]. The noncommutative thickening was introduced in [128]. The deformation quantization of Poisson manifolds was studied in [139]. For the basics of algebraic geometry of noncommutative rings, we refer the reader to the monographs [219] and [220]. For the *reconstruction theory* of A. L. Rosenberg, see the monograph [244]. A new interesting trend in noncommutative algebraic geometry can be found in the paper [148].

Bibliography

[1] S. A. Amitsur, A noncommutative Hilbert basis theorem and subrings of matrices, Trans. Am. Math. Soc. 149 (1970), 133–142.

[2] J. E. Andersen, Asymptotic faithfulness of the quantum $SU(n)$ representations of the mapping class groups, Ann. Math. 163 (2006), 347–368.

[3] D. V. Anosov, Geodesic flows on Riemann manifolds with negative curvature, Proc. Steklov Inst. 90 (1967).

[4] E. Artin, Über eine neue Art von L-Reihen, Abhandlungen aus dem Mathematischen Seminar der Hamburgerischen Universität, Bd. 3 (1924), 89-108.

[5] E. Artin, Theorie der Zöpfe, Abhandlungen aus dem Mathematischen Seminar der Hamburgerischen Universität, Bd. 4 (1925), 47-72.

[6] E. Artin, Theory of braids, Ann. Math. 48 (1947), 101–126.

[7] E. Artin, Geometric Algebra, Interscience, New York, 1957.

[8] M. Artin and W. F. Schelter, Graded algebras of global dimension 3, Adv. Math. 66 (1987), 171–216.

[9] M. Artin and M. van den Bergh, Twisted homogeneous coordinate rings, J. Algebra 133 (1990), 249–271.

[10] M. Artin and J.-L. Verdier, Seminar on étale cohomology of number fields, Woods Hole, 1964.

[11] M. Artin and J. J. Zhang, Noncommutative projective schemes, Adv. Math. 109 (1994), 228–287.

[12] M. Artin, J. Tate and M. van den Bergh, Some algebras associated to automorphisms of elliptic curves, in: The Grothendieck Festschrift, Vol. 1, Birkhäuser, Boston, 1990, pp. 33–85.

[13] W. Arveson, An Invitation to C^*-Algebras, GTM 38, Springer, 1976.

[14] M. Atiyah, K-Theory, Lecture Note Series by W. A. Benjamin, 1967.

[15] M. Atiyah, Global theory of elliptic operators, in: Proceedings of an International Conference on Functional Analysis and Related Topics, Univ. of Tokyo Press, 1970.

[16] M. Atiyah and I. Singer, The index of elliptic operators on compact manifolds, Bull. Am. Math. Soc. 69 (1963), 422–433.

[17] M. Atiyah and I. Singer, The index of elliptic operators, I, Ann. Math. 87 (1968), 484–530; II, Annals of Math. 87 (1968), 531–545; III, Annals of Math. 87 (1968), 546–604; IV, Annals of Math. 93 (1971), 119–138; V, Annals of Math. 93 (1971), 139–149.

[18] R. G. Ayoub and C. Ayoub, On the group ring of a finite abelian group, Bull. Aust. Math. Soc. 1 (1969), 245–261.

[19] R. Baer, Linear Algebra and Projective Geometry, Vol. II, Pure Appl. Math., Academic Press, New York, 1952.

[20] A. Baker, Transcendental Number Theory, Cambridge Univ. Press, 1975.

[21] M. Bauer, A characterization of uniquely ergodic interval exchange maps in terms of the Jacobi–Perron algorithm, Bol. Soc. Bras. Mat. 27 (1996), 109–128.

[22] P. Baum and A. Connes, Geometric K-theory for Lie groups and foliations, Enseign. Math. 46 (2000), 1–35.

[23] L. Bernstein, The Jacobi–Perron Algorithm, Its Theory and Applications, Lect. Notes in Math. 207, Springer, 1971.

[24] S. Bigelow, Braid groups are linear, J. Am. Math. Soc. 14 (2001), 471–486.

[25] S. Bigelow and R. Budney, The mapping class group of a genus two surface is linear, Algebraic Geom. Topol. 1 (2001), 699–708.

[26] J. S. Birman and H. M. Hilden, Mapping class groups of closed surfaces as covering spaces, in: Advances in the Theory of Riemann surfaces, Annals of Math. Studies Series 66, Princeton Univ. Press, 1971.

https://doi.org/10.1515/9783110788709-017

[27] M. Bhargava, Higher composition laws I: A new view on Gauss composition, and quadratic generalizations, Ann. Math. 159 (2004), 217–250.

[28] M. Bhargava, Higher composition laws II: On cubic analogues of Gauss composition, Ann. Math. 159 (2004), 865–886.

[29] M. Bhargava, Higher composition laws III: The parametrization of quartic rings, Ann. Math. 159 (2004), 1329–1360.

[30] M. Bhargava, Higher composition laws IV: The parametrization of quintic rings, Ann. Math. 167 (2008), 53–94.

[31] M. Bhargava, Higher composition laws and applications, in: Proceedings of the ICM, Madrid 2006, pp. 271–294.

[32] B. Blackadar, K-Theory for Operator Algebras, MSRI Publications, Springer, 1986.

[33] S. Bloch and K. Kato, L-functions and Tamagawa numbers of motives, in: The Grothendieck Festschrift, P. Cartier et al., eds, Progr. Math. 86, Birkhäuser, Boston, 1990, pp. 333–400.

[34] A. Bondal and D. Orlov, Reconstruction of a variety from the derived category and groups of autoequivalences, Compos. Math. 125 (2001), 327–344.

[35] A. Bondal and D. Orlov, Derived categories of coherent sheaves, in: Proceedings of the ICM, Vol. 2, Beijing 2002, pp. 47–56.

[36] A. Borel, Some finiteness properties of adele groups over number fields, Publ. Math. IHÉS 16 (1963), 5–30.

[37] A. Borel and J.-P. Serre, Théorèmes de finitude en cohomologie galoisienne, Comment. Math. Helv. 39 (1964), 111–164.

[38] Z. I. Borevich and I. R. Shafarevich, Number Theory, Acad. Press, 1966.

[39] J. B. Bost and A. Connes, Hecke algebras, type III factors and phase transition with spontaneous symmetry breaking in number theory, Sel. Math. 1 (1995), 411–457.

[40] R. Bowen and J. Franks, Homology for zero-dimensional nonwandering sets, Ann. Math. 106 (1977), 73–92.

[41] O. Bratteli, Inductive limits of finite dimensional C^*-algebras, Trans. Am. Math. Soc. 171 (1972), 195–234.

[42] O. Bratteli, P. E. T. Jorgensen and V. Ostrovskyı, Representation theory and numerical AF-invariants. The representations and centralizers of certain states on $\mathcal{O}_d$, Mem. Am. Math. Soc. 168 (797) (2004).

[43] B. W. Brock, N. D. Elkies and B. W. Jordan, Periodic continued fractions over S-integers in number fields and Skolem's p-adic method, Acta Arith. 197 (2021), 379–420.

[44] L. Brown, R. Douglas and P. Fillmore, Extensions of C^*-algebras and K-homology, Ann. Math. 105 (1977), 265–324.

[45] A. Carey, M. Marcolli and A. Rennie, Modular index invariants of Mumford curves, in: Noncommutative Geometry, Arithmetic, and Related Topics. Proceedings of the 21st Meeting of the Japan-U.S. Mathematics Institute (JAMI) Held at Johns Hopkins University, Baltimore, MD, USA, March 23–26, 2009. Johns Hopkins University Press, Baltimore, MD, 2011, pp. 31–73.

[46] L. Carlitz, Some binomial coefficient identities, Fibonacci Q. 4 (1966), 323–331.

[47] H. Cartan, Théorie de Galois pour les corps non commutatifs, Ann. Sci. Éc. Norm. Supér. 64 (1947), 59–77.

[48] H. Cartan and S. Eilenberg, Homological Algebra, Princeton Univ. Press, Princeton, 1956.

[49] J. W. S. Cassels, Diophantine equations with special reference to elliptic curves, J. Lond. Math. Soc. 41 (1966), 193–291.

[50] J. W. S. Cassels, Rational Quadratic Forms, Academic Press, 1978.

[51] A. J. Casson and S. A. Bleiler, Automorphisms of Surfaces after Nielsen and Thurston, Lond. Math. Soc. Student Texts 9, Cambridge, 1988.

[52] H. Cohen and X.-F. Roblot, Computing the Hilbert class field of real quadratic fields, Math. Comput. 69 (2000), 1229–1244.

[53] A. Connes, Une classification de facteurs de type III, Ann. Sci. Éc. Norm. Supér. 6 (1973), 133–252.

[54] A. Connes, Von Neumann algebras, in: Proceedings of the International Congress of Mathematics, Helsinki, 1978, pp. 97–109.

[55] A. Connes, A survey of foliations and operator algebras, in: Operator Algebras and Applications, Proc. Symp. Pure Math. 38, AMS, Providence, 1981.

[56] A. Connes, Noncommutative differential geometry, Publ. Math. IHÉS 62 (1985), 257–360.

[57] A. Connes, Géométrie Non Commutative, Paris, 1990.

[58] A. Connes, Noncommutative Geometry, Academic Press, 1994.

[59] A. Connes and M. Marcolli, Noncommutative Geometry, Quantum Fields and Motives, AMS Colloquium Publications, 2008.

[60] A. Connes and H. Moscovici, Cyclic cohomology, the Novikov conjecture and hyperbolic groups, Topology 29 (1990), 345–388.

[61] J. Cremona et al., LMFDB Collaboration: The L-functions and Modular Forms Database, 2013, http://www.lmfdb.org.

[62] J. Cuntz, Simple C^*-algebras generated by isometries, Commun. Math. Phys. 57 (1977), 173–185.

[63] J. Cuntz and W. Krieger, A class of C^*-algebras and topological Markov chains, Invent. Math. 56 (1980), 251–268.

[64] K. R. Davidson, C^*-Algebras by Example, Fields Inst. Monographs, AMS, Providence, RI, 1996.

[65] R. Dedekind, Über die Theorie der ganzen algebraischen Zahlen, in: Vorlesungen über Zahlentheorie, 1863.

[66] M. Dehn, Die Gruppe der Abbildungsklassen. (Das arithmetische Feld auf Flachen), Acta Math. 69 (1938), 135–206.

[67] P. Deligne, Travaux de Shimura, Séminaire Bourbaki, Lecture Notes in Math. 244, Springer, 1971, pp. 123–165.

[68] F. Diamond and J. Shurman, A First Course in Modular Forms, GTM 228, Springer, 2005.

[69] J. Dixmier, C^*-Algebras, North-Holland Publishing Company, 1977.

[70] R. G. Douglas, C^*-Algebra Extensions and K-Homology, Ann. of Math. Studies 95, Princeton Univ. Press, 1980.

[71] A. Douady and J. Hubbard, On the density of Strebel differentials, Invent. Math. 30 (1975), 175–179.

[72] S. Echterhoff, W. Lück, N. C. Phillips and S. Walters, The structure of crossed products of irrational rotation algebras by finite subgroups of $SL_2(\mathbf{Z})$, J. Reine Angew. Math. 639 (2010), 173–221.

[73] S. Eilenberg and S. MacLane, Natural isomorphisms in group theory, Proc. Natl. Acad. Sci. USA 28 (1942), 537–543.

[74] E. G. Effros, Dimensions and C^*-Algebras, Conf. Board of the Math. Sciences 46, Amer. Math. Soc., Providence, RI, 1981.

[75] E. G. Effros and C.-L. Shen, Approximately finite C^*-algebras and continued fractions, Indiana Univ. Math. J. 29 (1980), 191–204.

[76] G. A. Elliott, On the classification of inductive limits of sequences of semisimple finite-dimensional algebras, J. Algebra 38 (1976), 29–44.

[77] G. A. Elliott, On the K-theory of the C^*-algebra generated by a projective representation of $\mathbf{Z}^n$, Proc. Symp. Pure Math. 38 (1982), Part I, 177–180.

[78] G. Etesi, Gravity as a four dimensional algebraic quantum field theory, Adv. Theor. Math. Phys. 20 (2016), 1049–1082.

[79] G. Etesi, The universal von Neumann algebra of smooth four-manifolds, Adv. Theor. Math. Phys., to appear, 1712.01828.

[80] L. Euler, De usu novi algorithmi in Problemate Pelliano solvendo, Novi Comment. Acad. Sci. Imper. Petropol. 11 (1765).

[81] A. Fathi, F. Laudenbach and Z. Poénaru, Travaux de Thurston sur les surfaces, Séminaire Orsay, France, Astérisque 66–67 (1979).

[82] B. Farb and D. Margalit, A Primer on Mapping Class Groups, Princeton Math. Series 49, 2011.

[83] C. Farsi and N. Watling, C^*-algebras of dynamical systems of the non-commutative torus, Math. Scand. 75 (1994), 101–110.

[84] B. L. Feigin and A. V. Odesskii, Sklyanin's elliptic algebras, Funct. Anal. Appl. 23 (1989), 207–214.

[85] B. L. Feigin and A. V. Odesskii, Constructions of Sklyanin's elliptic algebras and quantum R-matrices, Funct. Anal. Appl. 27 (1993), 37–45.

[86] P. A. Fillmore, A User's Guide to Operator Algebras, Canadian Mathematical Society Monographs, J. Wiley & Sons, Inc., 1996.

[87] S. Fomin and A. Zelevinsky, Cluster algebras I: Foundations, J. Am. Math. Soc. 15 (2002), 497–529.

[88] S. Fomin, M. Shapiro and D. Thurston, Cluster algebras and triangulated surfaces, Part I: Cluster complexes, Acta Math. 201 (2008), 83–146.

[89] P. Furtwängler, Über das Verhalten der Ideale des Grundkörpers im Klassenkörper, Monatshefte Math. Phys. 27 (1916), 1–15.

[90] P. Freyd, D. Yetter, J. Hoste, W. Lickorish, K. Millet and A. Ocneanu, A new polynomial invariant of knots and links, Bull. Am. Math. Soc. 12 (1985), 239–246.

[91] S. Gelbart, An elementary introduction to the Langlands program, Bull. Am. Math. Soc. 10 (1984), 177–219.

[92] I. M. Gelfand and M. Naimark, On the embedding of normed rings into the ring of operators in Hilbert space, Mat. Sb. 12 (1943), 197–213.

[93] S. I. Gelfand and Yu.I. Manin, Homological Algebra, Algebra V, Encycl. Math. Sciences 38, Springer, 1994.

[94] J. G. Glimm, On a certain class of operator algebras, Trans. Am. Math. Soc. 95 (1960), 318–340.

[95] E. Golod and I. Shafarevich, On the class field tower, Izv. Akad. Nauk SSSR, Ser. Mat. 28 (1964), 261–272.

[96] R. E. Gompf, Stable diffeomorphism of compact 4-manifolds, Topol. Appl. 18 (1984), 115–120.

[97] W. H. Gottschalk and G. A. Hedlund, Topological Dynamics, AMS Colloquium Publications 36, 1955.

[98] D. F. Goldfeld, The class number of quadratic fields and the conjectures of Birch and Swinnerton-Dyer, Ann. Sc. Norm. Super. Pisa 3 (1976), 623–663.

[99] F. O. Gouvêa, P-Adic Numbers, Springer, 1993.

[100] J. M. Gracia-Bondia, J. C. Várilly and H. Figueroa, Elements of Noncommutative Geometry, Birkhäuser, 2000.

[101] B. H. Gross, Arithmetic on Elliptic Curves with Complex Multiplication, Lecture Notes Math. 776, Springer, 1980.

[102] A. Grothendieck, Standard conjectures on algebraic cycles, in: Algebraic Geometry, Internat. Colloq. Tata Inst. Fund. Res., Bombay, 1968.

[103] T. Hadfield, K-homology of the rotation algebras $\mathcal{A}_\theta$, Can. J. Math. 56 (2004), 926–944.

[104] R. Hain and E. Looijenga, Mapping class groups and moduli spaces of curves, Proc. Symp. Pure Math. 62(2) (1997), 97–142.

[105] D. Handelman, Positive matrices and dimension groups affiliated to C^*-algebras and topological Markov chains, J. Oper. Theory 6 (1981), 55–74.

[106] D. Handelman, Extensions for *AF* C^*-algebras and dimension groups, Trans. Am. Math. Soc. 271 (1982), 537–573.
[107] D. Handelman, Matrices of positive polynomials, Electron. J. Linear Algebra 19 (2009), 2–89.
[108] R. Hartshorne, Algebraic Geometry, GTM 52, Springer, 1977.
[109] A. Hatcher and D. McCullough, Finite Presentation of 3-Manifold Mapping Class Groups, Lecture Notes in Math. 1425, Springer, 1990, pp. 48–57.
[110] W. J. Harvey, Geometric structure of surface mapping class groups, in: Homological Group Theory, LMS Lect. Notes. 36, 1979, pp. 255–269.
[111] H. Hasse, Vorlesungen über Zahlentheorie, Springer, 1950.
[112] E. Hecke, Ueber die Konstruktion der Klassenkörper reeller quadratischer Körper mit Hilfe von automorphen Funktionen, Nachr. Ges. Wiss. Gött., Math.-Phys. Kl. 6 (1910), 619–623.
[113] E. Hecke, Bestimmung der Perioden gewisser Integrale durch die Theorie der Klassenkörper, Math. Z. 28 (1928), 708–727.
[114] G. Hemion, On the classification of homeomorphisms of 2-manifolds and the classification of 3-manifolds, Acta Math. 142 (1979), 123–155.
[115] D. Hilbert, Mathematical problems, Bull. Am. Math. Soc. 8 (1902), 437–479.
[116] D. Hilbert, Grundlagen der Geometrie, Teubner, Leipzig-Berlin, 1930.
[117] G. Hochschild, On the cohomology groups of an associative algebra, Ann. Math. 46 (1945), 58–67.
[118] G. Hochschild, B. Kostant and A. Rosenberg, Differential forms on regular affine algebras, Trans. Am. Math. Soc. 102 (1962), 383–408.
[119] H. Hopf, Über die Topologie der Gruppen-Mannigfaltigkeiten und ihrer Verallgemeinerungen, Ann. Math. 42 (1941), 22–52.
[120] J. Hubbard and H. Masur, Quadratic differentials and foliations, Acta Math. 142 (1979), 221–274.
[121] B. Hughes, Trees, ultrametrics, and noncommutative geometry, Pure Appl. Math. Q. 8 (2012), 221–312.
[122] D. Husemöller, Elliptic Curves, GTM 111, Springer, 1986.
[123] N. Jacobson, The fundamental theorem of the Galois theory for quasi-fields, Ann. Math. 41 (1940), 1–7.
[124] V. F. R. Jones, Index of subfactors, Invent. Math. 72 (1983), 1–25.
[125] V. F. R. Jones, A polynomial invariant for knots via von Neumann algebras, Bull. Am. Math. Soc. 12 (1985), 103–111.
[126] V. F. R. Jones, Subfactors and Knots, CBMS Regional Conference Series in Mathematics 80, AMS, Providence, 1991.
[127] S. Kamada, Braid and Knot Theory in Dimension Four, Mathematical Surveys and Monographs 95, AMS, 2002.
[128] M. Kapranov, Non-commutative geometry based on commutator expansions, J. Reine Angew. Math. 505 (1998), 73–118.
[129] M. Kapranov, Analogies between the Langlands correspondence and topological quantum field theory, in: Progress in Math. 131, Birkhäuser, 1995, pp. 119–151.
[130] G. Kasparov, The operator *K*-functor and extensions of C^*-algebras, Izv. Akad. Nauk SSSR, Ser. Mat. 44 (1980), 571–636.
[131] G. Kasparov, Operator *K*-theory and its applications: elliptic operators, group representations, higher signatures, C^*-algebras, In: Proceedings of ICM 83, Warsaw, 1984.
[132] Ch. Kassel, Quantum Groups, Graduate Texts in Mathematics 155, Springer, 1995.
[133] M. Khalkhali, Basic Noncommutative Geometry, EMS Series of Lectures in Mathematics, 2007.
[134] V. K. Kharchenko, Nekommutativnaya Teoriya Galua, Nauchnaya kniga, Novosibirsk, 1996.
[135] F. Kirwan, Complex Algebraic Curves, LMS Student Texts 23, 1992.
[136] A. W. Knapp, Elliptic Curves, Mathematical Notes 40, Princeton University Press, 1992.

[137] A. W. Knapp, Introduction to the Langlands program, Proc. Symp. Pure Math. 61 (1997), 245–302.
[138] N. Koblitz, Introduction to Elliptic Curves and Modular Forms, GTM 97, Springer, 1984.
[139] M. Kontsevich, Deformation quantization of Poisson manifolds, I, math.9709180.
[140] E. E. Kummer, Über die Zerlegung der aus Wurzeln der Einheit gebildeten complexen Zahlen in ihre Primfactoren, J. Reine Angew. Math. 35 (1847), 327–367.
[141] J. Kustermans and S. Vaes, The operator algebra approach to quantum groups, Proc. Natl. Acad. Sci. USA 97 (2000), 547–552.
[142] J. Kustermans and S. Vaes, Locally compact quantum groups in the von Neumann algebraic setting, Math. Scand. 92 (2003), 68–92.
[143] T. Y. Lam, Introduction to Quadratic Forms over Fields, Graduate Studies in Mathematics 67, AMS, 2005.
[144] G. Landi, An Introduction to Noncommutative Spaces and Their Geometries, Lecture Notes in Physics Monographs, Springer, 1997.
[145] R. P. Langlands, A letter to A. Weil, 10p, 1960's.
[146] R. P. Langlands, *L*-functions and automorphic representations, in: Proceedings of the ICM 1978, Helsinki, 1978, pp. 165–175.
[147] C. G. Latimer and C. C. MacDuffee, A correspondence between classes of ideals and classes of matrices, Ann. Math. 34 (1933), 313–316.
[148] O. A. Laudal, Noncommutative algebraic geometry, Rev. Mat. Iberoam. 19 (2003), 509–580.
[149] H. B. Lawson, Foliations, Bull. Am. Math. Soc. 80 (1974), 369–418.
[150] W. B. R. Lickorish, A representation of orientable combinatorial 3-manifolds, Ann. Math. 76 (1962), 531–540.
[151] D. Lind and B. Marcus, An Introduction to Symbolic Dynamics and Coding, Cambridge Univ. Press, 1995.
[152] S. MacLane, Categories for the Working Mathematician, Springer, 1971.
[153] J. Madore, An Introduction to Noncommutative Differential Geometry & its Applications, Cambridge Univ. Press, 1995.
[154] Yu.I. Manin, Topics in Noncommutative Geometry, Princeton Univ. Press, 1991.
[155] Yu.I. Manin, Real multiplication and noncommutative geometry, in: Legacy of Niels Hendrik Abel, Springer, 2004, pp. 685–727.
[156] M. Marcolli, Arithmetic Noncommutative Geometry, University Lecture Series 36, AMS, 2005.
[157] A. A. Markov, Über die freie Aquivalenz geschlossener Zöpfe, Math. Sb. 1 (1935), 73–78.
[158] B. Mazur, Arithmetic on curves, Bull. Am. Math. Soc. 14 (1986), 207–259.
[159] B. Mazur, Remarks on the Alexander polynomial, unpublished notes, c. 1964.
[160] B. Mazur, Notes on étale cohomology of number fields, Ann. Sci. Éc. Norm. Supér. 6 (1973), 521–552.
[161] H. Minkowski, Zur Geometrie der Zahlen, in: Verhandlungen des III internationalen Mathematiker-Kongresses in Heidelberg, Berlin, 1904, pp. 164–173.
[162] C. C. Moore and C. L. Schochet, Global Analysis on Foliated Spaces, MSRI Publ. 9, Cambridge Univ. Press, 2006.
[163] P. Morandi, Field and Galois Theory, GTM 167, Springer, 1996.
[164] G. D. Mostow, Strong Rigidity of Locally Symmetric Spaces, Annals of Mathematical Studies 78, Princeton Univ. Press, 1973.
[165] L. J. Mordell, On the rational solutions of the indeterminate equations of the third and fourth degrees, Proc. Camb. Philos. Soc. 21 (1922), 179–192.
[166] M. Morishita, Knots and Primes, An Introduction to Arithmetic Topology, Universitext, Springer, London, Dordrecht, Heidelberg, New York, 2012.
[167] L. Mosher, The classification of pseudo-Anosovs, in: Low-dimensional Topology and Kleinian Groups (Coventry/Durham) 1984, LMS Notes 112, 1986, pp. 13–75.

[168] D. Mumford, Tata Lectures on Theta I, Birkhäuser, 1983.

[169] D. Mundici, Farey stellar subdivisions, ultrasimplicial groups, and K_0 of AF C^*-algebras, Adv. Math. 68 (1988), 23–39.

[170] G. J. Murphy, C^*-Algebras and Operator Theory, Academic Press, 1990.

[171] F. J. Murray and J. von Neumann, On rings of operators, Ann. Math. 37 (1936), 116–229.

[172] A. Néron, Problèmes arithmétiques et géometriques rattachés à la notion de rang d'une courbe algébrique dans un corps, Bull. Soc. Math. Fr. 80 (1952), 101–166.

[173] J. Neukirch, Algebraic Number Theory, Springer, 1999.

[174] J. von Neumann, Continuous Geometry, Princeton Univ. Press, Princeton, 1960.

[175] J. Nielsen, Untersuchungen zur Topologie der zweiseitigen Flächen, I; II; III, Acta Math. 50 (1927), 189–358; 53 (1929), 1–76; 58 (1932), 87–167.

[176] I. Nikolaev, On complex and noncommutative tori, Proc. Am. Math. Soc. 134 (2006), 973–981.

[177] I. Nikolaev, On a desingularization of the moduli space of noncommutative tori, Proc. Am. Math. Soc. 136 (2008), 3769–3774.

[178] I. Nikolaev, Noncommutative geometry as a functor, Contemp. Math. 503 (2009), 151–157.

[179] I. Nikolaev, On a Teichmüller functor between the categories of complex tori and the Effros-Shen algebras, N.Y. J. Math. 15 (2009), 125–132.

[180] I. Nikolaev, Noncommutative geometry of algebraic curves, Proc. Am. Math. Soc. 137 (2009), 3283–3290.

[181] I. Nikolaev, Remark on the rank of elliptic curves, Osaka J. Math. 46 (2009), 515–527.

[182] I. Nikolaev, On a Cuntz-Krieger functor, N.Z. J. Math. 39 (2009), 165–169.

[183] I. Nikolaev, Langlands reciprocity for the even-dimensional noncommutative tori, Proc. Am. Math. Soc. 139 (2011), 4153–4162.

[184] I. Nikolaev, On the AF-algebra of a Hecke eigenform, Proc. Edinb. Math. Soc. 55 (2012), 207–213.

[185] I. Nikolaev, Operator algebras and conjugacy problem for the pseudo-Anosov automorphisms of a surface, Pac. J. Math. 261 (2013), 445–462.

[186] I. Nikolaev, Invariants of stationary AF-algebras and torsion subgroup of elliptic curves with complex multiplication, Missouri J. Math. Sci. 26 (2014), 23–32.

[187] I. Nikolaev, On traces of Frobenius endomorphisms, Finite Fields Appl. 25 (2014), 270–279.

[188] I. Nikolaev, AF-algebras and topology of mapping tori, Czechoslov. Math. J. 65 (2015), 1069–1083.

[189] I. Nikolaev, On a symmetry of complex and real multiplication, Hokkaido Math. J. 45 (2016), 43–51.

[190] I. Nikolaev, On cluster C^*-algebras, J. Funct. Spaces (2016), Article ID 9639875, 8 pp.

[191] I. Nikolaev, Real multiplication revisited, J. Gen. Lie Theory Appl. 10 (S2) (2016), Article ID 007, 5 p.

[192] I. Nikolaev, Tate curves and UHF-algebras, Indag. Math. 27 (2016), 383–391.

[193] I. Nikolaev, Riemann surfaces and AF-algebras, Ann. Funct. Anal. 7 (2016), 371–380.

[194] I. Nikolaev, Hyperbolic geometry, continued fractions and classification of the finitely generated totally ordered dimension groups, Houst. J. Math. 42 (2016), 167–178.

[195] I. Nikolaev, Remark on Weil's conjectures, Adv. Pure Appl. Math. 7 (2016), 213–221.

[196] I. Nikolaev, Noncommutative geometry of rational elliptic curves, Ann. Funct. Anal. 9 (2018), 202–209.

[197] I. Nikolaev, Mapping class groups are linear, in: Topological Algebras and Their Applications, A. A. Katz, ed., De Gruyter Proc. Math., De Gruyter, Berlin, 2018, pp. 193–200.

[198] I. Nikolaev, On algebraic values of function $\exp(2\pi i\theta + \log\log\varepsilon)$, Ramanujan J. 47 (2018), 417–425.

[199] I. Nikolaev, K-theory of cluster C^*-algebras, Banach Cent. Publ. 120 (2020), 23–34.

[200] I. Nikolaev, Sklyanin algebras revisited, Bol. Soc. Mat. Mex. 27 (2021), 1.

[201] I. Nikolaev, Untying knots in 4D and Wedderburn's Theorem, Commun. Algebra 49 (2021), 4133–4135.
[202] I. Nikolaev, Langlands reciprocity for C^*-algebras, in: Operator Theory, Functional Analysis and Applications 282, Birkhäsuer, 2021, pp. 515–528.
[203] I. Nikolaev, Dynamical ideals of non-commutative rings, Journal of Algebra and its Applications, to appear.
[204] I. Nikolaev, L-functions of noncommutative tori, 0912.4905.
[205] I. Nikolaev, On the rank conjecture, 1104.0609.
[206] I. Nikolaev, Remark on Serre C^*-algebras, 1208.2049.
[207] I. Nikolaev, Remark on arithmetic topology, 1706.06398.
[208] I. Nikolaev, K-theory of rational quadratic forms, 1711.00743.
[209] I. Nikolaev, Quantum dynamics of elliptic curves, 1808.03493.
[210] I. Nikolaev, Profinite mapping class groups, 1812.05390.
[211] I. Nikolaev, Arithmetic topology of 4-manifolds, 1907.03901.
[212] I. Nikolaev, Shafarevich–Tate groups of abelian varieties, 2005.09970.
[213] I. Nikolaev, K-theory of Etesi C^*-algebras, 2012.08237.
[214] I. Nikolaev, Noncommutative geometry of elliptic surfaces, 2106.11392.
[215] I. Nikolaev, Class field towers and minimal models, 2108.13470.
[216] P. W. Nowak and G. Yu, Large Scale Geometry, EMS Textbooks in Mathematics, 2012.
[217] A. B. Odesskii, Elliptic algebras, Usp. Mat. Nauk 57 (2002), 87–122.
[218] S. Otake, Orthogonal decompositions of integral trace forms of cyclotomic fields and their canonical forms over the ring of p-adic integers, J. Number Theory 134 (2014), 258–279.
[219] F. van Oystaeyen, Prime Spectrum of Non-commutative Rings, Lecture Notes in Mathematics 444, Springer, 1975.
[220] F. van Oystaeyen and A. Verschoren, Non-commutative Algebraic Geometry, Lecture Notes in Mathematics 887, Springer, 1980.
[221] G. Panti, Multidimensional continued fractions and a Minkowski function, Monatshefte Math. 154 (2008), 247–264.
[222] G. K. Pedersen, C^*-Algebras and Their Automorphism Groups, L.M.S. Monographs 14, Academic Press, 1979.
[223] R. C. Penner, The decorated Teichmüller space of punctured surfaces, Commun. Math. Phys. 113 (1987), 299–339.
[224] O. Perron, Grundlagen für eine Theorie des Jacobischen Kettenbruchalgorithmus, Math. Ann. 64 (1907), 1–76.
[225] O. Perron, Die Lehre von den Kettenbrüchen, Bd. 1, Teubner, 1954.
[226] R. S. Pierce, Associative Algebras, GTM 88, Springer, 1982.
[227] R. Piergallini, Four-manifolds as 4-fold branched covers of S^4, Topology 34 (1995), 497–508.
[228] N. C. Phillips, Crossed products of the Cantor set by free minimal actions of $\mathbf{Z}^d$, Commun. Math. Phys. 256 (2005), 1–42.
[229] M. V. Pimsner, Embedding some transformation group C^*-algebras into AF-algebras, Ergod. Theory Dyn. Syst. 3 (1983), 613–626.
[230] M. Pimsner and D. Voiculescu, Imbedding the irrational rotation C^*-algebra into an AF-algebra, J. Oper. Theory 4 (1980), 201–210.
[231] J. F. Plante, Foliations with measure preserving holonomy, Ann. Math. 102 (1975), 327–361.
[232] V. Platonov and A. Rapinchuk, Algebraic Groups and Number Theory, Academic Press, 1994.
[233] D. Poguntke, Simple quotients of group C^*-algebras for two step nilpotent groups and connected Lie groups, Ann. Sci. Éc. Norm. Supér. 16 (1983), 151–172.
[234] H. Poincaré, Sur les propriétés arithmetiques des courbes algébriques, J. Math. 5-e série, t. 7, fasc. III (1901), 160–253.

[235] B. Poonen and M. Stoll, The Cassels–Tate pairing on polarized abelian varieties, Ann. Math. 150 (1999), 1109–1149.

[236] K. Reidemeister, Knotentheorie, Ergebnisse der Mathematik und ihrer Grenzgebiete, Bd. 1, 1932, Chelsea, 1948.

[237] A. Reznikov, Three-manifolds class field theory (Homology of coverings for a nonvirtually b_1-positive manifold), Sel. Math. 3 (1997), 361–399.

[238] M. A. Rieffel, C^*-algebras associated with irrational rotations, Pac. J. Math. 93 (1981), 415–429.

[239] M. A. Rieffel, Non-commutative tori – a case study of non-commutative differentiable manifolds, Contemp. Math. 105 (1990), 191–211. Available http://math.berkeley.edu/~rieffel/.

[240] M. A. Rieffel and A. Schwarz, Morita equivalence of multidimensional noncommutative tori, Int. J. Math. 10 (1999), 289–299.

[241] P. M. Rodrigues and J. S. Ramos, Bowen–Franks groups as conjugacy invariants for T^n-automorphisms, Aequ. Math. 69 (2005), 231–249.

[242] J. Roe, Index Theory, Coarse Geometry, and Topology of Manifolds, CBMS Regional Series in Mathematics 90, AMS, Providence, 1996.

[243] M. Rørdam, F. Larsen and N. Laustsen, An Introduction to K-Theory for C^*-Algebras, London Mathematical Society Student Texts 49, Cambridge University Press, Cambridge, 2000.

[244] A. L. Rosenberg, Noncommutative Algebraic Geometry and Representations of Quantized Algebras, Mathematics and its Applications 330, Kluwer Academic Publishers, 1995.

[245] A. Rosenberg, Selected Papers on Noncommutative Geometry, New Prairie Press, 2014. ISBN 978-0-9915482-0-0.

[246] J. Rosenberg, C^*-algebras, positive scalar curvature, and the Novikov conjecture, Publ. Math. IHÉS 58 (1983), 197–212.

[247] K. Rubin, Tate–Shafarevich groups of elliptic curves with complex multiplication, in: Algebraic Number Theory - in honor of K. Iwasawa, Advanced Studies in Pure Mathematics 17, 1989, pp. 409–419.

[248] J. T. Schwartz, W^*-Algebras, Gordon and Breach, 1967.

[249] J. P. Serre, Fasceaux algébriques cohérents, Ann. Math. 61 (1955), 197–278.

[250] J. P. Serre, Complex multiplication, in: Algebraic Number Theory, J. W. S. Cassels and A. Fröhlich, eds, Acad. Press, 1967.

[251] J. P. Serre, Représentations Linéaires des Groupes Finis, Hermann, Paris, 1967.

[252] I. R. Shafarevich, Basic Notions of Algebra, Algebra I, Encycl. Math. Sci. 11, Springer-Verlag, 1990.

[253] G. Shimura, On elliptic curves with complex multiplication as factors of the Jacobians of modular function fields, Nagoya Math. J. 43 (1971), 199–208.

[254] G. Shimura, Class fields over real quadratic fields and Hecke operators, Ann. Math. 95 (1972), 130–190.

[255] T. Shioda, On elliptic modular surfaces, J. Math. Soc. Jpn. 24 (1972), 20–59.

[256] M. Schütt and T. Shioda, Mordell-Weil Lattices, Ergebnisse der Mathematik und ihrer Grenzgebiete 70, Springer, 2019.

[257] A. S. Sikora, Analogies between group actions on 3-manifolds and number fields, Comment. Math. Helv. 78 (2003), 832–844.

[258] J. H. Silverman, The Arithmetic of Elliptic Curves, GTM 106, Springer, 1985.

[259] J. H. Silverman, Advanced Topics in the Arithmetic of Elliptic Curves, GTM 151, Springer, 1994.

[260] J. H. Silverman and J. Tate, Rational Points on Elliptic Curves, Springer, 1992.

[261] E. K. Sklyanin, Some algebraic structures connected to the Yang–Baxter equations, Funct. Anal. Appl. 16 (1982), 27–34.

[262] E. K. Sklyanin, Some algebraic structures connected to the Yang–Baxter equations. Representations of quantum algebras, Funct. Anal. Appl. 17 (1982), 273–284.
[263] S. Smale, Differentiable dynamical systems, Bull. Am. Math. Soc. 73 (1967), 747–817.
[264] S. P. Smith and J. T. Stafford, Regularity of the four dimensional Sklyanin algebra, Compos. Math. 83 (1992), 259–289.
[265] J. T. Stafford and M. van den Bergh, Noncommutative curves and noncommutative surfaces, Bull. Am. Math. Soc. 38 (2001), 171–216.
[266] H. M. Stark, L-functions at $s = 1$. III. Totally real fields and Hilbert's twelfth problem, Adv. Math. 22 (1976), 64–84.
[267] W. A. Stein, Shafarevich–Tate groups of nonsquare order, in: Modular Curves and Abelian Varieties, Progr. Math. 224, Birkhäuser, Basel, 2004, pp. 277–289.
[268] K. Strebel, Quadratic Differentials, Springer, 1984.
[269] W. D. van Suijlekom, Noncommutative Geometry and Particle Physics, Mathematical Physics Studies, Springer, 2015.
[270] H. P. F. Swinnerton-Dyer, An application of computing to class field theory, in: Algebraic Number Theory, J. W. S. Cassels and A. Fröhlich, eds, Washington, D.C., 1967, pp. 280–291.
[271] M. Takesaki, Tomita's Theory of Modular Hilbert Algebras and its Applications, Lect. Notes in Mathematics 128, Springer-Verlag, 1970.
[272] J. T. Tate, WC-groups over p-adic fields, Sém. Bourbaki exp. 156 (1958), 265–277.
[273] J. T. Tate, Duality theorems in Galois cohomology over number fields, in: Proc. Intern. Cong. Stockholm, 1962, pp. 288–295.
[274] J. T. Tate, The arithmetic of elliptic curves, Invent. Math. 23 (1974), 179–206.
[275] W. P. Thurston, Three dimensional manifolds, Kleinian groups and hyperbolic geometry, Bull. Am. Math. Soc. 6 (1982), 357–381.
[276] W. P. Thurston, On the geometry and dynamics of diffeomorphisms of surfaces, Bull. Am. Math. Soc. 19 (1988), 417–431.
[277] T. Timmermann, An Invitation to Quantum Groups and Duality, EMS Textbooks in Mathematics, 2008.
[278] B. L. Tsygan, Homology of matrix Lie algebras over rings and the Hochschild homology, Usp. Mat. Nauk 38 (1983), 217–218.
[279] K. Uchida, Isomorphisms of Galois groups, J. Math. Soc. Jpn. 28 (1976), 617–620.
[280] J. C. Várilly, An Introduction to Noncommutative Geometry, EMS Series of Lectures in Mathematics, 2006.
[281] J. B. Wagoner, Strong shift equivalence theory and the shift equivalence problem, Bull. Am. Math. Soc. 36 (1999), 271–296.
[282] M. Waldschmidt, On a paper by Nikolaev, Ramanujan J. (2022).
[283] D. I. Wallace, Conjugacy classes of hyperbolic matrices in $SL(n, \mathbf{Z})$ and ideal classes in an order, Trans. Am. Math. Soc. 283 (1984), 177–184.
[284] J. H. Maclagan-Wedderburn, A theorem on finite algebras, Trans. Am. Math. Soc. 6 (1905), 349–352.
[285] A. Weil, Numbers of solutions of equations in finite fields, Bull. Am. Math. Soc. 55 (1949), 497–508.
[286] N. E. Wegge-Olsen, K-theory and C*-algebras, A Friendly Approach, Oxford Univ. Press, 1993.
[287] D. P. Williams, Crossed Products of C^*-Algebras, Math. Surveys and Monographs 134, Amer. Math. Soc., 2007.
[288] L. K. Williams, Cluster algebras: An introduction, Bull. Am. Math. Soc. 51 (2014), 1–26.
[289] S. L. Woronowicz, Compact matrix pseudogroups, Commun. Math. Phys. 111 (1987), 613–665.

Index

https://doi.org/10.1515/9783110788709-018

De Gruyter Studies in Mathematics

Volume 87
René Erlin Castillo
Fourier Meets Hilbert and Riesz. An Introduction to the Corresponding Transforms
ISBN 978-3-11-078405-3, e-ISBN 978-3-11-078409-1,
e-ISBN (ePUB) 978-3-11-078412-1

Volume 86
Flavia Smarrazzo, Alberto Tesei
Measure Theory and Nonlinear Evolution Equations
ISBN 978-3-11-055600-1, e-ISBN 978-3-11-055690-2,
e-ISBN (ePUB) 978-3-11-055604-9

Volume 85
Shai Dekel
Pointwise Variable Anisotropic Function Spaces on $\mathbb{R}^n$
ISBN 978-3-11-076176-4, e-ISBN 978-3-11-076179-5,
e-ISBN (ePUB) 978-3-11-076187-0

Volume 84
Marko Kostić
Selected Topics in Almost Periodicity, 2021
ISBN 978-3-11-076322-5, e-ISBN 978-3-11-076352-2,
e-ISBN (ePUB) 978-3-11-076360-7

Volume 83
Lydia Außenhofer, Dikran Dikranjan, Anna Giordano Bruno
Topological Groups and the Pontryagin-van Kampen Duality. An Introduction, 2021
ISBN 978-3-11-065349-6, e-ISBN 978-3-11-065493-6,
e-ISBN (ePUB) 978-3-11-065355-7

Volume 82
Friedrich Wagemann
Crossed Modules, 2021
ISBN 978-3-11-075076-8, e-ISBN 978-3-11-075095-9,
e-ISBN (ePUB) 978-3-11-075099-7

Volume 81
Ľubica Holá, Dušan Holý, Warren Moors
USCO and Quasicontinuous Mappings, 2021
ISBN 978-3-11-075015-7, e-ISBN 978-3-11-075018-8,
e-ISBN (ePUB) 978-3-11-075022-5

Printed and bound by CPI Group (UK) Ltd, Croydon, CR0 4YY
16/09/2026
14963052-0001